April 10–April 13, 2011
Salzburg, Austria

**Association for
Computing Machinery**

Advancing Computing as a Science & Profession

EuroSys'11

Proceedings of the **EuroSys 2011 Conference**

Sponsored by:
ACM SIGOPS

Supported by:
**VMWare, Google, Microsoft Research Cambridge,
University of Salzburg, Artist, Stadt: Salzburg, Land Salzburg**

Organized by:
University of Salzburg and EuroSys

Association for Computing Machinery

Advancing Computing as a Science & Profession

The Association for Computing Machinery
2 Penn Plaza, Suite 701
New York, New York 10121-0701

Notice to Past Authors of ACM-Published Articles
ACM intends to create a complete electronic archive of all articles and/or other material previously published by ACM. If you have written a work that has been previously published by ACM in any journal or conference proceedings prior to 1978, or any SIG Newsletter at any time, and you do NOT want this work to appear in the ACM Digital Library, please inform permissions@acm.org, stating the title of the work, the author(s), and where and when published.

ISBN: 978-1-4503-0634-8

Additional copies may be ordered prepaid from:

ACM Order Department
PO Box 30777
New York, NY 10087-0777, USA

Phone: 1-800-342-6626 (USA and Canada)
+1-212-626-0500 (Global)
Fax: +1-212-944-1318
E-mail: acmhelp@acm.org
Hours of Operation: 8:30 am – 4:30 pm ET

ACM Order Number 534113

Printed in the USA

Foreword

It is our great pleasure to present the *proceedings of the 6th ACM EuroSys Conference on Computer Systems – EuroSys 2011*. We are confident that this year's conference continues to grow the reputation and influence of EuroSys with a set of excellent papers. These cover a wide range of topics, from real-world configuration management and debugging to hardware reliability, from mobile devices to clouds, from fault tolerance to energy management.

This year's call for papers attracted a record number of 161 submissions, of which 78 came from North America, 49 from Europe, 25 from Asia, two each from South America and Australia, and one each from Central America and Africa. Seven papers were rejected for violating formatting requirements, which left 154 papers to be reviewed by the PC in two rounds with limited help from outside experts. In a first round of reviewing, each paper received a minimum of three reviews, leading to the elimination of 86 papers. Their authors were notified immediately, to maximise their opportunity to improve the paper for a possible re-submission to another conference.

The remaining 68 papers each received a minimum of three further reviews in the second round, after which reviews were sent to authors for rebuttal. For the vast majority of papers the reviews were quite consistent, reflecting the quality and commitment of the PC members. In a few cases where there was significant disagreement between reviews, further reviews were obtained before the PC meeting. The vast majority of reviews were written by the PC members themselves.

The *rebuttal process* was a first for EuroSys, and in a number of cases helped to provide valuable clarifications (including one case where a paper's strongest supporter downgraded their score after reading the rebuttal). A second novelty was the move to *double-blind reviewing*, where PC members did not know the authors of the papers they reviewed and discussed. Authors' identities were only revealed at the end of the PC meeting. There was strong consensus that this had worked well, that in the vast majority of cases PC members really did not know the origin of a paper and that this helped the PC to minimise bias.

At the meeting, the PC applied very high quality standards, and in the end selected 24 papers. A particularly rigorous standard was applied to papers authored by PC members, and only two of nine such papers were accepted. Given the high standard applied, it is particularly pleasing that ten European papers were accepted, giving Europe the highest acceptance rate of all geographic regions.

EuroSys 2011 takes place in Salzburg, Austria. The conference begins on April 10 with a day dedicated to workshops and tutorials held at the Faculty of Law of the University of Salzburg. In total, eight workshops and two tutorials have been selected to provide a forum for the latest research in multicore, cloud computing, embedded, security, and social network systems. The main conference is held on April 11–13 at the newly renovated Aula of the University of Salzburg across the street of the world-famous Festspielhaus.

We hope that you will find these proceedings interesting and thought provoking, and that those who attend the conference and its associated workshops and tutorials will find this a stimulating experience.

Christoph Kirsch
EuroSys 2011 General Chair
University of Salzburg

Gernot Heiser
EuroSys 2011 Program Chair
NICTA and *UNSW*

Table of Contents

Session 6: Hardcore OS

Session 7: Better Clouds

Session 8: Off the Beaten Path

Author Index

EuroSys 2011 Conference Organization

General Chair: Christoph Kirsch *(University of Salzburg, Austria)*

Program Chair: Gernot Heiser *(University of New South Wales, Australia)*

Finance Chair: Thomas Kalibera *(Charles University, Czech Republic)*

Workshop Chair: Herbert Bos *(Vrije Universiteit Amsterdam, The Netherlands)*

Tutorial Chair: Adam Dunkels *(Swedish Institute of Computer Science, Sweden)*

Poster Chair: Stergios Anastasiadis *(University of Ioannina, Greece)*

Publication Chair: Ana Sokolova *(University of Salzburg, Austria)*

Publicity Chair: Ben L. Titzer *(Google, USA)*

Sponsorship Co-chair: Marco Sanvido *(Pure Storage, USA)*

Sponsorship Co-chair: Michael Franz *(UC Irvine, USA)*

Local Arrangements Co-chair: Ana Sokolova *(University of Salzburg, Austria)*

Local Arrangements Co-chair: Andreas Haas *(University of Salzburg, Austria)*

Web Co-chair: Hannes Payer *(University of Salzburg, Austria)*

Web Co-chair: Silviu Craciunas *(University of Salzburg, Austria)*

Registration Chair: Petra Kirchweger *(University of Salzburg, Austria)*

Student Travel Grant Co-chair: Dejan Kostic *(EPFL, Switzerland)*

Student Travel Grant Co-chair: Yolande Berbers *(Katholieke Universiteit Leuven, Belgium)*

Program Committee: Ozalp Babaoglu *(Universita di Bologna, Italy)*
Andrew Baumann *(MSR Redmond, USA)*
Herbert Bos *(Vrije Universiteit Amsterdam, The Netherlands)*
Scott Brandt *(UCSC, USA)*
George Candea *(EPFL, Switzerland)*
Adam Dunkels *(Swedish Institute of Computer Science, Sweden)*
Dawson Engler *(Stanford University, USA)*
Sasha Fedorova *(Simon Fraser University, Canada)*
Steve Gribble *(University of Washington, USA)*
Robert Grimm *(NYU, USA)*
Andreas Haeberlen *(University of Pennsylvania, USA)*
Jacob Gorm Hansen *(VMWare, USA)*

Eurosys 2011 Sponsor & Supporters

Sponsor:

Supporters:

Platinum

Gold

Bronze

Organized by:

In cooperation with: USENIX

Keypad: An Auditing File System for Theft-Prone Devices

Roxana Geambasu   John P. John   Steven D. Gribble   Tadayoshi Kohno   Henry M. Levy

University of Washington

roxana, jjohn, gribble, yoshi, levy@cs.washington.edu

Abstract

This paper presents Keypad, an auditing file system for theft-prone devices, such as laptops and USB sticks. Keypad provides two important properties. First, Keypad supports fine-grained file auditing: a user can obtain *explicit evidence* that no files have been accessed *after* a device's loss. Second, a user can disable future file access after a device's loss, even in the *absence* of device network connectivity. Keypad achieves these properties by weaving together encryption and remote key storage. By encrypting files locally but storing encryption keys remotely, Keypad requires the involvement of an audit server with every protected file access. By alerting the audit server to refuse to return a particular file's key, the user can prevent new accesses after theft.

We describe the Keypad architecture, a prototype implementation on Linux, and our evaluation of Keypad's performance and auditing fidelity. Our results show that Keypad overcomes the challenges posed by slow networks or disconnection, providing clients with usable forensics and control for their (increasingly) missing mobile devices.

Categories and Subject Descriptors   D.4.6 [*Operating Systems*]: Security and Protection

General Terms   design, performance, security

Keywords   Keypad, auditing, file system, theft-prone

1. Introduction

Laptops, USB memory sticks, and other mobile computing devices greatly facilitate on-the-go productivity and the transport, storage, sharing, and mobile use of information. Unfortunately, their mobile nature and small form factors also make them highly susceptible to loss or theft. As example statistics, one in ten laptops is lost or stolen within a year of purchase [Nusca 2009], 600,000 laptops are lost

annually in U.S. airports alone [Ponemon Institute 2008], and dry cleaners in the U.K. found over 4,000 USB sticks in pockets in 2009 [Sorrel 2010]. The loss of such devices is most concerning for organizations and individuals storing confidential information, such as medical records, social security numbers (SSNs), and banking information.

Conventional wisdom suggests that standard encryption systems, such as BitLocker, PGP Whole Disk Encryption, and TrueCrypt, can protect confidential information. Unfortunately, encryption alone is sometimes insufficient to meet users' needs, for two reasons. First, traditional encryption systems can and do fail in the world of real users. As described in the seminal paper "Why Johnny Can't Encrypt" [Whitten 1999], security and usability are often at odds. Users find it difficult to create, remember, and manage passphrases or keys. As an example, a password-protected USB stick containing private medical information about prison inmates was lost along with a sticky note revealing its password [Savage 2009]. Encrypted file systems often rely on a locally stored key that is protected by a user's passphrase. User passphrases are known to be insecure; a recent study of consumer Web passwords found the most common one to be "123456" [Imperva 2010]. Finally, in the hands of a motivated data thief, devices are open to physical attacks on memory or cold-boot attacks [Halderman 2008] to retrieve passphrases or keys. Even physical attacks on TPMs and "tamper-resistant" hardware are possible [Anderson 1996, Robertson 2010].

Second, when encryption fails, it fails *silently*; an attacker might circumvent the encryption without the data owner ever learning of the access. The use of conventional encryption can therefore lead mobile device owners into a false sense of protection. For example, a hospital losing a laptop with encrypted patient information might not notify patients of its loss, even if the party finding the device has circumvented the encryption and accessed that information.

This paper presents the design, implementation, and evaluation of *Keypad*, a file system for loss- and theft-prone mobile devices that addresses these concerns. The principal goal of Keypad is to provide *explicit evidence* that protected data in a lost device either has or has not been exposed after loss. Specifically, someone who obtains a lost or stolen Keypad device *cannot* read or write files on the device without triggering the creation of access log entries on a remote server. This property holds even if the person finding the device also finds a note with the device's encryption password.

Keypad's forensic logs are detailed and fine grained. For example, a curious individual who finds a laptop at the coffee shop and seeks to learn its owner might register audit records for files in the home directory, but not for unaccessed confidential medical records also stored on the device. However, the professional data thief will register accesses to all of the specific confidential medical files that they view. Furthermore, Keypad lets device owners disable access to files on the mobile devices once they realize their devices have been lost or stolen, even if the devices have no network connectivity, such as USB memory sticks (in contrast to systems like Apple's MobileMe).

Keypad's basic technique is simple yet powerful: it tightly entangles the process of file access with logging on a *remote auditing server*. To do this, Keypad encrypts protected files with *file-specific* keys whose corresponding decryption keys are located on the server. Users never learn Keypad's decryption keys and thus they cannot choose weak passwords or accidentally reveal them; it is therefore computationally infeasible for an attacker to decrypt a file without leaving evidence in the log. When a file operation is invoked, Keypad logs the file operation remotely, temporarily downloads the key to access the file, and securely erases it shortly thereafter. Keypad is implemented on top of a traditional encrypted file system; obviously users should choose strong passwords (or use secure tokens, etc.) for that underlying file system, but Keypad provides a robust forensic trail of files accessed even if users choose weak passwords or the traditional system's keys are otherwise compromised.

While conceptually simple, making this vision practical presents significant technical challenges and difficult trade-offs. For example, neither the user nor Keypad can predict when a device will be lost or stolen. As a result, the system must provide both an accurate fine-grained forensic record, which is critical after loss, and acceptable performance, which is critical prior to loss.

The tension between performance and forensics is pervasive. As an example, consider the creation of a file. For forensic purposes, a naïve Keypad architecture might first pre-register newly created files and their corresponding keys with the remote server prior to writing any new data to those files. However, pre-registration would incur at least one full network round-trip, which could be problematic for some workloads over slow mobile networks, such as 3G or 4G. Delaying the registration is an obvious optimization, yet doing so would leave a loophole that a device thief could exploit to access files without triggering a log entry in the remote server. Overall, our experience demonstrates that we can achieve both forensic fidelity and acceptable performance by combining conventional systems techniques with

techniques from cryptography, including identity-based encryption [Boneh 2001, Shamir 1985].

We begin with a description of Keypad's motivation and goals in the following section. Keypad's architecture is presented in Section 3 and its implementation in Section 4. Section 5 provides a detailed evaluation of our prototype and Section 6 discusses its security. Section 7 reviews related work and we conclude in Section 8.

2. Motivation and Goals

Keypad is designed to increase assurances offered to owners of lost or stolen mobile devices. The mobile devices might have computational capabilities (e.g., laptops and phones) or might be simple storage devices (e.g., USB sticks). We view Keypad as particularly valuable to users storing personal or corporate documents, banking information, SSNs, medical records, and other highly sensitive data.

Examples. We provide two brief motivating examples. Alice is a businesswoman who carries a corporate laptop that stores documents containing trade secrets. Alice's IT department installs Keypad on the laptop, configuring it to track all accesses to files in her "corporate documents" folder. After returning to her hotel from a two-hour dinner, Alice notices that her laptop is missing. She immediately reports the loss to her IT department, which disables any future access to files in the corporate documents folder. The IT department also produces an audit log of all files accessed within the two-hour window since she last controlled her laptop, confirming that no sensitive files were accessed.

As a second example, at tax preparation time, Bob scans all of his tax documents, places them on a USB stick, encrypts it with a password, and physically hands the stick and password to his accountant. A few weeks later, Bob can no longer find his thumb drive and can't remember whether his accountant kept it or whether he lost it in the intervening weeks. Fortunately, Bob's stick was protected with Keypad and Bob uses a Web service provided by his drive manufacturer to view an audit log of all accesses to the drive. He sees that there were many accesses to his tax files over the previous week and he learns the IP addresses from which those accesses were made. Bob therefore places fraud alerts on his financial accounts and notifies the appropriate authorities.

In these scenarios, users benefit from additional advantages that Keypad has over traditional encrypted file systems. First, Keypad provides highly accurate, remotely readable forensic records of which files were accessed post-loss. If a file does not appear in those records, that suggests that no one accessed the file after device loss; if a file does appear in those records, this suggests that data was accessed and the owner should take appropriate mitigating actions. Second, by preventing key access, Keypad can prevent adversaries from accessing protected files post-loss, even in the absence of network connectivity, e.g., for a disconnected USB stick or an extracted laptop hard drive.

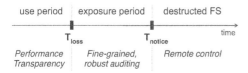

Figure 1. Timeline of theft/loss. This timeline shows the two critical events during the lifetime of a device: the device loss and the user noticing that the device has been lost. For each period, we enumerate the Keypad properties that matter in that period.

Goals. Figure 1 shows a high-level timeline of three periods in the life of a lost or stolen device, along with the properties the user requires during each. First is the normal use period during which the user has control of her device. The user loses control of the device at T_{loss}; however, the user may not know exactly when this occurs, so she must consider T_{loss} to be the last point at which she remembers having control. T_{notice} is the time at which the user realizes that she has lost her device, at which point she should take action. In our Alice scenario, the exposure period (T_{loss} to T_{notice}) is the full two-hour dinner window.

Our primary Keypad goal is to provide strong *audit security*. If an adversary gains control of a device and accesses a Keypad-protected file, at least one audit log entry should be produced on a remote audit server. Further, the adversary cannot tamper with the contents of the audit log or otherwise make it unavailable to the victim. Specifically, our goals are:

- *Robust auditing semantics:* Keypad must provide robust semantics by preventing unrecorded file accesses. To achieve this, the remote auditing server must observe data and metadata operations performed on the client.

- *Performance:* File access latency and throughput should be acceptable for Keypad-protected data. We mainly target office productivity and mobile workloads, rather than server- or engineering-oriented workloads. We also assume multiple network environments: at the office (LANs), at home (broadband), and on the road (3G or 4G). We seek minimal overhead at work or home, but will tolerate some increased latency in challenging mobile environments in exchange for Keypad's properties.

- *Fine granularity:* Keypad should produce detailed access logs of read and write accesses to individual Keypad-protected files. Administrators can control the granularity and coverage of these logs; e.g., configuring Keypad to produce audit logs for an entire file system or only for specific files identified as sensitive.

- *User transparency:* We assume that users are not technically sophisticated; therefore, Keypad's operation should be largely transparent to them and its auditing security should be independent of users' technical competence.

- *Remote access control:* The victim should be able to disable access to protected files after device loss, even if

the device has no network or computational capabilities. If an adversary has not yet accessed a protected file, then disabling access prevents any access to the file in the future. If an adversary has already accessed the file, we provide no guarantees about repeat accesses.

These goals mean that device owners will have accurate information about which files have been accessed post-loss. While we will consider optimizations that may introduce extra entries in the audit log, maintaining a *zero false-negative rate* is critical. If a file does not appear in the audit log, then one can confidently say that the file was not accessed. In addition, these audit goals must hold after T_{loss} even if an attacker uses his own software and hardware (and not Keypad) to access the files stored on the device.

We also have several non-goals for Keypad. First, we do not attempt to ensure the device's physical or software integrity after theft/loss. If a user recovers a lost device, he should assume that it has been tampered with, and inspect and reinstall the device from scratch to ensure that no keyloggers or malware have been installed. Second, Keypad deals with device theft/loss that is detectable by a user, and not with surreptitious attacks where an adversary might undetectedly access data on a user's device while he is away. This excludes evil-maid attacks from our threat model [Rutkowska 2009].

Third, Keypad ensures auditability and remote control solely at the file system interface level and below (e.g., the buffer cache). Auditability and control of clear-text data cached in applications' memories is out of Keypad's scope. Fourth, we do not seek to improve the confidentiality of protected files over traditional encryption. Instead, Keypad provides a secure audit log of file accesses if that traditional encryption fails. Finally, we do not guarantee that users can always access Keypad-protected files in the absence of network connectivity (which we consider increasingly rare, given ubiquitous cellular and WiFi networks). However, we do introduce a "paired-device" mechanism to mitigate the impact of disconnection while still maintaining auditability.

3. Keypad Design

Keypad augments encrypted file systems with two properties: auditability and remote data control. The basic idea is simple yet powerful. Keypad: (1) encrypts each file with its own symmetric key, (2) stores all keys on a remote audit service, (3) downloads the key for a file each time it is accessed, and (4) destroys the key immediately after use. This approach supports our auditability and remote data control goals. By configuring the audit service to log all storage accesses, we obtain fine-grained auditability; by disabling all keys associated with a stolen device on the service, we prevent further data access.

Despite its simplicity, designing a practical file system to achieve our goals poses three challenges. First is performance: each file access requires a blocking network request,

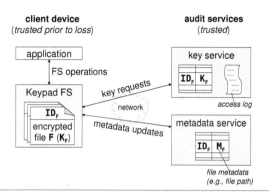

client device *(trusted prior to loss)*

application

FS operations

Keypad FS

ID_F

encrypted file **F** $(\mathbf{K_F})$

key requests

network

metadata updates

audit services *(trusted)*

key service

ID_F | K_F

access log

metadata service

ID_F | M_F

file metadata *(e.g., file path)*

Figure 2. Keypad Architecture. Each file is encrypted with its own random symmetric key. Keys are stored remotely on a key service. To enable forensics, a (separate) metadata service stores file metadata.

which could harm application performance and responsiveness over high latency cellular networks. Second is disconnection: involving the network on all file accesses prohibits file use during network unavailability. While we treat this as an exception, we still wish to support disconnected operation. Third is metadata: an auditor requires user-friendly, up-to-date metadata for each key to interpret access logs appropriately. As will be shown, efficiently maintaining metadata is complex, but possible. This section shows how Keypad's design addresses these three challenges.

3.1 Keypad Overview

Figure 2 shows Keypad's architecture. On the client device, each file F has a unique identifier (called the *audit ID* – ID_F) stored in its header, and the file's data is encrypted with a unique symmetric key, K_F. A remote *key service* maintains the mappings between audit IDs and keys. When an application wants to read or write a file, Keypad looks up the file's audit ID in its header and requests the associated key from the service. Before responding to the request, the service durably logs the requested ID and a timestamp. This process ensures that after T_{notice}, the user will be able to identify all compromised audit IDs for which there is a log entry after T_{loss}.

In addition to the key service, Keypad contains a *metadata service* that maintains information needed by users to interpret the logs. The information (called *file metadata*) includes a file's path, the process that created it, and the file's extended attributes. The metadata and key services fulfill conceptually independent functions; they could be run by a single provider or by distinct providers. Using distinct providers helps to mitigate privacy concerns that could arise if a single party tracked all file access information. The key service sees only accesses to opaque IDs and keys, while the metadata service learns the file system's structure, but not the access patterns. Thus, privacy-concerned users can avoid exposing full audit information to any audit service by using different key and metadata providers.

To meet our goal of robust auditing semantics, Keypad must carefully manage file metadata. For example, when an application creates a new file with name G, Keypad: (1) locally allocates an ID_G for the file, (2) sends a request to the key service to create a new key K_G and bind it to ID_G, and (3) sends a request to the metadata service to register the name G with ID_G. While steps 2 and 3 can occur concurrently, Keypad must confirm that both requests complete before it allows access to the new file. This ensures that file metadata is associated with keys prior to T_{loss}, so that any compromised keys can be correlated with their metadata after T_{notice}.

Similarly, during a file's lifetime, Keypad must keep the service's metadata current to ensure that a user will have fresh information in case of compromise. For example, whenever an application renames a file, Keypad sends a metadata-update request to the metadata service. Keypad must ensure that a thief cannot overwrite the user's metadata with bogus information after theft. For this reason, we implement the metadata store as an append-only log.

3.2 Semantics and Challenges

Keypad provides users with strong auditing semantics at audit time (i.e., post T_{loss}). We formulate an *ideal* invariant describing these semantics as follows:

> For any file **F** with identifier **ID$_F$** that was accessed after $\mathbf{T_{loss}}$ the following properties hold:
>
> (1) the key service shows an **ID$_F$** log entry after $\mathbf{T_{loss}}$, and
>
> (2) the metadata service shows all metadata updates that occurred on **ID$_F$** before $\mathbf{T_{loss}}$.

For (2), the metadata server must contain the latest file metadata (such as file pathname or other attributes) that the user assigned to the file. For example, suppose a user has downloaded a blank IRS tax form into `/tmp/irs_form.pdf`, renamed it as `/home/prepared_taxes_2011.pdf`, and filled it with sensitive information. Then, at forensics time, the user will need to have this latest path available on the service side to interpret a compromise of the taxes file accurately. Hence, maintaining up-to-date service-side metadata is vital to enable meaningful forensics.

In theory, we could achieve semantics arbitrarily close to this ideal invariant. If Keypad downloaded a file's key *every time* a block in the file is accessed and erased the key from memory immediately after using it, then we would obtain the first part of the invariant. Similarly, if Keypad waited for *every* metadata update to be acknowledged by the metadata service before completing that operation on the local disk, then we would obtain the second part.

In practice, however, achieving the ideal invariant is challenging at best. If Keypad must wait a full network round-trip for every block access and for every metadata operation (e.g., `rename`), then the system would be unacceptably slow over high-latency networks. Similarly, disconnected access would be impossible. The remainder of this section describes a combination of new techniques and re-purposed traditional mechanisms that help overcome these challenges. While each technique slightly weakens the invariant, we believe that the semantics remain clear and easy to grasp, and that we achieve our goals in nearly all realistic cases.

3.3 Encryption Key Caching and Prefetching

Many of Keypad's critical-path operations are remote key-fetching requests, e.g., issued whenever an application performs a file `read` or `write`. The number of such key requests can be minimized using standard OS mechanisms, such as caching and prefetching. For instance, instead of erasing a key immediately after use, Keypad can cache it locally. Similarly, on access to a file F, Keypad can prefetch keys for other related files, such as those in the same directory. Key caching and prefetching remove key retrieval from the critical path of many file accesses, dramatically improving performance (Section 5).

While caching and prefetching are well understood, they have non-standard implications in our system. First, these techniques cause keys to accumulate in the device's memory, affecting what users can deduce from the audit log of a lost device. Keys that are cached at time T_{loss} are susceptible to compromise: if an adversary can extract them from memory he can permanently remember those keys and bypass audit records for those files. The victim must thus make the worst-case assumption that all keys cached at T_{loss} are compromised. Second, key prefetching creates false positives in the audit log: some prefetched keys may not be used, although records for those keys will appear in the logs.

Keypad must therefore use caching and prefetching carefully to ensure good auditing semantics. For caching, we impose short lifetimes (T_{exp}) on keys and securely erase them at expiration. This bounds key accumulation in memory; the shorter the T_{exp}, the fewer keys will be exposed after T_{loss}. Experimentally, we find that key expirations as short as 100 seconds reap most of the performance benefit of caching, while exposing relatively few keys in memory at a given time. For prefetching, we designed a simple scheme to prefetch keys only when a file-scanning workload is detected (e.g., recursive file search or file hierarchy copying). This benefits file-system-heavy workloads where prefetching is the most useful, while maintaining high auditing precision for light workloads (e.g., interacting with a document). We discuss further prefetching alternatives in Section 4.

Key caching and prefetching alter Keypad's auditing semantics in a clear way: a user must now consider as compromised all files with audit records after $T_{loss} - T_{exp}$. Doing so ensures that the user will never experience false negatives. Hence, these techniques alter the invariant introduced in Section 3.1 in the following way: key and metadata service information must be present for any file F that was accessed after $T_{loss} - T_{exp}$. In Section 5.2 we quantify the effects of caching and prefetching on auditing.

3.4 Identity-Based Encryption for Metadata Updates

Metadata-update file system operations (such as file `create` and `rename`) account for a significant portion of file system operations in many workloads. For example, an OpenOffice file save invokes 11 file system operations, of which 7 are metadata operations that create and then rename temporary files. This large number of metadata operations would result in poor performance over slow networks if Keypad were to wait for an acknowledgement from the metadata service upon *every* metadata update before committing the update to disk, as required by our ideal auditing semantics. Figure 3a shows this scenario.

Overlapping local metadata updates with remote metadata service updates seems like a tempting optimization, however, it opens Keypad to possible attacks and frustrates our semantics. For example, consider a user who creates a new file called `/home/taxes_2011`, writes sensitive tax information inside, and closes the file and editing application. Suppose that due to network failures the create request does not reach the metadata service and therefore the service does not learn the new file's name. If a thief steals the device and reads the tax file ten minutes later, the access will produce an audit trail on the key service; however, no file metadata will be available for the user to interpret the log. Worse, the thief could block Keypad's metadata retries and send a bogus request to the service, e.g., declaring the new file's path as `/tmp/download` to mislead the user.

To respond to this challenge, Keypad leverages identity-based encryption (IBE) [Boneh 2001, Shamir 1985] in a way that both eliminates the network from the critical path of metadata updates and retains its strong auditing semantics. IBE allows a client to perform public-key encryption using any key string it chooses as the public key. A server called a private key generator (PKG) is required to generate the decryption key for the arbitrary public key. Most importantly for our use, the PKG need not know the public key string in advance, but the public key string must be provided to the PKG to learn the decryption key.

We modified Keypad to use IBE as follows. First, we add a level of indirection for file encryption keys. A file F's content is encrypted using a locally-generated random *data key* (denoted K_F^D) stored in the file's header. The data key is *itself* encrypted under the remote key, which in turn is stored on the key server. Section 4 provides more detail.

Second, Keypad's metadata service acts as a PKG, as shown in Figure 3b. When an application invokes a metadata operation (such as `rename`) for a file F, Keypad "locks" its encrypted data key K_F^D in the on-disk file header by encrypting it with IBE, using the new file's pathname as the public key string. While the metadata request is in flight, reads and writes can proceed *as long as* a copy of the file's cleartext data key K_F^D is cached in memory. Because files with metadata updates in flight are vulnerable to attacks, we reduce the key expiration time for such files to the bare minimum nec-

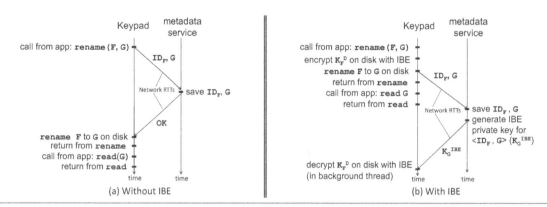

Figure 3. Timelines for handling metadata-update operations without IBE (a) and with IBE (b). The application is assumed to issue a `rename(F, G)` followed by a `read(G)`. Assuming that a copy of *F*'s decryption key is cached in memory, IBE allows overlap of accesses to *F* with the metadata service request until the cached key times out (1 second in our system).

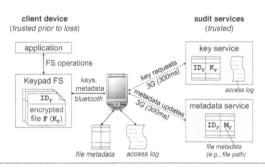

Figure 4. Paired-device architecture. By pairing a laptop with a mobile phone, Keypad supports disconnected operation and may even improve performance.

essary to hide network latencies on cellular networks. For example, our prototype expires cached keys with in-flight metadata updates in *one second*, minimizing attack opportunity. After the cached key times out, the file is essentially "locked" on disk by the IBE encryption, preventing subsequent file accesses until the metadata service confirms its success. On confirmation, the metadata service returns the IBE private key, allowing Keypad to "unlock" the file.

Suppose an attack or network failure prevents the service from registering the new metadata and subsequently the device is stolen. In the (extremely likely) case that the theft occurred more than one second after the user's rename request, the file's cached data key will have expired and the thief will need to obtain the IBE private key in order to unlock the file for access. As a result, the thief is forced to supply the *correct* file pathname to the metadata service if he desires to read the file; lying or avoiding the metadata update will prevent him from gaining access. Therefore, the thief cannot access the file without causing an audit record associated with correct and up-to-date metadata to be logged on the corresponding audit services.

3.5 Using Paired Devices for Disconnected Access

Although disconnected operations are assumed to be the exception rather than the rule, Keypad must still support them.

One option is to cache keys for an extended period of time and accumulate metadata registrations locally. However, this forces the user to give up auditability for the disconnected duration, which can be dangerous. Further, caching is not applicable to storage-only devices like USB sticks. To address this issue, we developed a *paired-device* extension to Keypad that supports disconnected operations without sacrificing auditability semantics.

Many of today's users carry multiple devices when they travel, such as a laptop as well as a smart phone or a tablet. These devices support short-range, low-latency networks, such as Bluetooth. The paired-device architecture, shown in Figure 4, uses a cell phone as a transparent extension of the Keypad key and metadata services. Keypad on the laptop is configured as usual, using strict caching, prefetching, and metadata registration policies to ensure fine-grained auditing. The phone is configured to hoard [Kistler 1991] any recently used keys, cache them until connectivity is restored, log any accesses and metadata updates to the local disk, and upload the logs when connectivity returns. If only the laptop is lost, the phone is used along with the audit service logs to provide a full audit trail. If the phone is stolen along with the laptop, then the audit service will list more files as exposed than if the laptop were stolen alone.

In addition to supporting (increasingly rare) disconnected cases, the paired-device architecture has another advantage: it can improve performance over slow mobile networks without sacrificing auditing. Because the laptop–phone link is relatively efficient, the paired phone can improve laptop performance by acting as a cache for it. Here the phone is configured to perform aggressive directory-level key prefetching and caching. On a key miss, the laptop contacts the phone via bluetooth and the phone returns the key, if available; otherwise the phone fetches the missed key and other related keys from the key service and returns the key to Keypad. Section 5 evaluates the performance improvement for this solution. As before, auditing properties are preserved if only the laptop is stolen. If both devices are stolen, then auditing is at a directory-level granularity.

3.6 Partial Coverage

Not all files necessarily require audit log entries. For example, as a trivial optimization we could exclude non-sensitive files such as binaries, libraries, and configuration files from Keypad's audited protection domain. In this scenario protected files are encrypted locally and their keys and metadata are stored remotely; unprotected files are (optionally) encrypted locally, but their encryption keys are derived from the user's login credentials.

The benefits of this optimization are obvious: Keypad's performance and availability costs are only incurred for protected files. There is also a risk: if a sensitive file is accidentally placed in an untracked file or directory, the audit logs will not reveal accesses to that sensitive data. One reasonable protection policy is to track accesses to any file in crucial directories, such as the user's home and temporary directory (e.g., `/home` and `/tmp` on Linux).

3.7 Summary

Keypad provides strong guarantees to its users. If a protected file is accessed, then at least one record related to that access will appear in the remote audit logs, and up-to-date metadata about the file will be available online. As we have shown, one challenge Keypad faces is preserving this strong property while overcoming the performance impact of communicating with remote services in the critical path of file accesses. We introduced a series of novel techniques to meet this challenge. Though some of these techniques have an impact on the quality of the information in the audit logs, we show in Section 5.2 that this impact is small.

4. Implementation

We implemented a Keypad prototype including the client-side Keypad filesystem, the key service, and the metadata service as shown in Figure 2. All components are coded in C++ and communicate using encrypted XML-RPC with persistent connections. Our client-side Keypad file system is an extension of EncFS [EncFS 2008], an open-source block-level encrypted file system based on FUSE [Fuse 2004]. EncFS encrypts all files, directories, and names under a single volume key, which is stored on disk encrypted under the user's password. Keypad extends EncFS in two ways. First, we modified EncFS to encrypt each file with its own per-file key. The single volume key is still used, however, to protect file headers and the file system's namespace, e.g., file and directory names. Second, Keypad stores all file keys on a remote key server and maintains up-to-date metadata on a metadata server. To support forensic analysis we built a simple Python tool; given a T_{loss} timestamp and an expiration time, T_{exp}, the tool reconstructs a full-fidelity audit report of all accesses after $T_{loss} - T_{exp}$, including full path names and access timestamps.

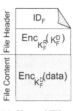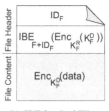

(a) Keypad File. (b) IBE-Locked File.

Figure 5. Keypad File Formats. Keypad on-disk file structure for the normal case (a) and the IBE-locked case (b).

Keypad File Structure. Figure 5(a) shows the internal structure of a Keypad file F, which consists of two regions: the file's header and its content. The file's header is fixed size and is encrypted using EncFS' volume key. For the file's content, our implementation adds a level of indirection for encryption keys to support techniques such as IBE efficiently. Specifically, file F's content is encrypted using a 256-bit random *data key*, denoted K_F^D. The data key is stored in the file's header encrypted under the *remote key*, denoted K_F^R. The remote key is stored on the key server and is identified by the file's audit ID (ID_F), which is a randomly generated 192-bit integer that is stored in the file's header along with the encrypted data key. This internal file structure is transparent to applications, which see only the decrypted contents of a file.

FS Operations. Keypad intercepts and alters two types of EncFS operations: file-content operations (`read`, `write`) and metadata-update operations (`create`, `rename` for files or directories). When an application accesses file content, Keypad: (1) looks up the file's audit ID from its header, (2) retrieves the remote key K_F^R, either from the local cache or the key service, (3) decrypts the data key K_F^D using K_F^R, (4) caches K_F^D temporarily, and (5) decrypts/encrypts the data using K_F^D.

When an application creates or updates file metadata, Keypad: (1) locks the data key using IBE, if enabled, and (2) sends the new metadata to the metadata service. The metadata is the file's path reported as a tuple of the form `directoryID/filename`. The names of Keypad directories are also kept current on the metadata service. While our current prototype applies IBE for file metadata update operations (e.g., file `create`, `rename`), it does not apply it to directory metadata operations (e.g., `mkdir` or directory `rename`), although this should be possible to add.

Key Expiration. Keypad caches keys for a limited time to improve performance. A background thread purges expired keys from the cache. If a key has been reused during its expiration period, the thread requests the key from the key service again, causing an audit record to be appended to the access log for that audit ID. If a response arrives before the key expires, the key's expiration time is updated in the cache, otherwise the key is removed. As a result, absent network failures, keys in Keypad never expire while in use. This ensures that long-term file accesses, such as playing a movie, will not exhibit hiccups due to remote-key fetching.

Key Prefetching. Key prefetching attempts to anticipate future file accesses by requesting file keys before the files are accessed. For our prototype, we sought a simple policy that would have both reasonable performance and little impact on auditability. We have experimented with two policies: (1) a random-prefetch scheme that prefetches random keys from the local directory upon every key-cache miss and (2) a full-directory-prefetch scheme that prefetches all keys in a directory when it detects that the directory is being scanned by an application. Our experiments indicated that the latter policy provided equally good performance, while incurring fewer false positives in the audit logs. Hence, our Keypad prototype uses it by default. The intuition behind our full-directory prefetch design is to avoid producing false positives for targeted workloads (such as interacting with a document, viewing a video, etc.) and to improve performance for scanning workloads (such as grepping through the files in a directory or copying a directory). Our full-directory-prefetch scheme avoids recursive prefetches to ensure that any false positives are triggered by real accesses to (related) files in the same directory. While other more effective prefetching policies may exist, our results show that our full-directory-prefetch policy, combined with our caching policies, reduce the number of blocking key requests to a point where the performance bottleneck shifts from blocking key requests to metadata requests (see Section 5).

IBE. To avoid blocking for metadata-update requests, our prototype implements IBE-based metadata registration, using an open-source IBE package [Boneh 2002]. On a metadata-update operation, Keypad locks the file until the metadata service confirms the receipt of the new file path; however, file operations can proceed for a one-second window, as previously described, to absorb the registration latency. Figure 5(b) shows the structure of an IBE-locked file. Its encrypted data key is further encrypted using IBE under a public key consisting of the file's path (directoryID/ filename) and the audit ID (ID_F). Embedding ID_F into the public key strongly binds ID_F and the path together at the metadata server. Handling updates for other types of file metadata functions (such as setfattr) works similarly, although our current prototype only supports pathnames as metadata.

Android-Based Paired-Device Prototype. We implemented a prototype of the paired-device architecture (Figure 4) using the Google Nexus One phone. A simple daemon (431 lines of Python) on the phone accepts key requests from the laptop over Bluetooth, saves accesses to a local database, responds to the laptop, and uploads access and metadata information to Keypad servers in bulk over wireless.

5. Evaluation

This section quantifies Keypad's performance and auditing quality. Keypad must be fast enough to preserve the usability of desktop and mobile applications, even in the face of adverse network conditions (e.g., 3G), while providing high quality auditing.

For our experiments, we used an eight-core 2GHz x86 machine running Linux 2.6.31 as our client. Our key service and name service daemons ran on 8 core 2.6GHz servers with 24GB of RAM, connected via gigabit Ethernet. We used Linux's traffic control utility to emulate different network latencies. We did not emulate different bandwidth constraints, however, Keypad's bandwidth requirements are very low. During a 12-day period in which one of our authors used Keypad continuously, average Keypad bandwidth was under 5 kb/s, with occasional spikes up to 45 kb/s.

Throughout the evaluation, we emulate the following RTTs for various networks: 0.1ms RTT for a LAN, 2ms RTT for a wireless LAN (WLAN), 25ms RTT for broadband, 125ms RTT for a DSL network, and 300ms RTT for a 3G cellular network. To illustrate network latency effects on Keypad performance, we often use examples from extreme network conditions, such as fast LANs and slow 3G networks, even though popular mobile connections today rely on WLAN and 4G.

5.1 Performance

To understand where the time goes for Keypad operations, we microbenchmarked file content (read and write) and metadata (create, rename, and mkdir) operations. Our measurements included client, server, and network latencies, as well as latency contributions for EncFS and Keypad.

Figure 6(a) shows the latency of file read and write operations for two cases: key-cache misses, which must fetch the key from the the server, and key-cache hits, which use a locally cached key. For each case we show data for two extreme networks: a fast 0.1ms-RTT LAN and a slow 300ms-RTT 3G network. The results show that misses are expensive on both networks, but for different reasons. On a LAN, the network is insignificant, but Keypad adds to the base EncFS time due to the XML-RPC marshalling overhead. On 3G, network latency dominates. When the key-cache hits, both the network and marshalling costs are eliminated; a file read with a cached key is only 0.01ms slower than the base EncFS read time of 0.337ms. This shows the importance of key caching to avoid misses, which we accomplish by carefully choosing our expiration and prefetching policies.

Figure 6(b) shows the latency of file metadata update operations. For create and rename, we show latency with and without IBE; mkdir is shown only without IBE, since it does not benefit from this optimization in our prototype. Without IBE, metadata update latency is driven primarily by network RTT: file creation takes 1.618ms on a LAN, and 302ms over 3G. With IBE, metadata update latency is inde-

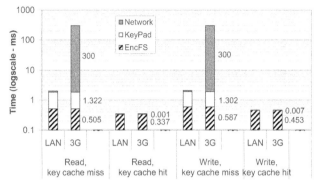

(a) FS Content Operations: `read`, `write`.

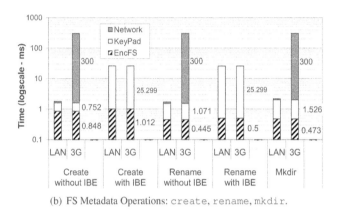

(b) FS Metadata Operations: `create`, `rename`, `mkdir`.

Figure 6. File Operation Latency. The latency of Keypad (a) content and (b) metadata-update operations. For each, we show the time spent in EncFS code, Keypad client and server code, and on the network. Labels on the graph show the latency for each component in the 3G 300ms RTT case. Results are averaged over 10 trials with a warm disk buffer cache.

pendent of network delay and is dominated by the computational cost of IBE itself. The figure shows that IBE meets its goal of improving performance of metadata updates over 3G. While IBE would add overhead for a LAN, it is unnecessary and would be disabled in the LAN environment.

5.1.1 Optimizations

We now demonstrate the effectiveness of our optimizations on a challenging workload: Apache compilation. While this workload is not characteristic of mobile devices, its complex nature make it ideal for evaluating the impact of our optimizations. In Section 5.1.2, we extend our evaluation to more typical workloads for mobile devices. As baselines, the Apache compilation takes 112s using the unmodified EncFS encrypted file system (i.e., with encryption but without auditing) and 63s on `ext3` (i.e., without encryption or auditing). Because Keypad enhances EncFS, the fair baseline comparison for Keypad is EncFS, and not `ext3`.

In what follows, we inspect the effect of optimizations as we enable one optimization after the other. We begin

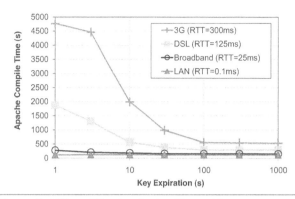

Figure 7. Effect of Key Expiration Time. This graph shows the effect of key expiration without any other optimizations enabled. A 100-s key expiration time is nearly optimal, and achieves compilation times of 115s, 153s, 292s, and 551s over a LAN, Broadband, DSL, and 3G, respectively. For comparison, the Apache compilation takes 112s on the unmodified EncFS and 63s on `ext3`.

by showing the effect of purely key caching with no other optimizations, then we add prefetching, then IBE, and finally we add the paired-device optimization.

Key Caching and Expiration. Key caching is crucial to performance. Even a cache with one-second expiration time has significant impact: 18% improvement on a LAN and 4.9x on 3G, relative to no caching at all. Figure 7 shows additional improvements for Apache compilation time as expirations are lengthened beyond one second. No optimizations other than caching are enabled here. Our results suggest that short expiration times are sufficient to extract nearly all the benefits. For LAN, Broadband, or DSL latencies, an expiration of 10s or so is optimal. Over 3G, a 100s key expiration time achieves all the benefit and provides 8.6x improvement over 1s (from 79.4 minutes down to 9.2 minutes). In comparison to EncFS, Keypad's performance degradation for 100s expiration times is already small over a LAN (5.3% overhead over EncFS), while for the other network types, further optimizations are required for performance.

Note that a 100s timeout is extremely small. To benefit from cached keys, a thief needs to steal the device within 100 seconds of the user's last access. Even in such cases, the user will know which files were exposed. We therefore believe that we can achieve both good performance and accurate auditing with these parameters.

Directory-Key Prefetching. Key caching alone avoids many key service requests: of the 75,744 reads and writes in the Apache compilation, only 486 involve the server when using a 100s expiration time. Directory-key prefetching avoids additional server requests. Prefetching a directory key on the first, third, or tenth miss in a directory results in 101, 249, and 424 key-cache misses, which translates into 63.3%, 24.1%, and 2.4% improvements, respectively, over not using directory-key prefetching over 3G. We adopted a prefetch-on-third-miss policy to strike a good balance between per-

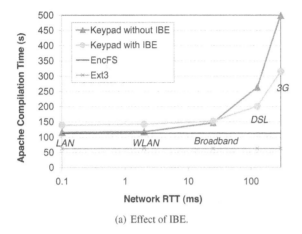

(a) Effect of IBE.

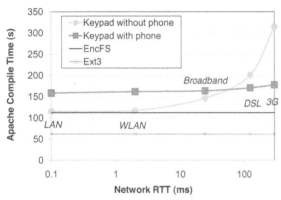

(b) Effect of Device Pairing.

Figure 8. Effect of IBE and Device-pairing Optimizations. (a) Effect of applying the IBE optimization atop a 100-s key caching policy and a third-miss prefetching policy; no device pairing is used here. (b) Effect of applying the device-pairing optimization atop the optimization setup in (a).

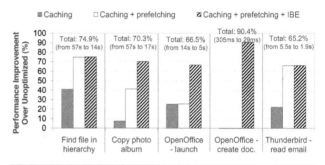

Figure 9. Impact of Optimizations on Various Applications. Impact of three of the optimizations on an emulated 3G network; labels indicate the total performance improvement when using all three optimizations over the unoptimized case, as well as the absolute numbers for the unoptimized and optimized.

formance and auditing quality (which is evaluated in Section 5.2). Over fast networks, such as a LAN and WLAN, the prefetch-on-third-miss policy coupled with 100-s key caching results in negligible performance overheads compared to EncFS: 2.8% for LAN and 4.3% for WLAN. Over slower networks, especially 3G, other smarter prefetching policies may improve performance by further eliminating blocking key requests. However, we find that with our simple prefetching policy, the dominating runtime component now becomes the blocking metadata requests (932 blocking metadata requests compared to the 249 blocking key requests). We next focus on optimizing metadata requests.

IBE. IBE tolerates the latency of metadata service requests over slow mobile networks. Figure 8(a) shows the impact of IBE on Apache compilation as a function of network RTT. As we see in the figure, IBE provides dramatic improvements on high-latency networks, including 3G- and 4G-class networks. For example, IBE improves the benchmark's per-

formance on 3G by 36.9%. The crossover for IBE is around 25ms, i.e., it should be used only for networks with RTTs over 25ms and disabled otherwise. As mentioned above, for faster networks, such as LANs or WLANs, IBE is not even necessary, as Keypad's overhead is already negligible after applying key caching and prefetching.

The Paired Device. Our paired device design is aimed at facilitating disconnected operation, but it can also provide performance benefits for high-latency network environments. Figure 8(b) shows the effect on the Apache workload of using a paired device as a caching proxy for key and metadata services. Two conclusions can be reached from the figure. First, performance for disconnected operation over Bluetooth should be similar to or better than that of a broadband connection (the latencies are similar). Second, pairing with another device is always beneficial for performance over cellular networks, because most operations only traverse the lower latency Bluetooth link. Obviously the paired device should not be used if fast networks are available, where Keypad is already efficient enough compared to EncFS.

5.1.2 Office-Oriented Workloads

Figure 9 shows the impact of our optimizations on more typical office-oriented workloads. We add optimizations incrementally, reporting additional improvement as more optimizations are added. The labels on top of each bar group show the total improvement with all three optimizations enabled. Different workloads benefit the most from different optimizations, depending primarily on the relative frequency of those operations. For example, caching and prefetching are important for a read-intensive workload such as a recursive grep ("Find file in hierarchy"). IBE provides large improvements for workloads that create files ("OpenOffice – create doc"). For mixed content/metadata workloads, such as copying a photo album across directories, all optimizations are important.

To better understand performance across many applications, we benchmarked the time to perform a number of pop-

Application	Task	EncFS	Keypad				
			LAN (RTT=0.1ms)	WLAN (RTT=2ms)	Broadband (RTT=25ms)	DSL (RTT=125ms)	3G (RTT=300ms)
OpenOffice Word Processor	Launch	0.5	0.5 | 0.5	0.6 | 0.6	1.3 | 1.3	2.7 | 2.7	4.6 | 4.6
	New document	0.0	0.0 | 0.0	0.0 | 0.0	0.0 | 0.0	0.0 | 0.1	0.0 | 0.3
	Save as	1.4	1.4 | 1.4	1.4 | 1.4	1.5 | 1.5	1.6 | 1.8	2.0 | 2.3
	Open	1.7	1.7 | 1.7	1.8 | 1.8	2.0 | 2.2	2.1 | 4.0	2.1 | 7.5
	Quit	0.1	0.1 | 0.1	0.1 | 0.1	0.3 | 0.4	0.4 | 0.7	0.4 | 1.2
Firefox	Launch	3.7	3.7 | 3.7	3.8 | 3.8	4.4 | 4.4	6.0 | 6.0	8.8 | 8.8
	Save a page	0.7	0.7 | 0.7	0.7 | 0.7	0.7 | 0.8	0.9 | 1.5	1.3 | 2.8
	Load bookmark	4.5	4.5 | 4.5	4.5 | 4.5	4.5 | 4.6	4.5 | 5.0	4.5 | 5.7
	Open tab	0.2	0.2 | 0.2	0.2 | 0.2	0.2 | 0.2	0.2 | 0.4	0.2 | 0.8
	Close tab	0.0	0.0 | 0.0	0.0 | 0.0	0.0 | 0.0	0.0 | 0.1	0.0 | 0.3
Thunderbird	Launch	1.3	1.3 | 1.3	1.3 | 1.3	1.4 | 1.4	2.0 | 2.0	3.1 | 3.1
	Read email	0.3	0.4 | 0.4	0.4 | 0.4	0.5 | 0.6	1.0 | 1.5	1.9 | 2.5
	Quit	0.2	0.2 | 0.2	0.2 | 2.2	0.2 | 0.4	0.2 | 1.3	0.2 | 2.9
Evince PDF Viewer	Launch	0.1	0.1 | 0.1	0.1 | 0.1	0.1 | 0.1	0.1 | 0.1	0.1 | 0.4
	Open document	0.1	0.1 | 0.1	0.1 | 0.1	0.1 | 0.1	0.2 | 0.2	0.4 | 0.4
	Quit	0.0	0.0 | 0.0	0.0 | 0.0	0.0 | 0.0	0.0 | 0.0	0.0 | 0.0

x \| y: x = time with warm key cache
y = time with cold key cache

Table 1. Typical Application Performance Over Keypad. For Keypad, we show both warm and cold key-cache times, separated by a \|.

ular tasks using EncFS and Keypad over several emulated networks (Table 1). For Keypad, we show both warm and cold key-cache times. A user will likely experience both, but with well-chosen key expiration times many operations will be absorbed by a warm cache.

From a user's perspective, Keypad performs roughly identically to EncFS over fast networks, such as a LAN and a wireless LAN. Hence, while at the office, the user should never feel our file system's presence, whether its key cache is warm or cold. With only a few exceptions, the user should perceive similar application performance over broadband with Keypad and the unmodified EncFS. Over mobile networks, the user may notice some application slowdown, especially after extended periods of inactivity.

The table and our own experience confirm that application launches are particularly expensive over 3G networks, as they often encounter a cold cache and many file system interactions. Keypad could optimize launch by profiling applications and prefetching needed keys; other file systems, such as NTFS, perform similar special-case optimizations to speed up application launch.

5.1.3   Comparison to Other File Systems

A networked file system might be an alternative to Keypad; instead of just storing keys remotely, all file system content would be remote. NFS provides a reasonably fair comparison to Keypad, since its short-term caching might provide audit properties comparable to ours. In contrast, for AFS and Coda, their long-term, coarse granularity caching policies might interfere more with precise audit semantics.

Figure 10 shows the relative performance of Keypad to (remote) NFSv3 and (local) EncFS for Apache compilation. We configured NFS with asynchronous batched writes and its default caching policy; this improves its performance, but

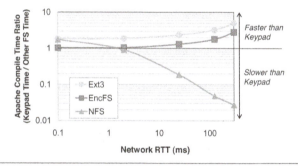

Figure 10. Comparison to EncFS and NFS.

would have some impact on auditing. Note that for these experiments, as before, we emulated different network RTTs but we did not constrain network bandwidth; thus, our results are upper bounds of NFS performance. Over actual 3G links, NFS performance would be significantly degraded because of wireless bandwidth constraints.

With LAN latencies, Keypad's performance is almost identical to EncFS with only a 2.78% increase in runtime, but worse than NFS, with a 75% increase. For reference, the unmodified EncFS itself is 71% slower compared to NFS with LAN-like latencies. As RTT grows, NFS degrades significantly. Even with an RTT of 2ms, NFS is 8.8% *slower* than Keypad, while for 3G network latencies of 300ms, NFS is 36.4x slower than Keypad! In contrast, Keypad is only 2.7x slower than EncFS over a 300ms network.

On large-RTT networks, NFS impacts interactivity. For example, launching OpenOffice over NFS with 3G latency takes 50.6 seconds, loading a bookmark in Firefox takes 27.6 seconds, and opening an email in Thunderbird takes 12.5 seconds, which we believe is unacceptable performance for these user-facing tasks.

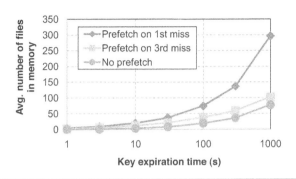

Figure 11. Effect of Optimization on Auditability. The average number of keys that reside in memory at any point in time, under various key expiration times and prefetching policies.

5.1.4 Anecdotal Experience

Anecdotally, one co-author used Keypad continuously to protect his laptop's $HOME and /tmp directories over a 12-day period, with an emulated 300ms client-to-server latency. Overall, the experience was positive: in most cases, there was no noticeable performance impact. Some activities, such as file system intensive CVS checkouts or recursive copies, were slower but usable. Other more typical activities, such as browsing the Web, editing documents, and exchanging email, had no noticeable performance degradation.

5.2 Auditing Properties

We now evaluate our optimizations' impact on auditability.

In-memory Key Sets. As described in Section 3.3, keys for recently-accessed or prefetched files stay in memory for their expiration period T_{exp}. This is not an issue for a thief who steals a passive storage device, such as a USB stick. For a laptop, because a thief can theoretically access cached-key files without triggering a server-side audit log, users must consider all files whose keys were retrieved between $T_{loss} - T_{exp}$ and T_{loss} as compromised. The size of this set at any point in time depends on the user's workload and on the aggressiveness of the caching and prefetching schemes.

To quantify this issue we used a trace gathered during our twelve-day deployment experience (Section 5.1.4) to calculate the impact of various optimizations on auditability. Figure 11 shows the size of the in-memory key set at any point in time averaged over use periods, for different key expiration times and prefetching policies. The graph shows that for reasonable key expiration and prefetching strategies, the average number of in-memory keys is small. For example, with a 100-second key expiration time and a prefetch-directory-on-third-miss strategy, on average there are 38 keys in memory at any instant. This is a small number and furthermore we observed that most of these keys exist as a side-effect of prefetching; i.e., they are files in the same directory as a file that was accessed by a user or program.

False Positives. Prefetching affects forensics by introducing false positives in the audit log. The rate of false posi-

tives depends on the prefetching policy as well as the thief's workload, since false positives only concern time post-T_{loss}. In the absence of an accepted "thief workload," we created a few scenarios that a thief might follow. Our goal was to gauge the impact of various prefetching policies on the rate of false positives, as a thief tries to find sensitive information on a captured device. We investigated three scenarios: (1) the thief launches Thunderbird, reads a few emails, browses folders, and searches for emails with a particular keyword; (2) he launches a document editor and looks at a few files; and (3) he inspects the history, bookmarks, cookies, and passwords in a Firefox window. For these workloads, our default prefetch policy (prefetch directory keys on the 3rd miss) leads to the following ratios between false positives and total accessed keys: 3:30, 6:67, and 0:12 for our Thunderbird, document editor, and Firefox workloads, respectively. Audit precision is high for these scenarios.

We have also discovered bad scenarios; if the thief navigates to a web page in Firefox, loading several files from the cache directory causes Keypad to prefetch the entire directory. While this causes several false positives, the user correctly learns that activity happened in the Firefox cache directory. Even in such cases, the auditing implications of our non-recursive prefetching policy are minimal, since all false positives are localized to one directory.

5.3 Summary

We measured the performance of our Keypad prototype on several workloads. Our measurement results and our experience using the system show that Keypad meets its goals of adding little overhead in the office or home environment, while remaining highly usable over cellular networks, such as 3G. Overall, our results show that with properly parameterized optimizations, Keypad can provide good performance while also maintaining good auditing fidelity. Furthermore, with current and future improvements of cellular network connectivity (e.g., 4G), we expect Keypad to have even better performance.

6. Security Analysis

Keypad is designed to provide strong audit guarantees for encrypted file systems if the first layer of defense, encryption with a password or cryptographic token, is breached. Keypad can additionally destroy the ability to read files after a mobile device is reported lost or stolen. Although we evaluated security properties extensively inline above, we now return for a unified discussion.

Context and Threat Model. We designed Keypad assuming that individuals who find or steal a mobile device range in sophistication, degree of planning, and interest. Curious individuals may insert a found USB stick into their computer, enter the password on the attached sticky note, and browse through a few files trying to find the device owner. Petty thieves may grab laptops opportunistically but have no real interest in accessing confidential files. Corporate spies may plan and execute device theft carefully, with the goal of accessing confidential files before the victim reports the device missing. We refer to all such individuals as "attackers."

Because a user has no way of knowing the motivation and skill of a potential attacker, Keypad assumes the worst. We assume that an attacker has full access to the lost device's hardware (for laptops and USB sticks) and software (for laptops). The attacker can perform cold-boot attacks on laptops, install new software, and manipulate or sever the device's network traffic. The attacker can also perform lower-level activities, such as physically extracting the hard drive from a laptop or memory from a USB stick and interrogating it with custom hardware. However, we do not consider attacks in which the adversary gains control of the device, modifies it, and returns it to the victim without his knowledge (see our non-goals discussion in Section 2). Any attacker with control over a device while in the user's possession could mount a slew of malicious attacks outside the scope of a forensic file system, ranging from online data exfiltration to the installation of password key loggers. Botnets and other forms of malware are therefore also explicitly outside our threat model.

Analysis. We begin with the premise that the audit servers are trusted and secure. The key and metadata servers are trusted to maintain accurate logs, and they are assumed to incorporate strong defenses to adversarial comprise, routinely back up their state, and have their own audit mechanisms. Neither the key server nor the metadata server is, however, fully trusted with the private information about a user's file access patterns prior to T_{loss}; accessing that information requires collusion between both servers or the device owner's invocation of the Keypad post-loss audit mechanisms. The unavailability of servers can deny access to files; for highly sensitive data, we argue that users would prefer unavailability over the potential for unaudited future file disclosure. Further, although not implemented in our prototype, the communications between the Keypad file system and the servers should be encrypted to ward off attackers who intercept network communications prior to device theft. The keys must change every T_{exp} seconds to ensure that an attacker who extracts the current network encryption key from the device cannot decrypt past intercepted data.

Consider now an attacker who obtains a lost or stolen Keypad device. If the device is cold, such as a powered-down laptop or a USB stick, then any successful attempt to access a protected file must generate at least one log record on the Keypad audit servers. This is true whether the attacker uses the Keypad file system or his own hardware or software to perform the access. All of Keypad's mechanisms – the storing of K_F^R on remote servers, the entangling of the metadata server and key server states to ensure consistency, and our method for using IBE – enforce this property. Additionally, the selection of 192-bit audit IDs at random makes it infea-

sible for an attacker to request information about valid audit IDs from the key and metadata servers prior to physically obtaining the protected device; such requests are additionally thwarted by authenticating the device to the servers.

Attackers who obtain warm, computational devices – such as running or hibernated laptops – may seek to violate the properties of Keypad by directly accessing the device's memory. Cached keys K_F^R should be evicted from memory upon device hibernation, and such evictions should be recorded on the audit servers. For fully running devices, we must assume that an attacker has accessed any file with an audit log entry after $T_{loss} - T_{exp}$. Although Keypad's focus is on providing file system auditing, a forensic analyst must also acknowledge that applications may have sensitive data in memory. A conservative analyst might use various heuristics to identify potentially vulnerable cleartext data. For example, he might mark as compromised any file opened since the device's last boot or hibernation, events that could be recorded on the audit servers. A potentially better future solution to this problem might be to employ encrypted memory technology [Provos 2000], possibly coupled with auditing.

Most importantly, even against an attacker who obtains warm computational devices, Keypad preserves the following invariant: if an analyst does not mark a file as accessed, then one can confidently conclude that the file has indeed not been accessed by an attacker. Finally, because entries in the key service are identified per-device, the service can deny access to all relevant keys if a device is reported missing.

For completeness, we must also consider an attacker who attempts to generate spurious entries in the remote audit logs. While such spurious entries might complicate the task of a forensic analyst, an attacker cannot use such actions to hide their actual accesses of confidential data.

7. Related Work

Keypad is related to previous work in three areas: (1) theft-protection systems, (2) data-protection systems, and (3) distributed file systems.

Theft-Protection Systems. Commercial and research theft-protection systems, such as Apple's MobileMe and Adeona [Ristenpart 2008], rely on software running on a device that can monitor file accesses, report device locations and file accesses to a remote server, and delete files upon request. A determined attacker can circumvent these protections and analyze the device's media using his own hardware, without the associated monitoring software installed. Keypad provides strong forensic and data-destruction capabilities even against thieves who use their own hardware and software to attack a Keypad-protected file system.

Data-Protection Systems. Encrypted file systems exist in academia (e.g., [Blaze 1993]) and industry (e.g., BitLocker, PGP Whole Disk, TrueCrypt). None provide remote auditing capabilities, therefore a security breach may go undetected. Keypad's forensic and data-destruction capabilities are orthogonal to work increasing the resilience of encrypted file systems to breach. Keypad can compose with new advances in encrypted file systems, providing both stronger barriers to access and a forensic trail if that barrier is breached.

ZIA [Corner 2002] and follow-on work [Corner 2003] protect files on a device with transient authentication. ZIA users wear small tokens that broadcast their presence. When a token is near a protected device, the device decrypts; when the token leaves, the device re-encrypts. Protection is lost if an attacker obtains both the device and the token, with no forensic guarantees. Keypad does not require a paired device, but if one is used, Keypad still provides a forensic trail of accesses even if both are lost or stolen. Keypad could be combined with ZIA for additional defense in depth.

Keypad's remote key-escrow architecture has been used frequently in the past to achieve a number of security and privacy goals. First, capture-resilient cryptography [MacKenzie 2001] uses a key server to prevent dictionary attacks against login passwords on stolen devices, as well as to enable remote wipe-out. Second, location-aware encryption [Studer 2010] uses a remote key server to dynamically adapt a device's data protection level based on its location. While the device is at a trusted location (e.g., at its owner's home), the server provides the decryption key; when the device is at an unknown or untrusted location, the server will require the user to enter a special password to return obtain the decryption key. Third, assured-delete systems, such as the Ephemerizer [Perlman 2005], revocable backup systems [Boneh 1996], and the Vanish distributed-trust self-destructing data system [Geambasu 2009] adopt the key-escrow architecture to ensure the deletion of sensitive data stored in backup systems or on Web services. Keypad resembles all of these systems in its remote key-escrow architecture and its secondary goal: post-theft data destruction. It differs from these systems in its primary goal: fine-grained auditability of mobile device data accesses.

In general, today's data-protection systems differ from our system in that they focus on data exposure *prevention*, whereas Keypad focuses on data exposure *detection* should prevention systems fail. In that sense, they should be considered as complementary rather than competitors.

Networked File Systems. Work in distributed file systems has aimed at providing shared and available remote storage (e.g., [Howard 1988, Lee 1996, Sandberg 1985]). Bayou [Peterson 1997] and Coda [Mummert 1995] support mobility, disconnected operation and data consistency. Coda's disconnected operation [Kistler 1991] relies on data caching, whereas Keypad uses device pairing, coupled with key caching, to support offline accesses. Coda supports encrypted communication but not storage. LBFS [Muthitacharoen 2001] uses compression to reduce latency for interactive file access over slow wide-area networks. SFS [Fu 2002, Mazieres 1999] is a network file system that supports secure network file transfers, avoiding the need for distributed key infrastructure by embedding public keys in file pathnames. SFS is concerned with secure communication, not with protecting a user's stored data from theft; it does not encrypt data on disk and does not support auditing.

In general, these systems do not support encryption and auditing. While they could be modified to support both on the server, there are significant performance issues, e.g., streaming an NFS-hosted video over 3G or wireless is slow and expensive. Finally, all of these systems are concerned with the transfer of *file data* between a client and server; in contrast, Keypad is concerned with *key management* and the transfer of encryption keys between a file system and a remote key server. Keypad is unique in its support for (and integration of) encryption and audit logging; it demonstrates the advantage of separating encryption and key management to enforce auditing for mobile device data.

8. Conclusions

This paper described Keypad, an auditing file system for loss- and theft-prone devices. Unlike basic disk encryption, Keypad provides users with evidence that sensitive data either was or was not accessed following the disappearance of a device. If data was accessed, Keypad gives the user an audit log showing which directories and files were touched. It also allows users to disable file access on lost devices, even if the device has been disconnected from the network or its disk has been removed. Keypad achieves its goals through the integration of encryption, remote key management, and auditing. Our measurements and experience demonstrate that Keypad is usable and effective for common workloads on today's mobile devices and networks.

9. Acknowledgements

We offer thanks to our shepherd Mahadev Satyanarayanan and the anonymous reviewers for their valuable comments on the paper. This work was supported by NSF grants CNS-0846065, CNS-0627367, and CNS-1016477, the Google Fellowship in Cloud Computing, the Alfred P. Sloan Research Fellowship, the Torode Family Career Development Professor, the Wissner-Slivka Chair, and a gift from Nortel Networks.

References

[Anderson 1996] Ross Anderson and Markus Kuhn. Tamper resistance: A cautionary note. In *Proceedings of the 2nd USENIX Workshop on Electronic Commerce (WOEC '96)*, 1996.

[Blaze 1993] Matt Blaze. A cryptographic file system for UNIX. In *Proceedings of the 1st ACM Conference on Computer and Communications Security (CCS '93)*, 1993.

[Boneh 2001] Dan Boneh and Matthew K. Franklin. Identity-based encryption from the Weil pairing. In *Proceedings of the 21st Annual International Cryptology Conference on Advances in Cryptology (CRYPTO '01)*, 2001.

[Boneh 2002] Dan Boneh, Matthew K. Franklin, Ben Lynn, Matt Pauker, Rishi Kacker, and Gene Tsudik. Identity-based encryption download. http://crypto.stanford.edu/ibe/download.html, 2002.

[Boneh 1996] Dan Boneh and Richard Lipton. A revocable backup system. In *Proceedings of the 6th USENIX Security Symposium*, 1996.

[Corner 2002] Mark D. Corner and Brian D. Noble. Zero-interaction authentication. In *Proceedings of the Eighth Annual International Conference on Mobile Computing and Networking (MobiCom '02)*, 2002.

[Corner 2003] Mark D. Corner and Brian D. Noble. Protecting applications with transient authentication. In *Proceedings of the First International Conference on Mobile Systems, Applications, and Services (MobiSys '03)*, 2003.

[EncFS 2008] EncFS. EncFS encrypted filesystem. `http://www.arg0.net/encfs`, 2008.

[Fu 2002] Kevin Fu, M. Frans Kaashoek, and David Mazieres. Fast and secure distributed read-only file system. *ACM Transactions on Computer Systems (TOCS)*, 20(1):1–24, 2002.

[Fuse 2004] Fuse. Filesystem in userspace. `http://fuse.sourceforge.net/`, 2004.

[Geambasu 2009] Roxana Geambasu, Tadayoshi Kohno, Amit Levy, and Henry M. Levy. Vanish: Increasing data privacy with self-destructing data. In *Proceedings of the 18th USENIX Security Symposium*, 2009.

[Halderman 2008] J. Alex Halderman, Seth D. Schoen, Nadia Heninger, William Clarkson, William Paul, Joseph A. Calandrino, Ariel J. Feldman, Jacob Appelbaum, and Edward W. Felten. Lest we remember: Cold boot attacks on encryption keys. In *Proceedings of the 17th USENIX Security Symposium*, 2008.

[Howard 1988] John H. Howard, Michael L. Kazar, Sherri G. Menees, David A. Nichols, M. Satyanarayanan, Robert N. Sidebotham, and Michael J. West. Scale and performance in a distributed file system. *ACM Transactions on Computer Systems (TOCS)*, 6(1):51–81, 1988.

[Imperva 2010] Imperva. Consumer password worst practices. `http://www.imperva.com/docs/WP_Consumer_Password_Worst_Practices.pdf`, 2010.

[Kistler 1991] James J. Kistler and M. Satyanarayanan. Disconnected operation in the Coda file system. In *Proceedings of the 13th ACM Symposium on Operating System Principles (SOSP '91)*, 1991.

[Lee 1996] Edward K. Lee and Chandramohan A. Thekkath. Petal: Distributed virtual disks. In *Proceedings of the 7th International Conference on Architectural Support for Programming Languages and Operating Systems (ASPLOS '96)*, 1996.

[MacKenzie 2001] Philip MacKenzie and Michael K. Reiter. Delegation of cryptographic servers for capture-resilient devices. In *Proceedings of the 8th ACM Conference on Computer and Communications Security (CCS '01)*, 2001.

[Mazieres 1999] David Mazieres, Michale Kaminsky, M. Frans Kaashoek, and Emmett Witchel. Separating key management from file system security. In *Proceedings of the 17th ACM Symposium on Operating Systems Principles (SOSP '99)*, 1999.

[Mummert 1995] Lily B. Mummert, Maria R. Ebling, and M. Satyanarayanan. Exploiting weak connectivity for mobile file access. In *Proceedings of 15th ACM Symposium on Operating Systems Principles (SOSP '95)*, 1995.

[Muthitacharoen 2001] Athicha Muthitacharoen, Benjie Chen, and David Mazieres. A low-bandwidth network file system. In *Proceedings of the 18th ACM Symposium on Operating Systems Principles (SOSP '01)*, 2001.

[Nusca 2009] Andrew Nusca. How to: Keep your laptop from being stolen. `http://www.zdnet.com/`, February 2009.

[Perlman 2005] Radia Perlman. File system design with assured delete. In *Proceedings of the 3rd IEEE International Security in Storage Workshop (SISW '05)*, 2005.

[Peterson 1997] Karin Peterson, Mike J. Spreitzer, Douglas B. Terry, Marvin M. Theimer, and Alan J. Demers. Flexible update propagation for weakly consistent replication. In *Proceedings of the 16th ACM Symposium on Operating Systems Principles (SOSP '97)*, 1997.

[Ponemon Institute 2008] Ponemon Institute. Airport insecurity: The case of lost and missing laptops; U.S. and EMEA result. `http://www.ponemon.org/data-security`, 2008.

[Provos 2000] Niels Provos. Encrypting virtual memory. In *Proceedings of the 9th USENIX Security Symposium*, 2000.

[Ristenpart 2008] Thomas Ristenpart, Gabriel Maganis, Arvind Krishnamurthy, and Tadayoshi Kohno. Privacy-preserving location tracking of lost or stolen devices: Cryptographic techniques and replacing trusted third parties with DHTs. In *Proceedings of the 17th USENIX Security Symposium*, 2008.

[Robertson 2010] Jordan Robertson. `http://www.usatoday.com/tech/news/computersecurity/2010-02-08-security-chip-pc-hacked_N.htm`, 2010.

[Rutkowska 2009] Joanna Rutkowska. Evil maid goes after TrueCrypt! `http://theinvisiblethings.blogspot.com/2009/10/evil-maid-goes-after-truecrypt.html`, 2009.

[Sandberg 1985] Russel Sandberg, David Goldberg, Steve Kleiman, Dan Walsh, and Bob Lyon. Design and implementation of the Sun network file system. In *Proceedings of the USENIX Annual Technical Conference*, 1985.

[Savage 2009] Michael Savage. NHS 'loses' thousands of medical records. `http://www.independent.co.uk/news/uk/politics/nhs-loses-thousands-of-medical-records-1690398.html`, 2009.

[Shamir 1985] Adi Shamir. Identity-based cryptosystems and signature schemes. In *Proceedings of the 5th Annual International Cryptology Conference on Advances in Cryptology (CRYPTO '85)*, 1985.

[Sorrel 2010] Charlie Sorrel. Brits send 4,500 USB sticks to the cleaners. `http://www.wired.com/`, 2010.

[Studer 2010] Ahren Studer and Adrian Perrig. Mobile user location-specific encryption (MULE): Using your office as your password. In *Proceedings of the 3rd ACM Conference on Wireless Network Security (WiSec '10)*, 2010.

[Whitten 1999] Alma Whitten and J.D. Tygar. Why Johnny can't encrypt: a usability evaluation of PGP 5.0. In *Proceedings of the 8th USENIX Security Symposium*, 1999.

Database Engines on Multicores,
Why Parallelize When You Can Distribute? *

Tudor-Ioan Salomie   Ionut Emanuel Subasu   Jana Giceva   Gustavo Alonso

Systems Group, Computer Science Department
ETH Zurich, Switzerland
{tsalomie, subasu, gicevaj, alonso}@inf.ethz.ch

Abstract

Multicore computers pose a substantial challenge to infrastructure software such as operating systems or databases. Such software typically evolves slower than the underlying hardware, and with multicore it faces structural limitations that can be solved only with radical architectural changes. In this paper we argue that, as has been suggested for operating systems, databases could treat multicore architectures as a distributed system rather than trying to hide the parallel nature of the hardware. We first analyze the limitations of database engines when running on multicores using MySQL and PostgreSQL as examples. We then show how to deploy several replicated engines within a single multicore machine to achieve better scalability and stability than a single database engine operating on all cores. The resulting system offers a low overhead alternative to having to redesign the database engine while providing significant performance gains for an important class of workloads.

Categories and Subject Descriptors   H.2.4 [*Information Systems*]: DATABASE MANAGEMENT—Systems

General Terms   Design, Measurement, Performance

Keywords   Multicores, Replication, Snapshot Isolation

1. Introduction

Multicore architectures pose a significant challenge to existing infrastructure software such as operating systems [Baumann 2009; Bryant 2003; Wickizer 2008], web servers [Veal

2007], or database engines [Hardavellas 2007; Papadopoulos 2008].

In the case of relational database engines, and in spite of the intense research in the area, there are still few practical solutions that allow a more flexible deployment of databases over multicore machines. We argue that this is the result of the radical architectural changes that many current proposals imply. As an alternative, in this paper we describe a solution that works well in a wide range of use cases and requires no changes to the database engine. Our approach is intended neither as a universal solution to all use cases nor as a replacement to a much needed complete redesign of the database engine. Rather, it presents a new architecture for database systems in the context of multicores relying on well known distributed system techniques and proving to be widely applicable for many workloads.

1.1 Background and Trends

Most commercial relational database engines are based on a decades old design optimized for disk I/O bottlenecks and meant to run on single CPU computers. Concurrency is achieved through threads and/or processes with few of them actually running simultaneously. Queries are optimized and executed independently of each other with synchronization enforced through locking of shared data structures. All these features make the transition to modern hardware platforms difficult.

It is now widely accepted that modern hardware, be it multicore, or many other developments such as flash storage or the memory-CPU gap, create problems for current database engine designs. For instance, locking has been shown to be a major deterrent for scalability with the number of cores [Johnson 2009a] and the interaction between concurrent queries when updates or whole table scans are involved can have a severe impact on overall performance [Unterbrunner 2009]. As a result, a great deal of work proposed either ways to modify the engine or to completely redesign the architecture. Just to mention a few examples, there are proposals to replace existing engines with pure main memory scans [Unterbrunner 2009]; to use dynamic

---

* This work is partly funded by the Swiss National Science Foundation under the programs ProDoc Enterprise Computing and NCCR MICS

programming optimizations to increase the degree of parallelism for query processing [Han 2009]; to use *helper cores* to efficiently pre-fetch data needed by working threads [Papadopoulos 2008]; to modularize the engine into a sequence of stages, obtaining a set of self-contained modules, which improve data locality and reduce cache problems [Harizopoulos 2005]; or to remove locking contention from the storage engine [Johnson 2009a,b]. Commercially, the first engines that represent a radical departure from the established architecture are starting to appear in niche markets. This trend can be best seen in the several database appliances that have become available (e.g., TwinFin of IBM/Netezza, and SAP Business Datawarehouse Accelerator; see [Alonso 2011] for a short overview).

1.2 Results

Inspired by recent work in *multikernel* operating systems [Baumann 2009; Liu 2009; Nightingale 2009; Wentzlaff 2009], our approach deploys a database on a multicore machine as a collection of distributed replicas coordinated through a middleware layer that manages consistency, load balancing, and query routing. In other words, rather than redesigning the engine, we partition the multicore machine and allocate an unmodified database engine to each partition, treating the whole as a distributed database.

The resulting system, *Multimed*, is based on techniques used in LANs as part of computer clusters in the Ganymed system [Plattner 2004], adapted to run on multicore machines. Multimed uses a primary copy approach (the *master database*) running on a subset of the cores. The master database receives all the update load and asynchronously propagates the changes to *satellite databases*. The satellites store copies of the database and run on non overlapping subsets of the cores. These satellites are kept in sync with the master copy (with some latency) and are used to execute the read only load (queries). The system guarantees global consistency in the form of snapshot isolation, although alternative consistency guarantees are possible.

Our experiments show that a minimally optimized version of Multimed exhibits both higher throughput with lower response time and more stable behavior as the number of cores increase than standalone versions of PostgreSQL and MySQL for standard benchmark loads (TPC-W).

1.3 Contribution

The main contribution of Multimed is to show that a share-nothing design similar to that used in clusters works well in multicore machines. The big advantage of such an approach is that the database engine does not need to be modified to run in a multicore machine. The parallelism offered by multicore is exploited through the combined performance of a collection of unmodified engines rather than through the optimization of a single engine modified to run on multiple cores. An interesting aspect of Multimed is that each engine is restricted in the number of cores and the amount of memory it can use. Yet, the combined performance of several engines is higher than that of a single engine using all the cores and all the available memory.

Like any database, Multimed is not suitable for all possible use cases but it does support a wide range of useful scenarios. For TPC-W, Multimed can support all update rates, from the browsing and shopping mix to the ordering mix, with only a slight loss of performance for the ordering mix. For business intelligence and data warehouse loads, Multimed can offer linear scalability by assigning more satellites to the analytical queries.

Finally, in this paper we show a few simple optimizations to improve the performance of Multimed. For instance, partial replication is used to reduce the memory footprint of the whole system. Many additional optimizations are possible over the basic design, including optimizations that do not require modification of the engine, (e.g., data placement strategies, and specialization of the satellites through the creation of indexes and data layouts tailored to given queries).

The paper is organized as follows. In the next section we motivate Multimed by analyzing the behavior of PostgreSQL and MySQL when running on multicore machines. The architecture and design of Multimed are covered in section 3, while section 4 discusses in detail our experimental evaluation of Multimed. Section 6 discusses related work and section 7 concludes the paper.

2. Motivation

To explore the behavior of traditional architectures in more detail, we have performed extensive benchmarks over PostgreSQL and MySQL (open source databases that we can easily instrument and where we can map bottlenecks to concrete code sequences). Our analysis complements and confirms the results of similar studies done on other database engines over a variety of multicore machines [Hardavellas 2007; Johnson 2009b; Papadopoulos 2008].

The hardware configuration and database settings used for running the following experiments are described in section 4.1. The values for L2 cache misses and context switches were measured using a runtime system profiler [OProfile].

2.1 Load interaction

Conventional database engines assign threads to operations and optimize one query at a time. The execution plan for each query is built and optimized as if the query would run alone in the system. As a result, concurrent transactions can significantly interfere with each other. This effect is minor in single CPU machines where real concurrency among threads is limited. In multicores, the larger number of hardware contexts leads to more transactions running in parallel which in turn amplifies load interaction.

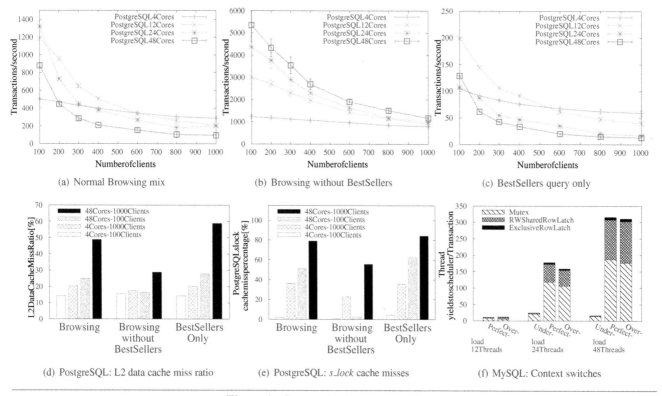

(a) Normal Browsing mix

(b) Browsing without BestSellers

(c) BestSellers query only

(d) PostgreSQL: L2 data cache miss ratio

(e) PostgreSQL: *s_lock* cache misses

(f) MySQL: Context switches

Figure 1. Current databases on multicore

We have investigated load interaction in both PostgreSQL and MySQL using the *Browsing* mix of the TPC-W Benchmark (see below for details on the experimental setup).

PostgreSQL's behavior with varying number of clients and cores is shown for the *Browsing* mix in figure 1(a); for all other queries in the mix except *BestSellers* in figure 1(b); and for the *BestSellers* query only in figure 1(c).

For the complete mix (figure 1(a)), we observe a clear performance degradation with the number of cores. We traced the problem to the *BestSellers* query, an analytical query that is performing scans and aggregation functions over the three biggest tables in the database. On one hand the query locks a large amount of resources and, while doing this, causes a large amount of context switches. On the other hand all the concurrent queries have to wait until the *BestSellers* query releases the locked resources. When this query is removed from the mix, figure 1(b), the throughput increases by almost five times and now it actually improves with the number of cores. When running the *BestSellers* query alone (figure 1(c)), we see a low throughput due to the interference among concurrently running queries and, again, low performance as the number of cores increases.

The interesting aspect of this experiment is that *BestSellers* is a query and, as such, is not doing any updates. The negative load interaction it causes arises from the competition for resources, which becomes worse as the larger number of cores allows us to start more queries concurrently.

Similar effects have been observed in MySQL, albeit for loads involving full table scans [Unterbrunner 2009]. Full table scans require a lot of memory bandwidth and slow down any other concurrent operation, providing another example of negative load interaction that becomes worse as the number of cores increases.

2.2 Contention

One of the reasons why loads interact with each other is contention. Contention in databases is caused mainly by concurrent access to locks and synchronization primitives.

To analyze this effect in more detail, we have profiled PostgreSQL while running the *BestSellers* query. The individual run time for this query, running alone in the system, is on average less than 80ms, indicating that there are no limitations in terms of indexing and data organization.

Figure 1(d) shows the L2 data cache misses for the full Browsing mix, the Browsing mix without the *BestSellers* and the *BestSellers* query alone. The L2 data cache miss ratio was computed using the expression below based on measured values for L2 cache misses, L2 cache fills and L2 requests (using the CPU performance counters). We have done individual measurements for each CPU core, but as there are no significant differences between the cores, we used the averaged values of the measured metrics.

$$L2DC\_Miss\_Ratio = \frac{100 \times L2Cache\_Misses}{(L2Cache\_Fills + L2Requests)}$$

With more clients and cores, we see a high increase in cache misses for the workloads containing the *BestSellers* query. We have traced this behavior to the "s_lock" (spin lock) function, which is used in PostgreSQL to control access to the shared buffers data structures (held in shared memory). Every time a lock can not be acquired, a context switch takes place, forcing an update of the L2 cache.

Figure 1(e) shows that the time spent on the "s_lock" function increases with both clients and cores, only when the *BestSellers* query is involved. We would expect to see an increase with the number of clients but not with more cores. Removing again the *BestSellers* query from the mix, we observe that it is indeed the one that causes PostgreSQL to waste CPU cycles on the "s_lock" function as the number of cores increases. Finally, looking at the "s_lock" while running only the *BestSellers* query we see that it dictates the behavior of the entire mix.

The conclusion from these experiments is that, as the number of cores and clients increase, the contention on the shared buffers significantly degrades performance: more memory leads to more data under contention, more cores just increase the contention. This problem that has also been observed by Boyd-Wickizer [2010].

The performance of MySQL for the *Browsing* mix with different number of cores and clients is shown in figure 5. MySQL's InnoDB storage engine acts as a queuing system: it has a fixed number of threads that process client requests (storage engine threads). If more client requests arrive than available threads, MySQL will buffer them until the previous ones have been answered. In this way MySQL is not affected by the number of clients but it shows the same pathological behavior as PostgreSQL with the number of cores: more cores result in lower throughput and higher response times.

While running this experiment, we monitored the times a thread had to yield to the OS due to waits for a lock or a latch. Figure 1(f) presents the number of thread yields per transaction for different loads on the system.

Running one storage engine thread for each CPU core available to MySQL, we looked at three scenarios: under-load (a total of 12 clients), perfect-load (same number of clients as storage engine threads) and over-load (200 concurrent clients). Running on 12 cores, we see very few thread yields per transaction taking place. This indicates that for this degree of multi-programming MySQL has no intrinsic problems. Adding extra cores and placing enough load as to fully utilize the storage engine threads (perfect load and over load scenarios), we see that the number of thread yields per transaction significantly increases. We also observe that the queuing effect in the system does not add extra thread yields. With increasing cores, the contention of acquiring a mutex or a latch increases exponentially.

Of the possible causes for the OS thread yields, we observe less than half are caused by the latches that MySQL's InnoDB storage engine uses for row level locking. The rest

are caused by mutexes that MySQL uses throughout its entire code. This implies that there is not a single locking bottleneck, but rather a problem with locking across the entire code-base, making it difficult to change the system so that it does not become worse with the number of cores.

In the case of the *BestSellers* query, MySQL does not show the same performance degradation issues due to the differences in engine architectures. MySQL has scalability problems with an increasing number of hardware contexts due to the synchronization primitives and contention over shared data structures.

2.3 Our approach

Load interaction is an intrinsic feature of existing database engines that can only become worse with multicore. Similarly, fixing all synchronization problems in existing engines is a daunting task that probably requires major changes to the underlying architecture. The basic insight of Multimed is that we can alleviate the problems of load interaction and contention by separating the load and using the available cores as a pool of distributed resources rather than as a single parallel machine.

Unlike existing work that focuses on optimizing the access time to shared data structures [Hardavellas 2007; Johnson 2009b] or aims at a complete redesign of the engine [Harizopoulos 2005; Unterbrunner 2009], Multimed does not require code modifications on the database engine. Instead, we use replicated engines each one of them running on a non-overlapping subset of the cores.

3. The Multimed System

Multimed is a platform for running replicated database engines on multicore machines. It is independent of the database engine used, its main component being a middleware layer that coordinates the execution of transactions across the replicated database engines. The main roles of Multimed are: (i) mapping database engines to hardware resources, (ii) scheduling and routing transactions over the replicated engines and (iii) communicating with the client applications.

3.1 Architectural overview

From the outside, Multimed follows the conventional client-server architecture of database engines. Multimed's client component is implemented as a JDBC Type 3 Driver. Internally, Multimed (figure 2) implements a master-slave replication strategy but offers a single system image, i.e., the clients see a single consistent system. The master holds a primary copy of the data and is responsible for executing all updates. Queries (the read only part of the load) run on the satellites. The satellites are kept up to date by asynchronously propagating *WriteSets* from the master. To provide a consistent view, queries can be scheduled to run on a satellite node only after that satellite has done all the updates executed by the master prior to the beginning of the query.

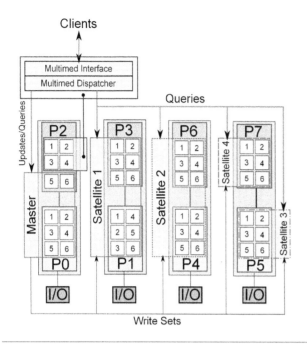

Figure 2. A possible deployment of Multimed

The main components of Multimed are the *Communication* component, the *Dispatcher* and the *Computational Nodes*. The Communication component implements an asynchronous server that allows Multimed to process a high number of concurrent requests. Upon receiving a transaction, Multimed routes the transaction to one of its *Computational Nodes*, each of which coordinates a database engine. The routing decision is taken by the *Dispatcher* subsystem.

With this architecture, Multimed achieves several goals. First, updates do not interfere with read operations as the updates are executed in the master and reads on the satellites; second, the read-only load can be separated across replicas so as to minimize the interference of heavy queries with the rest of the workload.

3.2 How Multimed works

We now briefly describe how Multimed implements replication, which is done by adapting techniques of middleware based replication [Plattner 2004] to run in a multicore machine. In a later section we explore the optimizations that are possible in this context and are not available in network based systems.

3.2.1 Replication model

Multimed uses lazy replication [Gray 1996] between its master and satellite nodes but guarantees a consistent view to the clients. The master node is responsible for keeping a durable copy of the database which is guaranteed to hold the latest version of the data. All the update transactions are executed on the master node as well as any operation requiring special features such as stored procedures, triggers, or user defined functions.

The satellite nodes hold replicas of the database. These replicas might not be completely up to date at a given moment but they are continuously fed with all the changes done at the master. A satellite node may hold a full or a partial replica of the database. Doing full replication has the advantage of not requiring knowledge of the data allocation for query routing. On the downside, full replication can incur both higher costs in keeping the satellites up to date due to larger update volumes, and lower performance because of memory contention across the replicas. In the experimental section we include an evaluation of partial replication but all the discussions on the architecture of Multimed are done on the basis of full replication to simplify the explanation.

Each time an update transaction is committed, the master commits the transaction locally. The master propagates changes as a list of rows that have been modified. A satellite enqueues these update messages and applies them in the same order as they were executed at the master node.

Multimed enforces snapshot isolation as a consistency criterion (see [Daudjee 2006] for a description of snapshot isolation and other consistency options such as session consistency in this type of system). In snapshot isolation queries are guaranteed to see all changes that have been committed at the time the transaction started, a form of multiversion concurrency control found today in database engines such as Oracle, SQLServer, or PostgreSQL. When a query enters the system, the *Dispatcher* needs to decide where it can run the transaction (i.e., to which node it should bind it, which may involve some small delay until a copy has all necessary updates) and if multiple options are available, which one to choose. Note that the master node is always capable of running any query without any delay and can be used as a way to minimize latency if that is an issue for particular queries.

Within each database, we rely on the snapshot isolation consistency of the underlying database engine. This means that an update transaction will not interfere with read transactions, namely the update transaction is applied on a different "snapshot" of the database and once it is committed, the shadow version of the data is applied on the active one. In this way, Multimed can schedule queries on replicas at the same time they are being updated.

3.2.2 WriteSet extraction and propagation

In order to capture the changes caused by an update transaction, Multimed uses row-level *insert*, *delete* and *update* triggers in the master database on the union of the tables replicated in all the satellites. The triggers fire every time a row is modified and the old and new versions are stored in the context of the transaction. All the changed rows, with their previous and current versions, represent the WriteSet of a transaction. In our system, this mechanism is implemented using SQL triggers and server side functions. This is the only mechanism specific to the underlying database but it is a standard feature in today's engines.

```
input  : Connection con, Server Count Number scn
WriteSet ws ← con.getWriteSet();
if ws.getStatementCount() > 0 then
    synchronized lock_object
        con.commit();
        ws.setSCN(scn.atomicIncrement());
        foreach satellite sat do sat.enqueue(ws);
    end
else con.commit();
```
Algorithm 1: WriteSet Propagation

When an update transaction is committed on the master node, the WriteSet is extracted (by invoking a server side function), parsed and passed on to each satellite for enqueuing in its update queue, following algorithm 1.

A total order is enforced by Multimed over the commit order of the transactions on the master node. This total order needs to be enforced so that it can be respected on the satellite nodes as well. This might be a performance bottleneck for the system, but as we show in the experimental section, the overhead induced by WriteSet extraction and by enforcing a total order over the commits of updates is quite small. In practice, Multimed introduces a small latency in starting a query (while waiting for a suitable replica) but it can execute many more queries in parallel and, often, the execution of each query is faster once started. Thus, the result is a net gain in performance.

3.3 Multimed's components

The main components of Multimed are the computational nodes, the dispatcher, the system model and the client communication interface.

3.3.1 Computational nodes

A *Computational Node* is an abstraction over a set of hardware (CPUs, memory, etc.) and software (database engine and stored data, connection pools, queues, etc) resources. Physically, it is responsible for forwarding queries and updates to its database engine.

Each *Computational Node* runs in its own thread. It is in charge of (i) executing queries; (ii) executing updates and (iii) returning results to the clients. Each transaction is bound to a single *Computational Node* (can be the master) which has the capability of handling all the requests that arrive in the context of a transaction.

Multimed has one master *Computational Node* and any number of satellite *Computational Nodes*. Upon arrival, queries are assigned a timestamp and dispatched to the first satellite available that has all the updates committed up to that timestamp (thereby enforcing snapshot isolation). Satellite nodes do not need to have any durability guarantees (i.e., do not need to write changes to disk). In the case of failure of a satellite, no data is lost, as everything is durably committed by the master *Computational Node*.

3.3.2 Dispatcher

The Multimed *Dispatcher* binds transactions to a Computational Node. It routes update transactions to the master node, leaving the read transactions to the satellites. The differentiation of the two types of transactions can be done based on the transaction's *readOnly* property (from the JDBC API).

The Dispatcher balances load by choosing the most lightly loaded satellite from among those that are able to handle the transaction. The ability of a satellite to handle a transaction is given by the freshness of the data it holds and by the actual data present (in the case of partial replication). The load can be the number of active transactions, the CPU usage, average run time on this node, etc. When no capable satellite is found, the *Dispatcher* waits until it can bind to a satellite with the correct update level or it may choose to bind the transaction to the master node.

3.3.3 System model

The system model describes the configuration of all *Computational Nodes*. It is used by the *Dispatcher* when processing client requests.

The *System Model* currently defines a static partitioning of the underlying software and hardware resources (dynamic partitioning is left for future work as it might involve reconfiguring the underlying database). It is used at start-time to obtain a logical and physical description of all the *Computational Nodes* and of the required connection settings to the underlying databases. It also describes the replication scheme in use, specifying what data is replicated where, the tables that are replicated, and where transactions need to be run.

3.3.4 Communication component

The communication subsystem, on the server side, has been implemented based on *Apache Mina 2.0* [Apache Mina]. The communication interface is implemented as an asynchronous server using *Java NIO* libraries. Upon arrival, each client request is passed on to a *Computational Node* thread for processing. We have implemented the server component of the system as a non-blocking message processing system so as to be able to support more concurrent client connections than existing engines. This is important to take advantage of the potential scalability of Multimed as often the management of client connections is a bottleneck in database engines (see the results for PostgreSQL above).

3.4 System optimizations

Multimed can be configured in many different ways and accepts a wide range of optimizations. In this paper we describe a selection to illustrate how Multimed can take advantage of multicore systems in ways that are not possible in conventional engines.

On the communication side, the messages received by the server component from the JDBC Type 3 Driver are small (under 100 bytes). By default, multiple messages will be packed together before being sent (based on Nagle's algorithm [Peterson 2000]), increasing the response time of a request. We disabled this by setting the TCP_NODELAY option on the Java sockets, reducing the RTT for messages by a factor of 10 at the cost of a higher number of packets on the network.

On the server side, all connections from the *Computational Nodes* to the database engines are done through JDBC Type 4 Drivers (native protocols) to ensure the best performance. Using our own connection pool increases performance as no wait times are incurred for creating/freeing a database connection.

At the *Dispatcher* level, the binding between an external client connection and an internal database connection is kept for as long as possible. This binding changes only when the JDBC *readOnly* property of the connection is modified.

For the underlying satellite node database engines, we can perform database engine-specific tweaks. For instance, for the PostgreSQL satellites, we turned off the synchronous commit of transactions and increased the time until these reach the disk. Consequently, the PostgreSQL specific options like *fsync*, *full_page_writes* and *synchronous_commit* were set to *off*, the *commit_delay* was set to its maximum limit of $100,000\mu s$, and the *wal_writer_delay* was set to $10,000ms$. Turning off the synchronous commit of the satellites does not affect the system, since they are not required for durability. Similar optimizations can be done with MySQL although in the experiments we only include the delay writes option. In the experimental section, we consider three optimization levels:

C0 implements full data replication on disk for all satellites. This is the naïve approach, where we expect performance gains from the reduced contention on the database's synchronization primitives, but also higher disk contention.

C1 implements full data replication in main memory for all satellites, thereby reducing the disk contention.

C2 implements partial or full data replication in main memory for the satellites and transaction routing at the *Dispatcher*. This approach uses far less memory than *C1*, but requires a-priori knowledge of the workload to partition the data adequately (satellites will be specialized for running only given queries). For the case of the 20GB database used in our experiments, a specialized replica containing just the tables needed to run the *BestSellers* query needs only 5.6GB thus allowing us to increase the number of in-memory satellite nodes. For CPU-bound use cases this approach allows us to easily scale to a large number of satellite nodes, and effectively push the bottleneck to the maximum disk I/O that the master database can use.

4. Evaluation

In this section we compare Multimed with conventional database engines running on multicore. We measure the throughput and response time of each system while running on a different number of cores, clients, and different database sizes. We also characterize the overhead and applicability of Multimed under different workloads. Aiming at a fair comparison between a traditional DBMS and Multimed, we used the TPC-W benchmark, which allows us to quantify the behavior under different update loads.

4.1 Setup

All the experiments were carried out on a four way AMD Opteron Processor 6174 with 48 cores, 128GB of RAM and two 146GB 15k RPM Seagate® Savvio® disks in RAID1.

Each AMD Magny Cours CPU consists of two dies, with 6 cores per die. Each core has a local L1 (128KB) and L2 cache (512KB). Each die has a shared L3 cache (12MB). The dies within a CPU are connected with two HyperTransport (HT) links between each other, each one of them having two additional HT links.

For the experiments with three and five satellites, each satellite was allocated entirely within a CPU, respectively within a die, to avoid competition for the cache. In the experiments with ten satellites, partial replication was used, making the databases smaller. In this case, each satellite was allocated on four cores for a total of 3 satellites per socket. Two of these satellites are entirely within a die and the third spawns two dies within the same CPU. Due to the small size of the replicas (the point we want to make with partial replication), we have not encountered cache competition problems when satellites share the L3 cache.

The hard disks in our machine prevented us from exploring more write intensive loads. In practice, network attached storage should be used, thereby allowing Multimed to support workloads with more updates. Nevertheless, the features and behavior of Multimed can be well studied in this hardware platform. A faster disk would only change at which point the the master hits the I/O bottleneck, improving the performance of Multimed even further.

The operating system used is a 64-bit Ubuntu 10.04 LTS Server, running PostgreSQL 8.3.7, MySQL 5.1 and Sun Java SDK 1.6.

4.2 Benchmark

The workload used is the TPC-W Benchmark over datasets of 2GB and 20GB. Each run consists of having the clients connect to the database and issue queries and updates, as per the specifications of the TPC-W mix being run. Clients issue queries for a time period of 30 minutes, without think times. Each experiment runs on a fresh copy of the database, so that dataset evolution does not affect the measurements. For consistent results, the memory and threading parameters of PostgreSQL and MySQL are fixed to the same values for both the standalone and Multimed systems.

The clients are emulated by means of 10 physical machines. This way more than 1000 clients can load the target system without incurring overheads due to contention on the client side. Clients are implemented in Java and are used to

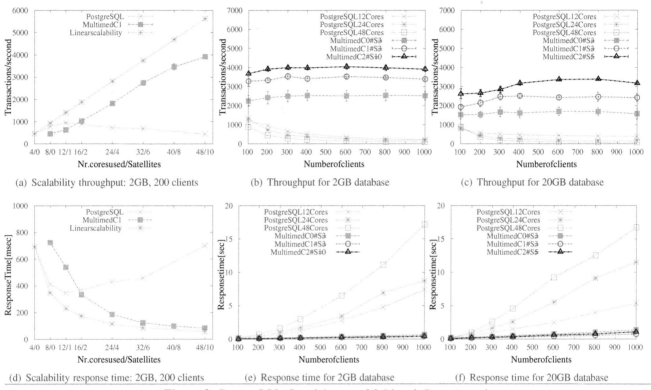

Figure 3. PostgreSQL: Standalone vs. Multimed, *Browsing* mix

emit the workload as well as to measure the throughput and response time.

The TPC-W benchmark specifies three workload mixes: TPC-W *Browsing* (10% updates), TPC-W *Shopping* (20% updates) and TPC-W *Ordering* (50% updates). Out of these three, we focus on the Browsing and Shopping mixes. The Ordering mix is disk intensive and hits an I/O bottleneck before any proper CPU usage is seen.

The TPC-W benchmark specifies both an application and a database level. We implemented only the database level, as this is the point of interest for this work. Due to the lack of the application level, some features required for correctly implementing the benchmark had to be emulated at the database level. For example the shopping cart, which should reside in the web server's session state, is present in our implementation as a table in the database. In order to limit the side effects of holding the shopping cart in the database, an upper bound is placed on the number of entries that it can hold, equal to the maximum number of concurrent clients.

We have done extensive tests on Multimed, trying to find the optimal configuration to use in the experiments. The number of cores on which the satellites and the master nodes are deployed can be adjusted. Also, the number of cores allocated for Multimed's middleware code can be configured. In the experiments below we mention the number of satellites (#S) and the optimization (C0-C2) that were used.

4.3 PostgreSQL: Standalone vs. Multimed version

This section compares the performance of PostgreSQL and Multimed running on top of PostgreSQL.

4.3.1 Query intensive workload

Figures 3(a) and 3(d) present the scalability of PostgreSQL compared to Multimed $C1$, in the case of the 2GB database, and 200 clients. The x-axis shows the number of cores used by both Multimed and PostgreSQL, as well as the number of satellites coordinated by Multimed. Both the throughput (figure 3(a)) and the response time (figure 3(d)) show that the TPC-W Browsing mix places a lot of pressure on standalone PostgreSQL, causing severe scalability problems with the number of cores. Multimed running on 4 cores, the master node on 4 cores, and each satellite on 4 cores scales up almost linearly to a total of 40 cores (equivalent of 8 satellites). The limit is reached when the disk I/O bound is hit: all queries run extremely fast, leaving only update transactions in the system to run longer, and facing contention on the disk. The gap between the linear scalability line and Multimed's performance is constant, being caused by the computational resources required by Multimed's middleware.

4.3.2 Increased update workload

Figures 3(b) and 3(c) present the throughput of PostgreSQL (running on different number of cores) and of Multimed (running with different configurations), as the number of clients increases. Note that PostgreSQL has problems in

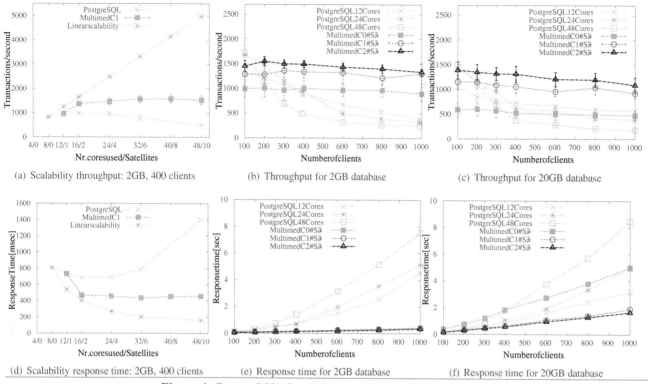

(a) Scalability throughput: 2GB, 400 clients   (b) Throughput for 2GB database   (c) Throughput for 20GB database

(d) Scalability response time: 2GB, 400 clients   (e) Response time for 2GB database   (f) Response time for 20GB database

Figure 4. PostgreSQL: Standalone vs. Multimed, *Shopping* mix

scaling with the number of clients issuing the workload, and its performance at 48 cores is lower than at 12.

For both dataset sizes, Multimed (at all optimization levels) outperforms the standalone version of PostgreSQL. The $C0$ optimization level for Multimed shows higher error bars, as all satellites are going concurrently to disk, in order to persist updates. Switching to the $C1$ optimization level, we reduce the contention on disk, by using more main memory. We see an improvement of more than 1000 transactions per second between the naïve $C0$ and optimized $C1$ versions of Multimed. Finally, switching to the less generic optimization level $C2$, Multimed accommodates more satellites in the available memory, and can take advantage of the available computational resources, until a disk I/O limit is hit. Using the $C2$ optimization, the problem of load interaction is also solved by routing the "heavy", analytical, queries to different satellite nodes, offloading the other nodes in the system. In all these experiments we have used static routing.

Note the fact that Multimed retains a steady behavior with increasing number of concurrent clients (up to 1000), without exhibiting performance degradation. Looking at the corresponding response times, even under heavy load, Multimed's response time is less than 1 second, indicating that Multimed is not only solving the problems of load interaction, but also the client handling limitations of PostgreSQL. For the Shopping mix, standalone PostgreSQL's performance is slightly better than for the Browsing mix due to the reduced number of heavy queries.

Figures 4(a) and 4(d) show that even in the case of the Shopping mix, PostgreSQL can not scale with the number of available cores, on the 2GB database, with 400 clients. Multimed scales up to 16 cores (2 satellites), at which point the disk becomes a bottleneck. Multimed's performance stays flat with increasing cores, while that of PostgreSQL drops.

Figures 4(b) and 4(c) show that PostgreSQL can not scale with the number of clients for this workload either, regardless of the database size. In the case of Multimed, for a small number of clients, all queries run very fast, leaving the updates to compete for the master node. Past 150 clients, the run time of queries increases and the contention on the master node is removed, allowing Multimed to better use the available satellites. We again observe that Multimed's behavior is steady and predictable with increasing load.

Using the $C0$ optimization level and for a low number of clients, Multimed performs worse than PostgreSQL, especially on the 20GB database, although it is more stable with the number of clients. With more updates in the system and with all of the satellites writing to disk, Multimed is blocked by I/O. As in the previous case, the $C1$ optimization solves the problem: standard deviation is reduced and the throughput increases. The $C2$ optimization, at the same number of satellites, also gives the system a performance gain as fewer WriteSets need to be applied on the satellites (they run faster).

Both in the case of a query intensive workload (Browsing mix) and in the case of increased update workload (Shopping

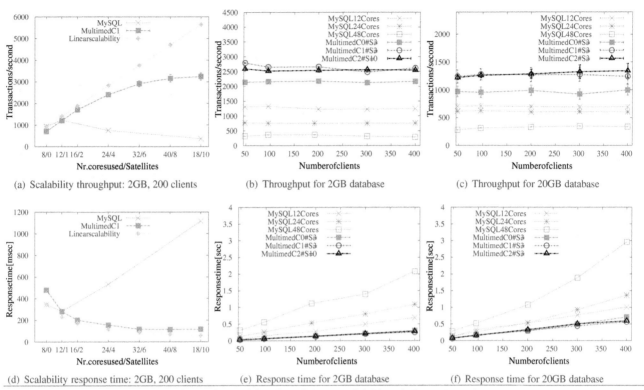

(a) Scalability throughput: 2GB, 200 clients   (b) Throughput for 2GB database   (c) Throughput for 20GB database

(d) Scalability response time: 2GB, 200 clients   (e) Response time for 2GB database   (f) Response time for 20GB database

Figure 5. MySQL: Standalone vs. Multimed, *Browsing* mix

mix), PostgreSQL does not scale with the number of cores or with the number of clients, regardless of the database size. PostgreSQL's inability to scale with the number of clients is due to the fact that for each new client a new process is spawned on the server. This might lead to the conclusion that the number of processes is far greater than what the operating system and the hardware can handle. This is disproved by Multimed, which can cope with 1000 clients in spite of the limitations of PostgreSQL. The problem in this case is not the large number of processes in the system, but rather the inability of a single PostgreSQL engine to handle high concurrency. Since Multimed splits the number of clients over a set of smaller sized satellites, it reduces the contention in each engine, resulting in a higher throughput

4.4 MySQL: Standalone vs. Multimed version

In this section we compare standalone MySQL to Multimed running on top of MySQL computational nodes.

For MySQL, we have done our experiments using its InnoDB storage engine. This engine is the most stable and used storage engine available for MySQL. However it has some peculiar characteristics: (i) it acts as a queuing system, allowing just a fixed number of concurrent threads to operate over the data (storage engine threads); (ii) it is slower than the PostgreSQL engine for disk operations. In all the results presented below, the number of cores available for MySQL is equal to the number of storage engine threads. Being a queuing system, MySQL will not show a degrada-

tion in throughput with the number of clients, but rather exhibits linear increase in response time. For this reason, the experiments for MySQL only go up to 400 clients.

4.4.1 Query intensive workload

Figures 5(a) and 5(d) present the ability of the standalone engine, and of Multimed running on top of it, to scale with the amount of computational resources, in the case of the 2GB database and 200 clients. The x-axis, as before, indicates the total number of cores available for MySQL and Multimed, as well as the number of satellites coordinated by Multimed. Each satellite runs on 4 cores.

In the case of the TPC-W Browsing mix, we notice that MySQL does not scale with the number of cores. Figure 5(a) shows that MySQL performs best at 12 cores. Adding more cores increases contention and performance degrades.

The same conclusion can be seen in the throughput and response time plots for both the 2GB and 20GB datasets (figures 5(b) and 5(c)), that show the performance of MySQL (running on different number of cores) and of Multimed (running on different configurations) with increasing clients. Since the behavior is independent of the dataset, we conclude that the contention is not caused by a small dataset, but rather by the synchronization primitives (i.e., mutexes) that are used by MySQL throughout its entire code.

In contrast, Multimed scales with the number of cores. Figure 5(a) shows that on the 2GB dataset, Multimed scales up to 6 satellites, at which point the disk I/O becomes the

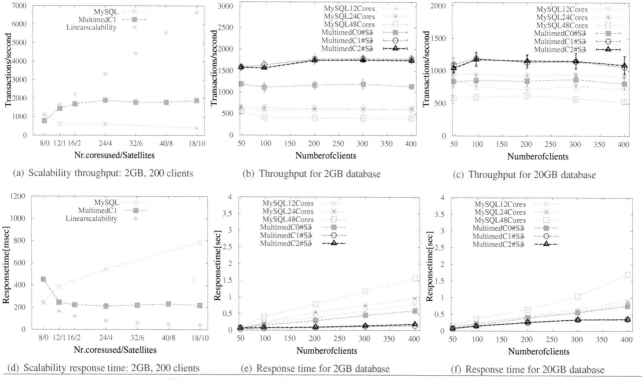

(a) Scalability throughput: 2GB, 200 clients

(b) Throughput for 2GB database

(c) Throughput for 20GB database

(d) Scalability response time: 2GB, 200 clients

(e) Response time for 2GB database

(f) Response time for 20GB database

Figure 6. MySQL: Standalone vs. Multimed, *Shopping* mix

bottleneck in the system, and the throughput and response times are flat. The fact that Multimed on top of PostgreSQL scaled in the same test up to 8 satellites corroborates the fact the PostgreSQL's storage engine is faster than MySQL's InnoDB for this workload.

The three configurations that we have run for Multimed show that by replicating data, Multimed can outperform standalone MySQL by a factor of 2, before it reaches the disk I/O bound. The *C*0 configuration of Multimed shows a behavior similar to standalone MySQL's best run. Removing this contention on disk from Multimed, by switching to its *C*1 configuration, increases performance. The *C*2 optimization does not yield better performance than *C*1. The system is already disk bound and load interaction does not influence MySQL for this workload. To improve performance here, a faster disk or lower I/O latency would be needed.

4.4.2 Increased update workload

The scalability plot (figure 6(a)), shows that MySQL performs best at 8 cores. With more cores performance degrades, confirming that contention is the bottleneck, not disk I/O. Multimed scales up to 16 cores, at which point the throughput flattens confirming that the disk becomes the bottleneck.

Figures 6(b) and 6(c) show that on larger datasets data contention decreases, allowing standalone MySQL to perform better. On the 2GB database, Multimed brings an im-

provement of 3x. In the case of the 20GB database, Multimed achieves a 1.5x improvement.

5. Discussion

Multimed adapts techniques that are widely used in database clusters. As a database replication solution, Multimed inherits many of the characteristics of replicated systems and database engines. In this section we discuss such aspects to further clarify the effectiveness and scope of Multimed.

5.1 Overhead

The main overhead introduced by Multimed over a stand alone database is latency. Transactions are intercepted by Multimed before being forwarded either to the master or to a satellite. In the case the transactions go to a satellite, there might be further delay while waiting for a satellite with the correct snapshot.

Multimed works because, for a wide range of database loads, such an increase in latency is easily compensated by the reduction in contention between queries and the increase in the resources available for executing each query. Although satellite databases in Multimed have fewer resources, they also have less to do. For the appropriate workloads, Multimed is faster because it separates loads across databases so that each can answer fewer queries faster than a large database can answer all the queries.

5.2 Loads supported

There is no database engine that is optimal for all loads [Stonebraker 2008]. Multimed is a replication based solution and, hence, it has a limitation in terms of how many updates can be performed as all the updates need to be done at the master. Although this may appear a severe limitation, it is not so in the context of database applications.

As the experiments above show, Multimed provides substantial performance improvements for the TPC-W browsing and shopping mixes. For the ordering mix, with a higher rate of updates, Multimed offers similar performance as the single database since the bottleneck in both cases is the disk. Multimed can be used to linearly scale read dominated loads such as those found in business intelligence applications and data warehousing. For instance, it is possible to show linear scale up of Multimed by simply assigning more satellites to complex analytical queries. As a general rule, the more queries and the more complex the queries, the better for Multimed.

Workloads with high update rates and without complex read operations are less suitable for Multimed – and indeed any primary copy replication approach – because the master becomes the bottleneck (regardless of why it becomes the bottleneck: CPU, memory, or disk). In cluster based replication, this problem is typically solved by simply assigning a larger machine to the master. Multimed can likewise be configured to mitigate this bottleneck with a larger allocation of resources (cores, memory) to the master.

5.3 Configuration

Tuning and configuring databases is a notoriously difficult problem. Some commercial database engines are known to provide thousands of tuning knobs. In fact, a big part of the impetus behind the autonomic computing initiative of a few years ago was driven by the need to automatically tune databases.

Similarly, tuning Multimed requires knowledge of database loads, knowledge of the engines used, and quite a bit of experimentation to find the right settings for each deployment. The advantage of Multimed over a stand alone database is that the number of global tuning knobs is less as each element of the system needs to be tailored to a specific load. The master can be tuned for writes, the satellites for reads. It is even possible to configure Multimed so that a satellite answers only specific queries (for instance, particularly expensive or long running ones) and then optimize the data placement, indexes, and configuration of that satellite for that particular type of query.

In terms of the interaction with the operating system and the underlying architecture, Multimed can be configured using simple rules: allocate contiguous cores to the same satellites, restrict the memory available to each satellite to that next to the corresponding cores, prevent satellites from interfering with each other when accessing system resources (e.g., cascading updates instead of updating all satellites at the same time), etc. Such rules are architecture specific but rather intuitive. Note as well that Multimed is intended to run in a database server. These are typically powerful machines with plenty of memory, often fast networks and even several network cards, and SAN/NAS storage rather than local disks. The more main memory is available, the faster the I/O, and the more cores, the more possibilities to tune Multimed to the application at hand and the bigger performance gains to be obtained from Multimed.

5.4 Optimizations

The version of Multimed presented in the paper does not include any sophisticated optimizations since the goal was to show that the basic concept works. There are, however, many possible optimizations over the basic system.

WriteSet extraction is currently done through triggers. Depending on the system and load, this can become a bottleneck (although note that the master where the trigger runs no longer handles any query, so it has extra capacity to run the triggers). An alternative is to extract the changes directly from the log buffers (which requires access to the database internals but has been done before in a number of systems). Another option is to capture the changes from the log, a common solution in many database replication products such as Data Streams of Oracle.

As mentioned, using a SAN/NAS storage will significantly speed up Multimed because the I/O bottleneck can be greatly reduced. Similarly, as has been done in some commercial systems like Oracle Exadata, one can use SSDs as caches between the remote disks and the server to minimize the I/O impact. In the same way that these solutions speed up traditional databases, Multimed will be faster using them.

The extra communication cost incurred by Multimed is very small compared to the total run time of each transaction. Even so, we have presented in section 3.4 a couple of tweaks that we have performed in order to reduce it even more.

Like any replicated database, Multimed requires additional resources for every copy of the database. In a multicore machine, the number of satellites that can be used depends on the available memory. The situation can be improved by using partial replication as we have shown in this paper. If each satellite contains only part of the original database, then the available memory allows for the creation of more satellites (subject to the deployment constraints discussed above). Multimed offers many options for exploring such settings: satellites can be tailored for answering certain queries (which only need the data necessary to answer those queries), or satellites with special indexes for given workloads. Satellites could also be enhanced with user defined functions, offering a way to extend the functionality of system without having to modify the master database. Several such optimizations have been shown to work well in cluster based system and can easily be adapted to run in Multimed [Plattner 2008].

6. Related work

Multimed builds on work done in database replication and multikernel operating systems.

6.1 Database Replication

Multimed uses many ideas from database replication strategies developed during the last decade starting with the work on Postgres-R [Kemme 2000]. Of the many existing systems [Cecchet 2008], however, not all approaches are suitable for multicore machines. For instance, many middleware database replication solutions use group communication to coordinate the copies [Bettina 2003; Elnikety 2006]. These solutions require a multi-master configuration and rely on full replication. As our experiments have shown, running the satellites in main memory and being able to implement partial replication is a great boost in performance. Nevertheless, some of the innovative applications pursued with these systems could also be implemented using the Multimed approach within a single multicore machine rather than on a cluster. An example is the work on tolerating byzantine failures using heterogeneous replicas [Vandiver 2007] or commercial version used in data warehousing [Xkoto].

The approaches closer to the design of Multimed are those relying on single master replication [Daudjee 2006; Plattner 2004] and that can support specialized satellites and partial replication. This type of design is starting to be widely used in cloud computing, for instance, in the Microsoft SQL Azure database [Campbell 2010].

6.2 Multikernel Operating Systems

The problems that multicore creates in system software either because of the increasing number of cores [Agarwal 2007; Borkar 2007] or their potential heterogeneity [Hill 2008; Kumar 2004] are by now well known. This has triggered a lot of activity in the area of operating systems to address these problems. For instance, [Wickizer 2008] proposes a new exokernel based operating system, Corey, which tries to manage the complexity of multicore machines by moving the responsibility into the application space. Disco [Bugnion 1997] and Cellular Disco [Govil 2000], make the case for resource partitioning by running a virtualization layer over a shared memory architecture, allowing the execution of multiple commodity operating systems, and treating multicore as a cluster. Finally, [Baumann 2009; Nightingale 2009; Wentzlaff 2009] make a clean statement that multicore machines should be viewed as distributed systems and adequate algorithms and communication models should be employed.

The work in Multimed borrows many of these concepts and applies them to the special (and architecturally very different) case of database engines.

7. Conclusions

In this paper we address the problem of making databases run efficiently on multicore machines. Multimed, the system we present in the paper, represents a departure from existing work in that it solves the problem for a wide range of loads without having to modify the engine. Instead, it uses existing databases in a replicated configuration and deploys them over a multicore machine as if the multicore machine were a distributed system. As shown in the evaluation, Multimed exhibits better and more stable performance on multicore architectures than PostgreSQL and MySQL.

A key aspect of Multimed is that it is independent of the database engine and it will benefit from current hardware developments, something that is not always the case for alternative approaches. Multimed will get better as the number of cores increases, as more main memory is available, through network attached storage, and by using SSD/Flash storage. In addition, it is in a better position to cope with the impending heterogeneity of multicore machines by allowing asymmetric replicas of the database that can be specialized to the characteristics of the underlying cores.

8. Acknowledgments

We would like to thank the reviewers for their helpful comments and insights.

In particular, we would like to thank Rebecca Isaacs from Microsoft Research for her help in preparing the final version of the paper.

References

[Agarwal 2007] A. Agarwal and M. Levy. The kill rule for multicore. In *IEEE DAC '07*, pages 750–753, San Diego, California, 2007.

[Alonso 2011] G. Alonso, D. Kossmann, and T. Roscoe. SwissBox: An Architecture for Data Processing Appliances. In *CIDR '11*, pages 32–37, Asilomar, California, 2011.

[Apache Mina] Apache Mina. http://mina.apache.org/.

[Baumann 2009] A. Baumann, P. Barham, P. Dagand, T. Harris, R. Isaacs, S. Peter, T. Roscoe, A. Schüpbach, and A. Singhania. The multikernel: a new OS architecture for scalable multicore systems. In *ACM SOSP '09*, pages 29–44, Big Sky, Montana, 2009.

[Bettina 2003] K. Bettina. *Future directions in distributed computing*. Springer-Verlag, 2003.

[Borkar 2007] S. Borkar. Thousand core chips: a technology perspective. In *IEEE DAC '07*, pages 746–749, San Diego, California, 2007.

[Boyd-Wickizer 2010] S. Boyd-Wickizer, A. Clements, Y. Mao, A. Pesterev, M.F. Kaashoek, R. Morris, and N. Zeldovich. An Analysis of Linux Scalability to Many Cores. In *USENIX OSDI '10)*, pages 1–8, Vancouver, Canada, 2010.

[Bryant 2003] R. Bryant and J. Hawkes. Linux Scalability for Large NUMA Systems. In *Linux Symposium*, pages 83–96, Ottawa, Canada, 2003.

[Bugnion 1997] E. Bugnion, S. Devine, K. Govil, and M. Rosenblum. Disco: Running Commodity Operating Systems on Scalable Multiprocessors. *ACM TOCS '97*, 15(4):412–447, 1997.

[Campbell 2010] D.G. Campbell, G. Kakivaya, and N. Ellis. Extreme scale with full SQL language support in Microsoft SQL Azure. In *ACM SIGMOD Conference*, pages 1021–1024, 2010.

[Cecchet 2008] E. Cecchet, G. Candea, and A. Ailamaki. Middleware-based database replication: the gaps between theory and practice. In *ACM SIGMOD 08*, pages 739–752, Vancouver, Canada, 2008.

[Daudjee 2006] K. Daudjee and K. Salem. Lazy Database Replication with Snapshot Isolation. In *VLDB '06*, pages 715–726, Seoul, Korea, 2006.

[Elnikety 2006] S. Elnikety, S. Dropsho, and F. Pedone. Tashkent: uniting durability with transaction ordering for high-performance scalable database replication. In *ACM EuroSys '06*, pages 117–130, Leuven, Belgium, 2006.

[Govil 2000] K. Govil, D. Teodosiu, Y. Huang, and M. Rosenblum. Cellular Disco: resource management using virtual clusters on shared-memory multiprocessors. *ACM TOCS 2000*, 18(3):229–262, 2000.

[Gray 1996] J. Gray, P. Helland, P. O'Neil, and D. Shasha. The dangers of replication and a solution. In *ACM SIGMOD '96*, pages 173–182, Montreal, Canada, 1996.

[Han 2009] W.S. Han and J. Lee. Dependency-aware reordering for parallelizing query optimization in multi-core CPUs. In *ACM SIGMOD '09*, pages 45–58, Providence, Rhode Island, 2009.

[Hardavellas 2007] N. Hardavellas, I. Pandis, R. Johnson, N. Mancheril, A. Ailamaki, and B. Falsafi. Database Servers on Chip Multiprocessors: Limitations and Opportunities. In *CIDR '07*, pages 79–87, Asilomar, California, 2007.

[Harizopoulos 2005] S. Harizopoulos and A. Ailamaki. StagedDB: Designing database servers for modern hardware. In *IEEE ICDE '05*, pages 11–16, Tokyo, Japan, 2005.

[Hill 2008] M.D. Hill and M.R. Marty. Amdahl's Law in the Multicore Era. In *IEEE HPCA '08*, volume 41, pages 33–38, Salt Lake City, Utah, 2008.

[Johnson 2009a] R. Johnson, M. Athanassoulis, R. Stoica, and A. Ailamaki. A new look at the roles of spinning and blocking. In *DaMoN '09*, pages 21–26, Providence, Rhode Island, 2009.

[Johnson 2009b] R. Johnson, I. Pandis, N. Hardavellas, A. Ailamaki, and B. Falsafi. Shore-MT: a scalable storage manager for the multicore era. In *ACM EDBT '09*, pages 24–35, Saint Petersburg, Russia, 2009.

[Kemme 2000] B. Kemme and G. Alonso. Don't Be Lazy, Be Consistent: Postgres-R, A New Way to Implement Database Replication. In *VLDB 2000*, pages 134–143, Cairo, Egypt, 2000.

[Kumar 2004] R. Kumar, D.M. Tullsen, P. Ranganathan, N. Jouppi, and K. Farkas. Single-ISA Heterogeneous Multi-Core Architectures for Multithreaded Workload Performance. In *IEEE ISCA '04*, volume 32, pages 64–75, Munich, Germany, 2004.

[Liu 2009] R. Liu, K. Klues, S. Bird, S. Hofmeyr, K. Asanović, and J. Kubiatowicz. Tessellation: space-time partitioning in a manycore client OS. In *USENIX HotPar'09*, pages 10–10, Berkley, California, 2009.

[Nightingale 2009] E.B. Nightingale, O. Hodson, R. McIlroy, C. Hawblitzel, and G. Hunt. Helios: heterogeneous multiprocessing with satellite kernels. In *ACM SOSP '09*, pages 221–234, Big Sky, Montana, 2009.

[OProfile] OProfile. http://oprofile.sourceforge.net/.

[Papadopoulos 2008] K. Papadopoulos, K. Stavrou, and P. Trancoso. HelperCore DB: Exploiting multicore technology to improve database performance. In *IEEE IPDPS '08*, pages 1–11, Miami, Florida, 2008.

[Peterson 2000] L. Peterson and B.S. Davie. *Computer networks: a systems approach*. Morgan Kaufmann Publishers Inc., 2000.

[Plattner 2004] C. Plattner and G. Alonso. Ganymed: Scalable Replication for Transactional Web Applications. In *ACM Middleware*, pages 155–174, Toronto, Canada, 2004.

[Plattner 2008] C. Plattner, G. Alonso, and M. Tamer Özsu. Extending DBMSs with satellite databases. *The VLDB Journal*, 17 (4):657–682, 2008.

[Stonebraker 2008] M. Stonebraker. Technical perspective - One size fits all: an idea whose time has come and gone. *Commun. ACM*, 51(12):76, 2008.

[Unterbrunner 2009] P. Unterbrunner, G. Giannikis, G Alonso, D. Fauser, and D. Kossmann. Predictable performance for unpredictable workloads. *Proc. VLDB Endow.*, 2(1):706–717, August 2009.

[Vandiver 2007] B. Vandiver, H. Balakrishnan, B. Liskov, and S. Madden. Tolerating byzantine faults in transaction processing systems using commit barrier scheduling. In *ACM SOSP '07*, pages 59–72, Stevenson, Washington, 2007.

[Veal 2007] B. Veal and A. Foong. Performance scalability of a multi-core web server. In *ACM ANCS '07*, pages 57–66, Orlando, Florida, 2007.

[Wentzlaff 2009] D. Wentzlaff and A. Agarwal. Factored operating systems (fos): The case for a scalable operating system for multicores. *ACM Operating Systems Review*, pages 76–85, 2009.

[Wickizer 2008] Silas B. Wickizer, H. Chen, Y Mao, M. Frans Kaashoek, R. Morris, A. Pesterev, L. Stein, M. Wu, et al. Corey: An Operating System for Many Cores. In *USENIX OSDI '08*, San Diego, California, 2008.

[Xkoto] Xkoto. http://www.xkoto.com/.

DEPSKY: Dependable and Secure Storage in a Cloud-of-Clouds

Alysson Bessani   Miguel Correia   Bruno Quaresma   Fernando André   Paulo Sousa *

University of Lisbon, Faculty of Sciences, Portugal

{bessani,mpc}@di.fc.ul.pt   {bmmrq84,andr3.fm,pjsousa}@gmail.com

Abstract

The increasing popularity of cloud storage services has lead companies that handle critical data to think about using these services for their storage needs. Medical record databases, power system historical information and financial data are some examples of critical data that could be moved to the cloud. However, the reliability and security of data stored in the cloud still remain major concerns. In this paper we present DEPSKY, a system that improves the availability, integrity and confidentiality of information stored in the cloud through the encryption, encoding and replication of the data on diverse clouds that form a cloud-of-clouds. We deployed our system using four commercial clouds and used Planet-Lab to run clients accessing the service from different countries. We observed that our protocols improved the perceived availability and, in most cases, the access latency when compared with cloud providers individually. Moreover, the monetary costs of using DEPSKY on this scenario is twice the cost of using a single cloud, which is optimal and seems to be a reasonable cost, given the benefits.

Categories and Subject Descriptors D.4.5 [*Operating Systems*]: Reliability–Fault-tolerance; C.2.0 [*Computer-Communication Networks*]: General–Security and protection; C.2.4 [*Distributed Systems*]: Distributed applications

General Terms Algorithms, Measurement, Performance, Reliability, Security

Keywords Cloud computing, Cloud storage, Byzantine quorum systems

1. Introduction

The increasing maturity of cloud computing technology is leading many organizations to migrate their IT infrastructure and/or adapting their IT solutions to operate completely or partially in the cloud. Even governments and companies that maintain critical infrastructures (e.g., healthcare, telcos) are adopting cloud computing as a way of reducing costs [Greer 2010]. Nevertheless, cloud computing has limitations related to security and privacy, which should be accounted for, especially in the context of critical applications.

This paper presents DEPSKY, a dependable and secure storage system that leverages the benefits of cloud computing by using a combination of diverse commercial clouds to build a *cloud-of-clouds*. In other words, DEPSKY is a virtual storage cloud, which is accessed by its users by invoking operations in several individual clouds. More specifically, DEPSKY addresses four important limitations of cloud computing for data storage in the following way:

Loss of availability: temporary partial unavailability of the Internet is a well-known phenomenon. When data is moved from inside of the company network to an external datacenter, it is inevitable that service unavailability will be experienced. The same problem can be caused by denial-of-service attacks, like the one that allegedly affected a service hosted in Amazon EC2 in 2009 [Metz 2009]. DEPSKY deals with this problem by exploiting replication and diversity to store the data on several clouds, thus allowing access to the data as long as a subset of them is reachable.

Loss and corruption of data: there are several cases of cloud services losing or corrupting customer data. For example, in October 2009 a subsidiary of Microsoft, Danger Inc., lost the contacts, notes, photos, etc. of a large number of users of the Sidekick service [Sarno 2009]. The data was recovered several days later, but the users of Ma.gnolia were not so lucky in February of the same year, when the company lost half a terabyte of data that it never managed to recover [Naone 2009]. DEPSKY deals with this problem using Byzantine fault-tolerant replication to store data on several cloud services, allowing data to be retrieved correctly even if some of the clouds corrupt or lose data.

Loss of privacy: the cloud provider has access to both the data stored in the cloud and metadata like access patterns. The provider may be trustworthy, but malicious insiders are a wide-spread security problem. This is an especial concern in applications that involve keeping private data like health records. An obvious solution is the customer encrypting the

---

* Now at Maxdata Informática, Portugal.

EuroSys'11, April 10–13, 2011, Salzburg, Austria.
Copyright © 2011 ACM 978-1-4503-0634-8/11/04... $10.00

data before storing it, but if the data is accessed by distributed applications this involves running protocols for key distribution (processes in different machines need access to the cryptographic keys). DEPSKY employs a secret sharing scheme and erasure codes to avoid storing clear data in the clouds and to improve the storage efficiency, amortizing the replication factor on the cost of the solution.

Vendor lock-in: there is currently some concern that a few cloud computing providers become dominant, the so called vendor lock-in issue [Abu-Libdeh 2010]. This concern is specially prevalent in Europe, as the most conspicuous providers are not in the region. Even moving from one provider to another one may be expensive because the cost of cloud usage has a component proportional to the amount of data that is read and written. DEPSKY addresses this issue in two ways. First, it does not depend on a single cloud provider, but on a few, so data access can be balanced among the providers considering their practices (e.g., what they charge). Second, DEPSKY uses erasure codes to store only a fraction (typically half) of the total amount of data in each cloud. In case the need of exchanging one provider by another arises, the cost of migrating the data will be at most a fraction of what it would be otherwise.

The way in which DEPSKY solves these limitations does not come for free. At first sight, using, say, four clouds instead of one involves costs roughly four times higher. One of the key objectives of DEPSKY is to reduce this cost, which in fact it does to about two times the cost of using a single cloud. This seems to be a reasonable cost, given the benefits.

The key insight of the paper is that these limitations of individual clouds can be overcome by using a *cloud-of-clouds* in which the operations (read, write, etc.) are implemented using a set of *Byzantine quorum systems protocols*. The protocols require *diversity* of location, administration, design and implementation, which in this case comes directly from the use of different commercial clouds [Vukolic 2010]. There are protocols of this kind in the literature, but they either require that the servers execute some code [Cachin 2006, Goodson 2004, Malkhi 1998a;b, Martin 2002], not possible in storage clouds, or are sensible to contention (e.g., [Abraham 2006]), which makes them difficult to use for geographically dispersed systems with high and variable access latencies. DEPSKY overcomes these limitations by not requiring code execution in the servers (i.e., storage clouds), but still being efficient by requiring only two communication round-trips for each operation. Furthermore, it leverages the above mentioned mechanisms to deal with data confidentiality and reduce the amount of data stored in each cloud.

In summary, the main contributions of the paper are:

1. The DEPSKY system, a storage cloud-of-clouds that overcomes the limitations of individual clouds by using an efficient set of Byzantine quorum system protocols, cryptography, secret sharing, erasure codes and the diversity that comes from using several clouds. The DEPSKY

protocols require at most two communication round-trips for each operation and store only approximately half of the data in each cloud for the typical case.

2. A set of experiments showing the costs and benefits (both monetary and in terms of performance) of storing updatable data blocks in more than one cloud. The experiments were made during one month, using four commercial cloud storage services (Amazon S3, Windows Azure, Nirvanix and Rackspace) and PlanetLab to run clients that access the service from several places worldwide.

2. Cloud Storage Applications

Examples of applications that can benefit from DEPSKY are the following:

Critical data storage. Given the overall advantages of using clouds for running large scale systems, many governments around the globe are considering the use of this model. Recently, the US government announced its interest in moving some of its computational infrastructure to the cloud and started some efforts in understanding the risks involved in doing these changes [Greer 2010]. The European Commission is also investing in the area through FP7 projects like TCLOUDS [tcl 2010].

In the same line of these efforts, there are many critical applications managed by companies that have no interest in maintaining a computational infrastructure (i.e., a datacenter). For these companies, the cloud computing pay-per-use model is specially appealing. An example would be power system operators. Considering only the case of storage, power systems have data historian databases that store events collected from the power grid and other subsystems. In such a system, the data should be always available for queries (although the workload is mostly write-dominated) and access control is mandatory.

Another critical application that could benefit from moving to the cloud is a unified medical records database, also known as electronic health record (EHR). In such an application, several hospitals, clinics, laboratories and public offices share patient records in order to offer a better service without the complexities of transferring patient information between them. A system like this has been being deployed in the UK for some years [Ehs 2010]. Similarly to our previous example, availability of data is a fundamental requirement of a cloud-based EHR system, and privacy concerns are even more important.

All these applications can benefit from a system like DEPSKY. First, the fact that the information is replicated on several clouds would improve the data availability and integrity. Moreover, the DEPSKY-CA protocol (Section 3) ensures the confidentiality of stored data and therefore addresses some of the privacy issues so important for these applications. Finally, these applications are prime examples of cases in which the extra costs due to replication are affordable for the added quality of service.

Content distribution. One of the most surprising uses of Amazon S3 is content distribution [Henry 2009]. In this scenario, users use the storage system as distribution points for their data in such a way that one or more producers store the content on their account and a set of consumers read this content. A system like DEPSKY that supports dependable updatable information storage can help this kind of application when the content being distributed is dynamic and there are security concerns associated. For example, a company can use the system to give detailed information about its business (price, available stock, etc.) to its affiliates with improved availability and security.

Future applications. Many applications are moving to the cloud, so, it is possible to think of new applications that would use the storage cloud as a back-end storage layer. Systems like databases, file systems, objects stores and key-value databases can use the cloud as storage layer as long as caching and weak consistency models are used to avoid paying the price of cloud access on every operation.

3. The DEPSKY System

This section presents the DEPSKY system. It starts by presenting the system architecture, then defines the data and system models, the two main algorithms (DEPSKY-A and DEPSKY-CA), and a set of auxiliary protocols.

3.1 DEPSKY Architecture

Figure 1 presents the architecture of DEPSKY. As mentioned before, the clouds are storage clouds *without the capacity of executing users' code*, so they are accessed using their standard interface without modifications. The DEPSKY algorithms are implemented as a software library in the clients. This library offers an *object store* interface [Gibson 1998], similar to what is used by parallel file systems (e.g., [Ghemawat 2003, Weil 2006]), allowing reads and writes in the back-end (in this case, the untrusted clouds).

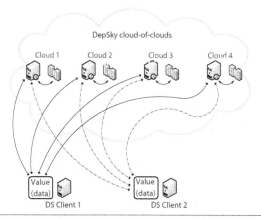

Figure 1. Architecture of DEPSKY (w/ 4 clouds, 2 clients).

3.2 Data Model

The use of diverse clouds requires the DEPSKY library to deal with the heterogeneity of the interfaces of each cloud provider. An aspect that is specially important is the format of the data accepted by each cloud. The data model allow us to ignore these details when presenting the algorithms.

Figure 2 presents the DEPSKY data model with its three abstraction levels. In the first (left), there is the *conceptual data unit*, which corresponds to the basic storage object with which the *algorithms* work (a register in distributed computing parlance [Lamport 1986, Malkhi 1998a]). A data unit has a unique name (X in the figure), a version number (to support updates on the object), verification data (usually a cryptographic hash of the data) and the data stored on the data unit object. In the second level (middle), the conceptual data unit is implemented as a *generic data unit* in an *abstract storage cloud*. Each generic data unit, or *container*, contains two types of files: a signed metadata file and the files that store the data. Metadata files contain the version number and the verification data, together with other informations that applications may demand. Notice that a data unit (conceptual or generic) can store several versions of the data, i.e., the container can contain several data files. The name of the metadata file is simply *metadata*, while the data files are called *value<Version>*, where *<Version>* is the version number of the data (e.g., *value1*, *value2*, etc.). Finally, in the third level (right) there is the *data unit implementation*, i.e., the container translated into the specific constructions supported by each cloud provider (Bucket, Folder, etc.).

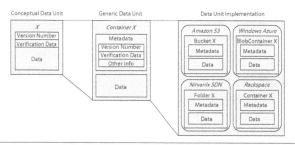

Figure 2. DEPSKY data unit and the 3 abstraction levels.

The data stored on a data unit can have arbitrary size, and this size can be different for different versions. Each data unit object supports the usual object store operations: creation (create the container and the metadata file with version 0), destruction (delete or remove access to the data unit), write and read.

3.3 System Model

We consider an *asynchronous distributed system* composed by three types of parties: writers, readers and cloud storage providers. The latter are the clouds 1-4 in Figure 1, while writers and readers are roles of the clients, not necessarily different processes.

Readers and writers. Readers can fail arbitrarily, i.e., they can crash, fail intermittently and present any behavior. Writers, on the other hand, are only assumed to fail by crashing. We do not consider that writers can fail arbitrarily because, even if the protocol tolerated inconsistent writes in the replicas, faulty writers would still be able to write wrong values in data units, effectively corrupting the state of the application that uses DEPSKY. Moreover, the protocols that tolerate malicious writers are much more complex (e.g., [Cachin 2006, Liskov 2006]), with active servers verifying the consistency of writer messages, which cannot be implemented on general storage clouds (Section 3.4).

All writers of a data unit *du* share a common private key $K_{r_w}^{du}$ used to sign some of the data written on the data unit (function $sign(DATA, K_{r_w}^{du})$), while readers of *du* have access to the corresponding public key $K_{u_w}^{du}$ to verify these signatures (function $verify(DATA, K_{u_w}^{du})$). This public key can be made available to the readers through the storage clouds themselves. Moreover, we assume also the existence of a collision-resistant *cryptographic hash function* H.

Cloud storage providers. Each cloud is modeled as a *passive storage entity* that supports five operations: *list* (lists the files of a container in the cloud), *get* (reads a file), *create* (creates a container), *put* (writes or modifies a file in a container) and *remove* (deletes a file). By passive storage entity, we mean that no protocol code other than what is needed to support the aforementioned operations is executed. We assume that access control is provided by the system in order to ensure that readers are only allowed to invoke the list and get operations.

Since we do not trust clouds individually, we assume they can fail in a Byzantine way [Lamport 1982]: data stored can be deleted, corrupted, created or leaked to unauthorized parties. This is the most general fault model and encompasses both malicious attacks/intrusions on a cloud provider and arbitrary data corruption (e.g., due to accidental events like the Ma.gnolia case). The protocols require a set of $n = 3f + 1$ storage clouds, at most f of which can be faulty. Additionally, the quorums used in the protocols are composed by any subset of $n - f$ storage clouds. It is worth to notice that this is the minimum number of replicas to tolerate Byzantine servers in asynchronous storage systems [Martin 2002].

The register abstraction provided by DEPSKY satisfies *regular semantics*: a read operation that happens concurrently with a write can return the value being written or the object's value before the write [Lamport 1986]. This semantics is both intuitive and stronger than the eventual consistency of some cloud-based services [Vogels 2009]. Nevertheless, if the semantics provided by the underlying storage clouds is weaker than regular, then DEPSKY's semantics is also weaker (Section 3.10).

Notice that our model hides most of the complexity of the distributed storage system employed by the cloud provider to manage the data storage service since it just assumes that the service is an object storage system prone to Byzantine faults that supports very simple operations. These operations are accessed through RPCs (Remote Procedure Calls) with the following failure semantics: the operation keeps being invoked until an answer is received or the operation is canceled (possibly by another thread, using a *cancel_pending* special operation to stop resending a request). This means that we have at most once semantics for the operations being invoked. This is not a problem because all storage cloud operations are idempotent, i.e., the state of the cloud becomes the same irrespectively of the operation being executed only once or more times.

3.4 Protocol Design Rationale

Quorum protocols can serve as the backbone of highly available storage systems [Chockler 2009]. There are many protocols for implementing Byzantine fault-tolerant (BFT) storage [Cachin 2006, Goodson 2004, Hendricks 2007, Liskov 2006, Malkhi 1998a;b, Martin 2002], but most of them require that the servers execute some code, a functionality not available on storage clouds. This leads to a key difference between the DEPSKY protocols and these classical BFT protocols: *metadata and data are written in separate quorum accesses*. Moreover, supporting multiple writers for a register (a data unit in DEPSKY parlance) can be problematic due to the lack of server code able to verify the version number of the data being written. To overcome this limitation we implement a single-writer multi-reader register, which is sufficient for many applications, and we provide a lock/lease protocol to support several concurrent writers for the data unit.

There are also some quorum protocols that consider individual storage nodes as passive shared memory objects (or disks) instead of servers [Abraham 2006, Attiya 2003, Chockler 2002, Gafni 2003, Jayanti 1998]. Unfortunately, most of these protocols require many steps to access the shared memory, or are heavily influenced by contention, which makes them impractical for geographically dispersed distributed systems such as DEPSKY due to the highly variable latencies involved. The DEPSKY protocols require two communication round-trips to read or write the metadata and the data files that are part of the data unit, independently of the existence of faults and contention.

Furthermore, as will be discussed latter, many clouds do not provide the expected consistency guarantees of a disk, something that can affect the correctness of these protocols. The DEPSKY protocols provide *consistency-proportional semantics*, i.e., the semantics of a data unit is as strong as the underling clouds allow, from eventual to regular consistency semantics. We do not try to provide atomic (linearizable) semantics due to the fact that all known techniques require server-to-server communication [Cachin 2006], servers sending update notifications to clients [Martin 2002] or write-backs [Goodson 2004, Malkhi 1998b]. None of these mechanisms is implementable using general-purpose storage clouds.

To ensure confidentiality of stored data on the clouds without requiring a key distribution service, we employ a *secret sharing scheme* [Shamir 1979]. In this scheme, a special party called dealer distributes a secret to n players, but each player gets only a share of this secret. The main properties of the scheme is that at least $f + 1 \leq n$ different shares of the secret are needed to recover it and that no information about the secret is disclosed with f or less shares. The scheme is integrated on the basic replication protocol in such way that each cloud receives just a share of the data being written, besides the metadata. This ensures that no individual cloud will have access to the data being stored, but that clients that have authorization to access the data will be granted access to the shares of (at least) $f + 1$ different clouds and will be able to rebuild the original data.

The use of a secret sharing scheme allows us to integrate confidentiality guarantees to the stored data without using a key distribution mechanism to make writers and readers of a data unit share a secret key. In fact, our mechanism reuses the access control of the cloud provider to control which readers are able to access the data stored on a data unit.

If we simply replicate the data on n clouds, the monetary costs of storing data using DEPSKY would increase by a factor of n. In order to avoid this, we compose the secret sharing scheme used on the protocol with an *information-optimal erasure code algorithm*, reducing the size of each share by a factor of $\frac{n}{f+1}$ of the original data [Rabin 1989]. This composition follows the original proposal of [Krawczyk 1993], where the data is encrypted with a random secret key, the encrypted data is encoded, the key is divided using secret sharing and each server receives a block of the encrypted data and a share of the key.

Common sense says that for critical data it is always a good practice to not erase old versions of the data, unless we can be certain that we will not need them anymore [Hamilton 2007]. An additional feature of our protocols is that old versions of the data are kept in the clouds.

3.5 DEPSKY-A– Available DepSky

The first DEPSKY protocol is called DEPSKY-A, and improves the availability and integrity of cloud-stored data by replicating it on several providers using quorum techniques. Algorithm 1 presents this protocol. Due to space constraints we encapsulate some of the protocol steps in the functions of the first two rows of Table 1. We use the '.' operator to denote access to metadata fields, e.g., given a metadata file m, $m.ver$ and $m.digest$ denote the version number and digest(s) stored in m. We use the '+' operator to concatenate two items into a string, e.g., "*value*"+*new_ver* produces a string that starts with the string "value" and ends with the value of variable *new_ver* in string format. Finally, the *max* function returns the maximum among a set of numbers.

The key idea of the *write algorithm* (lines 1-13) is to first write the value in a quorum of clouds (line 8), then write the

Function	Description
queryMetadata(du)	obtains the correctly signed file metadata stored in the container du of $n - f$ out-of the n clouds used to store the data unit and returns it in an array.
writeQuorum(du, name, value)	for every cloud $i \in \{0, ..., n-1\}$, writes the *value[i]* on a file named *name* on the container du in that cloud. Blocks until it receives write confirmations from $n - f$ clouds.
H(*value*)	returns the cryptographic hash of *value*.

Table 1. Functions used in the DEPSKY-A protocols.

Algorithm 1: DEPSKY-A

1 **procedure** DepSkyAWrite(du,value)
2 **begin**
3   **if** $max\_ver_{du} = 0$ **then**
4       $m \longleftarrow queryMetadata(du)$
5       $max\_ver_{du} \longleftarrow \max(\{m[i].ver : 0 \leq i \leq n-1\})$
6     $new\_ver \longleftarrow max\_ver_{du} + 1$
7     $v[0 .. n-1] \longleftarrow value$
8     $writeQuorum(du,$"value"$+new\_ver, v)$
9     $new\_meta \longleftarrow \langle new\_ver, \mathrm{H}(value) \rangle$
10    $sign(new\_meta, K_{r_w})$
11    $v[0 .. n-1] \longleftarrow new\_meta$
12    $writeQuorum(du,$"metadata"$, v)$
13    $max\_ver_{du} \longleftarrow new\_ver$

14 **function** DepSkyARead(du)
15 **begin**
16    $m \longleftarrow queryMetadata(du)$
17    $max\_id \longleftarrow i : m[i].ver = \max(\{m[i].ver : 0 \leq i \leq n-1\})$
18    $v[0 .. n-1] \longleftarrow \bot$
19   **parallel for** $0 \leq i < n-1$ **do**
20      $tmp_i \longleftarrow cloud_i.get(du,$ "value" $+m[max\_id].ver)$
21   **if** $\mathrm{H}(tmp_i) = m[max\_id].digest$ **then** $v[i] \longleftarrow tmp_i$
22   **wait until** $\exists i : v[i] \neq \bot$
23   **for** $0 \leq i \leq n-1$ **do** $cloud_i.cancel\_pending()$
24   **return** $v[i]$

---

corresponding metadata (lines 12). This order of operations ensures that a reader will only be able to read metadata for a value already stored in the clouds. Additionally, when a writer does its first writing in a data unit du (lines 3-5, max_ver_{du} is initialized as 0), it first contacts the clouds to obtain the metadata with the greatest version number, then updates the max_ver_{du} variable with the current version of the data unit.

The *read algorithm* just fetches the metadata files from a quorum of clouds (line 16), chooses the one with greatest version number (line 17) and reads the value corresponding to this version number and the cryptographic hash found in the chosen metadata (lines 19-22). After receiving the first reply that satisfies this condition the reader cancels the pending RPCs and returns the value (lines 22-24).

The rationale of why this protocol provides the desired properties is the following (proofs on the Appendix). Availability is guaranteed because the data is stored in a quorum of at least $n - f$ clouds and it is assumed that at most f clouds can be faulty. The read operation has to retrieve the value from only one of the clouds (line 22), which is always available because $(n - f) - f > 1$. Together with the data, signed metadata containing its cryptographic hash is also stored. Therefore, if a cloud is faulty and corrupts the data, this is detected when the metadata is retrieved.

3.6 DEPSKY-CA– Confidential & Available DepSky

The DEPSKY-A protocol has two main limitations. First, a data unit of size S consumes $n \times S$ storage capacity of the system and costs on average n times more than if it was stored in a single cloud. Second, it stores the data in cleartext, so it does not give confidentiality guarantees. To cope with these limitations we employ an information-efficient secret sharing scheme [Krawczyk 1993] that combines symmetric encryption with a classical secret sharing scheme and an optimal erasure code to partition the data in a set of blocks in such a way that (i.) $f + 1$ blocks are necessary to recover the original data and (ii.) f or less blocks do not give any information about the stored data[1].

The DEPSKY-CA protocol integrates these techniques with the DEPSKY-A protocol (Algorithm 2). The additional cryptographic and coding functions needed are in Table 2.

Function	Description
$generateSecretKey()$	generates a random secret key
$E(v,k)/D(e,k)$	encrypts v and decrypts e with key k
$encode(d,n,t)$	encodes d on n blocks in such a way that t are required to recover it
$decode(db,n,t)$	decodes array db of n blocks, with at least t valid, to recover d
$share(s,n,t)$	generates n shares in such a way that at least t of them are required to obtain any information about s
$combine(ss,n,t)$	combines shares on array ss of size n containing at least t correct shares to obtain the secret s

Table 2. Functions used in the DEPSKY-CA protocols.

The DEPSKY-CA protocol is very similar to DEPSKY-A with the following differences: (1.) the encryption of the data, the generation of the key shares and the encoding of the encrypted data on DepSkyCAWrite (lines 7-10) and the reverse process on DepSkyCARead (lines 30-31); (2.) the data stored in $cloud_i$ is composed by the share of the key $s[i]$ and the encoded block $e[i]$ (lines 12, 30-31); and (3.) $f + 1$ replies are necessary to read the data unit's current value instead of one on DEPSKY-A (line 28). Additionally, instead of storing a single digest on the metadata file, the writer generates and stores n digests, one for each cloud. These digests are accessed as different positions of the *digest* field of a metadata. If a key distribution infrastructure is available, or if readers and writer share a common key k, the secret

---
[1] Erasure codes alone cannot satisfy this confidentiality guarantee.

sharing scheme can be removed (lines 7, 9 and 31 are not necessary).

Algorithm 2: DEPSKY-CA

1   **procedure** DepSkyCAWrite(du,value)
2   **begin**
3   **if** $max\_ver_{du} = 0$ **then**
4     $m \longleftarrow queryMetadata(du)$
5     $max\_ver_{du} \longleftarrow \max(\{m[i].version : 0 \le i \le n-1\})$
6    $new\_ver \longleftarrow max\_ver_{du} + 1$
7    $k \longleftarrow generateSecretKey()$
8    $e \longleftarrow \mathsf{E}(value,k)$
9    $s[0 .. n-1] \longleftarrow share(k,n,f+1)$
10   $v[0 .. n-1] \longleftarrow encode(e,n,f+1)$
11   **for** $0 \le i < n-1$ **do**
12    $d[i] \longleftarrow \langle s[i], e[i] \rangle$
13    $h[i] \longleftarrow \mathsf{H}(d[i])$
14   $writeQuorum(du,\text{"value"}+new\_ver,d)$
15   $new\_meta \longleftarrow \langle new\_ver,h \rangle$
16   $sign(new\_meta,K_{r_w})$
17   $v[0 .. n-1] \longleftarrow new\_meta$
18   $writeQuorum(du,\text{"metadata"}.v)$
19   $max\_ver_{du} \longleftarrow new\_ver$

20   **function** DepSkyCARead(du)
21   **begin**
22    $m \longleftarrow queryMetadata(du)$
23    $max\_id \longleftarrow i : m[i].ver = \max(\{m[i].ver : 0 \le i \le n-1\})$
24    $d[0 .. n-1] \longleftarrow \perp$
25   **parallel for** $0 \le i \le n-1$ **do**
26     $tmp_i \longleftarrow cloud_i.get(du,\text{"value"}+m[max\_id].ver)$
27   **if** $\mathsf{H}(tmp_i) = m[max\_id].digest[i]$ **then** $d[i] \longleftarrow tmp_i$
28   **wait until** $|\{i : d[i] \neq \perp\}| > f$
29   **for** $0 \le i \le n-1$ **do** $cloud_i.cancel\_pending()$
30    $e \longleftarrow decode(d.e,n,f+1)$
31    $k \longleftarrow combine(d.s,n,f+1)$
32   **return** $\mathsf{D}(e,k)$

The rationale of the correctness of the protocol is similar to the one for DEPSKY-A (proofs also on the Appendix). The main differences are those already pointed out: encryption prevents individual clouds from disclosing the data; secret sharing allows storing the encryption key in the cloud without f faulty clouds being able to reconstruct it; the erasure code scheme reduces the size of the data stored in each cloud.

3.7 Read Optimization

The DEPSKY-A algorithm described in Section 3.5 tries to read the most recent version of the data unit from all clouds and waits for the first valid reply to return it. In the pay-per-use model this is far from ideal: even using just a single value, the application will be paying for n data accesses. A lower-cost solution is to use some criteria to sort the clouds and try to access them sequentially, one at time, until we obtain the desired value. The sorting criteria can be based on access monetary cost (cost-optimal), the latency of *queryMetadata* on the protocol (latency-optimal), a mix of the two or any other more complex criteria (e.g., an history of the latency and faults of the clouds).

This optimization can also be used to decrease the monetary cost of the DEPSKY-CA read operation. The main difference is that instead of choosing one of the clouds at a time to read the data, $f + 1$ of them are chosen.

3.8 Supporting Multiple Writers – Locks

The DEPSKY protocols presented do not support concurrent writes, which is sufficient for many applications where each process writes on its own data units. However, there are applications in which this is not the case. An example is a fault-tolerant storage system that uses DEPSKY as its backend object store. This system could have more than one node with the writer role writing in the same data unit(s) for fault tolerance reasons. If the writers are in the same network, a coordination system like Zookeeper [Hunt 2010] or DepSpace [Bessani 2008] can be used to elect a leader and coordinate the writes. However, if the writers are scattered through the Internet this solution is not practical without trusting the site in which the coordination service is deployed (and even in this case, the coordination service may be unavailable due to network issues).

The solution we advocate is a *low contention lock mechanism* that uses the cloud-of-clouds itself to maintain lock files on a data unit. These files specify which is the writer and for how much time it has write access to the data unit. The protocol is the following:

1. A process p that wants to be a writer (and has permission to be), first lists files on the data unit container on all n clouds and tries to find a zero-byte file called *lock-ID-T*. If such file is found on a quorum of clouds, $ID \neq p$ and the local time t on the process is smaller than $T + \Delta$, being Δ a safety margin concerning the difference between the synchronized clocks of all writers, someone else is the writer and p will wait until $T + \Delta$.

2. If the test fails, p can write a lock file called *lock-p-T* on all clouds, being $T = t + writer\_lease\_time$.

3. In the last step, p lists again all files in the data unit container searching for other lock files with $t < T + \Delta$ besides the one it wrote. If such file is found, p removes the lock file it wrote from the clouds and sleeps for a small random amount of time before trying to run the protocol again. Otherwise, p becomes the single-writer for the data unit until T.

Several remarks can be made about this protocol. First, the last step is necessary to ensure that two processes trying to become writers at the same time never succeed. Second, locks can be renewed periodically to ensure existence of a single writer at every moment of the execution. Moreover, unlocking can be easily done through the removal of the lock files. Third, the protocol requires synchronized clocks in order to employ leases and thus tolerate writer crashes. Finally,

this lock protocol is only *obstruction-free* [Herlihy 2003]: if several process try to become writers at the same time, it is possible that none of them are successful. However, due to the backoff on step 3, this situation should be very rare on the envisioned deployments for the systems.

3.9 Additional Protocols

Besides read, write and lock, DEPSKY provides other operations to manage data units. These operations and underlying protocols are briefly described in this section.

Creation and destruction. The creation of a data unit can be easily done through the invocation of the create operation in each individual cloud. In contention-prone applications, the creator should execute the locking protocol of the previous section before executing the first write to ensure it is the single writer of the data unit.

The destruction of a data unit is done in a similar way: the writer simply removes all files and the container that stores the data unit by calling *remove* in each individual cloud.

Garbage collection. As already discussed in Section 3.4, we choose to keep old versions of the value of the data unit on the clouds to improve the dependability of the storage system. However, after many writes the amount of storage used by a data unit can become too costly for the organization and thus some garbage collection is necessary. The protocol for doing that is very simple: a writer just lists all files *value<Version>* in the data unit container and removes all those with <Version> smaller than the oldest version it wants to keep in the system.

Cloud reconfiguration. Sometimes one cloud can become too expensive or too unreliable to be used for storing DEP-SKY data units. In this case we need a reconfiguration protocol to move the blocks from one cloud to another. The process is the following: (1.) the writer reads the data (probably from the other clouds and not from the one being removed); (2.) it creates the data unit container on the new cloud; (3.) executes the write protocol on the clouds not removed and the new cloud; (4.) deletes the data unit from the cloud being removed. After that, the writer needs to inform the readers that the data unit location was changed. This can be done writing a special file on the data unit container of the remaining clouds informing the new configuration of the system. A process will read this file and accept the reconfiguration if this file is read from at least $f + 1$ clouds.

3.10 Dealing with Weakly Consistent Clouds

Both DEPSKY-A and DEPSKY-CA protocols implement *single-writer multi-reader regular registers* if the clouds being accessed provide *regular semantics*. However, several clouds do not ensure this guarantee, but instead provide *read-after-write* or *eventual consistency* [Vogels 2009] for the data stored (e.g., Amazon S3 [Ama 2010]).

With a slight modification, our protocols can work with these weakly consistent clouds. The modification is very simple: repeat the data file reading from the clouds until the required condition is satisfied (receiving 1 or $f + 1$ data units, respectively in lines 22 and 28 of Algorithms 1 and 2). This modification ensures the read of a value described on a read metadata will be repeated until it is available.

This modification makes the DEPSKY protocols be *consistency-proportional* in the following sense: if the underlaying clouds provide regular semantics, the protocols provide regular semantics; if the clouds provide read-after-write semantics, the protocol satisfies read-after-write semantics; and finally, if the clouds provide eventually consistency, the protocols are eventually consistent. Notice that if the underlying clouds are heterogeneous in terms of consistency guarantees, DEPSKY ensures the weakest consistency among those provided. This comes from the fact that reading of a recently write value depends on the reading of the new metadata file, which, after a write is complete, will only be available eventually on weakly consistent clouds.

A problem with not having regular consistent clouds is that the lock protocol may not work correctly. After listing the contents of a container and not seeing a file, a process cannot conclude that it is the only writer. This problem can be minimized if the process waits a while between steps 2 and 3 of the protocol. However, the mutual exclusion guarantee will only be satisfied if the wait time is greater than the time for a data written to be seen by every other reader. Unfortunately, no eventually consistent cloud of our knowledge provides this kind of timeliness guarantee, but we can experimentally discover the amount of time needed for a read to propagate on a cloud with the desired coverage and use this value in the aforementioned wait. Moreover, to ensure some safety even when two writes happen in parallel, we can include a unique id of the writer (e.g., the hash of part of its private key) as the decimal part of its timestamps, just like is done in most Byzantine quorum protocols (e.g., [Malkhi 1998a]). This simple measure allows the durability of data written by concurrent writers (the name of the data files will be different), even if the metadata file may point to different versions on different clouds.

4. DEPSKY Implementation

We have implemented a DEPSKY prototype in Java as an application library that supports the read and write operations. The code is divided in three main parts: (1) data unit manager, that stores the definition and information of the data units that can be accessed; (2) system core, that implements the DEPSKY-A and DEPSKY-CA protocols; and (3) cloud providers drivers, which implement the logic for accessing the different clouds. The current implementation has 5 drivers available (the four clouds used in the evaluation - see next section - and one for storing data locally), but new drivers can be easily added. The overall implementation is about 2910 lines of code, being 1122 lines for the drivers.

The DEPSKY code follows a model of one thread per cloud per data unit in such a way that the cloud accesses can be executed in parallel (as described in the algorithms). All communications between clients and cloud providers are made over HTTPS (secure and private channels) using the REST APIs supplied by the storage cloud provider.

Our implementation makes use of several building blocks: RSA with 1024 bit keys for signatures, SHA-1 for cryptographic hashes, AES for symmetric cryptography, Shoenmakers' PVSS scheme [Schoenmakers 1999] for secret sharing with 192 bits secrets and the classic Reed-Solomon for erasure codes [Plank 2007]. Most of the implementations used come from the Java 6 API, while Java Secret Sharing [Bessani 2008] and Jerasure [Plank 2007] were used for secret sharing and erasure code, respectively.

5. Evaluation

In this section we present an evaluation of DEPSKY which tries to answer three main questions: *What is the additional cost in using replication on storage clouds? What is the advantage in terms of performance and availability of using replicated clouds to store data? What are the relative costs and benefits of the two DEPSKY protocols?*

The evaluation focus on the case of $n = 4$ and $f = 1$, which we expect to be the common deployment setup of our system for two reasons: (1.) f is the maximum number of faulty cloud storage providers, which are very resilient and so faults should be rare; (2.) there are currently not many more than four cloud storage providers that are adequate for storing critical data. Our evaluation uses the following cloud storage providers with their default configurations: Amazon S3, Windows Azure, Nirvanix and Rackspace.

5.1 Economical

Storage cloud providers usually charge their users based on the amount of data uploaded, downloaded and stored on them. Table 3 presents the cost in US Dollars of executing 10,000 reads and writes using the DEPSKY data model (with metadata and supporting many versions of a data unit) considering three data unit sizes: 100kb, 1Mb and 10Mb. This table includes only the costs of the operations being executed (invocations, upload and download), not the data storage, which will be discussed latter. All estimations presented on this section were calculated based on the values charged by the four clouds at September 25th, 2010.

In the table, the columns "DEPSKY-A", "DEPSKY-A opt", "DEPSKY-CA" e "DEPSKY-CA opt" present the costs of using the DEPSKY protocols with the read optimization respectively disabled and enabled. The other columns present the costs for storing the data unit (DU) in a single cloud.

The table shows that the cost of DEPSKY-A with $n = 4$ is roughly the sum of the costs of using the four clouds, as expected. However, if the read optimization is employed, the

Operation	DU Size	DEPSKY-A	DEPSKY-A opt	DEPSKY-CA	DEPSKY-CA opt	Amazon S3	Rackspace	Azure	Nirvanix
10kb Reads	100kb	0.64	0.14	0.32	0.14	0.14	0.21	0.14	0.14
	1Mb	6.55	1.47	3.26	1.47	1.46	2.15	1.46	1.46
	10Mb	65.5	14.6	32.0	14.6	14.6	21.5	14.6	14.6
10kb Writes	100kb	0.60	0.60	0.30	0.30	0.14	0.08	0.09	0.29
	1Mb	6.16	6.16	3.08	3.08	1.46	0.78	0.98	2.93
	10Mb	61.5	61.5	30.8	30.8	14.6	7.81	9.77	29.3

Table 3. Estimated costs per 10000 operations (in US Dollars). DEPSKY-A and DEPSKY-CA costs are computed for the realistic case of 4 clouds ($f = 1$). The "DEPSKY-A opt" and "DEPSKY-CA opt" setups consider the cost-optimal version of the protocols with no failures.

less expensive cloud cost dominates the cost of executing reads (only one out-of four clouds is accessed in fault-free executions). For DEPSKY-CA, the cost of reading and writing is approximately 50% of DEPSKY-A's due to the use of information-optimal erasure codes that make the data stored on each cloud roughly 50% of the size of the original data. The optimized version of DEPSKY-CA read also reduces this cost to half of the sum of the two less costly clouds due to its access to only $f + 1$ clouds in the best case. Recall that the costs for the optimized versions of the protocol account only for the best case in terms of monetary costs: reads are executed on the required less expensive clouds. In the worst case, the more expensive clouds will be used instead.

The storage costs of a 1Mb data unit for different numbers of stored versions is presented in Figure 3. We present the curves only for one data unit size because other size costs are directly proportional.

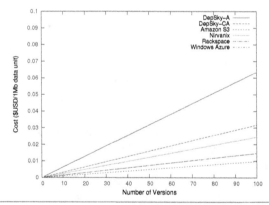

Figure 3. Storage costs of a 1Mb data unit for different numbers of stored versions.

The results depicted in the figure show that the cost of DEPSKY-CA storage is roughly half the cost of using DEPSKY-A and twice the cost of using a single cloud. This is no surprise since the storage costs are directly proportional to the amount of data stored on the cloud, and DEPSKY-A stores 4 times the data size, while DEPSKY-CA stores 2 times the data size and an individual cloud just stores a single copy of the data. Notice that the metadata costs are almost irrelevant when compared with the data size since its size is less than 500 bytes.

5.2 PlanetLab deployment

In order to understand the performance of DEPSKY in a real deployment, we used PlanetLab to run several clients accessing a cloud-of-clouds composed of popular storage cloud providers. This section explains our methodology and then presents the obtained results in terms of read and write latency, throughput and availability.

Methodology. The latency measurements were obtained using a logger application that tries to read a data unit from six different clouds: the four storage clouds individually and the two clouds-of-clouds implemented with DEPSKY-A and DEPSKY-CA.

The logger application executes periodically a *measurement epoch*, which comprises: read the data unit (DU) from each of the clouds individually, one after another; read the DU using DEPSKY-A; read the DU using DEPSKY-CA; sleep until the next epoch. The goal is to read the data through different setups within a time period as small as possible in order to minimize Internet performance variations.

We deployed the logger on eight PlanetLab machines across the Internet, on four continents. In each of these machines three instances of the logger were started for different DU sizes: 100kb (a measurement every 5 minutes), 1Mb (a measurement every 10 minutes) and 10Mb (a measurement every 30 minutes). These experiments took place during two months, but the values reported correspond to measurements done between September 10, 2010 and October 7, 2010.

In the experiments, the local costs, in which the protocols incur due to the use of cryptography and erasure codes, are negligible for DEPSKY-A and account for at most 5% of the read and 10% of the write latencies on DEPSKY-CA.

Reads. Figure 4 presents the 50% and 90% percentile of all observed latencies of the reads executed (i.e., the values below which 50% and 90% of the observations fell). The number of reads executed on each site is presented on the second column of Table 5.

Based on the results presented in the figure, several points can be highlighted. First, DEPSKY-A presents the best latency of all but one setups. This is explained by the fact that it waits for 3 out-of 4 copies of the metadata but only one of the data, and it usually obtains it from the best cloud available during the execution. Second, DEPSKY-CA latency is closely related with the second best cloud storage provider,

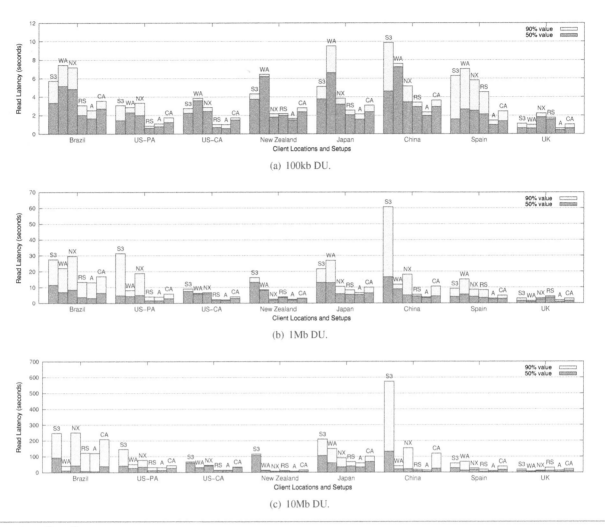

(a) 100kb DU.

(b) 1Mb DU.

(c) 10Mb DU.

Figure 4. $50^{th}/90^{th}$-percentile latency (in seconds) for 100kb, 1Mb and 10Mb DU read operations with PlanetLab clients located on different parts of the globe. The bar names are S3 for Amazon S3, WA for Windows Azure, NX for Nirvanix, RS for Rackspace, A for DEPSKY-A and CA for DEPSKY-CA. DEPSKY-CA and DEPSKY-A are configured with $n = 4$ and $f = 1$.

since it waits for at least 2 out-of 4 data blocks. Finally, notice that there is a huge variance between the performance of the cloud providers when accessed from different parts of the world. This means that no provider covers all areas in the same way, and highlight another advantage of the cloud-of-clouds: we can adapt our accesses to use the best cloud for a certain location.

The effect of optimizations. An interesting observation of our DEPSKY-A (resp. DEPSKY-CA) read experiments is that in a significant percentage of the reads the cloud that replied metadata faster (resp. the two faster in replying metadata) is not the first to reply the data (resp. the two first in replying the data). More precisely, in 17% of the 60768 DEPSKY-A reads and 32% of the 60444 DEPSKY-CA reads we observed this behavior. A possible explanation for that could be that some clouds are better serving small files (DEPSKY metadata is around 500 bytes) and not so good on serving large files (like the 10Mb data unit of some ex-

periments). This means that the read optimizations of Section 3.7 will make the protocol latency worse in these cases. Nonetheless we think this optimization is valuable since the rationale behind it worked for more than 4/5 (DEPSKY-A) and 2/3 (DEPSKY-CA) of the reads in our experiments, and its use can decrease the monetary costs of executing a read by a quarter and half, respectively.

Writes. We modified our logger application to execute writes instead of reads and deployed it on the same machines we executed the reads. We run it for two days in October and collected the logs, with at least 500 measurements for each location and data size. Due to space constraints, we do not present all these results, but illustrate the costs of write operations for different data sizes discussing only the observed results for an UK client. The 50% and 90% percentile of the latencies observed are presented in Figure 5.

The latencies in the figure consider the time of writing the data on all four clouds (file sent to 4 clouds, wait for only 3

Operation	DU Size	UK			US-CA		
		DEPSKY-A	DEPSKY-CA	Amazon S3	DEPSKY-A	DEPSKY-CA	Amazon S3
Read	100kb	189	135	59.3	129	64.9	31.5
	1Mb	808	568	321	544	306	104
	10Mb	1479	756	559	780	320	147
Write	100kb	3.53	4.26	5.43	2.91	3.55	5.06
	1Mb	14.9	26.2	53.1	13.6	19.9	25.5
	10Mb	64.9	107	84.1	96.6	108	34.4

Table 4. Throughput observed in kb/s on all reads and writes executed for the case of 4 clouds ($f = 1$).

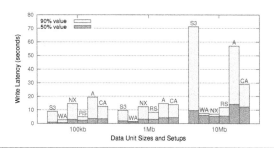

Figure 5. $50^{th}/90^{th}$-percentile latency (in seconds) for 100kb, 1Mb and 10Mb DU write operation for a PlanetLab client at the UK. The bar names are the same as in Figure 4. DEPSKY-A and DEPSKY-CA are configured with $n = 4$ and $f = 1$.

confirmations) and the time of writing the new metadata. As can be observed in the figure, the latency of a write is of the same order of magnitude of a read of a DU of the same size (this was observed on all locations). It is interesting to observe that, while DEPSKY's read latency is close to the cloud with best latency, the write latency is close to the worst cloud. This comes from the fact that in a write DEPSKY needs to upload data blocks on all clouds, which consumes more bandwidth at the client side and requires replies from at least three clouds.

Secret sharing overhead. As discussed in Section 3.6, if a key distribution mechanism is available, secret sharing could be removed from DEPSKY-CA. However, the effect of this on read and write latencies would be negligible since *share* and *combine* (lines 9 and 31 of Algorithm 2) account for less than 3 and 0.5 ms, respectively. It means that secret sharing is responsible for less than 0.1% of the protocols latency in the worst case[2].

Throughput. Table 4 shows the throughput in the experiments for two locations: UK and US-CA. The values are of the throughput observed by a single client, not by multiple clients as done in some throughput experiments. The table shows read and write throughput for both DEPSKY-A and DEPSKY-CA, together with the values observed from Amazon S3, just to give a baseline. The results from other locations and clouds follow the same trends discussed here.

By the table it is possible to observe that the read throughput decreases from DEPSKY-A to DEPSKY-CA and then to S3, at the same time that write throughput increases for this same sequence. The higher read throughput of DEPSKY when compared with S3 is due to the fact that it fetches the data from all clouds on the same time, trying to obtain the data from the fastest cloud available. The price to pay for this benefit is the lower write throughput since data should be written at least on a quorum of clouds in order to complete a write. This trade off appears to be a good compromise since reads tend to dominate most workloads of storage systems.

The table also shows that increasing the size of the data unit improves throughput. Increasing the data unit size from 100kb to 1Mb improves the throughput by an average factor of 5 in both reads and writes. By the other hand, increasing the size from 1Mb to 10Mb shows less benefits: read throughput is increased only by an average factor of 1.5 while write throughput increases by an average factor of 3.3. These results show that cloud storage services should be used for storing large chunks of data. However, increasing the size of these chunks brings less benefit after a certain size (1Mb).

Notice that the observed throughputs are at least an order of magnitude lower than the throughput of disk access or replicated storage in a LAN [Hendricks 2007], but the elasticity of the cloud allows the throughput to grow indefinitely with the number of clients accessing the system (according to the cloud providers). This is actually the main reason that lead us to not trying to measure the peak throughput of services built on top of clouds. Another reason is that the Internet bandwidth would probably be the bottleneck of the throughput, not the clouds.

Faults and availability. During our experiments we observed a significant number of read operations on individual clouds that could not be completed due to some error. Table 5 presents the *perceived availability* of all setups calculated as $\frac{reads\_completed}{reads\_tried}$ from different locations.

The first thing that can be observed from the table is that the number of measurements taken from each location is not the same. This happens due to the natural unreliability of PlanetLab nodes, that crash and restart with some regularity.

There are two key observations that can be taken from Table 5. First, DEPSKY-A and DEPSKY-CA are the two single setups that presented an availability of 1.0000 in almost all

---

[2] For a more comprensive discussion about the overhead imposed by Java secret sharing see [Bessani 2008].

Location	Reads Tried	DEPSKY-A	DEPSKY-CA	Amazon S3	Rackspace	Azure	Nirvanix
Brazil	8428	1.0000	0.9998	1.0000	0.9997	0.9793	0.9986
US-PA	5113	1.0000	1.0000	0.9998	1.0000	1.0000	0.9880
US-CA	8084	1.0000	1.0000	0.9998	1.0000	1.0000	0.9996
New Zealand	8545	1.0000	1.0000	0.9998	1.0000	0.9542	0.9996
Japan	8392	1.0000	1.0000	0.9997	0.9998	0.9996	0.9997
China	8594	1.0000	1.0000	0.9997	1.0000	0.9994	1.0000
Spain	6550	1.0000	1.0000	1.0000	1.0000	0.9796	0.9995
UK	7069	1.0000	1.0000	0.9998	1.0000	1.0000	1.0000

Table 5. The perceived availability of all setups evaluated from different points of the Internet.

locations[3]. Second, despite the fact that most cloud providers advertise providing 5 or 6 nines of availability, the perceived availability in our experiments was lower. The main problem is that outsourcing storage makes a company not only dependent on the provider's availability, but also on the network availability. This is a fact that companies moving critical applications to the cloud have to be fully aware.

6. Related Work

DEPSKY provides a single-writer multi-reader read/write register abstraction built on a set of untrusted storage clouds that can fail in an arbitrary way. This type of abstraction supports an updatable data model, requiring protocols that can handle multiple versions of stored data (which is substantially different than providing write-once, read-maybe archival storages such as the one described in [Storer 2007]).

There are many protocols for Byzantine quorums systems for register implementation (e.g., [Goodson 2004, Hendricks 2007, Malkhi 1998a, Martin 2002]), however, few of them address the model in which servers are passive entities that do not run protocol code [Abraham 2006, Attiya 2003, Jayanti 1998]. DEPSKY differentiates from them in the following aspects: (1.) it decouples the write of timestamp and verification data from the write of the new value; (2.) it has optimal resiliency ($3f + 1$ servers [Martin 2002]) and employs read and write protocols requiring two communication round-trips independently of the existence of contention, faults and weakly consistent clouds; finally, (3.) it is the first single-writer multi-reader register implementation supporting efficient encoding and confidentiality. Regarding (2.), our protocols are similar to others for fail-prone shared memory (or "disk quorums"), where servers are passive disks that may crash or corrupt stored data. In particular, Byzantine disk Paxos [Abraham 2006] presents a single-writer multi-reader regular register construction that requires two communication round-trips both for reading and writing in absence of contention. There is a fundamental difference between this construction and DEPSKY: it provides a weak liveness condition for the read protocol (termination only when there is a finite number of contending writes) while our protocol satisfies wait-freedom. An important consequence of this limitation is that reads may require several communication steps when contending writes are being executed. This same limitation appears on [Attiya 2003] that, additionally, does not tolerate writer faults. Regarding point (3.), it is worth to notice that several Byzantine storage protocols support efficient storage using erasure codes [Cachin 2006, Goodson 2004, Hendricks 2007], but none of them mention the use of secret sharing or the provision of confidentiality. However, it is not clear if information-efficient secret sharing [Krawczyk 1993] or some variant of this technique could substitute the erasure codes employed on these protocols.

Cloud storage is a hot topic with several papers appearing recently. However, most of these papers deal with the intricacies of implementing a storage infrastructure inside a cloud provider (e.g., [McCullough 2010]). Our work is closer to others that explore the use of existing cloud storage services to implement enriched storage applications. There are papers showing how to efficiently use storage clouds for backup [Vrable 2009], implement a database [Brantner 2008] or add provenance to the stored data [Muniswamy-Reddy 2010]. However none of these works provide guarantees like confidentiality and availability and do not consider a cloud-of-clouds.

Some works on this trend deal with the high-availability of stored data through the replication of this data on several cloud providers, and thus are closely related with DEPSKY. The HAIL (High-Availability Integrity Layer) protocol set [Bowers 2009] aggregates cryptographic protocols for proof of recoveries with erasure codes to provide a software layer to protect the integrity and availability of the stored data, even if the individual clouds are compromised by a malicious and mobile adversary. HAIL has at least three limitations when compared with DEPSKY: it only deals with static data (i.e., it is not possible to manage multiple versions of data), it requires that the servers run some code (opposite to DEPSKY, that uses the storage clouds as they are), and does not provide guarantee of confidentiality of the stored data. The RACS (Redundant Array of Cloud Storage) system [Abu-Libdeh 2010] employs RAID5-like techniques (mainly erasure codes) to implement high-available and storage-efficient data replication on diverse clouds. Differently from DEPSKY, the RACS system does not try to solve security problems of cloud storage, but instead deals with "economic failures" and vendor lock-in.

---

[3] This is somewhat surprising since we were expecting to have at least some faults on the client network that would disallow it to access any cloud.

In consequence, the system does not provide any mechanism to detect and recover from data corruption or confidentiality violations. Moreover, it does not provide updates of the stored data. Finally, it is worth to mention that none of these cloud replication works present an experimental evaluation with diverse clouds as it is presented in this paper.

There are several works about obtaining trustworthiness from untrusted clouds. Depot improves the resilience of cloud storage making similar assumptions to DEPSKY, that storage clouds are fault-prone black boxes [Mahajan 2010]. However, it uses a single cloud, so it provides a solution that is cheaper but does not tolerate total data losses and the availability is constrained by the availability of the cloud on top of which it is implemented. Works like SPORC [Feldman 2010] and Venus [Shraer 2010] make similar assumptions to implement services on top of untrusted clouds. All these works consider a single cloud (not a cloud-of-clouds), require a cloud with the ability to run code, and have limited support for cloud unavailability, which makes them different from DEPSKY.

7. Conclusion

This paper presents the design and evaluation of DEPSKY, a storage service that improves the availability and confidentiality provided by commercial storage cloud services. The system achieves these objectives by building a cloud-of-clouds on top of a set of storage clouds, combining Byzantine quorum system protocols, cryptographic secret sharing, erasure codes and the diversity provided by the use of several cloud providers. We believe DEPSKY protocols are in an unexplored region of the quorum systems design space and can enable applications sharing critical data (e.g., financial, medical) to benefit from clouds.

The paper also presents an extensive evaluation of the system. The key conclusion is that it provides confidentiality and improved availability at a cost roughly double of using a single cloud for a practical scenario, which seems to be a good compromise for critical applications.

Acknowledgments

We warmly thank our shepherd Scott Brandt and the Eu-roSys'11 reviewers for their comments on earlier versions of this paper. This work was partially supported by the EC through project TCLOUDS (FP7/2007-2013, ICT-257243), and by the FCT through the ReD (PTDC/EIA-EIA/109044/2008), RC-Clouds (PCT/EIA-EIA/115211/2009) and the Multian-nual and CMU-Portugal Programmes.

References

[Ama 2010] Amazon S3 FAQ: What data consistency model does amazon S3 employ? http://aws.amazon.com/s3/faqs/, 2010.

[tcl 2010] Project TCLOUDS – trustworthy clouds - privacy and resilience for Internet-scale critical infrastructure. http://www.tclouds-project.eu/, 2010.

[Ehs 2010] UK NHS Systems and Services. http://www.connectingforhealth.nhs.uk/, 2010.

[Abraham 2006] Ittai Abraham, Gregory Chockler, Idit Keidar, and Dahlia Malkhi. Byzantine disk Paxos: optimal resilience with Byzantine shared memory. *Distributed Computing*, 18(5):387–408, April 2006.

[Abu-Libdeh 2010] Hussam Abu-Libdeh, Lonnie Princehouse, and Hakim Weatherspoon. RACS: A case for cloud storage diversity. *Proc. of the 1st ACM Symposium on Cloud Computing*, pages 229–240, June 2010.

[Attiya 2003] Hagit Attiya and Amir Bar-Or. Sharing memory with semi-Byzantine clients and faulty storage servers. In *Proc. of the 22rd IEEE Symposium on Reliable Distributed Systems - SRDS 2003*, pages 174–183, October 2003.

[Bessani 2008] Alysson N. Bessani, Eduardo P. Alchieri, Miguel Correia, and Joni S. Fraga. DepSpace: a Byzantine fault-tolerant coordination service. In *Proc. of the 3rd ACM European Systems Conference – EuroSys'08*, pages 163–176, April 2008.

[Bowers 2009] Kevin D. Bowers, Ari Juels, and Alina Oprea. HAIL: a high-availability and integrity layer for cloud storage. In *Proc. of the 16th ACM Conference on Computer and Communications Security - CCS'09*, pages 187–198, 2009.

[Brantner 2008] Matthias Brantner, Daniela Florescu, David Graf, Donald Kossmann, and Tim Kraska. Building a database on S3. In *Proc. of the 2008 ACM SIGMOD International Conference on Management of Data*, pages 251–264, 2008.

[Cachin 2006] Christian Cachin and Stefano Tessaro. Optimal resilience for erasure-coded Byzantine distributed storage. In *Proc. of the Int. Conference on Dependable Systems and Networks - DSN 2006*, pages 115–124, June 2006.

[Chockler 2009] Gregory Chockler, Rachid Guerraoui, Idit Keidar, and Marko Vukolić. Reliable distributed storage. *IEEE Computer*, 42(4):60–67, 2009.

[Chockler 2002] Gregory Chockler and Dahlia Malkhi. Active disk Paxos with infinitely many processes. In *Proc. of the 21st Symposium on Principles of Distributed Computing – PODC'02*, pages 78–87, 2002.

[Feldman 2010] Ariel J. Feldman, William P. Zeller, Michael J. Freedman, and Edward W. Felten. SPORC: Group collaboration using untrusted cloud resources. In *Proc. of the 9th USENIX Symposium on Operating Systems Design and Implementation – OSDI'10*, pages 337–350, October 2010.

[Gafni 2003] Eli Gafni and Leslie Lamport. Disk Paxos. *Distributed Computing*, 16(1):1–20, 2003.

[Ghemawat 2003] Sanjay Ghemawat, Howard Gobioff, and Shun-Tak Leung. The Google file system. In *Proc. of the 19th ACM Symposium on Operating Systems Principles – SOSP'03*, pages 29–43, 2003.

[Gibson 1998] Garth Gibson, David Nagle, Khalil Amiri, Jeff Butler, Fay Chang, Howard Gobioff, Charles Hardin, Erik Riedel, David Rochberg, and Jim Zelenka. A cost-effective, high-bandwidth storage architecture. In *Proc. of the 8th Int. Con-

ference on Architectural Support for Programming Languages and Operating Systems - ASPLOS'98, pages 92–103, 1998.

[Goodson 2004] Garth Goodson, Jay Wylie, Gregory Ganger, and Micheal Reiter. Efficient Byzantine-tolerant erasure-coded storage. In *Proc. of the Int. Conference on Dependable Systems and Networks - DSN'04*, pages 135–144, June 2004.

[Greer 2010] Melvin Greer. Survivability and information assurance in the cloud. In *Proc. of the 4th Workshop on Recent Advances in Intrusion-Tolerant Systems – WRAITS'10*, 2010.

[Hamilton 2007] James Hamilton. On designing and deploying Internet-scale services. In *Proc. of the 21st Large Installation System Administration Conference – LISA'07*, pages 231–242, 2007.

[Hendricks 2007] James Hendricks, Gregory Ganger, and Michael Reiter. Low-overhead byzantine fault-tolerant storage. In *Proc. of the 21st ACM Symposium on Operating Systems Principles – SOSP'07*, pages 73–86, 2007.

[Henry 2009] Alyssa Henry. Cloud storage FUD (failure, uncertainty, and durability). Keynote Address at the 7th USENIX Conference on File and Storage Technologies, February 2009.

[Herlihy 2003] Maurice Herlihy, Victor Lucangco, and Mark Moir. Obstruction-free syncronization: double-ended queues as an example. In *Proc. of the 23th IEEE Int. Conference on Distributed Computing Systems - ICDCS 2003*, pages 522–529, July 2003.

[Hunt 2010] Patrick Hunt, Mahadev Konar, Flavio Junqueira, and Benjamin Reed. Zookeeper: Wait-free coordination for Internet-scale services. In *Proc. of the USENIX Annual Technical Conference – ATC 2010*, pages 145–158, June 2010.

[Jayanti 1998] Prasad Jayanti, Tushar Deepak Chandra, and Sam Toueg. Fault-tolerant wait-free shared objects. *Journal of the ACM*, 45(3):451–500, May 1998.

[Krawczyk 1993] Hugo Krawczyk. Secret sharing made short. In *Proc. of the 13th Int. Cryptology Conference – CRYPTO'93*, pages 136–146, August 1993.

[Lamport 1986] Leslie Lamport. On interprocess communication (part II). *Distributed Computing*, 1(1):203–213, January 1986.

[Lamport 1982] Leslie Lamport, Robert Shostak, and Marshall Pease. The Byzantine generals problem. *ACM Transactions on Programing Languages and Systems*, 4(3):382–401, July 1982.

[Liskov 2006] Barbara Liskov and Rodrigo Rodrigues. Tolerating Byzantine faulty clients in a quorum system. In *Proc. of the 26th IEEE Int. Conference on Distributed Computing Systems - ICDCS'06*, July 2006.

[Mahajan 2010] Prince Mahajan, Srinath Setty, Sangmin Lee, Allen Clement, Lorenzo Alvisi, Mike Dahlin, and Michael Walfish. Depot: Cloud storage with minimal trust. In *Proc. of the 9th USENIX Symposium on Operating Systems Design and Implementation – OSDI 2010*, pages 307–322, October 2010.

[Malkhi 1998a] Dahlia Malkhi and Michael Reiter. Byzantine quorum systems. *Distributed Computing*, 11(4):203–213, 1998.

[Malkhi 1998b] Dahlia Malkhi and Michael Reiter. Secure and scalable replication in Phalanx. In *Proc. of the 17th IEEE Symposium on Reliable Distributed Systems - SRDS'98*, pages 51–60, October 1998.

[Martin 2002] Jean-Philippe Martin, Lorenzo Alvisi, and Mike Dahlin. Minimal Byzantine storage. In *Proc. of the 16th Int. Symposium on Distributed Computing – DISC 2002*, pages 311–325, 2002.

[McCullough 2010] John C. McCullough, JohnDunagan, Alec Wolman, and Alex C. Snoeren. Stout: An adaptive interface to scalable cloud storage. In *Proc. of the USENIX Annual Technical Conference – ATC 2010*, pages 47–60, June 2010.

[Metz 2009] Cade Metz. DDoS attack rains down on Amazon cloud. *The Register*, October 2009. http://www.theregister.co.uk/2009/10/05/amazon_bitbucket_outage/.

[Muniswamy-Reddy 2010] Kiran-Kumar Muniswamy-Reddy, Peter Macko, and Margo Seltzer. Provenance for the cloud. In *Proc. of the 8th USENIX Conference on File and Storage Technologies – FAST'10*, pages 197–210, 2010.

[Naone 2009] Erica Naone. Are we safeguarding social data? Technology Review published by MIT Review, http://www.technologyreview.com/blog/editors/22924/, February 2009.

[Plank 2007] James S. Plank. Jerasure: A library in C/C++ facilitating erasure coding for storage applications. Technical Report CS-07-603, University of Tennessee, September 2007.

[Rabin 1989] Michael Rabin. Efficient dispersal of information for security, load balancing, and fault tolerance. *Journal of the ACM*, 36(2):335–348, February 1989.

[Sarno 2009] David Sarno. Microsoft says lost sidekick data will be restored to users. *Los Angeles Times*, Oct. 15th 2009.

[Schoenmakers 1999] Berry Schoenmakers. A simple publicly verifiable secret sharing scheme and its application to electronic voting. In *Proc. of the 19th Int. Cryptology Conference – CRYPTO'99*, pages 148–164, August 1999.

[Shamir 1979] Adi Shamir. How to share a secret. *Communications of ACM*, 22(11):612–613, November 1979.

[Shraer 2010] Alexander Shraer, Christian Cachin, Asaf Cidon, Idit Keidar, Yan Michalevsky, and Dani Shaket. Venus: Verification for untrusted cloud storage. In *Proc. of the ACM Cloud Computing Security Workshop – CCSW'10*, 2010.

[Storer 2007] Mark W. Storer, Kevin M. Greenan, Ethan L. Miller, and Kaladhar Voruganti. Potshards: Secure long-term storage without encryption. In *Proc. of the USENIX Annual Technical Conference – ATC 2007*, pages 143–156, June 2007.

[Vogels 2009] Werner Vogels. Eventually consistent. *Communications of the ACM*, 52(1):40–44, 2009.

[Vrable 2009] Michael Vrable, Stefan Savage, and Geoffrey M. Voelker. Cumulus: Filesystem backup to the cloud. *ACM Transactions on Storage*, 5(4):1–28, 2009.

[Vukolic 2010] Marko Vukolic. The Byzantine empire in the intercloud. *ACM SIGACT News*, 41(3):105–111, 2010.

[Weil 2006] Sage A. Weil, Scott A. Brandt, Ethan L. Miller, Darrell D. E. Long, and Carlos Maltzahn. Ceph: A scalable, high-performance distributed file system. In *Proc. of the 7th USENIX Symposium on Operating Systems Design and Implementation – OSDI 2006*, pages 307–320, 2006.

Protocols Correctness Proofs

This section presents correctness proofs of the DEPSKY-A and DEPSKY-CA protocols. The first lemma states that the auxiliary functions presented in Table 1 are wait-free.

Lemma 1 *A correct process will not block executing write-Quorum or queryMetadata.*

Proof: Both operations require $n - f$ clouds to answer the put and get requests. For *writeQuorum*, these replies are just *acks* and they will always be received since at most f clouds are faulty. For the *queryMetadata*, the operation is finished only if one metadata file is available. Since we are considering only non-malicious writers, a metadata written in a cloud is always valid and thus correctly signed using $K_{r_w}^{du}$. It means that a valid metadata file will be read from at least $n - f$ clouds and the process will choose one of these files and finish the algorithm. ■

The next two lemmas state that if a correctly signed metadata is obtained from the cloud providers (using *queryMetadata*) the corresponding data can also be retrieved and that the metadata stored on DEPSKY-A and DEPSKY-CA satisfy the properties of a regular register [Lamport 1986] (if the clouds provide this consistency semantics).

Lemma 2 *The value associated with the metadata with greatest version number returned by queryMetadata, from now on called* outstanding metadata, *is available on at least $f + 1$ non-faulty clouds.*

Proof: Recall that only valid metadata files will be returned by *queryMetadata*. These metadata will be written only by a non-malicious writer that follows the DepSkyAWrite (resp. DepSkyCAWrite) protocol. In this protocol, the data value is written on a quorum of clouds on line 8 (resp. line 14) of Algorithm 1 (resp. Algorithm 2), and then the metadata is generated and written on a quorum of clouds on lines 9-12 (resp. lines 15-18). Consequently, a metadata is only put on a cloud if its associated value was already put on a quorum of clouds. It implies that if a metadata is read, its associated value was already written on $n - f$ servers, from which at least $n - f - f \geq f + 1$ are correct. ■

Lemma 3 *The outstanding metadata obtained on an DepSkyARead (resp. DepSkyCARead) concurrent with zero or more DepSkyAWrites (resp. DepSkyCAWrites) is the metadata written on the last complete write or being written by one of the concurrent writes.*

Proof: Assuming that a client reads an outstanding metadata m, we have to show that m was written on the last complete write or is being written concurrently with the read. This proof can easily be obtained by contradiction. Suppose m was written before the start of the last complete write before the read and that it was the metadata with greatest version number returned from *queryMetadata*. This is clearly impossible because m was overwritten by the last complete write (which has a greater version number) and thus will never be selected as the outstanding metadata. It means that m can only correspond to the last complete write or to a write being executed concurrently with the read. ■

With the previous lemmas we can prove the wait-freedom of the DEPSKY-A and DEPSKY-CA registers, showing that their operations will never block.

Theorem 1 *All DEPSKY read and write operations are wait-free operations.*

Proof: Both Algorithms 1 and 2 use functions *queryMetadata* and *writeQuorum*. As show in Lemma 1, these operations can not block. Besides that, read operations make processes wait for the value associated with the outstanding metadata. Lemma 2 states that there are at least $f + 1$ correct servers with this data, and thus at least one of them will answer the RPC of lines 20 and 27 of Algorithms 1 and 2, respectively, with values that matches the digest contained on the metadata (or the different block digests in the case of DEPSKY-CA) and make $d[i] \neq \perp$, releasing itself from the barrier and completing the algorithm. ■

The last two theorems show that DEPSKY-A and DEPSKY-CA implement single-writer multi-reader regular registers.

Theorem 2 *A client reading a DEPSKY-A register in parallel with zero or more writes (by the same writer) will read the last complete write or one of the values being written.*

Proof: Lemma 3 states that the outstanding metadata obtained on line lines 16-17 of Algorithm 1 corresponds to the last write executed or one of the writes being executed. Lemma 2 states that the value associated with this metadata is available from at least $f + 1$ correct servers, and thus it can be obtained by the client on lines 19-22: just a single valid reply will suffice for releasing the barrier of line 22 and return the value. ■

Theorem 3 *A client reading a DEPSKY-CA register in parallel with zero or more writes (by the same writer) will read the last complete write or one of the values being written.*

Proof: This proof is similar with the one for DEPSKY-A. Lemma 3 states that the outstanding metadata obtained on lines 22-23 of Algorithm 2 corresponds to the last write executed or one of the writes being executed concurrently. Lemma 2 states that the values associated with this metadata are stored on at least $f + 1$ non-faulty clouds, and thus a reader can obtain them through the execution of lines 25-28: all non-faulty clouds will return their values corresponding to the outstanding metadata allowing the reader to decode the encrypted value, combine the key shares and decrypt the read data (lines 30-32), inverting the processing done by the writer on DepSkyCAWrite (lines 7-10). ■

Feature Consistency in Compile-Time–Configurable System Software *

Facing the Linux 10,000 Feature Problem

Reinhard Tartler, Daniel Lohmann, Julio Sincero, Wolfgang Schröder-Preikschat

Friedrich–Alexander University Erlangen–Nuremberg

{tartler, lohmann, sincero, wosch}@cs.fau.de

Abstract

Much system software can be configured at compile time to tailor it with respect to a broad range of supported hardware architectures and application domains. A good example is the Linux kernel, which provides more than 10,000 configurable features, growing rapidly.

From the maintenance point of view, compile-time configurability imposes big challenges. The configuration model (the selectable features and their constraints as presented to the user) and the configurability that is actually implemented in the code have to be kept in sync, which, if performed manually, is a tedious and error-prone task. In the case of Linux, this has led to numerous defects in the source code, many of which are actual bugs.

We suggest an approach to automatically check for configurability-related implementation defects in large-scale configurable system software. The configurability is extracted from its various implementation sources and examined for inconsistencies, which manifest in seemingly conditional code that is in fact unconditional. We evaluate our approach with the latest version of Linux, for which our tool detects 1,776 configurability defects, which manifest as dead/superfluous source code and bugs. Our findings have led to numerous source-code improvements and bug fixes in Linux: 123 patches (49 merged) fix 364 defects, 147 of which have been confirmed by the corresponding Linux developers and 20 as fixing a new bug.

* This work was partly supported by the German Research Foundation (DFG) under grant no. SCHR 603/7-1 and SFB/TR 89.

Categories and Subject Descriptors   D.4.7 [*Operating Systems*]: Organization and Design;   D.2.9 [*Management*]: Software configuration management

General Terms   Algorithms, Design, Experimentation, Management, Languages

Keywords   Configurability, Maintenance, Linux, Static Analysis, VAMOS

1. Introduction

I know of no feature that is always needed. When we say that two functions are almost always used together, we should remember that "almost" is a euphemism for "not".   DAVID L. PARNAS [1979]

Serving no user value on its own, system software has always been "caught between a rock and a hard place". As a link between hardware and applications, system software is faced with the requirement for variability to meet the specific demands of both. This is particularly true for operating systems, which ideally should be tailorable for domains ranging from small, resource-constrained embedded systems over network appliances and interactive workstations up to mainframe servers. As a result, many operating systems are provided as a software family [Parnas 1979]; they can (and have to) be configured at compile time to derive a concrete operating-system variant.

Configurability as a system property includes two separated – but related – aspects: *implementation* and *configuration*. Kernel developers implement configurability in the code; in most cases they do this by means of conditional compilation and the C preprocessor [Spinellis 2008], despite all the disadvantages with respect to understandability and maintainability ("#ifdef hell") this approach is known for [Liebig 2010, Spencer 1992]. Users configure the operating system to derive a concrete variant that fits their purposes. In simple cases they have to do this by (un-)commenting #define directives in some global configure.h file; however, many operating systems today come with an interactive configuration tool. Based on an internal model of features and constraints, this tool guides the user through the configuration process by a hierarchical / topic-oriented view on the available features. In fact, it performs implicit consistency checks with respect to the selected features, so that the outcome is always a valid configuration that represents a viable variant. In today's operating systems, this extra guidance is crucial because of the sheer enormity of available features: eCos, for instance, provides more than 700 features, which are configured with (and checked by) ECOSCONFIG [Massa 2002]; the Linux kernel is configured with KCONFIG and provides more than 10,000 (!) features. This is a lot of variability – and, as we show in this paper, the source of many bugs that could easily be avoided by better tool support.

Our Contributions

This article builds upon previous work. In [Sincero 2010], we have introduced the extraction of a source-code variability model from C Preprocessor (CPP)-based software, which represents a building block for this work. A short summary of this approach is presented in Section 3.2.3. In this paper, we extend that work by incorporating other sources of variability and automatically (cross-) checking them for configurability-related implementation defects in large-scale configurable system software. We evaluate our approach with the latest version of Linux. In summary, we claim the following contributions:

1. It is the first work that shows the problem with the increasing configurability in system software that causes serious maintenance issues. (Section 2.2)

2. It is the first work that checks for configurability-related implementation defects under the consideration of both symbolic *and* logic integrity. (Section 3.1)

3. It presents an algorithm to effectively slice very large configuration models, which are commonly found in system software. This greatly assists our crosschecking approach. (Section 3.2.2)

4. It presents a practical and scalable tool chain that has detected 1,776 configurability-related defects and bugs in Linux 2.6.35; for 121 of these defects (among them 22 confirmed new bugs) our fixes have already been merged into the mainline tree. (Section 4)

In the following, we first analyze the problem in further detail before presenting our approach in Section 3. We evaluate and discuss our approach in Section 4 and Section 5, respectively, and discuss related work in Section 6. The problem of configurability-related defects will be introduced in the context of Linux, which will also be the case study used throughout this paper. Our findings and suggestions, however, also apply to other compile-time configurable system software.

2. Problem Analysis

Linux today provides more than 10,000 configurable features, which is a lot of variability with respect to hardware platforms and application domains. The possibility to leave out functionality that is not needed (such as x86 PAE support in an Atom-based embedded system) and to choose between alternatives for those features that are needed (such as the default IO scheduler to use) is an important factor for its ongoing success in so many different application and hardware domains.

2.1 Configurability in Linux

The enormous configurability of the Linux kernel demands dedicated means to ensure the validity of the resulting Linux variants. Most features are not self-contained; instead, their possible inclusion is constrained by the presence or absence of other features, which in turn impose constraints on further features, and so on. In Linux, variant validity is taken care of by the KCONFIG tool chain, which is depicted in Figure 1:

❶ Linux employs the KCONFIG language to specify its configurable features together with their constraints. In version 2.6.35 a total of 761 *Kconfig files* with 110,005 lines of code define 11,283 features plus dependencies. We call the thereby specified variability the Linux **configuration space**.

The following KCONFIG lines, for instance, describe the (optional) Linux feature to include support for hot CPU plugging in an enterprise server:

```
config HOTPLUG_CPU
    bool "Support for hot-pluggable CPUs"
    depends on SMP && HOTPLUG
        && SYS_SUPPORTS_HOTPLUG_CPU
```

The HOTPLUG_CPU feature *depends on* general support for hot-pluggable hardware and must not be selected in a single-processor system.

❷ The KCONFIG configuration tool implicitly enforces all feature constraints during the interactive feature selection process. The outcome is, by construction, the description of a valid Linux **configuration variant**.

Technically, the output is a C header file (autoconf.h) and a Makefile (auto.make) that define a CONFIG_<FEATURE> preprocessor macro and make variable for every selected KCONFIG feature:

```
#define CONFIG_HOTPLUG_CPU 1
#define CONFIG_SMP 1
```

It's a convention that *all* and *only* KCONFIG flags are prefixed with CONFIG_.

❸ Features are implemented in the Linux source base. Whereas some coarse-grained features are enforced by including or excluding whole compilation units in the build process, the majority of features are enforced within the source files by means of the conditional compilation with

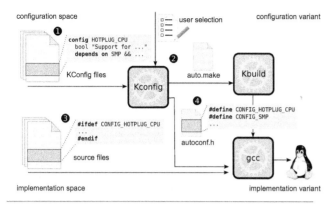

Figure 1. Linux build process (simplified).

the C preprocessor. A total of 27,166 source files contain 82,116 #ifdef blocks. We call the thereby implemented variability the Linux **implementation space**.

❹ The KBUILD utility drives the actual variant compilation and linking process by evaluating auto.make and embedding the configuration variant definition autoconf.h into every compilation unit via GCC's "forced include"[1] mechanism. The result of this process is a concrete Linux **implementation variant**.

2.2 The Issue

Overall, the configurability of Linux is defined by two separated, but related models: The configuration space defines the *intended* variability, whereas the implementation space defines the *implemented* variability of Linux. Given the size of both spaces – 110 kloc for the configuration space and 12 mloc for the implementation space in Linux 2.6.35 –, it is not hard to imagine that this is prone to inconsistencies, which manifest as configurability defects, many of which are bugs. We have identified two types of integrity issues, namely *symbolic* and *logic*, which we introduce in the following by examples from Linux:

Consider the following change, which corrects a simple feature misnaming (detected by our tool and confirmed as a bug) in the file kernel/smp.c[2]:

```
diff --git a/kernel/smp.c b/kernel/smp.c
--- a/kernel/smp.c
+++ b/kernel/smp.c

-#ifdef CONFIG_CPU_HOTPLUG
+#ifdef CONFIG_HOTPLUG_CPU
```

Patch 1. Fix for a symbolic defect

The issue, which was present in Linux 2.6.30, is an example of a **symbolic integrity violation**; the implementation space references a feature that does not exist in the configuration

spaces, so the actual implementation of the HOTPLUG_CPU feature is incomplete. This bug remained undetected in the kernel code base for more than six months. We cannot claim credit for detecting this particular bug (it had been reported to the respective developer just before we submitted our patch); however, we have found 116 similar defects caused by symbolic integrity violation that have been confirmed as *new*.

A symbolic integrity violation indicates a configuration–implementation space mismatch with respect to a feature *identifier*. However, consistency issues also occur at the level of feature *constraints*. Consider the following fix, which fixes what we call a **logic integrity violation**:

```
diff --git a/arch/x86/include/asm/mmzone_32.h
          b/arch/x86/include/asm/mmzone_32.h
--- a/arch/x86/include/asm/mmzone_32.h
+++ b/arch/x86/include/asm/mmzone_32.h
@@ -61,11 +61,7 @@ extern s8 physnode_map[];

 static inline int pfn_to_nid(unsigned long pfn)
 {
-#ifdef CONFIG_NUMA
   return((int) physnode_map[(pfn)
            / PAGES_PER_ELEMENT]);
-#else
-  return 0;
-#endif
 }

 /*
```

Patch 2. Fix for a logical defect

The patch itself does not look too complicated – the particularities of the issue it fixes stem from the context: In the source, the affected pfn_to_nid() function is nested within a larger code block whose presence condition is #ifdef CONFIG_DISCONTIGMEM. According to the KCONFIG model, however, the DISCONTIGMEM feature *depends on* the NUMA feature, which means that it also implies the selection of NUMA in any valid configuration. As a consequence, the #ifdef CONFIG_NUMA is superfluous; the #else branch is dead and both are removed by the patch. The patch has been confirmed as fixing a *new* defect by the respective Linux developers and is currently processed upstream for final acceptance into mainline Linux.

Compared to symbolic integrity violations, logic integrity violations are generally much more difficult to analyze and fix. So far we have fixed 38 logic integrity violations that have been confirmed as new defects.

Note that Patch 2 does not fix a real bug – it only improves the source-code quality of Linux by removing some dead code and superfluous #ifdef statements. Some readers might consider this as "less relevant cosmetical improvement"; however, such "cruft" (especially if it contributes to "#ifdef

---

[1] implemented by the -include command-line switch

[2] Shown in unified diff format. Lines starting with -/+ are being removed/added

version	features	#ifdef blocks	source files
2.6.12 (2005)	5338	57078	15219
2.6.20	7059	62873	18513
2.6.25	8394	67972	20609
2.6.30	9570	79154	23960
2.6.35 (2010)	11223	84150	28598
relative growth (5 years)	**110%**	**47%**	**88%**

Table 1. Growth of configurability in Linux

hell") causes long-term maintenance costs and impedes the general accessibility of the source.

2.3 Problem Summary

Overall, we find 1,316 symbolic + 460 logic integrity violations in Linux 2.6.35 – numbers that speak for themselves. The situation becomes more severe every day, given how quickly Linux is growing: Within the last five years, the number of configuration-conditional blocks in the source (#if blocks that test for some KCONFIG item) has grown by around fifty percent, the number of features (KCONFIG items) and source files have practically doubled (Table 1).

We think that configurability as a system property has to be seen as a significant (and so far underestimated) cause of software defects in its own respect.

3. The Approach

As pointed out in Section 2.2, many configurability-related defects are caused by inconsistencies that result from the fact that configurability is defined by two (technically separated, but conceptually related) models: the configuration space and the implementation space. The general idea of our approach is to extract all configurability-related information from both models into a common representation (a propositional formula), which is then used to cross-check the variability exposed within and across both models in order to find inconsistencies. We call these inconsistencies *configurability defects:*

> A **configurability defect** (short: **defect**) is a configuration-conditional **item** that is either **dead** (never included) or **undead** (always included) under the precondition that its **parent** (enclosing item) is included.

Examples for *items* in Linux are: KCONFIG options, build rules, and (most prominent) #ifdef blocks. The CONFIG_NUMA example discussed in Section 2.2 (see Figure 2) bears two *defects* in this respect: Block$_2$ is *undead* and Block$_3$ is *dead*. Defects can be further classified as:

Confirmed – a defect that has been confirmed as *unintentional* by the corresponding developers. If the defect has an effect on the binary code of at least one Linux implementation variant, we call it a **bug**.

Rule violation – a defect that, even though it breaks a generally accepted development rule, has been confirmed as *intentional* by the corresponding developers.

Patch 1 discussed in Section 2.2 fixes a *bug*, Patch 2 a *confirmed defect*. In the case of Linux a rule violation is usually the use of the CONFIG_ prefix for preprocessor flags that are not (yet) defined by KCONFIG. We will discuss the source of rule violations more thoroughly in Section 5.1.

3.1 Challenges in Analyzing Configurability Consistency – "What's wrong with GREP?"

Since version 2.6.23, Linux has included the (AWK/GREP-based) script checkkconfigsymbols.sh. This script is supposed to be used by maintainers to check for referential integrity between the KCONFIG model and the source code before committing their changes.

However, for Linux 2.6.30 this script reports 760 issues, among them the CONFIG_CPU_HOTPLUG issue discussed in Section 2.2, which remained in the kernel for more than six months. Apparently, kernel maintainers do *not* use this script systematically. While we can only speculate why this is the case, we have identified a number of shortcomings:

Accuracy. The output is disturbed by many **false positives**, defect reports that are not valid, but caused by some CONFIG_ macros being mentioned in a historical comment. We consider this as a constant annoyance that hinders the frequent employment of the script.

Performance. The script can only be applied on the complete source tree. On reasonable modern hardware (Intel quadcore with 2.83 GHz) it takes over 7 minutes until the output begins. We consider this as too long and too inflexible for integration into the daily incremental build process.

Coverage. Despite its verbosity, the script misses many *valid* defects. **False negatives** are caused, on the one hand, by logic integrity issues, like the CONFIG_NUMA example from Figure 2, as logic integrity is not covered at all. However, even many referential integrity issues are not detected – the script does not deal well with KCONFIG's tristate options (which are commonly used for loadable kernel modules). We consider this as a constant source of doubt with respect to the script's output.

The lack of accuracy causes a lot of noise in the output. This, and the fact that the script cannot be used during incremental builds, renders the script barely usable. Most of these shortcomings come from the fact that it does not actually parse and analyze the expressed variability, but just employs regular expressions to cross-match CONFIG_ identifiers. We conclude that the naïve GREP-based approach is (too) limited in this respect and that this problem has not been considered seriously in the past.

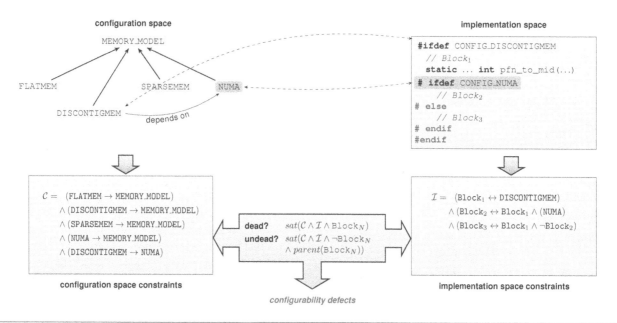

Figure 2. Our approach at a glance: The variability constraints defined by both spaces are extracted separately into propositional formulas, which are then examined against each other to find inconsistencies we call *configurability defects*.

3.2 Our Solution

Essential for the analysis of configurability problems is a common representation of the variability that spreads over different software artifacts. The idea is to individually convert each variability source (e.g., source files, KCONFIG, etc.) to a common representation in form of a sub-model and then combine these sub-models into a model that contains the whole variability of the software project. This makes it possible to analyze each sub-model as well their combination in order to reveal inconsistencies across sub-models.

Most of the constructs that model the variability both in the configuration and implementation spaces can be directly translated to propositional logic; therefore, propositional logic is our abstraction means of choice. As a consequence, the detection of configuration problems boils down to a satisfiability problem.

Linux (and many other systems) keep their configuration space ($\mathcal{C}$) and their implementation space ($\mathcal{I}$) separated. The variability model ($\mathcal{V}$) can be represented by the following boolean formula:

$$\mathcal{V} = \mathcal{C} \wedge \mathcal{I} \qquad (1)$$

$\mathcal{V} \mapsto \{0, 1\}$ is a boolean formula over all features of the system; $\mathcal{C}$ and $\mathcal{I}$ are the boolean formulas representing the constraints of the configuration and implementation spaces, respectively. Properly capturing and translating the variability of different artifacts into the formulas $\mathcal{C}$ and $\mathcal{I}$ is crucial for building the complete variability model $\mathcal{V}$. Once the model $\mathcal{V}$ is built we use it to search for defects.

With this model, we validate the implementation for configurability defects, that is, we check if the conditions for the presence of the block ($\texttt{Block}_N$) are fulfillable in the model $\mathcal{V}$. For example, consider Figure 2: The formula shown for dead blocks is satisfiable for $\texttt{Block}_1$ and $\texttt{Block}_2$, but not for $\texttt{Block}_3$. Therefore, $\texttt{Block}_3$ is considered to be *dead*; similarly the formula for undead blocks indicates that $\texttt{Block}_2$ is *undead*.

3.2.1 Challenges

In order to implement the solution sketch described above in practice for real-world large-scale system software, we face the following challenges:

Performance. As we aim at dealing with huge code bases, we have to guarantee that our tools finish in a reasonable amount of time. More importantly, we also aim at supporting programmers at development time when only a few files are of interest. Therefore, we consider the efficient check for variability consistency during incremental builds essential.

Flexibility. Projects that handle thousands of features will eventually contain *desired* inconsistencies with respect to their variability. Gradual addition or removal of features and large refactorings are examples of efforts that may lead to such inconsistent states within the lifetime of a project. Also, evolving projects may change their requirements regarding their variability descriptions. Therefore, a tool that checks for configuration problems should be flexible enough to incorporate information about desired issues in order to deliver precise and useful results; it should also minimize the number of false positives and false negatives.

Require: $\mathcal{S}$ initialized with an initial set of items
```
 1:  R = S
 2:  while S ≠ ∅ do
 3:      item = S.pop()
 4:      PC = presenceCondition(item)
 5:      for all i such that i ∈ PC do
 6:          if i ∉ R then
 7:              S.push(i)
 8:              R.push(i)
 9:          end if
10:      end for
11:  end while
12:  return R
```

Figure 3. Algorithm for configuration model slicing

In order to achieve both *performance* and *flexibility*, the implementation of our approach needs to take the particularities of the software project into account. Moreover, the precision of the configurability extraction mechanism has direct a impact on the rate of false positive and false negative reports. As many projects have developed their own, custom tools and languages to describe configuration variability, the configurability extraction needs to be tightly tailored.

In the following sections, we describe how we have approached these challenges to achieve good performance, flexibility, and, at the same time, a low number of false positives and false negatives.

3.2.2 Configuration Space

There are several strategies to convert configuration space models into boolean formulas [Benavides 2005, Czarnecki 2007]. However, due to the size of real models – the KCONFIG model contains more than 10,000 features –, the resulting boolean formulas become very complex. The search for a solution to problems that use such formulas may become intractable.

Therefore, we have devised an algorithm that implements *model slicing* for KCONFIG. This allows us to generate submodels from the original model that are smaller than the complete model. To illustrate, suppose we want to check if a specific block of the source code can be enabled by any valid user configuration. This is expressed by the satisfiability of the formula $\mathcal{V} \wedge \mathrm{Block}_N$. With a full model, the term $\mathcal{V}$ would contain all user-visible features as logical variables; for the Linux kernel it would have more than 10,000 variables. Nevertheless, not all features influence the solution for this specific problem. The key challenge is to find a sufficient – and preferably minimal – subset of features that can possibly influence the selection of the code block under analysis.

Our slicing algorithm for this purpose is depicted in Figure 3. The goal is to find the set of configuration items that can possibly affect the selection of one or more given initial items. (In our tool, which we will present in Section 4.1, this initial set of items will be taken from the #ifdef expressions.)

The basic idea is to check the presence conditions of each item for additional relevant items. Both *direct* and *indirect* dependencies from the initial set of features are thus taken into account such that the resulting set contains all features that can influence the features in the initial set.

In the first step (Line 1) the resulting set $\mathcal{R}$ is initialized with the list of input features. Then, the algorithm iterates until the working stack $\mathcal{S}$ is empty. In each iteration (Lines 2–11), a feature is taken from the stack and its *presence condition* is calculated through the function *presenceCondition(feature)*, which returns a boolean formula of the form *feature* $\rightarrow \varphi$. This formula represents the condition under which the feature can be enabled; φ is a boolean formula over the available features. Then, all features that appear in φ and have not already been processed (Line 6), are added to the working stack $\mathcal{S}$ and the result set $\mathcal{R}$. This algorithm always terminates; in the worst case, it will return all features and the slice will be exactly like the original model.

To implement our algorithm for Linux, we also have to implement the function *presenceCondition()* that takes the semantic details of the KCONFIG language into account. In a nutshell, the KCONFIG language supports the definitions of five types of features. Moreover, the features can have a number of attributes like prompts, direct and reverse dependencies, and default values. The presence condition of a feature is the set of conditions that must be met, so that either the user can select it or a default value is set automatically. Consider the following feature defined in the KCONFIG language:

```
config DISCONTIGMEM
        def_bool y
        depends on (!SELECT_MEMORY_MODEL &&
                ARCH_DISCONTIGMEM_ENABLE) ||
                DISCONTIGMEM_MANUAL
```

The presence condition for the feature DISCONTIGMEM is simply the selection of the feature itself and the expression of the depends on option. If a feature has several definitions of prompts and defaults, the feature implies the disjunction of the condition of each option that control its selection. The formal semantics of the KCONFIG language has been studied elsewhere [Berger 2010, Zengler 2010]; such formalisms describe in detail how to correctly derive the presence conditions.

3.2.3 Implementation Space

Many techniques [Baxter 2001, Hu 2000] have been proposed to translate the CPP semantics to boolean formulas. However, for our approach, we need to consider the language features of CPP that implement conditional compilation only. Therefore we devised an algorithm [Sincero 2010] that is tailored in this respect in order to be precise and have good performance. In short, our algorithm generates a boolean formula that describes a source file by means of its *conditional com-*

pilation structures; it therefore examines the CPP directives #ifdef, #ifndef, #if, #elif, #else, which are the constructs responsible for conditional compilation. As result, we receive a formula that describes the presence conditions for each conditional block. It thereby includes all flags (features) that appear in any conditional compilation expression as a logical variable. We build the presence condition $\mathcal{PC}$ of the conditional block b_i as follows:

$$\mathcal{PC}(b_i) = expr(b_i) \land noPredecessors(b_i) \land parent(b_i) \quad (2)$$

Let b_i be a conditional block. In order for this block to be selected, it is required that its expression $expr(b_i)$ evaluates to true, in #elif cascades none of its predecessors are selected $noPredecessors(b_i)$, and for nested blocks, its enclosing #ifdef block $parent(b_i)$ is selected. If all these conditions are met, then CPP will necessarily select this block. Additionally, also the reverse is true: if the CPP selects the block, all these presence conditions need to hold. This results in a biimplication: $b_i \leftrightarrow \mathcal{PC}(b_i)$. Therefore, the formula for a file with several blocks can be built as follows:

$$\mathcal{F}_u(\vec{f}, \vec{b}) = \bigwedge_{i=1..m} b_i \leftrightarrow \mathcal{PC}(b_i) \quad (3)$$

where $\vec{f}$ is a vector containing all flags present in the file, and $\vec{b}$ is a vector containing a variable for each block of the file. An example is shown on the right hand side of Figure 2: in the upper part we show the source code, and in the lower part we show the generated formula by our algorithm; note that in this example $\mathcal{F}_u([\texttt{DISCONTIGMEM}, \texttt{NUMA}], [\texttt{Block}_1, \texttt{Block}_2, \texttt{Block}_3]) = \mathcal{PC}(\texttt{Block}_1) \land \mathcal{PC}(\texttt{Block}_2) \land \mathcal{PC}(\texttt{Block}_3) = \mathcal{I}$

3.2.4 Crosschecking Among Variability Spaces

Our approach converts the different representations of variability to a common format so that we can check for inconsistencies, the configurability *defects*. Defects appear in two ways, either as **dead**, that is, unselectable blocks, or **undead**, that is, always present blocks. Both kinds of defects indicate code that is only seemingly conditional. They can be found within single models as presented in the previous two sections in isolation as well as across multiple models.

Within a single model we have **implementation-only** defects, which represent code blocks that cannot be selected regardless of the systems' selected features; the structure of the source file itself contains contradictions that impede the selection of a block. This can be determined by checking the satisfiability of the formula $sat(b_i \leftrightarrow \mathcal{PC}(b_i))$. **Configuration-only** defects represents features that are present in the configuration-space model but do not appear in any valid configuration of the model, which means that the presence condition of the feature is not satisfiable. We can check for such defects by solving: $sat(feature \rightarrow presenceCondition(feature))$.

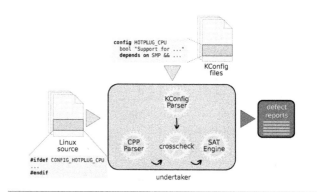

Figure 4. Principle of Operation

Across multiple models we have **configuration-implementation** defects, which occur when the rules from the configuration space contradict rules from the implementation space. We check for such defects by solving $sat((b_i \leftrightarrow \mathcal{PC}(b_i)) \land \mathcal{V})$. **Referential** defects are caused by a *missing feature* (m) that appears in either the configuration or the implementation space *only*. That is, $sat((b_i \leftrightarrow \mathcal{PC}(b_i)) \land \mathcal{V} \land \neg(m_1 \lor \cdots \lor m_n))$ is unsatisfiable.

Implementation-only defects have already been addressed in [Sincero 2010]; this paper focuses on the detection of configuration-implementation and referential defects in Linux. The defect analysis can be done using the *dead* and *undead* formulas as shown in the center of Figure 2.

We categorize all identified defects as either *logic* or *symbolic*. Logic defects are those that can only be found by determining the satisfiability of a complex boolean formula. Symbolic defects belong to a sub-group of referential defects where the expression of the analyzed block $exp(b_i)$ is an atomic formula.

4. Evaluation

In order to evaluate our approach, we have developed a prototype tool for Linux and a workflow to submit our results to the kernel developers. We started submitting our first patches in February 2010, at which time Linux version 2.6.33 has just been released. Most of our patches entered the mainline kernel tree during the merge window of version 2.6.36. In the following, we describe our tool and summarize the results.

4.1 Implementation for Linux

We named our tool UNDERTAKER, because its task is to identify (and eventually bury) dead and undead CPP-Blocks. Its basic principle of operation is depicted in Figure 4: The different sources of variability are parsed and transformed into propositional formulas. For CPP parsing, we use the BOOST::WAVE[3] parsing library; for proper parsing of the Kconfig files, we have hooked up in the original Linux KCONFIG implementation. The outcome of these parsers is fed into the crosscheck-

---
[3] http://www.boost.org

ing engine as described in Section 3.2.4 and solved using the PICOSAT[4] package. The tool itself is published as Free Software and available on our website.[5]

Our tool scans each `.c` and `.h` file in the source tree individually. Unlike the script `checkkonfigsymbols.sh` as discussed in Section 3.1, this allows developers to focus on the part of the source code they are currently working on and to get instant results for incremental changes. The results come as *defect reports* per file: For each file all configurability-related CPP blocks are analyzed for satisfiability, which yields the defect types described in the previous section. For instance, the report produced for the *configuration-implementation defect* from Figure 2 looks like this:

```
Found Kconfig related DEAD in arch/parisc/include/asm/mmzone.h,
line 40: Block B6 is unselectable, check the SAT formula.
```

Based on this information, the developer now revisits the KCONFIG files. The basis for the report is a formula that is falsified by our SAT solver. For this particular example the following formula was created:

```
1   #B6:arch/parisc/include/asm/mmzone.h:40:1:logic:undead
2   B2 &
3   !B6 &
4   (B0 <-> !_PARISC_MMZONE_H) &
5   (B2 <-> B0 & CONFIG_DISCONTIGMEM) &
6   (B4 <-> B2 & !CONFIG_64BIT) &
7   (B6 <-> B2 & !B4) &
8   (B9 <-> B0 & !B2) &
9   (CONFIG_64BIT -> CONFIG_PA8X00) &
10  (CONFIG_ARCH_DISCONTIGMEM_ENABLE -> CONFIG_64BIT) &
11  (CONFIG_ARCH_SELECT_MEMORY_MODEL -> CONFIG_64BIT) &
12  (CONFIG_CHOICE_11 -> CONFIG_SELECT_MEMORY_MODEL) &
13  (CONFIG_DISCONTIGMEM -> !CONFIG_SELECT_MEMORY_MODEL &
         CONFIG_ARCH_DISCONTIGMEM_ENABLE | CONFIG_DISCONTIGMEM_MANUAL) &
14  (CONFIG_DISCONTIGMEM_MANUAL -> CONFIG_CHOICE_11 &
         CONFIG_ARCH_DISCONTIGMEM_ENABLE) &
15  (CONFIG_PA8X00 -> CONFIG_CHOICE_7) &
16  (CONFIG_SELECT_MEMORY_MODEL -> CONFIG_EXPERIMENTAL |
         CONFIG_ARCH_SELECT_MEMORY_MODEL)
```

This formula can be deciphered easily by examining its parts individually. The first line shows an "executive summary" of the defect; here, Block B6, which starts in Line 40 in the file `arch/parisc/include/asm/mmzone.h`, inhibits a *logical* configuration defect in form of a block that cannot be unselected ("undead"). Lines 4 to 8 show the *presence conditions* of the corresponding blocks (cf. Section 3.2.3 and [Sincero 2010]); they all start with a block variable and by construction cannot cause the formula to be unsatisfiable. From the structure of the formula, we see that Block B2 is the enclosing block. Lines 9ff. contain the extracted implications from KCONFIG (cf. Section 3.2.2). In this case, it turns out that the KCONFIG implications from Line 9 to 16 show a *transitive* dependency from the KCONFIG item `CONFIG_DISCONTIGMEM` (cf. Block B2, Line 5) to the item `CONFIG_64BIT` (cf. Block B4, Line 6). This means that the KCONFIG selection has no impact on the evaluation of the #ifdef expression and the

code can thus be simplified. We have therefore proposed the following patch to the Linux developers[6]:

```
1   diff --git a/arch/parisc/include/asm/mmzone.h b/arch/parisc/include/
        asm/mmzone.h
2   --- a/arch/parisc/include/asm/mmzone.h
3   +++ b/arch/parisc/include/asm/mmzone.h
4   @@ -35,6 +35,1 @@ extern struct node_map_data node_data[];
5
6   -#ifndef CONFIG_64BIT
7   #define pfn_is_io(pfn) ((pfn & (0xf0000000UL >> PAGE_SHIFT)) == (0
        xf0000000UL >> PAGE_SHIFT))
8   -#else
9   -/* io can be 0xf0f0f0f0f0xxxxxx or 0xfffffffff0000000 */
10  -#define pfn_is_io(pfn) ((pfn & (0xf000000000000000UL >> PAGE_SHIFT))
        == (0xf000000000000000UL >> PAGE_SHIFT))
11  -#endif
```

Please note that this is one of the more complicated examples. Most of the defects reports have in fact only a few lines and are much easier to comprehend.

Results. Table 2 (upper half) summarizes the defects that UNDERTAKER finds in Linux 2.6.35, differentiated by subsystem. When counting defects in Linux, some extra care has to be taken with respect to architectures: Linux employs a separate KCONFIG-model per architecture that may also declare architecture-specific features. Hence, we need to run our defect analysis over every architecture and intersect the results. This prevents us from counting, for example, MIPS-specific blocks of the code as *dead* when compiling for x86. An exception of this rule is the code below `arch/`, which is architecture-specific by definition and checked against the configuration model of the respective architecture only.

Most of the 1,776 defects are found in `arch/` and `drivers/`, which together account for more than 75 percent of the configurability-related #ifdef-blocks. For these subsystems, we find more than three defects per hundred #ifdef-blocks, whereas for all other subsystems this ratio is below one percent (`net/` below two percent). These numbers support the common observation (e.g., [Engler 2001]) that "most bugs can be found in driver code", which apparently also holds for configurability-related defects. They also indicate that the problems induced by "#ifdef-hell" grow more than linearly, which we consider as a serious issue for the increasing configurability of Linux and other system software.

Even though the majority of defects (74%) are caused by symbolic integrity issues, we also find 460 logic integrity violations, which would be a lot harder to detect by "developer brainpower".

Performance. We have evaluated the performance of our tool with Linux 2.6.35. A full analysis of this kernel processes 27,166 source files with 82,116 configurability-conditional code blocks. This represents the information from the implementation space. The configuration space provides 761 KCONFIG files defining 11,283 features.

A *full* analysis that produces the results as shown in Table 2 takes around 15 minutes on a modern Intel quadcore with 2.83 GHz and 8 GB RAM. However, the implementation

---

[4] http://fmv.jku.at/picosat/

[5] http://vamos.informatik.uni-erlangen.de/trac/undertaker/

[6] http://lkml.org/lkml/2010/5/12/202

subsystem	#ifdefs	logic	symbolic	total
arch/	33757	345	581	926
drivers/	32695	88	648	736
fs/	3000	4	13	17
include/	7241	6	11	17
kernel/	1412	7	2	9
mm/	555	0	1	1
net/	2731	1	49	50
sound/	3246	5	10	15
virt/	53	0	0	0
other subsystems	601	4	1	5
$\sum$	85291	460	1316	1776
fix proposed		150 (1)	214 (22)	364 (23)
confirmed defect		38 (1)	116 (20)	154 (21)
confirmed rule-violation		88 (0)	21 (2)	109 (2)
pending		24 (0)	77 (0)	101 (0)

Table 2. Upper half: #ifdef blocks and defects per subsystem in Linux version 2.6.35; Lower half: acceptance state of defects (bugs) for which we have submitted a patch

< 0.5 s		93.69%
< 5 s		99.65%
< 30 s		100%

Figure 5. Processing time for 27,166 Linux source files

still leaves a lot of room for optimization: Around 70 percent of the consumed CPU time is *system* time, which is mostly caused by the fact that we fork() the SAT solver for every single #ifdef block.

Despite this optimization potential, the runtime of UNDER-TAKER is already appropriate to be integrated into (much more common) incremental Linux builds. Figure 5 depicts the file-based runtime for the Linux source base: Thanks to our slicing algorithm, 94 percent of all source files are analyzed in less than half a second; less than one percent of the source files take more than five seconds and only four files take between 20 and 30 seconds. The upper bound (29.1 seconds) is caused by kernel/sysctl.c, which handles a very high number of features; changes to this file often require a complete rebuild of the kernel anyway. For an incremental build that affects about a dozen files, UNDERTAKER typically finishes in less than six seconds.

4.2 Evaluation of Findings

To evaluate the quality of our findings, we have given our defect reports to two undergraduate students to analyze them, propose a change, and submit the patch upstream to the responsible kernel maintainers. Figure 6 depicts the whole workflow.

The first step is *defect analysis*: The students have to look up the source-code position for which the defect is reported and understand its particularities, which in the case of logical defects (as in the CONFIG_NUMA example presented in Figure 2) might also involve analyzing KCONFIG dependencies and further parts of the source code. This

information is then used to develop a *patch* that fixes the defect.

Based on the response to a submitted patch, we *improve and resubmit* and finally classify it (and the defects it fixes) in two categories: *accept (confirmed defect)* and *reject (confirmed rule violation)*. The latter means that the responsible developers consider the defect for some reason as *intended*; we will discuss this further in Section 5.1. As a matter of pragmatics, these defects are added into a local whitelist to filter them out in future runs.

In the period of February to July 2010, the students so far have submitted 123 patches. The submitted patches focus on the arch/ and driver/ subsystems and fix 364 out of 1,776 identified defects (20%). 23 (6%) of the analyzed and fixed defects were classified as bugs. If we extrapolate this defect/bug ratio to the remaining defects, we can expect to find another 80+ configurability-related bugs in the Linux kernel.

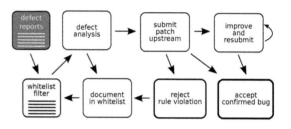

Figure 6. Based on the analysis of the defect reports, a patch is manually created and submitted to kernel developers. Based on the acceptance, we classify the defects that are fixed by our patch either as *confirmed rule violation* or *confirmed defect*.

Reaction of Kernel Maintainers. Table 3 lists the state of the submitted patches in detail; the corresponding defects are listed in Table 2 (lower half). In general, we see that our patches are well received: 87 out of 123 (71%) have been answered; more than 70 percent of them within less than one day, some even within minutes (Figure 7). We take this as indication that many of our patches are easy to verify and in fact appreciated.

Contribution to Linux. Table 3 also classifies the submitted patches as *critical* and *noncritical*, respectively. Critical patches fix *bugs*, that is, configurability defects that have an impact on the binary code. We did not investigate in detail the run-time observable effects of the 23 identified bugs. However, what can be seen from Table 3 is that the responsible developers consider them as worth fixing: 16 out of 17 (94%) of our critical patches have been answered; 9 have already been merged into Linus Torvalds' master git tree for Linux 2.6.36.

The majority of our patches fixes defects that affect the source code only, such as the examples shown in Section 2.2. However, even for these noncritical patches 57 out of 106 (54%) have already reached acknowledged state or better.

<1 hour	28.74%
<1 day	72.41%
<1 week	90.8%

Figure 7. Response time of 87 answered patches

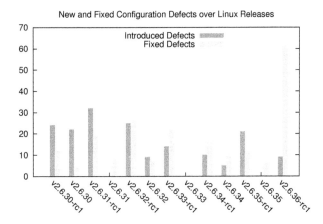

New and Fixed Configuration Defects over Linux Releases

Introduced Defects
Fixed Defects

Figure 8. Evolution of defect blocks over various Kernel versions. Most of our work was merged after the release of Linux version 2.6.35.

These patches clean up the kernel sources by removing 5,129 lines of configurability-related dead code and superfluous #ifdef statements ("cruft"). We consider this as a strong indicator that the Linux community is aware of the negative effects of configurability on the source-code quality and welcomes attempts to improve the situation.

Figure 8 depicts the impact of our work on a larger scale. To build this figure, we ran our tool on previous kernel versions and calculated the number of configurability defects that were *fixed* and *introduced* with each release. Most of our patches entered the mainline kernel tree during the merge window of version 2.6.36. Given that the patch submissions of two students have already made such a measurable impact, we expect that a consequent application of our approach, ideally directly by developers that work on new or existing code, could significantly reduce the problem of configurability-related consistency issues in Linux.

5. Discussion

Our findings have yielded a notable number of configurability defects in Linux. In the following, we discuss some potential causes for the introduction of defects and rule violations, threats to validity, and the broader applicability of our approach.

5.1 Interpretation of the Feedback

About 57 of the 123 submitted patches were accepted without further comments. We take this as indication that experts can easily verify the correctness of our submissions. Because of the distributed development of the Linux kernel, drawing

patch status	critical	noncritical	$\sum$
submitted	17	106	123
unanswered	1	35	36
ruleviolation	1	14	15
acknowledged	1	14	15
accepted	5	3	8
mainline	9	40	49

Table 3. *Critical* patches do have an effect on the resulting binaries (kernel and runtime-loadable modules). Noncritical patches remove text from the source code only.

the line between acknowledged and accepted (i.e., patches that have been merged for the next release), is challenging. We therefore count the 87 patches for which we received comments by Linux maintainers that maintain a public branch on the internet or are otherwise recognized in the Linux community as a confirmation that we identified a valid defect.

Causes for Defects. We have not yet analyzed the causes for defects systematically; doing this (e.g., using HERODOTOS [Palix 2010]) remains a topic for further research. However, we can already name some common causes, for which we need to consider how changes get integrated into Linux:

Logical defects are often caused by *copy and paste* (which confirms a similar observation in [Engler 2001]). Apparently code is often copied *together* with an enclosing #ifdef–#else block into a new context, where either the #ifdef or the #else branch is always taken (i.e., undead) and the counterpart is dead.

The most common source for symbolic defects is *spelling mistakes*, such as the CONFIG_HOTPLUG example in Patch 1. Another source for this kind of defects is *incomplete merges* of ongoing developments, such as architecture-specific code that is maintained by respective developer teams who maintain separate development trees and only submit hand-selected *patch series* for inclusion into the mainline. Obviously, this hand selection does not consider configurability-based defects – despite the recommendations in the patch submission guidelines:[7]

```
6: Any new or modified CONFIG options don't muck up the config menu.
7: All new Kconfig options have help text.
8: Has been carefully reviewed with respect to relevant Kconfig combinations. This is very hard to get right with testing -- brainpower pays off here.
```

Our approach provides a systematic, tool-based approach for this demanded checking of KCONFIG combinations.

Reasons for Rule Violations. On the other hand, we count 15 patches that were rejected by Linux maintainers. For all these patches, the respective maintainers confirmed the defects as valid (in one case even a bug!), but *nevertheless* prefer to keep them in the code. Reasons for this (besides *carelessness* and *responsibility uncertainties*) include:

---

[7] Documentation/SubmitChecklist in the Linux source.

Documentation. Even though all changes to the Linux source code are kept in the version control system (GIT), some maintainers have expressed their preference to keep outdated or unsupported feature implementations in the code in order to serve as a reference or template (e.g., to ease the porting of driver code to a newer hardware platform).

Out-of-tree development. In a number of cases, we find configurability-related items that are referenced from code in private development trees only. Keeping these symbolic defects in the kernel seems to be considered helpful for future code submission and review.

While it is debatable if all of the above are good reasons or not, of course we have to accept the maintainers preferences. The whitelist approach provides a pragmatic way to make such preferences explicit – so that they are no longer reported as defects, but can be addressed later if desired.

5.2 Threats to Validity

Accuracy. A strong feature of our approach is the accuracy with which configurability defects can be found. In our approach, false positives are conditional blocks that are falsely reported as unselectable. This means that there is a KCONFIG selection for which the code is *seen* by the compiler. By design, our approach operates *exact* and avoids this kind of error. Since by construction we avoid false positives (sans implementation bugs), the major threat to validity is the rate of false negatives, that is, the rate of the remaining, unidentified issues.

In fact, we have found for 2 (confirmed) defects explicit #error statements in the source that provoke compilation errors in case an *invalid* set of features has been selected. In our experiment, we classified these defects as confirmed rule violations. On top of that, we can find 28 similar #error statements in Linux 2.6.35. This indicates some distrust of developers in the variability declarations in KCONFIG, which our tool helps to mitigate by checking the effective constraints accurately.

Coverage. So far we do not consider the (discouraged) 509 #undef and #define CONFIG_ statements that we find in the code. However, these statements could possibly lead to incomplete results for only at most 4.51 percent of the 11,283 KCONFIG items.

Another restriction of the current implementation is that we do not yet analyze nonpropositional expressions in #ifdef statements, like comparisons against the integral value of some CONFIG_ flag. This affects about 2% out of 82,116 #ifdef blocks. We are currently looking into improving our implementation to reduce this number even further by rewriting the extracted constraints and process them using a satisfiability modulo theories (SMT) or constraint solving problem (CSP) engine.

An important, yet not considered source of feature constraints is the build system (makefiles). 91 percent of the Linux source files are feature-dependent, that is, they are not compiled at all when the respective feature has not been selected. Incorporation of these additional constraints into our approach is straight-forward: they can simply be added as further conjunctions to the variability model. These additional constraints could possibly restrict the variability even further, and thereby lead to false negatives.

Subtle semantic details and anachronisms of the KCONFIG language and implementation [Berger 2010, Zengler 2010] made our engineering difficult and contributed to the number of false negatives. At the time we conducted the experiment in Section 4, our implementation did not completely support the KCONFIG features *default value* and *select*. Meanwhile, we have fixed these issues in the undertaker, which increases the raw number of defects from 1,776 to 2,972.

In no case our approach resulted in a change that proposes to remove blocks that are used in production. However, in one case[8] we stumbled across old code that is useful with some additional debug-only patches that have never been merged. It turned out that the patches in question are no longer necessary in favor of the new tracing infrastructure. Our patch therefore has contributed to the removal of otherwise useless and potentially confusing code.

Despite all potential sources of false negatives: Compared to the 760 issues reported by the GREP-based approach (including many false positives!, see Section 3.1), our tool already finds more than twice as many defects. As our approach prevents false positives, this has to be considered as a *lower bound* for the number of configurability defects in Linux!

5.3 General Applicability of the Approach

Linux is the most configurable piece of software we are aware of, which made it a natural target to evaluate accuracy and scalability of our approach. However, the approach can be implemented for other software families as well, given there is some way to extract feature identifiers and feature constraints from all sources of variability. This is probably always the case for the implementation space (code), which is generally configured by CPP or some similar preprocessor. Extracting the variability from the configuration space is straight-forward, too, as long as features and constraints are described by some semi-formal model (such as KCONFIG). The configurability of eCos, for instance, is described in the configuration description language (CDL) [Massa 2002], whose expressiveness is comparable to KCONFIG.

KCONFIG itself is employed by more and more software families besides Linux. Examples include OpenWRT[9] or

---

[8] http://kerneltrap.org/mailarchive/linux-ext4/2010/2/8/6762333/thread

[9] http://www.openwrt.org

BusyBox.[10] For these software families our approach could be implemented with minimal effort.

However, even if the system software is configured by a simple configure script (such as FreeBSD), it would still be possible to extract feature identifiers and, hence, use our approach to detect symbolic configurability defects – which in the case of Linux account for 74 percent of all defects. Feature constraints, on the other hand, are more difficult to extract from configure files, as they are commonly given as human-readable comments only. A possible solution might be to employ techniques of natural language processing to automatically infer the constraints from the comments, similar to the approach suggested in [Tan 2007].

In a more general sense, our approach could be combined with other tools to make them *configurability aware*. For instance, modifications on in-kernel APIs and other larger refactorings are commonly done tool assisted (e.g., Padioleau [2008]). However, refactoring tools are generally not aware of code liveness and suggest changes in dead code. Our approach contributes to avoiding such useless work.

5.4 Variability-Aware Languages

The high relevance of static configurability for system software gives rise to the question if we are in need of better programming languages. Ideally, the language and compiler would directly support configurability (implementation and configuration), so that symbolic and semantic integrity issues can be prevented upfront by means of type-systems or at least be checked for at compile-time.

With respect to implementation of configurability it is generally accepted that CPP might not be the right tool for the job [Liebig 2010, Spencer 1992]. Many approaches have been suggested for a better separation of concerns in configurable (system) software, including, but not limited to: object-orientation [Campbell 1993], component models [Fassino 2002, Reid 2000], aspect-oriented programming (AOP) [Coady 2003, Lohmann 2009], or feature-oriented programming (FOP) [Batory 2004]. However, in the systems community we tend to be reluctant to adopt new programming paradigms, mostly because we fear unacceptable runtime overheads and immature tools. C++ was ruled out of the Linux kernel for exactly these reasons.[11] The authors certainly disagree here (in previous work with embedded operating systems we could show that C++ class composition [Beuche 1999] and AOP [Lohmann 2006] provide excellent means to implement overhead-free, fine-grained static configurability). Nevertheless, we have to accept CPP as the still de-facto standard for implementing static configurability in system software [Liebig 2010, Spinellis 2008].

With respect to modeling configurability, feature modeling and other approaches from the product line engineering

domain [Czarnecki 2000, Pohl 2005] provide languages and tooling to describe the variability of software systems, including systematic consistency checks. KCONFIG for Linux or CDL for eCos fit in here. However, what is generally missing is the bridge between the modeled and the implemented configurability. Hence tools like the UNDERTAKER remain necessary.

6. Related Work

Automated bug detection by examining the source code has a long tradition in the systems community. Many approaches have been suggested to extract rules, invariants, specifications, or even misleading source-code comments from the source code or execution traces [Engler 2001, Ernst 2000, Kremenek 2006, Li 2005, Tan 2007]. Basically, all of these approaches extract some *internal model* about what the code *should* look like/behave and then match this model against the reality to find defects that are potential bugs. For instance, iComment [Tan 2007] employs means of natural language processing to find inconsistencies between the programmer's intentions expressed in source-code comments and the actual implementation; Kremenek [2006] and colleagues use logic and probability to automatically infer specifications that can be checked by static bug-finding tools. However, none of the existing approaches takes *configurability* into account when inferring the internal model. In fact, the existing tools are more or less *configurability agnostic* – they either ignore configuration-conditional parts completely, fall back to simple heuristics, or have to be executed on preprocessed source code. Thereby, important information is lost. Our analysis framework could be combined with these approaches to make them configurability-aware and to systematically improve their coverage with respect to the (extremely high) number of Linux variants. However, we also think that configurability has to be understood as a significant source of bugs in its own respect. Our approach does just that.

A reason for the fact that existing source-code analysis tools ignore configurability (more or less) might be that conditionally-compiled code tends to be hard to analyze in real-world settings. Many approaches for analyzing conditional-compilation usage have been suggested, usually based on symbolic execution. However, even the most powerful symbolic execution techniques (such as KLEE [Cadar 2008]) would currently not scale to the size of the Linux kernel. Hence, several authors proposed to apply transformation systems to symbolically simplify CPP code with respect to configurability aspects [Baxter 2001, Hu 2000]. Our approach is technically similar in the sense that we also analyze only the configurability-related subset of CPP. However, by "simulating" the mechanics of the CPP using propositional formulas [Sincero 2010], we can more easily integrate (and check against) other sources of configurability, such as the configuration-space model.

---

[10] http://www.busybox.net

[11] *Trust me – writing kernel code in C++ is a BLOODY STUPID IDEA* LINUS TORVALDS [2004], http://www.tux.org/lkml/#s15-3

So far we have submitted 123 patches to the Linux community, which is a reasonably high number to confirm many observations of Guo [2009]: Patches for actively-maintained files are *a lot* more likely to receive responses. It really is worth the effort to figure out who is the principal maintainer (which is not always obvious) and to ensure that patches are easy reviewable and easy to integrate.

7. Summary and Conclusions

#ifdef's sprinkled all over the place are neither an incentive for kernel developers to delve into the code nor are they suitable for long-term maintenance.[12]

To cope with a broad range of application and hardware settings, system software has to be highly configurable. Linux 2.6.35, as a prominent example, offers 11,283 configurable features (KCONFIG items), which are implemented at compile time by 82,116 conditional blocks (#ifdef, #elif, ...) in the source code. The number of features has more than doubled within the last five years! From the maintenance point of view, this imposes big challenges, as the configuration model (the selectable features and their constraints) and the configurability that is *actually* implemented in the code have to be kept in sync. In the case of Linux, this has led to numerous inconsistencies, which manifest as dead #ifdef-blocks and bugs.

We have suggested an approach for automatic consistency checks for compile-time configurable software. Our implementation for Linux has yielded 1,776 configurability issues. Based on these findings, we so far have proposed 123 patches (49 merged, 8 accepted, 15 acknowledged) that fix 364 of these issues (among them 20 confirmed new bugs) and improve the Linux source-code quality by removing 5,129 lines of unnecessary #ifdef-code. The performance of our tool chain is good enough to be integrated into the regular Linux build process, which offers the chance for the Linux community to prevent configurability-related inconsistencies from the very beginning. We are currently finalizing out tools in this respect to submit them upstream.

The lesson to learn from this paper is that configurability has to be seen as a significant (and so far underestimated) cause of software defects in its own respect. Our work is meant as a call for attention on these problems – as well as a first attempt to improve on the situation.

Acknowledgments

We would like to thank our students Christian Dietrich and Christoph Egger for their enduring and admiring work on the implementation and evaluation. We thank Julia Lawall for inspiring conversations and her helpful comments on an early version of this work. Many thanks go to the anonymous reviewers for their constructive comments that helped to improve this paper a lot as well as to our shepherd Dawson Engler.

References

[Batory 2004] Don Batory. Feature-oriented programming and the AHEAD tool suite. In *Proceedings of the 26th International Conference on Software Engineering (ICSE '04)*, pages 702–703. IEEE Computer Society Press, 2004.

[Baxter 2001] Ira D. Baxter and Michael Mehlich. Preprocessor conditional removal by simple partial evaluation. In *Proceedings of the 8th Working Conference on Reverse Engineering (WCRE '01)*, page 281, Washington, DC, USA, 2001. IEEE Computer Society Press. ISBN 0-7695-1303-4.

[Benavides 2005] D. Benavides, A. Ruiz-Cortés, and P. Trinidad. Automated reasoning on feature models. In *Proceedings of the 17th International Conference on Advanced Information Systems Engineering (CAISE '05)*, volume 3520, pages 491–503, Heidelberg, Germany, 2005. Springer-Verlag.

[Berger 2010] Thorsten Berger and Steven She. Formal semantics of the CDL language. Technical note, University of Leipzig, 2010.

[Beuche 1999] Danilo Beuche, Abdelaziz Guerrouat, Holger Papajewski, Wolfgang Schröder-Preikschat, Olaf Spinczyk, and Ute Spinczyk. The PURE family of object-oriented operating systems for deeply embedded systems. In *Proceedings of the 2nd IEEE International Symposium on Object-Oriented Real-Time Distributed Computing (ISORC '99)*, pages 45–53, St Malo, France, May 1999.

[Cadar 2008] Cristian Cadar, Daniel Dunbar, and Dawson Engler. KLEE: Unassisted and automatic generation of high-coverage tests for complex systems programs. In *8th Symposium on Operating System Design and Implementation (OSDI '08)*, Berkeley, CA, USA, 2008. USENIX Association.

[Campbell 1993] Roy Campbell, Nayeem Islam, Peter Madany, and David Raila. Designing and implementing Choices: An object-oriented system in C++. *Communications of the ACM*, 36(9), 1993.

[Coady 2003] Yvonne Coady and Gregor Kiczales. Back to the future: A retroactive study of aspect evolution in operating system code. In Mehmet Akşit, editor, *Proceedings of the 2nd International Conference on Aspect-Oriented Software Development (AOSD '03)*, pages 50–59, Boston, MA, USA, March 2003. ACM Press.

[Czarnecki 2000] Krysztof Czarnecki and Ulrich W. Eisenecker. *Generative Programming. Methods, Tools and Applications.* Addison-Wesley, May 2000. ISBN 0-20-13097-77.

[Czarnecki 2007] Krzysztof Czarnecki and Andrzej Wasowski. Feature diagrams and logics: There and back again. In *Proceedings of the 11th Software Product Line Conference (SPLC '07)*, pages 23–34. IEEE Computer Society Press, Sept. 2007.

[Engler 2001] Dawson Engler, David Yu Chen, Seth Hallem, Andy Chou, and Benjamin Chelf. Bugs as deviant behavior: a general approach to inferring errors in systems code. In *Proceedings of the 18th ACM Symposium on Operating Systems Principles (SOSP '01)*, pages 57–72, New York, NY, USA, 2001. ACM Press.

---

[12] Linux maintainer Thomas Gleixner in his ECRTS '10 keynote *"Realtime Linux: academia v. reality"*. http://lwn.net/Articles/397422

[Ernst 2000] Michael D. Ernst, Adam Czeisler, William G. Griswold, and David Notkin. Quickly detecting relevant program invariants. In *Proceedings of the 22nd International Conference on Software Engineering (ICSE '00)*, pages 449–458, New York, NY, USA, 2000. ACM Press. ISBN 1-58113-206-9.

[Fassino 2002] Jean-Philippe Fassino, Jean-Bernard Stefani, Julia Lawall, and Gilles Muller. THINK: A software framework for component-based operating system kernels. In *Proceedings of the 2002 USENIX Annual Technical Conference*, pages 73–86. USENIX Association, June 2002.

[Guo 2009] Philip J. Guo and Dawson Engler. Linux kernel developer responses to static analysis bug reports. In *Proceedings of the 2009 USENIX Annual Technical Conference*, Berkeley, CA, USA, June 2009. USENIX Association. ISBN 978-1-931971-68-3.

[Hu 2000] Ying Hu, Ettore Merlo, Michel Dagenais, and Bruno Lagüe. C/C++ conditional compilation analysis using symbolic execution. In *Proceedings of the 16th IEEE International Conference on Software Maintainance (ICSM'00)*, page 196, Washington, DC, USA, 2000. IEEE Computer Society Press. ISBN 0-7695-0753-0.

[Kremenek 2006] Ted Kremenek, Paul Twohey, Godmar Back, Andrew Ng, and Dawson Engler. From uncertainty to belief: inferring the specification within. In *7th Symposium on Operating System Design and Implementation (OSDI '06)*, pages 161–176, Berkeley, CA, USA, 2006. USENIX Association.

[Li 2005] Zhenmin Li and Yuanyuan Zhou. PR-miner: automatically extracting implicit programming rules and detecting violations in large software code. In *Proceedings of the 10th European Software Engineering Conference and the 13th ACM Symposium on the Foundations of Software Engineering (ESEC/FSE '00)*, pages 306–315, New York, NY, USA, 2005. ACM Press. ISBN 1-59593-014-0.

[Liebig 2010] Jörg Liebig, Sven Apel, Christian Lengauer, Christian Kästner, and Michael Schulze. An analysis of the variability in forty preprocessor-based software product lines. In *Proceedings of the 32nd International Conference on Software Engineering (ICSE '10)*, New York, NY, USA, 2010. ACM Press.

[Lohmann 2009] Daniel Lohmann, Wanja Hofer, Wolfgang Schröder-Preikschat, Jochen Streicher, and Olaf Spinczyk. CiAO: An aspect-oriented operating-system family for resource-constrained embedded systems. In *Proceedings of the 2009 USENIX Annual Technical Conference*, pages 215–228, Berkeley, CA, USA, June 2009. USENIX Association. ISBN 978-1-931971-68-3.

[Lohmann 2006] Daniel Lohmann, Fabian Scheler, Reinhard Tartler, Olaf Spinczyk, and Wolfgang Schröder-Preikschat. A quantitative analysis of aspects in the eCos kernel. In *Proceedings of the ACM SIGOPS/EuroSys European Conference on Computer Systems 2006 (EuroSys '06)*, pages 191–204, New York, NY, USA, April 2006. ACM Press.

[Massa 2002] Anthony Massa. *Embedded Software Development with eCos*. New Riders, 2002. ISBN 978-0130354730.

[Padioleau 2008] Yoann Padioleau, Julia L. Lawall, Gilles Muller, and René Rydhof Hansen. Documenting and automating collateral evolutions in Linux device drivers. In *Proceedings of the ACM SIGOPS/EuroSys European Conference on Computer Systems 2008 (EuroSys '08)*, Glasgow, Scotland, March 2008.

[Palix 2010] Nicolas Palix, Julia Lawall, and Gilles Muller. Tracking code patterns over multiple software versions with Herodotos. In *Proceedings of the 9th International Conference on Aspect-Oriented Software Development (AOSD '10)*, pages 169–180, New York, NY, USA, 2010. ACM Press. ISBN 978-1-60558-958-9.

[Parnas 1979] David Lorge Parnas. Designing software for ease of extension and contraction. *IEEE Transactions on Software Engineering*, SE-5(2):128–138, 1979.

[Pohl 2005] Klaus Pohl, Günter Böckle, and Frank J. van der Linden. *Software Product Line Engineering: Foundations, Principles and Techniques*. Springer-Verlag, 2005. ISBN 978-3540243724.

[Reid 2000] Alastair Reid, Matthew Flatt, Leigh Stoller, Jay Lepreau, and Eric Eide. Knit: Component composition for systems software. In *4th Symposium on Operating System Design and Implementation (OSDI '00)*, pages 347–360, Berkeley, CA, USA, October 2000. USENIX Association.

[Sincero 2010] Julio Sincero, Reinhard Tartler, Daniel Lohmann, and Wolfgang Schröder-Preikschat. Efficient extraction and analysis of preprocessor-based variability. In *Proceedings of the 9th International Conference on Generative Programming and Component Engineering (GPCE '10)*, New York, NY, USA, 2010. ACM Press.

[Spencer 1992] Henry Spencer and Gehoff Collyer. #ifdef considered harmful, or portability experience with C News. In *Proceedings of the 1992 USENIX Annual Technical Conference*, Berkeley, CA, USA, June 1992. USENIX Association.

[Spinellis 2008] Diomidis Spinellis. A tale of four kernels. In Wilhem Schäfer, Matthew B. Dwyer, and Volker Gruhn, editors, *Proceedings of the 30th International Conference on Software Engineering (ICSE '08)*, pages 381–390, New York, NY, USA, May 2008. ACM Press. ISBN 987-1-60558-079-1.

[Tan 2007] Lin Tan, Ding Yuan, Gopal Krishna, and Yuanyuan Zhou. /*icomment: Bugs or bad comments?*/. In *Proceedings of the 21st ACM Symposium on Operating Systems Principles (SOSP '07)*, pages 145–158, New York, NY, USA, 2007. ACM Press. ISBN 978-1-59593-591-5.

[Zengler 2010] Christoph Zengler and Wolfgang Küchlin. Encoding the Linux kernel configuration in propositional logic. In Lothar Hotz and Alois Haselböck, editors, *Proceedings of the 19th European Conference on Artificial Intelligence (ECAI 2010) Workshop on Configuration 2010*, pages 51–56, 2010.

A Case for Scaling Applications to Many-core with OS Clustering

Xiang Song   Haibo Chen   Rong Chen   Yuanxuan Wang   Binyu Zang

Parallel Processing Institute, Fudan University

{xiangsong, hbchen, chenrong, yxwang1987, byzang}@fudan.edu.cn

Abstract

This paper proposes an approach to scaling UNIX-like operating systems for many cores in a backward-compatible way, which still enjoys common wisdom in new operating system designs. The proposed system, called Cerberus, mitigates contention on many shared data structures within OS kernels by clustering multiple commodity operating systems atop a VMM, and providing applications with the traditional shared memory interface. Cerberus extends a traditional VMM with efficient support for resource sharing and communication among the clustered operating systems. It also routes system calls of an application among operating systems, to provide applications with the illusion of running on a single operating system.

We have implemented a prototype system based on Xen/Linux, which runs on an Intel machine with 16 cores and an AMD machine with 48 cores. Experiments with an unmodified MapReduce application, dbench, Apache Web Server and Memcached show that, given the nontrivial performance overhead incurred by the virtualization layer, Cerberus achieves up to 1.74X and 4.95X performance speedup compared to native Linux. It also scales better than a single Linux configuration. Profiling results further show that Cerberus wins due to mitigated contention and more efficient use of resources.

Categories and Subject Descriptors D.4.7 [*Operating Systems*]: Organization and Design

General Terms Design

Keywords Multicore, Scalability, OS Clustering

1. Introduction

Scaling UNIX-like operating systems on shared memory multicore or multiprocessor machines has been a goal

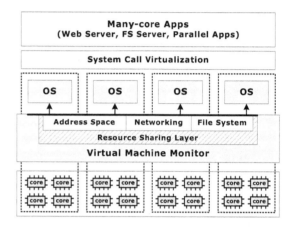

Figure 1: *Architecture overview of OS Clustering.*

of system researchers for a long time. Currently, there is a debate on the approach to scaling operating systems: designing new operating systems from scratch (e.g., Corey [Boyd-Wickizer 2008], Barrelfish [Baumann 2009] and fos [Wentzlaff 2008]); or continuing the traditional path of refining commodity kernels by iteratively eliminating bottlenecks using both traditional parallel programming skills or new data structures (e.g., RCU [McKenney 2002], Sloppy Counter [Boyd-Wickizer 2010]). However, with continual growth of the number of cores in a single machine and the still speculative structure of future many-core machines, there is currently no conclusion on the best long-term direction.

In this paper, we seek to add a point to the debate, by evaluating a middle ground between these two trends, motivated by the observation that commodity operating systems can scale well with a small number of CPU cores, and one virtual machine monitor (VMM) can effectively consolidate multiple operating systems. The proposed approach, called OS clustering (shown in Figure 1), is an operating system structuring strategy that attempts to provide a near- or middle-term solution to mitigate the scalability problem of commodity operating systems, yet without non-trivial testing efforts and possible backward compatibility issues in new operating system designs. The basic idea is clustering multiple commodity operating systems atop a VMM to serve one application, while providing the familiar *POSIX programming interface* to shared-memory applications. The resulting system, called Cerberus, supports existing many-core applications with little or no porting effort.

The goal of Cerberus is to make a bridge between two different directions (i.e., designing new OSes and refining commodity OSes) of scaling operating systems. On one hand, Cerberus incorporates some common wisdom in new operating system designs, such as state replication and message passing. On the other hand, Cerberus is designed based on reusing commodity operating systems, which means Cerberus may still share the benefits of improvements to commodity operating systems. It should be noted that Cerberus also comes at the cost of increased resource consumption due to the increased number of OS instances. However, for future many-core platforms with likely abundant resources, we believe it is worthwhile to trade resources for scalability.

In general, Cerberus could mitigate or avoid many instances of resource contention within a single operating system as well as in the VMM, due to the reduced number of CPU cores managed by a single operating system kernel. It is also easier for the inter-OS communication protocol to scale with the number of OS instances, rather than with the number of cores. Thus, contention within many subsystems could be mitigated for shared-memory multi-threaded and multiprocessing applications.

As well as state replication and distribution in operating systems and the VMM, Cerberus also retrofits some techniques in new OS designs back to commodity operating systems. Cerberus extends traditional system virtualization techniques with support for efficient resource sharing among the clustered operating systems. Specifically, the VMM is built with the *address range* support from Corey [Boyd-Wickizer 2008] to minimize the page fault costs for cross-OS memory sharing, which is critical for some memory-intensive applications. Further, to reduce contention for file accesses, Cerberus incorporates an efficient distributed file-system among clustered OSes, which optimizes local accesses while maintaining good performance for remote accesses.

Moreover, Cerberus incorporates a system call virtualization layer that allows processes/threads of an application to be executed in multiple operating systems, yet provides users with the illusion of running in a single operating system. This layer relies on both message passing and shared memory mechanisms to route system calls to specific operating systems and marshal the results, thus providing applications with a unified TCP/IP stack and file system. This layer uses the notion of "*SuperProcess*", which groups processes/threads in multiple operating systems, to manage the spawned processes/threads.

We have implemented a Cerberus prototype based on Xen-3.3.0 [Barham 2003] and Linux-2.6.18, which runs on an Intel machine with 16 cores and an AMD machine with 48 cores. The prototype adds 1800 lines of code to the Xen VMM and requires no code change to the Linux core kernel. A loadable kernel module and a user-level module are implemented to support the system call virtualization, which has 8,800 lines of code in total.

To measure the effectiveness of Cerberus, we have conducted several performance measurements and compared the performance of a shared memory MapReduce application, dbench [Tridgell 2010], Memcached and Apache web sever running on a single Linux (native Linux and virtualized Linux) and Cerberus with different number of VMs. Performance results show that though Cerberus incurs overhead for some primitives, it does provide better performance scalability. The performance speedup ranges from 1.74X to 4.95X over native Linux and from 1.37X to 11.62X compared to virtualized Linux on 48 cores. The profiling results using Oprofile [Levon 2004] and Xenprof [Menon 2005] indicate that Cerberus mitigates or avoids many instances of contention within both Xen and Linux.

In summary, the contributions of this paper are:

- A technique called OS clustering, which provides a backward-compatible way to scale existing shared memory applications on multicore machines;

- A set of mechanisms to enable efficient sharing of resources among clustered operating systems;

- The design and implementation of our prototype system Cerberus, as well as the evaluation of Cerberus using realistic application benchmarks, which demonstrate both the performance and scalability of our approach.

The rest of the paper is organized as follows. The next section relates Cerberus with previous work on OS scalability. Section 3 provides an overview on the challenges and approaches of Cerberus. Sections 4 and 5 present the design of the two major enabling parts of Cerberus, namely *Super-Process* and resource sharing. Then, section 6 describes the implementation details on Xen and Linux. The experimental results are shown in section 7. We present the discussion of the limitations and future work in section 8. Finally, we end this paper with a concluding remark in section 9.

2. Related Work

Improving the scalability of UNIX-like operating systems has been a longstanding goal of system researchers. This section relates Cerberus to other work in operating system scalability.

2.1 OS Structuring Strategies

Cerberus is influenced by much existing work on system virtualization, building new scalable OSes and refining existing OSes. Cerberus differs from existing work mainly in that it aims at improving performance scalability of existing applications by using a backward-compatible technique called OS clustering.

The idea of running multiple operating systems in a single machine is not new, but rather an inherent goal of system virtualization [Goldberg 1974]. For example, Disco [Bugnion 1997] (and its relative Cellular Disco [Govil 1999]) had run multiple virtual machines in the form of a virtual cluster to support distributed applications. Denali [Whitaker 2002] also safely multiplexes a large number of Internet services atop a lightweight virtual machine monitor. Cerberus puts these ideas into the context of multicore architecture, and more importantly supports efficiently running a contemporary shared-memory application with POSIX APIs on multiple clustered operating systems with little or no modification.

One viable way to scale operating systems is partitioning a hardware platform as a distributed system and distributing replicated kernels among partitioned hardware. Hive [Chapin 1995] uses a strategy called multicellular, which organizes an operating system as multiple independent kernels (i.e., cells), which communicates with each other for resource management to provide better reliability and scalability. Barrelfish [Baumann 2009] tries to scale applications on multicore system by using a multikernel model, which distributes replicated kernels on multiple cores and uses message-passing instead of shared-memory to maintain their consistency. The factored operating system [Wentzlaff 2008] argues that with the likely abundant cores, it would be more appropriate to space-multiplex cores instead of time-slicing them. Helios [Nightingale 2009] is an operating system that aims at bridging the heterogeneity of different processing units in a platform by using a satellite kernel, which provides the same abstractions across different processor architectures. Cerberus is also influenced by these systems in the use of replicated kernels and state, but retrofits the ideas to commodity operating systems to scale existing shared-memory applications.

Other work has focused on improving OS scalability by controlling or reducing sharing and improving data locality. Corey [Boyd-Wickizer 2008] is an exokernel [Engler 1995] style operating system that provides three new abstractions (share, address range and kernel core) for applications to explicitly control sharing of resources. K42 [Appavoo 2007] and its relatives (Tornado [Gamsa 1999] and Hurricane [Unrau 1995]) are designed to reduce contention and to improve locality for NUMA systems. Cerberus shares some similarities with the clustered objects of K42, but applies at a much higher level (complete operating systems).

2.2 Efforts in Commodity OSes

There are extensive studies on the scalability issues of commercial kernels and a lot of approaches are proposed to fix them. RCU [McKenney 2002], MCS lock [Mellor-Crummey 1991] and local runqueues [Aas 2005] are strategies that aim at reducing the contention on shared data structures. Recently, Boyd-Wickizer et al. [Boyd-Wickizer 2010] analyzed and fixed the scalability of many-core applications on Linux by refining the kernel, and improving applications' user-level design and use of kernel services. Cerberus also aims at improving scalability of applications on Linux, but runs multiple commodity OS instances to host one application instead of refining the core kernel.

In summary, the effort of Cerberus is complementary to the efforts of improving the scalability of existing commercial operating systems. With more scalable operating systems, Cerberus would require fewer operating systems to be clustered together to provide a scalable runtime environment.

3. Overview and Approaches

This section first discusses the challenges and illustrates the solutions to running a shared-memory application on multiple operating systems in a virtualized system. Then, we give an overview of the Cerberus architecture.

3.1 The Case for OS Clustering

To meet its design goals, Cerberus uses pervasive system virtualization. Rather than designing a new OS from scratch or fixing the internal mechanisms of commodity OSes, Cerberus clusters multiple commodity operating systems atop a VMM, and allows an application to run on multiple clustered operating systems with a shared-memory interface. Using multiple federated operating systems hosted by a VMM to run one application means that processes/threads belonging to one shared-memory application now run on multiple OSes. Hence, Cerberus uses a set of mechanisms to ensure system consistency.

Single Shared-Memory Interface: To avoid requiring a port of existing applications, it is critical to provide the existing shared-memory interface to applications. Usually, application programmers using traditional shared-memory APIs (e.g., POSIX) often make the assumption that their programs run within an operating system. Thus, a shared-memory application running in an operating system in the form of multiple processes and threads often expects to share a consistent view of system resources. These processes/threads also rely on the operating system interfaces and services to communicate with each other. For example, threads belonging to one process are expected to see the same address space and processes in one application have parent-children relations and use IPCs to notify each other.

To address these issues, Cerberus incorporates a system-call virtualization layer, which coordinates system calls in multiple clustered operating systems and marshals the results. Cerberus uses the notion of *SuperProcess* (section 4), which denotes a group of processes/threads executing in multiple operating systems. Each process/thread in a *SuperProcess* is called a *SubProcess*. The *SuperProcess* coordinates the control delivery and data communication of its *SubProcess*es, to maintain the system's consistency, and provide

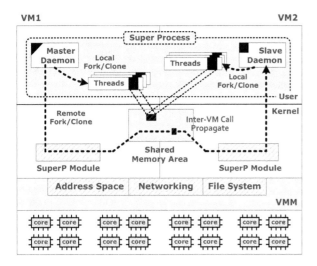

Figure 2: *The Cerberus System Architecture. Cerberus is composed of an extension in the VMM, which handles the resource sharing; and a kernel module in the guest OS, which manages the* SuperProcess *and inter-VM system calls.*

the application with the illusion of running on one operating system.

Efficient Resource Sharing: Another challenge is efficient sharing of resources among processes/threads crossing the operating system boundary, to still provide applications with a consistent view of system resources. Unfortunately, traditional VMMs are not built with support for sharing many resources between operating systems, but rather, enforce strong isolation among guest operating systems for security reasons.

Hence, Cerberus implements a resource-sharing layer in both the VMM and OSes, which supports efficient sharing of resources such as address spaces, networking and file systems. The resource-sharing layer exploits the fact that the clustered OSes share the hardware to coordinate accesses to shared resources. Cerberus uses both shared-memory and message-passing mechanisms to coordinate accesses and events among clustered operating systems. To serialize accesses to resources shared by multiple OS instances, it uses message-passing and lock-free mechanisms when necessary. For events among instances of different operating systems, Cerberus uses a two-level message queue to deliver these events. As the system state of an application is replicated and by default private, false sharing and unnecessary serialization can be significantly avoided.

3.2 System Architecture

The system architecture of Cerberus is shown in Figure 2. There are several virtual machines running atop a VMM. The VMM manages the underlying hardware resources and partitions the resources among the VMs. Currently, Cerberus requires the VMs to run the same operating system kernel for simplicity. Roughly speaking, Cerberus organizes the *Super-*

Process in the form of a coordinated distributed system, using both messages and shared memory. Multiple processes of one application run on multiple OSes in the form of a *SuperProcess*, which consists of one master daemon and multiple slave daemons. There is exactly one slave daemon for an application in VMs not running the master daemon. The master daemon is responsible for loading the initial parts of an application and creating the slave daemons. Afterwards, the master daemon works similarly to the slave daemon, according to the semantics of the application.

The daemons communicate with each other to decide which VMs should serve a process/thread creation request, to balance load among clustered OSes. To run a process in a VM other than the requesting VM, the *SuperProcess* daemon issues a *remote spawn*, which replicates the current running state to the target VM.

Cerberus also routes system calls using the SuperProcess *module* in each operating system, which is a loadable kernel module. The module intercepts system calls made by an application. To retain the semantics of system calls and a consistent view of the execution context, the module routes the system calls, as well as marshalling and translating the results.

Cerberus uses cross-VM message-passing mechanisms to handle communication between daemons in multiple VMs. A daemon uses the message-passing mechanism to send process/thread creation requests and signal remote processes/threads. There is also a shared-memory area for data communication among multiple VMs.

Currently, Cerberus decides the number of operating systems to run based on a user-specified heuristic for simplicity (the scalability limit of an application with certain number of cores). By default, Cerberus allocates a fixed, equal portion of resources to an operating system and lets the application decide the assignment of processes/threads to operating systems. Each operating system is pinned on a fixed number of cores.

In the following sections, we will describe the mechanisms in Cerberus to support *SuperProcess* (section 4) and efficient sharing of address space, file system and networking among clustered operating systems (section 5).

4. Supporting SuperProcess

This section describes the underlying design to support the *SuperProcess* abstraction, which provides applications with the illusion of running on a single operating system.

4.1 Remote Process/Thread Spawning

Cerberus uses techniques from traditional process checkpoint/restart mechanisms to support remote process spawning. As shown in Figure 3, Cerberus first checkpoints the state of the current running process, including the register state, memory mappings and opened files, among others. The checkpointed process state is then put into a shared

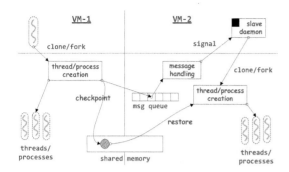

Figure 3: *The sequence of doing a remote fork/clone.*

memory area. To spawn a process on behalf of the requesting daemon in a different operating system, the daemon in the target OS first spawns a child process itself. Then, it retrieves the checkpointed state from the shared memory area, and restores it to the child process.

Currently, many applications use the threaded programming model. As all threads of a threaded application share the same binary image, Cerberus proactively creates a resident process (similar to the dispatcher in K42 [Krieger 2006]) for each clustered operating system, and maintains the consistency of each resident process by propagating changes to the application's global resources. For example, Cerberus automatically propagates memory mapping and unmapping requests in the issuing operating system to other clustered operating systems. Thus, for applications that create a large number of threads, the cost of remote spawning of threads is reduced, as the thread creation requests can be done locally by each resident process.

It should be noted that although creating a remote process or thread in Cerberus is more heavyweight than within a single OS, Cerberus supports parallel fork/clone that allows simultaneously creating processes/threads in multiple VMs, which amortizes the cost of a single operation.

4.2 Process Management

Cerberus relies mainly on the system call interception and redirection mechanisms to group processes distributed across multiple operating systems to provide correct semantics.

Cerberus virtualizes the process identity (such as the process ID), the parent-child relationship and the group information. To achieve this, Cerberus intercepts the system calls manipulating such information, translates the arguments before dispatching the operations, and marshals the results before returning to applications.

For process IDs, Cerberus maintains a global mapping table of the virtual process ID (seen by applications) and the physical ID (seen by the operating system). Cerberus thus relies on the virtual ID to maintain the process relationship. For example, the PID passed by the *kill* shell command will be translated by the Cerberus system call interception layer.

If the virtual PID belongs to the current operating system, the signal will be delivered to the process associated with the real PID. Otherwise, Cerberus will redirect the signal to the corresponding operating system. Cerberus also maintains a logical to physical CPU mapping table and provides the correct cores and operating systems to run threads and processes. For example, the pthread library provides interfaces to get and set the affinity (pthread_get(/set)_affinity) that obtain the set of cores on which a thread can run and assign specific threads to run on some cores, and Cerberus translates these calls.

4.3 Coordination of State Accesses

As the state of an application is shared or replicated among multiple clustered operating systems, Cerberus uses lock-free mechanisms and message passing to coordinate changes to the state from each OS. For some shared state among OSes, Cerberus uses compare-and-swap to allow each OS to eagerly access some replicated state such as page table pages. Upon a conflict, Cerberus rolls back the changes to state from one OS. For some shared data structures such as the virtual file descriptor table and inode table, Cerberus partitions these data structures to individual OSes, to avoid access serialization and cache ping-ponging, and uses message passing to coordinate the state.

Cerberus also implements an inter-VM notification mechanism that uses a hierarchical message-passing mechanism: when a process notifies processes in other VMs on the occurrences of certain events (e.g., signals, unmap requests), it first sends a message to the *SubProcess* in that VM. The *SubProcess* will queue the message marked with the type of the message and deliver the message to the appropriate VM. Then the receiver VM will send the corresponding event to the appropriate threads/processes. For example, for a futex [Franke 2002][1] call on the address of a remote thread, Cerberus will translate the address into the real address to monitor. On being notified by the local operating system about the change of the address, Cerberus will send a message to the receiving thread.

5. Supporting Resource Sharing

Cerberus supports the efficient sharing of address spaces, file systems and networks across the clustered operating systems, to provide a consistent view for applications.

5.1 Sharing Address Spaces

Cerberus identifies the range of shared address space by interpreting the application's semantics. An application running in a multi-threading mode should normally have its address space shared with all threads in a process. A forked process usually shares little with its parent. For a multi-

---

[1] A futex allows two entities to synchronize with each other using a shared memory location. The pthread mutex is implemented based on this mechanism.

threaded application, Cerberus maintains a global list of the shared *address ranges*. It intercepts the memory mapping requests (e.g., *mmap*) from each thread and updates the list accordingly. Cerberus creates a virtual memory mapping for that shared address range to let the page fault handler be aware of that address range. When handling a page fault, the Cerberus module first checks for a pending list entry and updates the virtual memory mapping before resolving the faulting address.

To efficiently share an address space across operating systems, Cerberus incorporates the *address range* abstraction from Corey [Boyd-Wickizer 2008]. This supports sharing a subset of the root page table by multiple guest VMs, according to the address range. The level of page table sharing might be changed according to the virtual memory mapping of an operating system. Cerberus also dynamically coalesces and splits the sharing of page tables according to the application's memory mapping requests. According to the list of shared address ranges, the page fault handler in the VMM will connect the page table of a shared address range in one VM's page table to that in other VMs, when there is a first access to that shared address range in those domains. Cerberus determines the level of sharing based on the size of the address range.

5.2 Sharing File Systems

Running a single application on multiple operating systems raises the problem of sharing files among processes in each clustered operating system. This is because each operating system will have its own file system and device driver, preventing a process from accessing files managed by another operating system. One intuitive approach would be the use of a networked file system managed by one operating system, with other operating systems as NFS clients to access files in the operating system running the NFS server. However, this creates two performance problems. First, all file accesses are now centralized to one operating system, which can easily make the accesses the new performance bottleneck. Second, there are some inherent performance overheads, as a networked file system usually has inferior performance compared to a local one. For example, recent measurements [Nightingale 2005, Zhao 2006] showed that NFS could be several times slower than a native file system such as ext3.

Fortunately, most files in many multiprocessing applications are usually accessed exclusively, with few opportunities to be accessed by multiple processes (except some non-performance-critical ones such as log files)[2]. Hence, Cerberus uses a hybrid approach of both networked and local file system, which seeks to give accesses to private files little

---

[2] For multi-threaded applications, applications usually map the files into memory using mmap and then threads can modify the memory-mapped file directly, which will be discussed in the following sections.

contention and high performance, while maintaining acceptable performance for shared files.

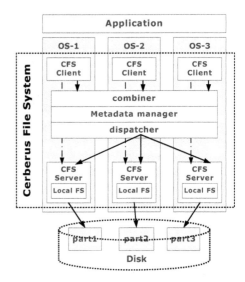

Figure 4: *Architecture of the Cerberus file system (CFS), which is organized as a mesh of networked file systems: each OS manages its local partition and exposes it to other OSes through the CFS client. Cerberus dispatches accesses to files and marshals the results according the managed metadata.*

Figure 4 shows the architecture of our approach, which forms of a mesh of networked file systems: each operating system manages a local partition and exposes it to other operating systems through an NFS-like interface; processes in such an operating system access private files directly in the local partition and access files in other partitions through the CFS client. To identify a file as shared or private, Cerberus maintains a mapping from each inode describing a file to the owner ID (e.g., virtual machine ID). As the metadata of files in each partition is maintained only by one operating system, Cerberus offers a similar metadata consistency and crash-recover model to native systems. It should be noted that the CFS implemented by Cerberus does not rely on network but rather on the virtual machine communication interfaces for communication and shared memory. This avoids redundant file data copies and the associated data exchange, and thus is more efficient than NFS [Zhao 2006]. Again, the sharing of a file between the CFS client and CFS server is done using the *address range* abstraction to minimize soft page faults.

To provide applications with a consistent view of the clustered file system, Cerberus intercepts accesses to the attributes or state of each file and directory, distributes accesses to each partition when necessary, and marshals the results before returning to user applications. Such operations (e.g., list a directory) are relatively costly compared to those in a single operating system. However, they are rare and usually occur in non-performance critical paths of applications.

5.3 Sharing File Contents

For multithreaded applications, it is common that the content of a file is shared by multiple threads. Thus, Cerberus supports the sharing of a file based on the *address space sharing* in Cerberus to maintain consistency for a file accessed by multiple operating systems. Cerberus uses memory mapped I/O (MMIO) to map a file into a shared address range, which is visible to all threads in multiple operating systems. Cerberus only allows the *SuperProcess* to access shared files using MMIO. To provide backward compatibility for applications using the traditional read/write APIs, Cerberus handles file I/O to shared files using a similar idea to that in Overshadow [Chen 2008], by translating file related I/Os to MMIOs. On the first read/write operation to the file, Cerberus maps the file in a shared address space using the *mmap* system call. Cerberus ensures that the buffer is mapped using the *MAP_SHARED* flag. Cerberus also ensures that the address range of the memory buffer is shared among clustered operating systems using the *address range* abstraction. Thus, changes from one operating system will be directly visible to other operating systems. Then, Cerberus emulates the read/write system calls by operating on the mmapped area.

To provide file-I/O semantics, Cerberus maintains a virtual file metadata structure that reflects the logical view of the files seen by a process. Cerberus also virtualizes the system calls that operate on the metadata of files. For example, the fseek system call will advance the file position maintained in the virtualized metadata and return the state in virtualized metadata for fstat-like system calls.

Note that this scheme is transparent to the in-kernel file systems and buffer cache management, as each buffer cache will have a consistent view of the file. The same piece of a file might be replicated among multiple buffer caches, causing wasted memory. However, multiple replicas also increase the concurrency of file access and avoid unnecessary contention.

5.4 Shared Networking Interfaces

To provide applications with a consistent view of networking interfaces, Cerberus exploits the fact that typical servers are usually equipped with multiple NICs, and each NIC is now hardware virtualizable (e.g., through SR-IOV [PCI-SIG 2010]). Hence, Cerberus directly assigns either virtualized NICs or physical NICs to each operating system for high performance. This could avoid contention on TCP/IP stacks if the operations are done on the local (virtual) NICs. To hide applications from such geographic distributions, Cerberus virtualizes the socket interface by intercepting related system calls and relies on the file descriptor virtualization described previously to manage socket descriptors. Cerberus maintains the (virtual) NIC information, and redirects calls that bind to a NIC if necessary. Cerberus then dispatches related operations (e.g., send, receive) to the VM that manages the NIC. The associated data will be exchanged using the shared memory area managed by Cerberus to avoid possible data copies.

6. Prototype Implementation

We have implemented Cerberus based on Xen to run multiple Linux instances with a single shared memory interface, using the shadow mode of page table management in Xen. The system call layer in Cerberus currently supports only a subset of the POSIX interface, but is sufficient to run many applications including shared-memory MapReduce applications, Apache, Memcached and file system benchmarks. For simplicity, Cerberus currently requires applications to be statically linked[3], and to link with a small piece of user-level code containing a few Cerberus-specific signal handlers, that handle remote requests such as futex and socket operations.

6.1 Inter-VM Message Passing

The inter-VM message passing mechanism is implemented by leveraging the cross-VM event channel mechanism in Xen. Cerberus creates a point-to-point event channel between each pair of clustered operating systems. The *SuperP* module inside each operating system has a handler to receive such cross-VM events and distribute them to the receivers. In the case of concurrent cross-VM events, each operating system maintains a cross-event queue to buffer the incoming events, and handles them in order. All cross-VM communication of Cerberus, such as futex and signal operations, uses this mechanism.

6.2 Memory Management

In Cerberus, the sharing of page tables is implemented in the shadow page tables, and by manipulating the P2M (physical-to-machine) table, thus is transparent to guest operating systems. We have also investigated an implementation of page sharing for Xen's direct mode (with writable page tables), with the aim of supporting para-virtualization. However, our preliminary results show that supporting writable page tables could result in significant changes to guest operating systems, as well as incurring non-trivial performance overhead.

On x86-64, Xen uses 4-levels of page tables and Cerberus supports sharing at the lower three levels (i.e., L1 – L3). Cerberus records the root page table page for an address range when the guest kernel connects an allocated page table page to the upper-level page table. When sharing a page table page among multiple OSes, one machine page might be accessed by multiple OSes, and thus might correspond to more than one guest-physical page in Xen. Hence, Cerberus creates a per-VM representation of each shared page table, but in an on-demand way. When a VM tries to write a page table page for the first time, Cerberus will create a represen-

---

[3] This will not increase much memory usage, as application code is shared by default.

tation of the page table page in that VM and map it to a single machine page by manipulating the P2M table, which maps guest physical memory to the host machine memory. Cerberus uses compare-and-swap to serialize updates to shared page table pages among multiple VMs: when a VM tries to update the shared page table, it uses a compare-and-swap to see if the entry has already been filled by other VMs, and frees the duplicated page table page if so.

Other than sharing page tables, Cerberus also needs to synchronize the virtual memory area (VMA) mappings across clustered VMs. As threads on different VMs have separate address spaces, they maintain their VMAs individually. Memory management system calls (e.g., mmap) on a single VM only change the VMA mappings of the threads in that VM. Thus, Cerberus intercepts most memory management system calls (e.g., mmap, mremap, mprotect, munmap and brk). Before handling the memory management system call, Cerberus will first force the VM to handle the virtual memory synchronization requests from other VMs. After finishing the call, Cerberus will allow the VM to propagate the system call to all other VMs in the system. This is done by adding a virtual memory synchronization request with appropriate parameters to the request queue of each receiver VM.

6.3 Cerberus File System

Inodes in a Cerberus file system (CFS) are divided into two kinds, namely local inodes and remote inodes. Local inodes describe files on a domain-local file system, and may be accessed directly. Remote inodes correspond to files stored on remote domains. A remote inode can be uniquely identified by its owner domain and its inode number in that domain. When a remote inode is created, CFS will keep track of this unique identifier. Each time a remote inode access is required, CFS will pack the inode identifier and other information into a message, and send it to remote domain via the inter-VM message passing mechanism.

Another data structure we track is the dentry. A dentry is an object describing relationships between inodes, and storing names of inodes. Unlike inodes, dentries in the original Linux file system do not have identifiers. To simplify remote dentry access, we assign a global identifier to each remote dentry. The dentry id is assigned in a lazy way, that is, only when a dentry is visited from a remote domain for the first time, will we assign a global identifier to it.

6.4 Virtualizing Networking

Cerberus virtualizes the socket interface by intercepting the related system calls. The socket operations are divided into two kinds, namely local and remote socket operations. We use virtual file descriptor numbers to distinguish the operations. Each virtual file descriptor number is associated with a virtual file descriptor. The virtual file descriptor describes the owner VM, the *responder* (a user-level daemon on the owner VM) and the real file descriptor corresponding to it.

When a process accesses a virtual file descriptor, Cerberus will first check the corresponding owner VM. If it is a local access, Cerberus just handles the request as in native Xen-Linux using the real file descriptor. Otherwise, Cerberus will send a remote socket operation request to the target VM, and let the *responder* handle the socket request. With this simple mechanism, Cerberus can currently support several socket-related operations (such as bind, listen, accept, read, write, select, sendto and recvmsg).

6.5 System Call Virtualization

We classify system calls into two types according to which system state they access. The first type includes system calls that only access local state or are stateless (e.g., get_systime). For such system calls, replicating calls among multiple OSes will not cause state consistency problems, and thus Cerberus does not need to handle them specially. The second type includes system calls that access and modify global state in the operating system (e.g., mmap). Cerberus needs to intercept this kind of system call, coordinate state changes, and marshal the results to support cross-VM interactions. To do interceptions, the Cerberus module modifies the system call table to change the function pointers of certain system call handlers to Cerberus-specific handlers during loading. When a system call is invoked, the Cerberus handler checks if it should be handled by Cerberus, and if so, invokes specific handlers provided by Cerberus.

We have currently virtualized 35 POSIX system calls (belonging to the second type) at either system call level or virtual file system level. They are divided into five categories: process/thread creation and exit (e.g., fork, clone, exec, exit, exit_group, getpid and getppid); thread communication (e.g., futex and signal); memory management (e.g., brk, mmap, munmap, mprotect and mremap); network operations (e.g., socket, connect, bind, accept, listen, select, sendto, recvfrom, shutdown and close); and file operations (e.g., open, read, write, mkdir, rmdir, close and readdir). We currently leave system calls related to security, realtime signals, debugging and kernel modules unhandled. In our experience, virtualizing a system call is usually not very difficult, as it mostly involves partitioning/marshaling the associated cross-process state. Table 1 gives some typical examples of how they are implemented.

6.6 Implementation Efforts

In total, the implementation adds 1,800 lines of code to Xen to support management of Cerberus and efficient sharing of data among *SubProcess* in multiple Linux instances. The support for system call interception, super-process and Cerberus file system is implemented as a loadable kernel module, which is comprised of 8,800 lines of code. It takes 1,250 lines of code to enable *SuperProcess* management. About 800 lines of code are used to support network virtualization and 750 lines of code to support the Cerberus file system. The Cerberus system call virtualization layer

Syscall	Approaches
clone	Cerberus first makes sure each VM has the *resident process*. Then, it queries the *SuperProcess* daemons for the target domain. Native clone is invoked for a local clone. Otherwise a remote clone request with the marshalled parameters (e.g., stack address) is sent. The resident process on the target domain then creates a new thread.
getpid	Cerberus returns a virtual pid to the caller. The virtual pid contains the domain id and the *SubProcess* number.
signal	Cerberus scans the mapping between virtual pid and process to find the target domain and process. A remote signal request is sent when necessary. The native signal call is then invoked on the receiver domain.
mmap	Cerberus first handles the VMA synchronization request, and then makes the native mmap call. Finally, it broadcasts the mmap result to other VMs.
accept	Cerberus checks the virtual fd table to get the owner domain of the fd. A remote accept request is sent when necessary. The accept operation is done by the corresponding *responder* with the real fd, and the resulting virtual fd of the created connection is sent back.
sendto	If the fd does not refer to a remote connection (either the socket is not established or it is a local connection), Cerberus will invoke the native sendto. Otherwise, Cerberus will query the virtual fd table to get the owner domain. A remote sendto request is sent and handled by the corresponding *responder* with the real fd.
mkdir	Cerberus first gets the global identifiers of the inode and dentry of the parent directory. If it is a local request, Cerberus passes it to the native file system. Otherwise, Cerberus gets the owner domain id and sends a remote CFS request with the global identifiers, the type of the new node (directory in this case) and the directory name. The owner domain creates the new child directory and sends the corresponding global identifier of the newly created inode and dentry back to the request domain.

Table 1: *System call implementation examples*

takes about 3000 lines of code, including marshaling multiple system calls (e.g., clone). The Cerberus system support code consists of 3000 lines, including the management of shared memory pool, cross-VM messages and process checkpointing and restoring (including 700 lines of code from Crak [Zhong 2001]).

7. Experimental Results

This section evaluates the potential costs and benefits in performance and scalability of Cerberus's approach to mitigating contention in operating system kernels.

7.1 Experimental Setup

The benchmarks used include histogram from the Phoenix testsuite [Ranger 2007][4], dbench 3.0.4 [Tridgell 2010], Apache web server 2.2.15 [Fielding 2002] and Memcached 1.4.5 [Fitzpatrick 2004].

Moreover, we present the costs of basic operations in Cerberus. We use OProfile to study the time distribution of *histogram*, *dbench*, *Apache* and *Memcached* on Cerberus, Xen-Linux and Linux.

Most experiments were conducted on an AMD 48-core machine with 8 6-core AMD 2.4 GHz Opteron chips. Each core has a separate 128 KByte L1 cache and a separate 512 KByte L2 cache. Each chip has a shared 8-way 6 MByte L3 cache. The size of physical memory is 128 GByte. We use Debian GNU/Linux 5.0, which is installed on a 147 GByte SCSI hard disk with the ext3 file system. There are a total of four network interface cards and each is configured with different IPs in a subnet. The input files and executable

for testing are stored in a separate 300 GByte SCSI hard disk with ext3 file system. The Apache and Memcached benchmarks were conducted on a 4 quad-core Intel machine with 8 NICs (as it has more NICs than the AMD machine), to reduce the bottlenecks from the NIC itself. Due to resource limitations, we can run up to 24 virtual machines on the AMD machine. All performance measurements were tested at least three times and we report the mean.

Cerberus is based on is Xen 3.3.0, which by default runs with the Linux kernel version 2.6.18. We thus use the kernel version 2.6.18 for the three measured systems. Xen-Linux uses the privileged domain (Dom0) in *direct paging mode*, for good performance. As Xen-3.3.0 can support at most 32 VCPUs for one VM, we only evaluate Xen-Linux with up to 32 cores.

We compare the performance and scalability of Cerberus with Linux. We also present the performance results of Xen-Linux to show the performance overhead incurred by the virtualization layer, as well as the performance benefit of Cerberus over typical virtualized systems. To investigate the performance gain of Cerberus, we used Oprofile and Xenoprof to collect the distribution of time of *histogram*, *dbench*, *Apache* and *Memcached* on Xen-Linux and Linux and that on Cerberus using the 2 core per-VM configuration. All profiling tests use the *CPU_CYCLE_UNHALTED* event as the performance counter event.

7.2 Cerberus Primitives

We also wrote a set of microbenchmarks to evaluate the cost of many primitives of Cerberus, to understand the basic cost underlying Cerberus.

Sending Signals: To evaluate the performance of the Cerberus signal mechanism, we use a micro-benchmark to test

---

[4] The reason we chose histogram is because it has severe performance scalability problems on our testing machine, which other programs don't exhibit.

	localhost	remote host
Native Linux	12.5ms	125.5ms
Xen-Linux	42.9ms	132.6ms
Cerberus local	43.1ms	131.8ms
Cerberus remote	87.1ms	154.7ms

Table 2: *Cost of ping-ponging one packet 1000 times*

	Intel	AMD
Native Linux	7.9ms	4.0ms
Xen-Linux	38.7ms	74.1ms
Cerberus local	43.1ms	72.3ms
Cerberus remote	25.8ms	45.0ms

Table 3: *Cost of ping-ponging 1000 signals*

the time it takes to send a signal using a ping-pong scheme (e.g., sending a signal to a process and that process sending a signal back to the originator) on both the Intel and AMD machines. Table 3 depicts the evaluation results. It can be seen that the virtualization layer introduces some overhead to the signal mechanism. However, sending a cross-VM signal takes less time than sending a local signal. There are two reasons: 1) The inter-VM message passing mechanism is efficient; 2) Sending a signal to a remote process only needs to forward the request to the target VM, so signaling the target process and executing the sender process can be done in parallel.

Primitive	Config	Time
remote fork	1 process	5.40 ms
	24 processes	31.77 ms
remote clone	1 thread	3.21 ms
	24 threads	30.79 ms

Table 4: *The costs of fork and clone in Cerberus*

Remote Fork and Clone: The first and second columns of Table 4 show the cost of spawning 1 processe/thread on a remote VM in the AMD machine with 2 VMs and concurrently spawning 24 processes/threads on remote VMs in the AMD machine with 24 VMs. Cerberus suffers from some overhead due to checkpointing, transferring and restoring process/thread state from the issuing VM to the receiving VMs. However, with increasing numbers of VMs, the steps of creating remote threads can be processed in parallel. This helps to reduce some overhead of creating threads as shown in the table.

Inter-VM Message Passing: To evaluate the costs of inter-VM message passing, we pass a message between VMs in order using a ping-pong scheme, e.g., sending that message to a VM and the VM responds by sending a message back to the sender. The time for one round-trip is around

10.24 μs within the same chip and 11.34 μs between chips, which we believe is modest and acceptable.

Reading a File with CFS: To evaluate the performance of (CFS), we write a micro-benchmark to test the time it costs to read the beginning portion of a simple file on the AMD machine. We generate one hundred files with random content, clear the buffer cache, read the first ten bytes of each file, and then calculate the average execution time. The result shows that one read operation on a native Xen-Linux file takes 6.47 μs, and one read operation on a local CFS file takes 6.52 μs, while on a remote CFS file it will cost 17.81 μs. The performance of local read operations on the CFS is close to that of native Xen-Linux file system. However, remote read operations introduce some performance overhead.

Sending and Receiving Packets: To evaluate the performance of the Cerberus network system, we use a micro-benchmark to test the time for sending and receiving network packets, using a ping-pong scheme on the Intel machine. The micro-server establishes a network connection with the client and creates a child to handle the following requests. The client will send an 8 byte string to the server through a socket connection (localhost/remote host) to trigger the test. Table 2 depicts the evaluation results. It shows the execution time of ping-ponging one message 1000 times under different configurations. It can be seen that the virtualization layer introduces some overhead for sending and receiving packets, while forwarding a packet in Cerberus introduces more overhead. However, if the connection is from a remote host, the overhead of the packet forwarding is below 25% compared to native Linux.

7.3 Performance Results

For *histogram* and *dbench*, we ran each workload on the AMD 48-core machine under Xen-Linux and Linux with the number of cores increasing from 2 to 48. For Cerberus, we evaluate two configurations, which run one and two cores for each VM (Cerberus-1core and Cerberus-2cores), running on a different number of cores, increasing from 2 to 48. When running one virtual machine on two cores, we configured each VM with cores that have minimal communication costs (e.g., sharing the L3 cache). For the Apache web server and Memcached service benchmarks, we run each workload on the Intel 16-core machine under Cerberus, Xen-Linux and Linux with different number of cores, increasing from 2 to 16. As both applications require a relatively large number of NICs, we did not test them on 48-core AMD machine. During the Apache and Memcached tests, we setup one instance of the web server on each core, which accepts service requests from clients running on a pool of 16 dual-core machines (32 clients for Apache and 64 clients for Memcached).

Histogram: Figure 5 shows the performance and scalability of histogram processing 4 GByte of data on Cerberus, native Linux and Xen-Linux. All input data is held in an in-memory tmpfs to avoid applications being bottlenecked by

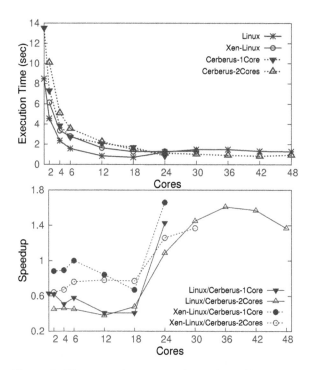

Figure 5: *The execution time and speedup of histogram on Cerberus compared to those on Linux and Xen-Linux under two configurations: which use 1 and 2 cores/domain accordingly.*

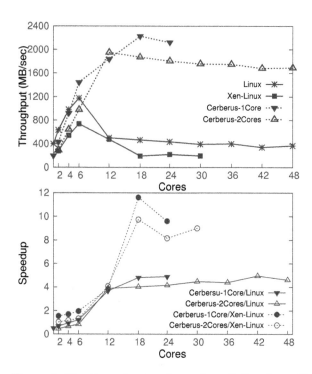

Figure 6: *The throughput and speedup of dbench on Cerberus compared to those on Linux and Xen-Linux under two configurations: which use 1 and 2 cores/domain accordingly.*

disk I/O. Cerberus performs significantly worse than Linux for a small number of cores, due to the performance overhead in shadow page management and the inherent virtualization overhead. However, as the number of cores increases, the execution time of histogram eventually decreases and outperforms native Linux. The speedup of Cerberus over Xen-Linux is around 51% on 24 cores for the one core per-VM configuration, and 30% on 30 cores for the two cores per-VM configuration. The speedup over Linux is around 43% on 24 cores for one core per-VM, and 37% on 48 cores for two cores per-VM. The performance of two cores per-VM is worse than that of one core per-VM, due to the increased contention on the shadow page table inside Xen. The speedup of Cerberus degrades a little (57% vs. 37%) from 42 cores to 48 cores, probably because the costs of creating threads and communication increases, thus the benefit degrades.

Table 5 shows the top 3 hottest functions in the profiling report of the *histogram* benchmark. Linux suffers from contention in __up_read and __down_read_trylock due to memory management. Xen-Linux spends most of its time in address 0x0 (/vmlinux-unknown) when the number of cores exceeds eight[5], which might be used for a para-virtualized kernel to interact with the hypervisor. However, Cerberus does not encounter contention in Linux and Xen-Linux, the time spent

in the lock-free implementation (cmpxchg) increases a little with the increasing number of cores.

dbench: Figure 6 depicts the throughput and speedup of dbench on Cerberus over Xen-Linux and Linux. The throughput of dbench on Xen-Linux and Linux degrades dramatically when the number of cores increases from 6 to 12 and degrades slightly afterwards. By contrast, though the throughput of Cerberus is worse than that on Linux for a small number of cores (1-6), its throughput scales well to 18 cores and 12 cores for the one and two core per-VM configuration. It appears that dbench has reached its extreme throughput here and has no further space for improvement. Starting from 12 cores or 18 cores, the throughput degrades slightly due to the increased process creation and inter-VM communication costs. Again, the one core per-VM configuration is slightly better than the two cores per-VM configuration, due to the per-VM lock on the shadow page table. In total, the speedup is 4.89X for the one core per-VM configuration on 24 cores, 4.95X for 42 cores, and 4.61X for 48 cores.

Table 6 shows the top 3 hottest functions in the profiling report of *dbench* benchmark. We ignore the portion of samples related to *mwait_idle*, as it means the CPU has nothing to do. From the table we can see that Linux and Xen-Linux both spend substantial time in ext3 file system operations, which may be the reason for poor scalability. On the other hand, Cerberus does not encounter such scalability problems, but is slightly affected by the shadow paging mode.

---

[5] The profiling results are obtained through Xenoprof using the *CPU_CYCLE_UNHALTED* event

The evaluation on histogram and dbench also shows that these applications poorly utilize multicore resources when the number of cores reaches a certain level. This indicates that horizontally allocating more cores to such applications may not be a good idea. Instead, allocating a suitable amount of cores to such applications could result in better utilization and performance tradeoff.

Threads	Top 3 Functions	Percent
Linux		
48	__up_read	38.6%
	__down_read_trylock	35.9%
	calc_hist	8.3%
1	calc_hist	81.2%
	find_busiest_group	0.06%
	page_fault	0.03%
Xen-Linux		
32	/vmlinux-unknown	70.9%
	calc_hist	11.6%
	__handle_mm_fault	3.2%
1	calc_hist	60.3%
	__handle_mm_fault	3.6%
	sh_gva_to_gfn__guest_4	2.7%
Cerberus		
2/VM	calc_hist	22.5%
	sh_x86_emulate_cmpxchg__guest_2	8.9%
	/xen-unknown	8.3%

Table 5: *The summary of the top 3 hottest functions in histogram benchmark profiling*

Threads	Top 3 Functions	Percent
Linux		
48	ext3_test_allocatable	66.6%
	bitmap_search_next_usable_block	18.2%
	journal_dirty_metadata	0.02%
1	/lib/libc-2.7.so	20.7%
	copy_user_generic	14.1%
	__d_lookup	0.03%
Xen-Linux		
32	ext3_test_allocatable	59.7%
	bitmap_search_next_usable_block	17.7%
	/vmlinux-unknown	5.99%
1	/lib/libc-2.7.so	13.7%
	copy_user_generic	9.9%
	__d_lookup	4.1%
Cerberus		
2/VM	sh_x86_emulate_cmpxchg__guest_2	11.2%
	/xen-unknown	8.67%
	sh_x86_emulate_write__guest_2	5.2%

Table 6: *The summary of the top 3 hottest functions in dbench benchmark profiling*

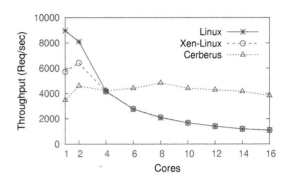

Figure 7: *The per-core throughput of Apache on Cerberus compared to those on Linux and Xen-Linux*

Apache Web Server: Figure 7 shows the per-core throughput of Apache on the Intel 16-core machine under Cerberus, Xen-Linux and Linux. There are a total of eight NICs, and each is configured with a different IP in a subnet. We run one web server instance on each core and share one NIC between two web servers. The throughput of Apache on Linux significantly degrades with the growing number of cores. When evaluating Cerberus, we directly assign 8 NICs to 8 different VMs (using PCI passthrough). The per-core throughput of 16 cores is only 1085 requests/sec for Linux, which is 12.1% of that on 1 core. By contrast, the throughput of Cerberus is quite stable. Although Cerberus performs worse than Linux for a small number (1-2) of cores (4603 vs. 8118 on 2 cores), it outperforms Linux when the number of cores exceeds 4 and scales nearly linearly. Cerberus achieves a speedup of 3.49X and 3.53X over Linux and Xen-Linux (3833 vs. 1099 and 1085).

The profiling of *Apache* shows that more CPU time is spent idle with the increasing number of cores used to host web servers, and there is some load imbalance. Oprofile shows that the same server instance takes 2.57X more CPU cycles under 1-core configuration than that under 16-core configuration. The same scenario also appears in Xen-Linux(2.39X). This may be caused by contention in the network layer in Linux and Xen-Linux. However, Cerberus does not encounter such a problem and can fully utilize its CPU resources. This evaluation shows that Cerberus could also avoid some imbalance caused by Linux, and achieve more efficient use of resources.

Memcached: Figure 8 shows the average throughput of Memcached server on the Intel 16-core machine under Cerberus, Xen-Linux and Linux. The configuration is similar to Apache. We run one Memcached server instance on each core and share one card by two servers listening to different UDP ports. The throughput of Memcached server on Linux significantly degrades when the number of cores exceeds 4. By contrast, the throughput of Cerberus does not degrade until the Memcached instances start to provide service on the same VM, as two instances affect each other heavily. However Cerberus still outperforms Xen-Linux and Linux.

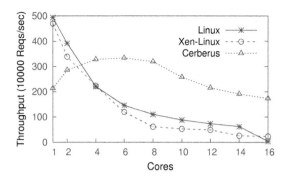

Figure 8: *The per-core throughput of Memcached on Cerberus compared to that on Linux and Xen-Linux*

The profiling of *Memcached* shows that many CPU cycles are spent polling network events. Further per-CPU profiling shows that a few Memcached instances spend much time in the *ep_poll_callback* and *task_rq_lock* functions, and seem to block other instances.

We also evaluated *histogram*, *dbench*, *Apache* and *Memcached* on other Linux versions (Linux 2.6.26, the standard kernel for Debian GNU/Linux, and Linux 2.6.35, the newest stable kernel). Only the scalability of *histogram* improves in Linux 2.6.35. Others still suffer from heavy contention, and have similar performance and scalability.

Performance of Different Configurations: We also measured the performance of different cores per-VM using 48 cores. As shown in Table 7, the performance actually degrades when the number of cores per-VM increases. The degradation is especially significant due to heavy contention on shadow page table management, with the execution time increasing more than 12X (10.624s vs. 0.860s) when the number of cores per-VM increases from 2 to 8. The evaluation shows that Cerberus does not rely on the scalability of the VMM and can also mitigate performance scalability problems within the VMM when configured properly.

#Cores/VM	Histogram(sec)	Dbench(MB/sec)
2	0.860	2123.6
4	1.130	1805.0
8	10.624	1273.8

Table 7: *Performance of histogram and dbench with different number of cores per-VM*

Performance Comparison with Xen-Linux Shadow Mode: As Cerberus is based on the shadow mode of Xen-Linux, we also give a performance comparison with Xen-Linux for reference. We used Domain0 in shadow mode and direct mode running 32 virtual cores to run histogram and dbench. Due to heavy contention in shadow mode, Xen-Linux experiences extremely bad performance, spending about 246.54s on histogram, and has only 3.4 MB/s throughput for dbench. Yet, for the direct mode, the execution time for histogram is 1.80s and the throughput is 246.54 MB/s for

dbench. Hence, when running parallel workloads on multiple cores, it should be better to use direct mode rather than shadow mode. The performance evaluation also shows that Cerberus could not only mitigate the contention within operating systems, but also reduce the contention from multiple cores accessing the shared state owned by a single virtual machine (i.e., shadow page management).

8. Discussion and Future Work

Though Cerberus has demonstrated the applicability of scaling applications with OS clustering, There are still ample optimization and research opportunities remaining. We describe our current limitations as well as possible extensions.

Viability of Our Approach: Our approach is not a panacea to the scalability of applications on multicore, but is only effective in specific scenarios where applications themselves have good parallelism and do not have intensive communication. Specifically, Cerberus might not show performance advantages in the following scenarios. First, applications that clone a number of short-lived, intensively-communicating threads/processes will probably not benefit from our approach, due to the relatively expensive cost of message passing and thread creation. Second, as remote network and remote file introduce overhead in Cerberus, applications with frequent remote resource access might experience degraded performance. Finally, applications with frequent small-size memory mapping operations (e.g., mmap, mremap) will stress the current synchronization mechanism for virtual memory in Cerberus and might have some performance degradation.

Application Cooperation: To retain application transparency, Cerberus relies on some relatively expensive operations (such as inter-VM fork/clone) to support cross-OS execution of an application. In our future work, we would like to investigate ways of adding some appropriate application programming interfaces and libraries to let applications cooperate with Cerberus, thus further reducing the performance overhead. For example, it would be interesting to let user applications explicitly specify which address space range should share the page table, to avoid unnecessary serialization and contention. Moreover, in a fork-intensive application, it would be beneficial for applications to direct Cerberus on which parts need to be checkpointed.

Hardware-assisted Virtualization: Currently, Cerberus is implemented based on hardware platform without hardware-assisted virtualization, thus come with the associated (usually non-trivial) overhead of virtualization. However, hardware-assisted virtualization techniques such as Intel VT-x and AMD SVM with extended page tables or nested page tables are commercially available. Our future work includes incorporating hardware-assisted virtualization to reduce the virtualization overhead, thus further enlarging the performance benefits of Cerberus.

Fault Tolerance: Currently, Cerberus does not provide fault tolerance to applications. While running applications on multiple VMs, it would be desiable when one process fails, processes in other VMs could take over the tasks and proceed as if the failure never happened. However, in Cerberus, if one process of a *SuperProcess* failed in one VM, it is uncertain what would happen to other processes on the other VMs.

9. Conclusions

Scaling operating systems on many-core systems is a critical issue for researchers and developers to fully harness the likely abundant future processing resources. This paper has presented Cerberus, a system that runs a single many-core application on multiple commodity operating systems, yet provides applications with the illusion of running on a single operating system. Cerberus has the potential to mitigate the pressure of applications on the efficiency of operating systems managing resources on many cores. Cerberus is enabled by retrofitting a number of new design techniques back to commodity operating systems to mitigate contention and to support efficient resource sharing. A system call virtualization layer coordinates accesses from process instances in clustered operating systems to ensure state consistency. Experiments with four applications on a 48-core AMD machine and a 16-core Intel machine show that Cerberus outperforms native Linux for a relatively large number of cores, and also scales better than Linux.

10. Acknowledgments

We thank our shepherd Andrew Baumann and the anonymous reviewers for their detailed and insightful comments. This work was funded by China National Natural Science Foundation under grant numbered 61003002, a grant from the Science and Technology Commission of Shanghai Municipality numbered 10511500100, China National 863 program numbered 2008AA01Z138, a research grant from Intel as well as a joint program between China Ministry of Education and Intel numbered MOE-INTEL-09-04, Fundamental Research Funds for the Central Universities in China and Shanghai Leading Academic Discipline Project (Project Number: B114).

References

[Aas 2005] Josh Aas. Understanding the Linux 2.6.8.1 CPU scheduler. http://joshaas.net/linux/linux_cpu_scheduler.pdf, Februry 2005.

[Appavoo 2007] Jonathan Appavoo, Dilma Da Silva, Orran Krieger, Marc Auslander, Michal Ostrowski, Bryan Rosenburg, Amos Waterland, Robert W. Wisniewski, Jimi Xenidis, Michael Stumm, and Livio Soares. Experience distributing objects in an SMMP OS. *TOCS*, 25(3):6, 2007.

[Barham 2003] Paul Barham, Boris Dragovic, Keir Fraser, Steven Hand, Tim Harris, Alex Ho, Rolf Neugebauer, Ian Pratt, and Andrew Warfield. Xen and the art of virtualization. In *Proc. SOSP*, pages 164–177, 2003.

[Baumann 2009] Andrew Baumann, Paul Barham, Pierre-Evariste Dagand, Tim Harris, Rebecca Isaacs, Simon Peter, Timothy Roscoe, Adrian Schuepbach, and Akhilesh Singhania. The multikernel: A new OS architecture for scalable multicore systems. In *Proc. SOSP*, 2009.

[Boyd-Wickizer 2008] Silas Boyd-Wickizer, Haibo Chen, Rong Chen, Yandong Mao, Frans Kaashoek, Robert Morris, Aleksey Pesterev, Lex Stein, Ming Wu, Yuehua Dai, Yang Zhang, and Zheng Zhang. Corey: An operating system for many cores. In *Proc. OSDI*, 2008.

[Boyd-Wickizer 2010] Silas Boyd-Wickizer, Austin Clements, Yandong Mao, Aleksey Pesterev, M. Frans Kaashoek, Robert Morris, and Nickolai Zeldovich. An analysis of Linux scalability to many cores. In *Proc. OSDI*, 2010.

[Bugnion 1997] E. Bugnion, S. Devine, and M. Rosenblum. DISCO: running commodity operating systems on scalable multiprocessors. In *Proc. SOSP*, pages 143–156, 1997.

[Chapin 1995] John Chapin, Mendel Rosenblum, Scott Devine, Tirthankar Lahiri, Dan Teodosiu, and Anoop Gupta. Hive: Fault Containment for Shared-Memory Multiprocessors. In *Proc. SOSP*, 1995.

[Chen 2008] X. Chen, T. Garfinkel, E.C. Lewis, P. Subrahmanyam, C.A. Waldspurger, D. Boneh, J. Dwoskin, and D.R.K. Ports. Overshadow: a virtualization-based approach to retrofitting protection in commodity operating systems. In *Proc. ASPLOS*, pages 2–13, 2008.

[Engler 1995] Dawson R. Engler, M. Frans Kaashoek, and James W. O'Toole. Exokernel: An operating system architecture for application-level resource management. In *Proc. SOSP*, pages 251–266, 1995.

[Fielding 2002] RT Fielding and G. Kaiser. The Apache HTTP server project. *Internet Computing, IEEE*, 1(4):88–90, 2002.

[Fitzpatrick 2004] B. Fitzpatrick. Distributed caching with memcached. *Linux journal*, 2004.

[Franke 2002] H. Franke, R. Russell, and M. Kirkwood. Fuss, Futexes and Furwocks: Fast Userlevel Locking in Linux. In *Proceedings of the Ottawa Linux Symposium*, 2002.

[Gamsa 1999] Ben Gamsa, Orran Krieger, Jonathan Appavoo, and Michael Stumm. Tornado: maximizing locality and concurrency in a shared memory multiprocessor operating system. In *Proc. OSDI*, 1999.

[Goldberg 1974] R.P. Goldberg. Survey of virtual machine research. *IEEE Computer*, 7(6):34–45, 1974.

[Govil 1999] Kinshuk Govil, Dan Teodosiu, Yongqiang Huang, and Mendel Rosenblum. Cellular disco: resource management using virtual clusters on shared-memory multiprocessors. In *Proc. SOSP*, pages 154–169, 1999.

[Krieger 2006] O. Krieger, M. Auslander, B. Rosenburg, R.W. Wisniewski, J. Xenidis, D. Da Silva, M. Ostrowski, J. Appavoo, M. Butrico, M. Mergen, et al. K42: building a complete operating system. *ACM SIGOPS Operating Systems Review*, 40 (4):145, 2006.

[Levon 2004] John Levon. *OProfile Manual*. Victoria University of Manchester, 2004. http://oprofile.sourceforge.net/doc/.

[McKenney 2002] Paul E. McKenney, Dipankar Sarma, Andrea Arcangeli, Andi Kleen, Orran Krieger, and Rusty Russell. Read-copy update. In *Proceedings of Linux Symposium*, pages 338–367, 2002.

[Mellor-Crummey 1991] John M. Mellor-Crummey and Michael L. Scott. Algorithms for scalable synchronization on shared-memory multiprocessors. *ACM Transaction on Computer Systems*, 9(1):21–65, 1991. ISSN 0734-2071.

[Menon 2005] A. Menon, J.R. Santos, Y. Turner, G.J. Janakiraman, and W. Zwaenepoel. Diagnosing performance overheads in the Xen virtual machine environment. In *Proc. VEE*, 2005.

[Nightingale 2005] E.B. Nightingale, P.M. Chen, and J. Flinn. Speculative execution in a distributed file system. In *Proc. SOSP*, pages 191–205, 2005.

[Nightingale 2009] E.B. Nightingale, O. Hodson, R. McIlroy, C. Hawblitzel, and G. Hunt. Helios: Heterogeneous multiprocessing with satellite kernels. In *Proc. SOSP*, 2009.

[PCI-SIG 2010] PCI-SIG. Single-root I/O virtualization specifications. http://www.pcisig.com/specifications/iov/single_root/, 2010.

[Ranger 2007] C. Ranger, R. Raghuraman, A. Penmetsa, G. Bradski, and C. Kozyrakis. Evaluating mapreduce for multi-core and multiprocessor systems. In *Proc. HPCA*, 2007.

[Tridgell 2010] A. Tridgell. Dbench filesystem benchmark. http://samba.org/ftp/tridge/dbench/, 2010.

[Unrau 1995] R.C. Unrau, O. Krieger, B. Gamsa, and M. Stumm. Hierarchical clustering: A structure for scalable multiprocessor operating system design. *The Journal of Supercomputing*, 9(1): 105–134, 1995.

[Wentzlaff 2008] D. Wentzlaff and A. Agarwal. Factored Operating Systems (fos): The Case for a Scalable Operating System for Multicores. *Operating System Review*, 2008.

[Whitaker 2002] A. Whitaker, M. Shaw, and S.D. Gribble. Scale and performance in the Denali isolation kernel. In *Proc. OSDI*, 2002.

[Zhao 2006] X. Zhao, A. Prakash, B. Noble, and K. Borders. Improving Distributed File System Performance in Virtual Machine Environments. Technical report, CSE-TR-526-06. University of Michigan, 2006.

[Zhong 2001] H. Zhong and J. Nieh. CRAK: Linux checkpoint/restart as a kernel module. *Technical Report CUCS-014-01, Department of Computer Science, Columbia University*, 2001.

Refuse to Crash with Re-FUSE

Swaminathan Sundararaman

Computer Sciences Department,
University of Wisconsin-Madison
swami@cs.wisc.edu

Laxman Visampalli

Qualcomm Inc.
laxmanv@qualcomm.com

Andrea C. Arpaci-Dusseau
Remzi H. Arpaci-Dusseau

Computer Sciences Department,
University of Wisconsin-Madison
{dusseau,remzi}@cs.wisc.edu

Abstract

We introduce Re-FUSE, a framework that provides support for restartable user-level file systems. Re-FUSE monitors the user-level file-system and on a crash transparently restarts the file system and restores its state; the restart process is completely transparent to applications. Re-FUSE provides transparent recovery through a combination of novel techniques, including request tagging, system-call logging, and non-interruptible system calls. We tested Re-FUSE with three popular FUSE file systems: NTFS-3g, SSHFS, and AVFS. Through experimentation, we show that Re-FUSE induces little performance overhead and can tolerate a wide range of file-system crashes. More critically, Re-FUSE does so with minimal modification of existing FUSE file systems, thus improving robustness to crashes without mandating intrusive changes.

Categories and Subject Descriptors   D.0 [*Software*]: General—File system Reliability

General Terms   Reliability, Fault tolerance, Performance

Keywords   FUSE, Restartable, User-level File Systems

1. Introduction

File system deployment remains a significant challenge to those developing new and interesting file systems designs [Ganger 2010]. Because of their critical role in the long-term management of data, organizations are sometimes reluctant to embrace new storage technology even though said innovations may address current needs. Similar problems exist in industry, where venture capitalists are loathe to fund storage startups, as it is well known that it takes three to

Eurosys'11,   April 10-13, Salzburg, Austria.

five years for storage products to "harden" and thus become ready for real commercial usage [Vahdat 2010].

One reason for this reluctance to adopt new technology is that unproven software often still has bugs in it, beyond those that are discovered through testing [Lu 2008]. Such "heisenbugs" [Gray 1987] often appear only in deployment, are hard to reproduce, and can lead to system unavailability in the form of a crash.

File system crashes are harmful for two primary reasons. First, when a file system crashes, manual intervention is often required to repair any damage and restart the file system; thus, crashed file systems stay down for noticeable stretches of time and decrease availability dramatically, requiring costly human time to repair. Second, crashes give users the sense that a file system "does not work" and thus decrease the chances for adoption.

To address this problem, we introduce Restartable FUSE (Re-FUSE), a restartable file system layer built as an extension to the Linux FUSE user-level file system infrastructure [Sourceforge 2010a]. Nearly 200 FUSE file systems have already been implemented [Sourceforge 2010b, Wikipedia 2010], indicating that the move towards user-level file systems is significant. In this work, we add a transparent restart framework around FUSE which hides many file-system crashes from users; Re-FUSE simply restarts the file system and user applications continue unharmed.

Restart with Re-FUSE is based on three basic techniques. The first is *request tagging*, which differentiates activities that are being performed on the behalf of concurrent requests; the second is *system-call logging*, which carefully tracks the system calls issued by a user-level file system and caches their results; the third is *non-interruptible system calls*, which ensures that no user-level file-system thread is terminated in the midst of a system call. Together, these three techniques enable Re-FUSE to recover correctly from a crash of a user-level file system by simply re-issuing the calls that the FUSE file system was processing when the crash took place; no user-level application using a user-level file system will notice the failure, except perhaps for a small drop in performance during the restart. Additional performance optimizations, including *page versioning* and *socket buffering*, are employed to lower the overheads of logging and recovery mechanisms.

We evaluate Re-FUSE with three popular file systems, NTFS-3g, SSHFS, and AVFS, which differ in their data-access mechanisms, on-disk structures, and features. Less than ten lines of code were added to each of these file systems to make them restartable, showing that the modifications required to use Re-FUSE are minimal. We tested these file systems with both micro- and macro-benchmarks and found that performance overhead during normal operation is minimal. Moreover, recovery time after a crash is small, on the order of a few hundred milliseconds in our tests.

Overall, we find that Re-FUSE successfully detects and recovers from a wide range of fail-stop and transient failures. By doing so, Re-FUSE increases system availability, as most crashes no longer make the entire file system unavailable for long periods of time. Re-FUSE thus removes one critical barrier to the deployment of future file-system technology.

The rest of the paper is organized as follows. Section 2 gives an overview of FUSE and user-level file systems. Section 3 discusses the essentials of a restartable user-level file system framework. Section 4 presents Re-FUSE, and Section 5 describes the three modified FUSE file systems. Section 6 evaluates the robustness and performance of Re-FUSE. Section 7 concludes the paper.

2. FUSE Background

Before delving into Re-FUSE, we first present background on the original FUSE system. We discuss the rationale for such a framework and present its basic architecture.

2.1 Rationale

FUSE was implemented to bridge the gap between features that users want in a file system and those offered in kernel-level file systems. Users want simple yet useful features on top of their favorite kernel-level file systems. Examples of such features are encryption, de-duplication, and accessing files inside archives. Users also want simplified file-system interfaces to access systems like databases, web servers, and new web services such as Amazon S3. The simplified file-system interface obviates the need to learn new tools and languages to access data. Such features and interfaces are lacking in many popular kernel-level file systems.

Kernel-level file-system developers may not be open to the idea of adding all of the features users want in file systems for two reasons. First, adding a new feature requires a significant amount of development and debugging effort [Zadok 2000]. Second, adding a new feature in a tightly coupled system (such as a file system) increases the complexity of the already-large code base. As a result, developers are likely only willing to include functionality that will be useful to the majority of users.

FUSE enables file systems to be developed and deployed at user level and thus simplifies the task of creating a new file system in a number of ways. First, programmers no longer need to have an in-depth understanding of kernel internals (e.g., memory management, VFS, block devices, and network layers). Second, programmers need not understand how these kernel modules interact with others. Third, programmers can easily debug user-level file systems using standard debugging tools such as gdb [GNU 2010] and valgrind [Nethercote 2007]. All of these improvements combine to allow developers to focus on the features they want in a particular file system.

In addition to Linux, FUSE has been developed for FreeBSD [Creo 2010], Solaris [Open Solaris 2010], and OS X [Google Code 2010] operating systems. Though most of our discussion revolves around the Linux version of FUSE, the issues faced herein are likely applicable to FUSE within other systems.

2.2 Architecture

To better understand how FUSE file systems are different than traditional kernel-level file systems, we begin by giving a bit of background on how kernel-level file-systems are structured. In the majority of operating systems, requests to file systems from applications begin at the system-call layer and eventually are routed to the proper underlying file system through a virtual file system (VFS) layer [Kleiman 1986]. The VFS layer provides a unified interface to implement file systems within the kernel, and thus much common code can be removed from the file systems themselves and performed instead within the generic VFS code. For example, VFS code caches file-system objects, thus greatly improving performance when objects are accessed frequently.

FUSE consists of two main components: the *Kernel File-system Module* (KFM) and a user-space library *libfuse* (see Figure 1). The KFM acts as a pseudo file system and queues *application requests* that arrive through the VFS layer. The libfuse layer exports a simplified file-system interface that each user-level file system must implement and acts as a liaison between user-level file systems and the KFM.

A typical application request is processed as follows. First, the application issues a system call, which is routed through VFS to the KFM. The KFM queues this application request (e.g., to read a block from a file) and puts the calling thread to sleep. The user-level file system, through the libfuse interface, retrieves the request off of the queue and begins to process it; in doing so, the user-level file system may issue a number of system calls itself, for example to read or write the local disk, or to communicate with a remote machine via the network. When the request processing is complete, the user-level file system passes the result back through libfuse, which places it within a queue, where the KFM can retrieve it. Finally, the KFM copies the result into the page cache, wakes the application blocked on the request, and returns the desired data to it. Subsequent accesses to the same block will be retrieved from the page cache, without involving the FUSE file system.

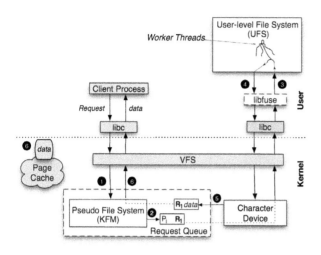

Figure 1. FUSE Framework. *The figure presents the FUSE framework. The user-level file system (in solid white box) is a server process that uses libfuse to communicate with the Kernel-level FUSE Module (KFM). The client process is the application process invoking operations on the file system. File-system requests are processed in the following way: (1) the application sends a request through the KFM via the VFS layer; (2) the request gets tagged and is put inside the request queue; (3) the user-level file-system worker thread dequeues the request; (4) the worker services the request and returns the response; (5) the response is added back to the queue; (6) finally, the KFM copies the data into the page cache before returning it to the application.*

Unlike kernel file systems, where the calling thread executes the bulk of the work, FUSE has a *decoupled* execution model, in which the KFM queues application requests and a separate user-level file system process handles them. As we will see in subsequent sections, this decoupled model is useful in the design of Re-FUSE. In addition, FUSE uses multithreading to improve concurrency in user-level file systems. Specifically, the libfuse layer allows user-level file-systems to create worker threads to concurrently process file-system requests; as we will see in subsequent sections, such concurrency will complicate Re-FUSE.

The caching architecture of FUSE is also of interest. Because the KFM pretends to be a kernel file system, it must create in-memory objects for each user-level file system object accessed by the application. Doing so improves performance greatly, as in the common case, cached requests can be serviced without consulting the user-level file system.

3. Restartable User-Level File Systems

In this section, we discuss the essentials of a restartable user-level file system framework. We present our goals, and then discuss both our assumptions of the fault model as well as assumptions we make about typical FUSE file systems. We conclude by discussing some challenges a restartable system must overcome, as well as some related approaches.

3.1 Goals

We now present our goals in building a restartable file-system framework for FUSE. Such a framework should have the following four properties:

Generic: A gamut of user-level file-systems exist today. These file systems have varied underlying data-access mechanisms, features, and reliability guarantees. Ideally, the framework should enable any user-level file system to be made restartable with little or no changes.

Application-Transparent: We believe it is difficult for applications using a user-level file system to handle file-system crashes. Expecting every application developer to change and recompile their code to work with a restartable file-system framework is likely untenable. Thus, any restartable framework should be completely transparent to applications and hide failures from them.

Lightweight: FUSE already has significant overheads compared to kernel-level file systems. This additional overhead is attributed to frequent context switching from user to kernel and back as well as extra data copying [Rajgarhia 2010]. Thus, adding significant overhead on top of already slower file-systems is not palatable; a restartable framework should strive to minimize or remove any additional overheads.

Consistent: User-level file systems use different underlying systems (such as databases, web servers, file systems, etc.) to access and store their data. Each of these systems provide different consistency guarantees. The restartable framework should function properly with whatever underlying consistency mechanisms are in use.

3.2 The Fault Model

Faults in a user-level file-system impact availability. A fault could occur due to developer mistakes, an incomplete implementation (such as missing or improper error handling), or a variety of other reasons. On a fault, a user-level file system becomes unavailable until it is restarted.

We believe that user-level file systems are likely to be less reliable than kernel-level file systems, due to a number of factors. First, unlike kernel-level file systems, most user-level file systems are written by novice programmers. Second, no common testing infrastructure exists to detect problems; as a result, systems are likely not stress-tested as carefully as kernel file systems are before release. Finally, no FUSE documentation exists to inform user-level file-system developers about the errors, corner cases, and failure scenarios that a file system should handle.

Our goal is to tolerate a broad class of faults that occur due to programming mistakes and transient changes in the environment. Examples of sources of such faults include sloppy or missing error handling, temporary resource unavailability, memory corruption, and memory leaks. Given the relative inexperience of the developers of many user-level file systems, it is hard to eliminate such failures.

The subset of these failures we seek to address are those that are "fail-stop" and transient [Qin 2005, Swift 2004, Zhou 2006]. In these cases, when such faults are triggered, the system crashes quickly, before ill effects such as permanent data loss can arise; upon retry, the problem is unlikely to re-occur. Faulty error-handling code and certain programming bugs are thus avoided on restart, as the fault that caused these errors to manifest does not take place again.

As with many systems, our goal is not to handle faults caused by basic logic errors and fail-silent bugs. Avoiding logic errors is critical to the correct operation of the file-system; we believe that such bugs should (and likely would) be detected and eliminated during development. On the other hand, fail-silent bugs are more problematic, as they do not crash the system but silently corrupt the in-memory state of the file system. Such corruption could slowly propagate to other components in the system (e.g., the page cache); recovery from such faults is difficult if not impossible. To the best of our knowledge, all previous restartable solutions make the same fail-stop and transient assumption that we make [Candea 2004, David 2008, Qin 2005, Sundararaman 2010, Swift 2003; 2004].

In our failure model, we assume that user-level file-system crashes are due to transient, fail-stop faults. We also assume that all the other components (i.e., the operating system, FUSE itself, and any remote host) work correctly. We believe it is reasonable to make this assumption as the rest of the components that the user-level file system interacts with (i.e., kernel components) are more rigorously tested and used by a larger number of users.

3.3 The User-level File-System Model

To design a restartable framework for FUSE, we must first understand how user-level file systems are commonly implemented; we refer to these assumptions as our *reference model* of a user-level file system.

It is infeasible to examine all FUSE file systems to obtain the "perfect" reference model. Thus, to derive a reference model, we instead analyze six diverse and popular file systems. Table 1 presents details on each of the six file systems we chose to study. NTFS-3g and ext2fuse each are kernel-like file systems "ported" to user space. AVFS allows programs to look inside archives (such as tar and gzip) and TagFS allows users to organize documents using tags inside existing file systems. Finally, SSHFS and HTTPFS allow users to mount remote file systems or websites through the SSH and HTTP protocols, respectively. We now discuss the properties of the reference file-system model.

Simple Threading Model: A single worker thread is responsible for processing a file-system request from start to finish, and only works on a single request at any given time. Amongst the reference-model file systems, only NTFS-3g is single-threaded by default; the rest all operate in multi-threaded mode.

File System	Category	LOC	Downloads
NTFS-3g	block-based	32K	N/A
ext2fuse	block-based	19K	40K
AVFS	pass-through	39K	70K
TagFS	pass-through	2K	400
SSHFS	network-based	4K	93K
HTTPFS	network-based	1K	8K

Table 1. Reference Model File Systems.

Request Splitting: Each request to a user-level file system is eventually translated into one or more system calls. For example, an application-level write request to a NTFS-3g file-system is translated to a sequence of block reads and writes where NTFS-3g reads in the meta-data and data blocks of the file and writes them back after updating them.

Access Through System Calls: Any external calls that the user-level file system needs to make are issued through the system-call interface. These requests are serviced by either the local system (e.g., the disk) or a remote server (e.g., a web server); in either case, system calls are made by the user-level file system in order to access such services.

Output Determinism: For a given request, the user-level file system always performs the same sequence of operations. Thus, on replay of a particular request, the user-level file system outputs the same values as the original invocation [Altekar 2009].

Synchronous Writes: Both dirty data and meta-data generated while serving a request are immediately written back to the underlying system. Unlike kernel-level file systems, a user-level file system does not buffer writes in memory; doing so makes a user-level file system stateless, a property adhered to by many user-level file systems in order to afford a simpler implementation.

Our reference model clearly does not describe all possible user-level file-system behaviors. The FUSE framework does not impose any rules or restrictions on how one should implement a file system; as a result, it is easy to deviate from our reference model, if one desires. We discuss this issue further at the end of Section 4.

3.4 Challenges

FUSE in its current form does not tolerate any file-system mistakes. On a user-level file system crash, the kernel cleans up the resources of the killed file-system process, which forces FUSE to abort all new and in-flight requests of the user-level file system and return an error (a "connection abort") to the application process. The application is thus left responsible for handling failures from the user-level file system. FUSE also prevents any subsequent operations on the crashed file system until a user manually restarts it. As a result, the file system remains unavailable to applications during this process. Three main challenges exist in restarting user-level file systems; we now discuss each in detail.

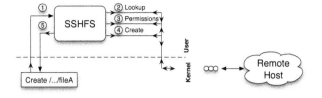

Figure 2. SSHFS Create Operation. *The figure shows a simplified version of SSHFS processing a create request. The number within the gray circle indicates the sequence of steps SSHFS performs to complete the operation. The FUSE, application process, and network components of the OS are not shown for simplicity.*

Generic Recovery Mechanism: Currently there are hundreds of user-level file systems and most of them do not have in-built crash-consistency mechanisms. Crash consistency mechanisms such as journaling or snapshotting could help restore file-system state after a crash. Adding such mechanisms would require significant implementation effort, not only for user-level file-systems but also to the underlying data-management system. Thus, any recovery mechanism should not depend upon the user-level file system itself in order to perform recovery.

Synchronized State: Even if a user-level file system has some in-built crash-consistency mechanism, leveraging such a mechanism could still lead to a disconnect between application perceived file-system state and the state of the recovered file system. This discrepancy arises because crash-consistency mechanisms group file-system operations into a single transaction and periodically commit them to the disk; they are designed only for power failures and not for soft crashes. Hence, on restart, a crash-consistency mechanism only ensures that the file system is restored back to the last known consistent state, which results in a loss of updates that occurred between the last checkpoint and the crash. As applications are not killed on a user-level file-system crash, the file-system state recovered after a crash may not be the same as that perceived by applications. Thus, any recovery mechanism must ensure that the file system and application eventually realize the same view of file system state.

Residual State: The non-idempotent nature of system calls in user-level file systems can leave *residual state* on a crash. This residual state prevents file systems from recreating the state of partially-completed operations. Both undo or redo of partially completed operations through the user-level file system thus may not work in certain situations. The create operation in SSHFS is a good example of such an operation. Figure 2 shows the sequence of steps performed by SSHFS during a create request. SSHFS can crash either before file create (Step 4) or before it returns the result to the FUSE module (Step 5). Undo would incorrectly delete a file if it was already present at the remote host if the crash happened before Step 4; redo would incorrectly return an error to the application if it crashed before Step 5. Thus any recovery mechanism must properly handle residual state.

3.5 Existing Solutions

There has been a great deal of research on restartable systems. Solutions such as CuriOS [David 2008], Rx [Qin 2005], and Microreboot [Candea 2004] help restart and recover application processes from crashes. These solutions require significant implementation effort to both the file system and underlying data-access system and also have high performance overheads. For example, CuriOS heavily instruments file-system code to force the file system to store its state in a separate address space. On restart, CuriOS uses the stored state to rebuild in-memory file-system state, but does not take care of on-disk consistency.

Solutions that use either roll-back [Hitz 1994] or roll-forward [Hagmann 1987, Sweeney 1996, Ts'o 2002] do not work well for user-level file systems. The residual state left by non-idempotent operations coupled with utilization of an underlying data-access system (such as a database) prevent proper recovery using these techniques.

Our earlier work on Membrane [Sundararaman 2010] shows how to restart kernel-level file systems. However, the techniques developed therein are highly tailored to the in-kernel environment and have no applicability to the FUSE context. Thus, a new FUSE-specific approach is warranted.

4. Re-FUSE: Design and Implementation

Re-FUSE is designed to transparently restart the affected user-level file system upon a crash, while applications and the rest of the operating system continue to operate normally. In this section, we first present an overview of our approach. We then discuss how Re-FUSE anticipates, detects, and recovers from faults. We conclude with a discussion of how Re-FUSE leverages many existing aspects of FUSE to make recovery simpler, and some limitations of our approach.

4.1 Overview

The main challenge for Re-FUSE is to restart the user-level file system without losing any updates, while also ensuring the restart activity is both lightweight and transparent. File systems are *stateful*, and as a result, both in-memory and on-disk state needs to be carefully restored after a crash. Three types of work must be done by the system to ensure correct recovery. First is *anticipation*, which is the additional work that must be done during the normal operation of a file system to prepare the file system for a failure. The second is *detection*, which notices a problem has occurred. The third component, *recovery*, is the additional work performed after a failure is detected to restore the file system back to its fully-operational mode.

Unlike existing solutions, Re-FUSE takes a different approach to restoring the consistency of a user-level file system after a file-system crash. After a crash, most existing systems rollback their state to a previous checkpoint and attempt to restore the state by re-executing operations from the beginning [Candea 2004, Qin 2005, Sundararaman 2010]. In contrast, Re-FUSE does not attempt to rollback to a consistent state, but rather continues forward from the inconsistent state towards a new consistent state. Re-FUSE does so by allowing partially-completed requests to continue executing from where they were stopped at the time of the crash. This action has the same effect as taking a snapshot of the user-level file system (including on-going operations) just before the crash and resuming from the snapshot during the recovery.

Most of the complexity and novelty in Re-FUSE comes in the fault anticipation component of the system. We now discuss this piece in greater detail, before presenting the more standard detection and recovery protocols.

4.2 Fault Anticipation

In anticipation of faults, Re-FUSE must perform a number of activities in order to ensure it can properly recover once the said fault arises. Specifically, Re-FUSE must track the progress of application-level file-system requests in order to continue executing them from their last state once a crash occurs. The inconsistency in file-system state is caused by partially-completed operations at the time of the crash; fault anticipation must do enough work during normal operation in order to help the file system move to a consistent state during recovery.

To create light-weight continuous snapshots of a user-level file system, Re-FUSE fault anticipation uses three different techniques: request tagging, system-call logging, and uninterruptible system calls. Re-FUSE also optimizes its performance through page versioning. We now discuss each of these in detail.

4.2.1 Request Tagging

Tracking the progress of each file-system request is difficult in the current FUSE implementation. The decoupled execution model of FUSE combined with request splitting at the user-level file system makes it hard for Re-FUSE to correlate an application request with the system calls performed by a user-level file system to service said application request.

Request tagging enables Re-FUSE to correlate application requests with the system calls that each user-level file system makes on behalf of the request. As the name suggests, request tagging transparently adds a request ID to the task structure of the file-system process (i.e., worker thread) that services it.

Re-FUSE instruments the libfuse layer to automatically set the ID of the application request in the task structure of the file-system thread whenever it receives a request from the KFM. Re-FUSE adds an additional attribute to the task structure to store the request ID. Any system call that the thread issues on behalf of the request thus has the ID in its task structure. On a system call, Re-FUSE inspects the tagged request ID in the task structure of the process to correlate the system call with the original application request. Re-FUSE also uses the tagged request ID in the task structure of the file-system process to differentiate system calls made by the

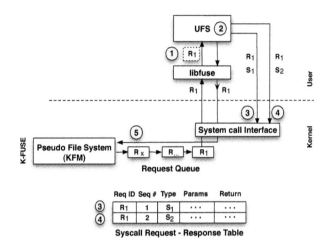

Figure 3. Request Tagging and System-call Logging. *The figure shows how Re-FUSE tracks the progress of individual file-system request. When KFM queues the application requests (denoted by R with a subscript). Re-FUSE tracks the progress of the request in the following way: (1) the request identifier is transparently attached to the task structure of the worker thread at the libfuse layer; (2) the user-level file system worker thread issues one or more system calls (denoted by S with a subscript) while processing the request; (3 and 4) Re-FUSE (at the system call interface) identifies these calls through the request ID in the caller's task structure and logs the input parameters along with the return value; (5) the KFM, upon receiving the response from the user-level file system for a request, deletes its entries from the log.*

user-level file system from other processes in the operating system. Figure 3 presents these steps in more detail.

4.2.2 System-Call Logging

Re-FUSE checkpoints the progress of individual application requests inside the user-level file system by logging the system calls that the user-level file system makes in the context of the request. On a restart, when the request is re-executed by the user-level file system, Re-FUSE returns the results from recorded state to mimic its execution.

The logged state contains the type, input arguments, and the response (return value and data), along with a request ID, and is stored in a hash table called the *syscall request-response table*. This hash table is indexed by the ID of the application request. Figure 3 shows how system-call logging takes place during regular operations.

Re-FUSE maintains the number of system calls that a file-system process makes to differentiate between user-level file-system requests to the same system call with identical parameters. For example, on a create request, NTFS-3g reads the same meta-data block multiple times between other read and write operations. Without a sequence number, it would be difficult to identify its corresponding entry in the syscall request-response table. Additionally, the sequence number also serves as a sanity check to verify that the system calls happen in the same order during replay. Re-FUSE removes the entries of the application request from the hash table when the user-level file system returns the response to the KFM.

4.2.3 Non-interruptible System Calls

The threading model in Linux prevents this basic logging approach from working correctly. Specifically, the threading model in Linux forces all threads of a process to be killed when one of the thread terminates (or crashes) due to a bug. Moreover, the other threads are killed independent of whether they are executing in user or kernel mode. Our logging approach only works if the system call issued by the user-level file system finishes completely, as a partially-completed system call could leave some residual state inside the kernel, thus preventing correct replay of in-flight requests.

To remedy this problem, Re-FUSE introduces the notion of *non-interruptible system calls*. Such a system call provides the guarantee that if a system call starts executing a request, it continues until its completion. Of course, the system call can still complete by returning an error, but the worker thread executing the system call cannot be killed prematurely when one of its sibling threads is killed within the user-level file-system. In other words, by using non-interruptible system calls, Re-FUSE allows a user-level file-system thread to continue to execute a system call to completion even when another file-system thread is terminated due to a crash.

Re-FUSE implements non-interruptible system calls by changing the default termination behavior of a thread group in Linux. Specifically, Re-FUSE modifies the termination behavior in the following way: when a thread abruptly terminates, Re-FUSE allows other threads in the group to complete whatever system call they are processing until they are about to return the status (and data) to the user. Re-FUSE then terminates said threads after logging their responses (including the data) to the syscall request-response table.

Re-FUSE eagerly copies input parameters to ensure that the crashed process does not infect the kernel. Lazy copying of input parameters to a system call in Linux could potentially corrupt the kernel state as non-interruptible system calls allow other threads to continue accessing the process state. Re-FUSE prevents access to corrupt input arguments by eagerly copying in parameters from the user buffer into the kernel and also by skipping COPY_FROM_USER and COPY_TO_USER functions after a crash. It is important to note that the process state is never accessed within a system call except for copying arguments from the user to the kernel at the beginning. Moreover, non-interruptible system calls are enforced only for user-level file system processes (i.e., only for processes that have a FUSE request ID set in their task structure). As a result, other application processes remain unaffected by non-interruptible system calls.

4.2.4 Performance Optimizations

Logging responses of read operations has high overheads in terms of both time and space as we also need to log the data returned with each read request. To reduce these overheads, instead of storing the data as part of the log records, Re-FUSE implements *page versioning*, which can greatly improve performance. Re-FUSE first tracks the pages accessed (and also returned) during each read request and then marks them as copy-on-write. The operating system automatically creates a new version whenever a subsequent request modifies the previously-marked page. The copy-on-write flag on the marked pages is removed when the response is returned back from the user-level file system to the KFM layer. Once the response is returned back, the file-system request is removed from the request queue at the KFM layer and need not be replayed back after a crash.

Page versioning does not work for network-based file systems, which use socket buffers to send and receive data. To reduce the overheads of logging read operations, Re-FUSE also caches the socket buffers of the file-system requests until the request completes.

4.3 Fault Detection

Re-FUSE detects faults in a user-level file-system through file-system crashes. As discussed earlier, Re-FUSE only handles faults that are both transient and fail-stop. Unlike kernel-level file systems, detection of faults in a user-level file system is simple. The faults Re-FUSE attempts to recover crash the file-system as soon as they are triggered (see Section 3.2). Re-FUSE inspects the return value and the signal attached to the killed file-system process to differentiate between regular termination and a crash.

Re-FUSE currently only implements a lightweight fault-detection mechanism. Fault detection can be further hardened in user-level file systems by applying techniques used in other systems [Cowan 1998, Necula 2005, Zhou 2006]. Such techniques can help to automatically add checks (by code or binary instrumentation) to crash file systems more quickly when certain types of bugs are encountered (e.g., out-of-bounds memory accesses).

4.4 Fault Recovery

The recovery subsystem is responsible for restarting and restoring the state of the crashed user-level file system. To restore the in-memory state of the crashed user-level file system, Re-FUSE leverages the information about the file-system state available through the KFM. Recovery after a crash mainly consists of the following steps: cleanup, re-initialize, restore the in-memory state of the user-level file system, and re-execute the in-flight file-system requests at the time of the crash. The decoupled execution model in the FUSE preserves application state on a crash. Hence, application state need not be restored. We now explain the steps in the recovery process in detail.

The operating system automatically cleans up the resources used by a user-level file system on a crash. The file system is run as a normal process with no special privileges by the FUSE. On a crash, like other killed user-level processes, the operating system cleans up the resources of the file system, obviating the need for explicit state clean up.

Re-FUSE holds an extra reference on the FUSE device file object owned by the crashed process. This file object is the gateway to the request queue that was being handled by the crashed process and KFM's view of the file system. Instead of doing a new mount operation, the file-system process sends a restart message to the KFM to attach itself to the old instance of the file system in KFM. This action also informs the KFM to initiate the recovery process for the particular file system.

The in-memory file-system state required to execute file-system requests is restored using the state cached inside the kernel (i.e., the VFS layer). Re-FUSE then exploits the following property: an access on a user-level file-system object through the KFM layer recreates it. Re-FUSE performs a lookup for each of the object cached in the VFS layer, which recreates the corresponding user-level file-system object in memory. Re-FUSE also uses the information returned in each call to point the cached VFS objects to the newly created file-system object. It is important to note that lookups do not recreate all file-system objects but only those required to re-execute both in-flight and new requests. To speed up recovery, Re-FUSE looks up file-system objects lazily.

Finally, Re-FUSE restores the on-disk consistency of the user-level file-system by re-executing in-flight requests. To re-execute the crashed file-system requests, a copy of each request that is available in the KFM layer is put back on the request queue for the restarted file system. For each replayed request, the FUSE request ID, sequence number of the external call, and input arguments are matched with the entry in the syscall request-response table and if they match correctly, the cached results are returned to the user-level file system. If the previously encountered fault is transient, the user-level file system successfully executes the request to completion and returns the results to the waiting application.

On an error during recovery, Re-FUSE falls back to the default FUSE behavior, which is to crash the user-level file system and wait for the user to manually restart the file system. An error could be due to a non-transient fault or a mismatch in one or more input arguments in the replayed system call (i.e., violating our assumptions about the reference file-system model). Before giving up on recovering the file system, Re-FUSE dumps useful debugging information about the error for the file-system developer.

4.5 Leveraging FUSE

The design of FUSE simplifies the recovery process in a user-level file system for the following four reasons. First, in FUSE, the file-system is run as a stand-alone user-level process. On a file-system crash, only the file-system process is killed and other components such as FUSE, the operating system, local file system, and even a remote host are not corrupted and continue to work normally.

Second, the decoupled execution model blocks the application issuing the file-system request at the kernel level (i.e., inside KFM) and a separate file-system process executes the request on behalf of the application. On a crash, the decoupled execution model preserves application state and also provides a copy of file-system requests that are being serviced by the user-level file system.

Third, requests from applications to a user-level file system are routed through the VFS layer. As a result, the VFS layer creates an equivalent copy of the in-memory state of the file system inside the kernel. Any access (such as a lookup) to the user-level file system using the in-kernel copy recreates the corresponding in-memory object.

Finally, application requests propagated from KFM to a user-level file system are always idempotent (i.e., this idempotency is enforced by the FUSE interface). The KFM layer ensures idempotency of operations by changing all relative arguments from the application to absolute arguments before forwarding it to the user-level file system. The idempotent requests from the KFM allow requests to be re-executed without any side effects. For example, the read system call does not take the file offset as an argument and uses the current file offset of the requesting process; the KFM converts this relative offset to an absolute offset (i.e., an offset from beginning of the file) during a read request.

4.6 Limitations

Our approach is obviously not without limitations. First, one of the assumptions that Re-FUSE makes for handling non-idempotency is that operations execute in the same sequence every time during replay. If file systems have some internal non-determinism, additional support would be required from the remote (or host) system to undo the partially-completed operations of the file system. For example, consider block allocation inside a file system. The block allocation process is deterministic in most file systems today; however, if the file system randomly picked a block during allocation, the arguments to the subsequent replay operations (i.e., the block number of the bitmap block) would change and thus could potentially leave the file system in an inconsistent state.

Re-FUSE does not currently support all I/O interfaces. For example, file systems cannot use mmap to write back data to the underlying system as updates to mapped files are not immediately visible through the system-call interface. Similarly, page versioning does not work in direct-I/O mode; Re-FUSE needs the data to be cached within the page cache.

Multi-threading can also limit the applicability of Re-FUSE. For example, multi-threading in block-based file systems could lead to race conditions during replay of in-flight requests and hence data loss after recovery. Different threading models could also involve multiple threads to handle a single request. For such systems, the FUSE request ID needs

Component	Original	Added	Modified
libfuse	9K	250	8
KFM	4K	750	10
Total	**13K**	**1K**	**18**

FUSE Changes

Component	Original	Added	Modified
VFS	37K	3K	0
MM	28K	250	1
NET	16K	60	0
Total	**81K**	**3.3K**	**1**

Kernel Changes

Table 2. Implementation Effort. *The table presents the code changes required to transform FUSE and Linux 2.6.18 into their restartable counterparts.*

to be explicitly transferred between the (worker) threads so that the operating system can identify the FUSE request ID for which the corresponding system call is issued.

The file systems in our reference model do not cache data in user space, but user-level file systems certainly could do so to improve performance (e.g., to reduce the disk or network traffic). For such systems, the assumption about the completion of requests (by the time the response is written back) would be broken and result in lost updates after a restart. One solution to handle this issue is to add a commit protocol to the request-handling logic, where in addition to sending a response message back, the user-level file system should also issue a commit message after the write request is completed. Requests in the KFM could be safely thrown away from the request queue only after a commit message is received from the user-level file system. In the event of a crash, all cached requests for which the commit message has not been received will be replayed to restore file-system state. For multi-threaded file systems, Re-FUSE would also need to maintain the execution order of requests to ensure correct replay. Moreover, if a user-level file system internally maintains a special cache (for some reason), for correct recovery, the file system would need to to explicitly synchronize the contents of the cache with Re-FUSE.

4.7 Implementation Statistics

Our Re-FUSE prototype is implemented in Linux 2.6.18 and FUSE 2.7.4. Table 2 shows the code changes done in both FUSE and the kernel proper. For Re-FUSE, around 3300 and 1000 lines of code were added to the Linux kernel and FUSE, respectively. The code changes in libfuse include request tagging, fault detection, and state restoration; changes in KFM center around support for recovery. The code changes in the VFS layer correspond to the support for system-call logging, and modifications in the MM and NET modules correspond to page versioning and socket-buffer caching respectively.

5. Re-FUSE File Systems

Re-FUSE is not entirely transparent to user-level file systems. We briefly describe the minor changes required in the three file systems employed in this work.

NTFS-3g: NTFS-3g reads a few key metadata pages into memory during initialization, just after the creation of the file system, and uses these cached pages to handle subsequent requests. However, any changes to these key metadata pages are immediately written back to disk while processing requests. On a restart of the file-system process, NTFS-3g would again perform the same initialization process. However, if we allow the process to read the current version of the metadata pages, it could potentially access inconsistent data and may thus fail. To avoid this situation, we return the oldest version of the metadata page on restart, as the oldest version points to the version that existed before the handling of a particular request (note that NTFS-3g operates in single-threaded mode).

AVFS: To make AVFS work with Re-FUSE, we simply increment the reference count of open files and cache the file descriptor so that we can return the same file handle when it is reopened again after a restart.

SSHFS: To make SSHFS work correctly with Re-FUSE, we made the following changes to SSHFS. SSHFS internally generates its own request IDs to match the responses from the remote host with waiting requests. The request IDs are stored inside SSHFS and are lost on a crash. After restart, on replay of an in-flight request, SSHFS generates new request IDs which could be different than the old ones. In order to match new request IDs with the old ones, Re-FUSE uses the FUSE request ID tagged in the worker thread along with the sequence number. Once requests are matched, Re-FUSE correctly returns the cached response. Also, to mask the SSHFS crash from the remote server, Re-FUSE holds an extra reference count on the network socket, and re-attaches it to the new process that is created. Without this action, upon a restart, SSHFS would start a new session, and the cached file handle would not be valid in the new session.

6. Evaluation

We now evaluate Re-FUSE in the following three categories: generality, robustness, and performance. Generality helps to demonstrate that our solution can be easily applied to other file systems with little or no change. Robustness helps show the correctness of Re-FUSE. Performance results help us analyze the overheads during regular operations and during a crash to see if they are acceptable.

All experiments were performed on a machine with a 2.2 GHz Opteron processor, two 80GB WDC disks, and 2GB of memory running Linux 2.6.18. We evaluated Re-FUSE with FUSE (2.7.4) using NTFS-3g (2009.4.4), AVFS (0.9.8), and SSHFS (2.2) file systems. For SSHFS, we use public-key authentication to avoid typing the password on restart.

File System	Original	Added	Modified
NTFS-3g	32K	10	1
AVFS	39K	4	1
SSHFS	4K	3	2

Table 3. Implementation Complexity. *The table presents the code changes required to transform NTFS-3g, AVFS and SSHFS into their restartable counterparts.*

6.1 Generality

To show Re-FUSE can be used by many user-level file systems, we chose NTFS-3g, AVFS, and SSHFS. These file systems are different in their underlying data access mechanism, reliability guarantees, features, and usage. Table 3 shows the code changes required in each of these file systems to work with Re-FUSE.

From the table, we can see that file-system specific changes required to work with Re-FUSE are minimal. To each user-level file system, we have added less than 10 lines of code, and modified a few more. Some of these lines were added to daemonize the file system and to restart the process in the event of a crash. A few further lines were added or modified to make recovery work properly, as discussed previously in Section 5.

6.2 Robustness

To analyze the robustness of Re-FUSE, we use fault injection. We employ both controlled and random fault-injection to show the inability of current user-level file systems to tolerate faults and how Re-FUSE helps them.

The injected faults are fail-stop and transient. These faults try to mimic some of the possible crash scenarios in user-level file systems. We first run the fault injection experiments on a vanilla user-level file system running over FUSE and then compare the results by repeating them over the adapted user-level file system running over Re-FUSE both with and without kernel modifications. The experiments without the kernel modifications are denoted by *Restart* and those with the kernel changes are denoted by *Re-FUSE*. We include the restart column to show that, without the kernel support, simple restart and replay of in-flight operations does not work well for FUSE.

6.2.1 Controlled Fault Injection

We employ controlled fault injection to understand how user-level file systems react to failures. In these experiments, we exercise different file-system code paths (e.g., `create()`, `mkdir()`, etc.) and crash the file system by injecting transient faults (such as a null-pointer dereference) in these code paths. We performed a total of 60 fault-injection experiments for all three file systems; we present the user-visible results.

User-visible results help analyze the impact of a fault both at the application and the file-system level. We choose *application state*, *file-system consistency*, and *file-system state*

as the user-visible metrics of interest. Application state indicates how a fault affects the execution of the application that uses the user-level file system. File-system consistency indicates if a potential data loss could occur as a result of a fault. File-system state indicates if a file system can continue servicing subsequent requests after a fault.

Table 4 summarizes the results of our fault-injection experiments. The caption explains how to interpret the data in the table. We now discuss the major observations and the conclusions of our fault-injection experiments.

First, we analyze the vanilla versions of the file systems running on vanilla FUSE and a vanilla Linux kernel. The results are shown in the leftmost result column in Table 4. We observe that the vanilla versions of user-level file systems and FUSE do a poor job in hiding failures from applications. In all experiments, the user-level file system is unusable after the fault; as a result, applications have to prematurely terminate their requests after receiving an error (a "software-caused connection abort") from FUSE. Moreover, in 40% of the cases, crashes lead to inconsistent file system state.

Second, we analyze the usefulness of fault-detection and simple restart at the KFM *without* any explicit support from the operating system. The second result column (denoted by Restart) of Table 4 shows the result. We observe that a simple restart of the user-level file system and replay of in-flight requests at the KFM layer ensures that the application completes the failed operation in the majority of the cases (around 60%). It still cannot, however, re-execute a significant amount (around 40%) of partially-completed operations due to the non-idempotent nature of the particular file-system operation. Moreover, an error is wrongly returned to the application and the crashes leave the file system in an inconsistent state.

Finally, we analyze the usefulness of Re-FUSE that includes restarting the crashed user-level file system, replaying in-flight requests, and has support from the operating system for re-executing non-idempotent operations (i.e., all the support described in Section 4). The results of the experiments are shown in the rightmost column of Table 4. From the table, we can see that all faults are handled properly, applications successfully complete the operation, and the file system is always left in a consistent state.

6.2.2 Random Fault Injection

In order to stress the robustness of our system, we use random fault injection. In the random fault-injection experiments, we arbitrarily crash the user-level file system during different workloads and observe the user-visible results. The sort, Postmark, and OpenSSH macro-benchmarks are used as workloads for these experiments; each is described further below. We perform the experiments on the vanilla versions of the user-level file systems, FUSE and Linux kernel, and on the adapted versions of the user-level file systems that run with Re-FUSE.

Table 4

Operation	NTFS_fn	Regular Application?	Regular FS:Consistent?	Regular FS:Usable?	Restart Application?	Restart FS:Consistent?	Restart FS:Usable?	Re-Fuse Application?	Re-Fuse FS:Consistent?	Re-Fuse FS:Usable?
		NTFS-3g								
create	fuse_create	×	×	×	e	×	√	√	√	√
mkdir	fuse_create	×	×	×	e	×	√	√	√	√
symlink	fuse_create	×	×	×	e	×	√	√	√	√
link	link	×	×	×	e	×	√	√	√	√
rename	link	×	×	×	e	×	√	√	√	√
open	fuse_open	×	√	×	√	√	√	√	√	√
read	fuse_read	×	√	×	√	√	√	√	√	√
readdir	fuse_readdir	×	√	×	√	√	√	√	√	√
readlink	fuse_readlink	×	√	×	√	√	√	√	√	√
write	fuse_write	×	×	×	√	×	√	√	√	√
unlink	delete	×	×	×	e	×	√	√	√	√
rmdir	inode_sync	×	×	×	e	×	√	√	√	√
truncate	fuse_truncate	×	×	×	√	√	√	√	√	√
utime	inode_sync	×	√	×	√	√	√	√	√	√
SSHFS_fn										
create	open_common	×	√	×	e	√	√	√	√	√
mkdir	mkdir	×	√	×	e	√	√	√	√	√
symlink	symlink	×	√	×	e	√	√	√	√	√
rename	rename	×	√	×	e	√	√	√	√	√
open	open_common	×	√	×	√	√	√	√	√	√
read	sync_read	×	√	×	√	√	√	√	√	√
readdir	getdir	×	√	×	√	√	√	√	√	√
readlink	readlink	×	√	×	√	√	√	√	√	√
write	write	×	√	×	√	√	√	√	√	√
unlink	unlink	×	√	×	e	√	√	√	√	√
rmdir	rmdir	×	√	×	e	√	√	√	√	√
truncate	truncate	×	√	×	√	√	√	√	√	√
chmod	chmod	×	√	×	√	√	√	√	√	√
stat	getattr	×	√	×	√	√	√	√	√	√
AVFS_fn										
create	mknod	×	×	×	e	×	√	√	√	√
mkdir	mkdir	×	×	×	e	×	√	√	√	√
symlink	symlink	×	×	×	e	×	√	√	√	√
link	link	×	×	×	e	×	√	√	√	√
rename	rename	×	×	×	e	×	√	√	√	√
open	open	×	√	×	√	√	√	√	√	√
read	read	×	√	×	√	√	√	√	√	√
readdir	readdir	×	√	×	√	√	√	√	√	√
readlink	readlink	×	√	×	√	√	√	√	√	√
write	write	×	×	×	√	×	√	√	√	√
unlink	unlink	×	×	×	e	×	√	√	√	√
rmdir	rmdir	×	×	×	e	×	√	√	√	√
truncate	truncate	×	×	×	√	×	√	√	√	√
chmod	chmod	×	√	×	√	√	√	√	√	√
stat	getattr	×	√	×	√	√	√	√	√	√

Table 4. Fault Study. *The table shows the affect of fault injections on the behavior of NTFS-3g, SSHFS and AVFS, respectively. Each row presents the results of a single experiment, and the columns show (in left-to-right order) the intended operation, the file system function that was fault injected, how it affected the application, whether the file system was consistent after the fault, and whether the file system was usable for other operations. Various symbols are used to condense the presentation. For application behavior, "√": application observed successful completion of the operation; "×": application received the error "software caused connection abort"; "e": application incorrectly received an error.*

Table 5

File System + Re-FUSE	Injected Faults	Sort (Survived)	OpenSSH (Survived)	Postmark (Survived)
NTFS-3g	100	100	100	100
SSHFS	100	100	100	100
AVFS	100	100	100	100

Table 5. Random Fault Injection. *The table shows the affect of randomly injected crashes on the three file systems supported with Re-FUSE. The second column refers to the total number of random (in terms of the crash point in the code) crashes injected into the file system during the span of time it is serving a macro-benchmark. The crashes are injected by sending the signal SIGSEGV to the file system process periodically. The right-most three columns indicate the number of survived crashes by the re-inforced file systems during each macro-benchmark. We do not include the results of the experiments on the vanilla versions of these file systems in the table; those file systems remain unusable after the first crash even though we inject the crash at varied time-points during the workload.*

We use three commonly-used macro-benchmarks to help analyze file-system robustness (and later, performance). Specifically, we utilize the sort utility, Postmark [Katcher 1997], and OpenSSH [Sourceforge 2010c]. The sort benchmark represents data-manipulation workloads, Postmark represents I/O-intensive workloads, and OpenSSH represents user-desktop workloads.

Table 5 presents the result of our study. From the table, Re-FUSE ensures that the application continues executing through the failures, thus making progress. We also found that a vanilla user-level file system with no support for fault handling cannot tolerate crashes (not shown in the table).

In summary, both from controlled and random fault injection experiments, we clearly see the usefulness of Re-FUSE in recovering from user-level file system crashes. In a standard environment, a user-level file system is always unusable after the crash and applications using the user-level file system are killed. Moreover, in many cases, the file system is also left in an inconsistent state. In contrast, Re-FUSE, upon detecting a user-level file system crash, transparently restarts the crashed user-level file system and restores it to a consistent and usable state. It is important to understand that even though Re-FUSE recovers cleanly from both controlled and random faults, it is still limited in its applicability (i.e., Re-FUSE only works for faults that are both fail-stop and transient and for file systems that strictly adhere to the reference file-system model described in Section 3.3).

6.3 Performance

Though fault-tolerance is our primary goal, we also evaluate the performance of Re-FUSE in the context of regular operations and recovery time.

Benchmark	ntfs	ntfs+ Re-FUSE	overhead %	sshfs	sshfs+ Re-FUSE	Overhead %	avfs	avfs+ Re-FUSE	Overhead %
Sequential read	9.2	9.2	0.0	91.8	91.9	0.1	17.1	17.2	0.6
Sequential write	13.1	14.2	8.4	519.7	519.8	0.0	17.9	17.9	0.0
Random read	150.5	150.5	0.0	58.6	59.5	1.5	154.4	154.4	0.0
Random write	11.3	12.4	9.7	90.4	90.8	0.4	53.2	53.7	0.9
Create	20.6	23.2	12.6	485.7	485.8	0.0	17.1	17.2	0.6
Delete	1.4	1.4	0.0	2.9	3.0	3.4	1.6	1.6	0.0

Table 6. Microbenchmarks. *This table compares the execution time (in seconds) for various benchmarks for restartable versions of ntfs-3g, sshfs, avfs (on Re-FUSE) against their regular versions on the unmodified kernel. Sequential reads/writes are 4 KB at a time to a 1-GB file. Random reads/writes are 4 KB at a time to 100 MB of a 1-GB file. Create/delete copies/removes 1000 files each of size 1MB to/from the file system respectively. All workloads use a cold file-system cache.*

Benchmark	ntfs	ntfs+ Re-FUSE	Overhead %	sshfs	sshfs+ Re-FUSE	Overhead %	avfs	avfs+ Re-FUSE	Overhead %
Sort	133.5	134.2	0.5	145.0	145.2	0.1	129.0	130.3	1.0
OpenSSH	32.5	32.5	0.0	55.8	56.4	1.1	28.9	29.3	1.4
PostMark	112.0	113.0	0.9	5683	5689	0.1	141.0	143.0	1.4

Table 7. Macrobenchmarks. *The table presents the performance (in seconds) of different benchmarks running on both standard and restartable versions of ntfs-3g, sshfs, and avfs. The sort benchmark (CPU intensive) sorts roughly 100MB of text using the command-line sort utility. For the OpenSSH benchmark (CPU+I/O intensive), we measure the time to copy, untar, configure, and make the OpenSSH 4.51 source code. PostMark (I/O intensive) parameters are: 3000 files (sizes 4KB to 4MB), 60000 transactions, and 50/50 read/append and create/delete biases.*

6.3.1 Regular Operations

We now evaluate the performance of Re-FUSE. Specifically, we measure the overhead of our system during regular operations and also during user-level file system crashes to see if a user-level file system running on Re-FUSE has acceptable overheads. We use both micro- and macro-benchmarks to evaluate the overheads during regular operation.

Micro-benchmarks help analyze file-system performance for frequently performed operations in isolation. We use sequential read/write, random read/write, create, and delete operations as our micro benchmarks. These operations exercise the most frequently accessed code paths in file systems. The caption in Table 6 describes our micro-benchmark configuration in more detail. We also use the previously-described macro-benchmarks sort, Postmark, and OpenSSH; the caption in Table 7 describes the exact configuration parameters for our experiments.

Table 6 and Table 7 show the results of micro- and macro-benchmarks respectively. From the tables, we can see that for both micro- and macro-benchmarks, Re-FUSE has minimal overhead, often less than 3%. The overheads are small due to in-memory logging and our optimization through page versioning (or socket buffer caching in the context of SSHFS). These results show that the additional reliability Re-FUSE achieves comes with negligible overhead for common file-system workloads, thus removing one important barrier of adoption for Re-FUSE.

File System	Vanilla Total Time (s)	Re-FUSE Total Time (s)	Re-FUSE Restart Time (ms)
NTFS-3g	133.5	134.45	65.54
SSHFS	145.0	145.4	255.8
AVFS	129.0	130.7	6.0

Table 8. Restart Time in Re-FUSE. *The table shows the impact of a single restart on the restartable versions of the file systems. The benchmark used is sort and the restart is triggered approximately mid-way through the benchmark.*

6.3.2 Recovery Time

We now measure the overhead of recovery time in Re-FUSE. Recovery time is the time Re-FUSE takes to restart and restore the state of the crashed user-level file system. To measure the recovery-time overhead, we ran the sort benchmark for ten times and crashed the file system half-way through each run. Sort is a good benchmark for testing recovery as it makes many I/O system calls and both reads and updates file-system state.

Table 8 shows the elapsed time and the average time Re-FUSE spent in restoring the crashed user-level file system state. The restoration process includes restart of the user-level file-system process and restoring its in-memory state. From the table, we can see that the restart time is in the order of a few milliseconds. The application also does not see any observable increase in its execution time due to the user-level file-system crash.

7. Conclusions

"Failure is not falling down but refusing to get up."
–Chinese Proverb

Software imperfections are common and are a fact of life especially for code that has not been well tested. Even though user-level file systems crashes are isolated from the operating system by FUSE, the reliability of individual file systems has not necessarily improved. File systems still remain unavailable to applications after a crash. Re-FUSE embraces the fact that failures sometimes occur and provides a framework to transparently restart crashed file systems.

We develop a number of new techniques to enable efficient and correct user-level file system restartability. In particular, request tagging allows Re-FUSE to differentiate between concurrently-serviced requests; system-call logging enables Re-FUSE to track (and eventually, replay) the sequence of operations performed by a user-level file system; non-interruptible system calls ensure that user-level file-system threads move to a reasonable state before file system recovery begins. Through experiments, we demonstrate that our techniques are reasonable in their performance overheads and effective at detection and recovery from a certain class of faults.

In the future, much work can be done to enhance Re-FUSE. More file systems can be ported to use it, and more experience with the real pitfalls of running a file system within such a framework can be obtained. It is unlikely developers will ever build the "perfect" file system; Re-FUSE presents one way to tolerate these imperfections.

Acknowledgments

We thank the anonymous reviewers and Steve Gribble (our shepherd) for their feedback and comments, which have substantially improved the content and presentation of this paper. We also thank Sriram Subramanian, Ashok Anand, Sankaralingam Panneerselvam, Mohit Saxena, and Asim Kadav for their comments on earlier drafts of the paper.

This material is based upon work supported by the National Science Foundation under the following grants: CCF-0621487, CNS-0509474, CNS-0834392, CCF-0811697, CCF-0811697, CCF-0937959, as well as by generous donations from NetApp and Google.

Any opinions, findings, and conclusions or recommendations expressed in this material are those of the authors and do not necessarily reflect the views of the National Science Foundation or other institutions.

References

[Altekar 2009] Gautam Altekar and Ion Stoica. Odr: output-deterministic replay for multicore debugging. In *Proceedings of the 22nd ACM Symposium on Operating Systems Principles (SOSP '07)*, pages 193–206, Big Sky, Montana, October 2009.

[Candea 2004] George Candea, Shinichi Kawamoto, Yuichi Fujiki, Greg Friedman, and Armando Fox. Microreboot – A Technique for Cheap Recovery. In *Proceedings of the 6th Symposium on Operating Systems Design and Implementation (OSDI '04)*, pages 31–44, San Francisco, California, December 2004.

[Cowan 1998] Crispin Cowan, Calton Pu, Dave Maier, Heather Hinton, Jonathan Walpole, Peat Bakke, Steve Beattie, Aaron Grier, Perry Wagle, and Qian Zhang. StackGuard: Automatic adaptive detection and prevention of buffer-overflow attacks. In *Proceedings of the 7th USENIX Security Symposium (Sec '98)*, San Antonio, Texas, January 1998.

[Creo 2010] Creo. Fuse for FreeBSD. http://fuse4bsd.creo.hu/, 2010.

[David 2008] Francis M. David, Ellick M. Chan, Jeffrey C. Carlyle, and Roy H. Campbell. CuriOS: Improving Reliability through Operating System Structure. In *Proceedings of the 8th Symposium on Operating Systems Design and Implementation (OSDI '08)*, San Diego, California, December 2008.

[Ganger 2010] Greg Ganger. File System Virtual Machines. http://www.pdl.cmu.edu/FSVA/index.shtml, 2010.

[GNU 2010] GNU. The GNU Project Debugger. http://www.gnu.org/software/gdb, 2010.

[Google Code 2010] Google Code. MacFUSE. http://code.google.com/p/macfuse/, 2010.

[Gray 1987] Jim Gray. Why Do Computers Stop and What Can We Do About It? In *6th International Conference on Reliability and Distributed Databases*, June 1987.

[Hagmann 1987] Robert Hagmann. Reimplementing the Cedar File System Using Logging and Group Commit. In *Proceedings of the 11th ACM Symposium on Operating Systems Principles (SOSP '87)*, Austin, Texas, November 1987.

[Hitz 1994] Dave Hitz, James Lau, and Michael Malcolm. File System Design for an NFS File Server Appliance. In *Proceedings of the USENIX Winter Technical Conference (USENIX Winter '94)*, San Francisco, California, January 1994.

[Katcher 1997] Jeffrey Katcher. PostMark: A New File System Benchmark. Technical Report TR-3022, Network Appliance Inc., October 1997.

[Kleiman 1986] Steve R. Kleiman. Vnodes: An Architecture for Multiple File System Types in Sun UNIX. In *Proceedings of the USENIX Summer Technical Conference (USENIX Summer '86)*, pages 238–247, Atlanta, Georgia, June 1986.

[Lu 2008] Shan Lu, Soyeon Park, Eunsoo Seo, and Yuanyuan Zhou. Learning from Mistakes — A Comprehensive Study on Real World Concurrency Bug Characteristics. In *Proceedings of the 13th International Conference on Architectural Support for Programming Languages and Operating Systems (ASPLOS XIII)*, Seattle, Washington, March 2008.

[Necula 2005] George C. Necula, Jeremy Condit, Matthew Harren, Scott McPeak, and Westley Weimer. CCured: Type-Safe

Retrofitting of Legacy Software. *ACM Transactions on Programming Languages and Systems*, 27(3), May 2005.

[Nethercote 2007] Nicholas Nethercote and Julian Seward. Valgrind: a framework for heavyweight dynamic binary instrumentation. *SIGPLAN Not.*, 42(6):89–100, 2007. ISSN 0362-1340.

[Open Solaris 2010] Open Solaris. Fuse on Solaris. http://hub.opensolaris.org/bin/view/Project+fuse/, 2010.

[Qin 2005] Feng Qin, Joseph Tucek, Jagadeesan Sundaresan, and Yuanyuan Zhou. Rx: Treating Bugs As Allergies. In *Proceedings of the 20th ACM Symposium on Operating Systems Principles (SOSP '05)*, Brighton, United Kingdom, October 2005.

[Rajgarhia 2010] Aditya Rajgarhia and Ashish Gehani. Performance and extension of user space file systems. In *SAC '10: Proceedings of the 2010 ACM Symposium on Applied Computing*, pages 206–213, New York, NY, USA, 2010. ACM.

[Sourceforge 2010a] Sourceforge. AVFS: A Virtual Filesystem. http://sourceforge.net/projects/avf/, 2010.

[Sourceforge 2010b] Sourceforge. File systems using FUSE. http://sourceforge.net/apps/mediawiki/fuse/index.php?title=FileSystems, 2010.

[Sourceforge 2010c] Sourceforge. OpenSSH. http://www.openssh.com/, 2010.

[Sundararaman 2010] Swaminathan Sundararaman, Sriram Subramanian, Abhishek Rajimwale, Andrea C. Arpaci-Dusseau, Remzi H. Arpaci-Dusseau, and Michael M. Swift. Membrane: Operating System Support for Restartable File Systems. In *Proceedings of the 8th USENIX Symposium on File and Storage Technologies (FAST '10)*, San Jose, California, February 2010.

[Sweeney 1996] Adan Sweeney, Doug Doucette, Wei Hu, Curtis Anderson, Mike Nishimoto, and Geoff Peck. Scalability in the XFS File System. In *Proceedings of the USENIX Annual Technical Conference (USENIX '96)*, San Diego, California, January 1996.

[Swift 2003] Michael M. Swift, Brian N. Bershad, and Henry M. Levy. Improving the Reliability of Commodity Operating Systems. In *Proceedings of the 19th ACM Symposium on Operating Systems Principles (SOSP '03)*, Bolton Landing, New York, October 2003.

[Swift 2004] Michael M. Swift, Brian N. Bershad, and Henry M. Levy. Recovering device drivers. In *Proceedings of the 6th Symposium on Operating Systems Design and Implementation (OSDI '04)*, pages 1–16, San Francisco, California, December 2004.

[Ts'o 2002] Theodore Ts'o and Stephen Tweedie. Future Directions for the Ext2/3 Filesystem. In *Proceedings of the USENIX Annual Technical Conference (FREENIX Track)*, Monterey, California, June 2002.

[Vahdat 2010] Amin Vahdat. VCs loathe to fund storage startups. Personal Communication, October 2010.

[Wikipedia 2010] Wikipedia. Filesystem in Userspace. http://en.wikipedia.org/wiki/Filesystem_in_Userspace, 2010.

[Zadok 2000] E. Zadok and J. Nieh. FiST: A language for stackable file systems. In *Proc. of the Annual USENIX Technical Conference*, pages 55–70, San Diego, CA, June 2000. USENIX Association.

[Zhou 2006] Feng Zhou, Jeremy Condit, Zachary Anderson, Ilya Bagrak, Rob Ennals, Matthew Harren, George Necula, and Eric Brewer. SafeDrive: Safe and Recoverable Extensions Using Language-Based Techniques. In *Proceedings of the 7th Symposium on Operating Systems Design and Implementation (OSDI '06)*, Seattle, Washington, November 2006.

Increasing Performance in Byzantine Fault-Tolerant Systems with On-Demand Replica Consistency[*]

Tobias Distler   Rüdiger Kapitza

Friedrich–Alexander University Erlangen–Nuremberg

{distler,rrkapitz}@cs.fau.de

Abstract

Traditional agreement-based Byzantine fault-tolerant (BFT) systems process all requests on all replicas to ensure consistency. In addition to the overhead for BFT protocol and state-machine replication, this practice degrades performance and prevents throughput scalability. In this paper, we propose an extension to existing BFT architectures that increases performance for the default number of replicas by optimizing the resource utilization of their execution stages.

Our approach executes a request on only a selected subset of replicas, using a *selector* component co-located with each replica. As this leads to divergent replica states, a selector on-demand updates outdated objects on the local replica prior to processing a request. Our evaluation shows that with each replica executing only a part of all requests, the overall performance of a Byzantine fault-tolerant NFS can be almost doubled; our prototype even outperforms unreplicated NFS.

Categories and Subject Descriptors D.4.7 [*Organization and Design*]: Distributed Systems; C.4 [*Performance of Systems*]: Fault Tolerance

General Terms Design, Performance, Reliability

Keywords Byzantine Failures; Performance

1. Introduction

Today's information society heavily depends on computer-provided services. Byzantine fault tolerance (BFT) based on replicated state machines is a general approach to make these services tolerate a wide spectrum of faults, including hardware failures, software crashes, and malicious attacks.

In general, the architecture of an agreement-based BFT system can be divided into two stages [Yin 2003]: *agreement* and *execution*. The agreement stage is responsible for imposing a total order on client requests; the execution stage processes the requests on all replicas, preserving the order determined by the agreement stage to ensure consistency.

Up to now, a number of protocol optimizations and architecture variants [Castro 1999, Correia 2004, Hendricks 2007, Kotla 2007; 2004, Wester 2009] have been proposed to improve the performance of both stages. In all these cases, the performance increase achieved is based on minimizing certain parts of the overhead introduced by BFT replication (e. g., request ordering [Correia 2004, Kotla 2007] or state-machine replication [Kotla 2004]). However, as all these systems ensure replica consistency by executing all requests on all replicas, the maximum throughput achievable for the fault-tolerant service is bounded by the throughput of a single replica; that is, the corresponding non–fault-tolerant unreplicated service.

In this paper, we present an extension to existing BFT state-machine–replication architectures that increases the upper throughput bound of an agreement-based BFT service (in the absence of faults) beyond the throughput of the corresponding unreplicated service. Our evaluation shows that our extension also reduces the response time under load. The approach exploits the fact that for systems tolerating f faults, $f + 1$ *identical* replies provided by different replicas prove a reply correct. In accordance with other authors [Hendricks 2007, Kotla 2007, Wester 2009, Wood 2011], we assume faults to be rare. Therefore, we execute each request on only a subset of $f + 1$ replicas during normal-case operation. In case of a fault, additional replicas process the request.

The subset of replicas to execute a request is selected for each request individually, based on the service-state variables accessed by the request. In particular, we divide the service state into *objects* and assign each object to $f + 1$ replicas. With different objects being assigned to different subsets of replicas, request execution is distributed across all replicas. Assuming uniform object distribution and uniform object access, each replica only executes $\frac{f+1}{n}$ of all requests, with n being the number of replicas. As a result, the

---

[*] This work was partially supported by the German Research Council (DFG) under grant no. KA 3171/1.

upper throughput bound T of the overall system increases to $T' = \frac{n}{f+1}T$. For a standard agreement-based BFT architecture with $3f + 1$ execution replicas and $f = 1$, this doubles the throughput bound. Our evaluation of a Byzantine fault-tolerant version of NFS shows that for medium and high workloads the load reduction at the execution stage outweighs the overhead introduced by request ordering and state-machine replication. As a result, the BFT service outperforms its non–fault-tolerant unreplicated equivalent.

Replica states in our system are not completely identical at all times but may differ across replicas, depending on the requests a replica has executed. However, instances of the same object are consistent across all $f + 1$ assigned replicas as they all process requests accessing the object in the same order. As clients may send requests accessing objects assigned to different replicas, a *selector* module co-located with every service replica provides *on-demand replica consistency (ODRC)*. The selector ensures that all objects possibly read or modified by a request are consistent with the current service state by the time the request is executed; objects outside the scope of a request are unaffected and may remain outdated.

Applying ODRC reduces the costs of keeping replicas consistent to the extent actually needed in order to provide consistent replies to client requests. In consequence, this approach improves resource utilization of replica execution stages, as replicas for the most part only execute requests that produce replies the client really needs to make progress. While other systems [Distler 2011, Wood 2011] have been proposed that minimize the number of replicas by utilizing the idea of providing only $f + 1$ replies in the absence of faults, the main goal of ODRC is to increase performance for the default number of replicas (i. e., $3f + 1$ in traditional BFT systems, such as PBFT [Castro 1999]).

We show that ODRC can be integrated into existing agreement-based BFT state-machine–replication architectures by introducing a single module between the agreement stage and the execution stage of a replica. Apart from that, our approach solely relies on existing mechanisms. In particular, this paper makes the following contributions:

- It presents an extension to existing agreement-based BFT state-machine–replication architectures that improves performance for the default number of replicas using ODRC (Sections 3 and 4).

- It outlines service integration for ODRC for a Byzantine fault-tolerant NFS service (Section 6) and a BFT variant of ZooKeeper (Section 7).

- It evaluates the impact of ODRC in the absence as well as the presence of faults (Sections 6 and 7).

In addition, Section 2 presents our system model. Section 5 describes important optimizations. Section 8 discusses the implications of applying on-demand replica consistency. Section 9 presents related work, and Section 10 concludes.

2. System Model and Assumptions

This section presents our system model and defines the BFT system properties required for ODRC.

2.1 General System Model

We assume the standard system model used for BFT state-machine replication [Castro 1999, Kotla 2007; 2004, Rodrigues 2001, Yin 2003] that comprises the possibility of replicas and clients behaving arbitrarily. Nodes may operate at different speeds. They are linked by a network that may fail to deliver messages, corrupt and/or delay them, or deliver them out of order. Our system is safe under this asynchronous model. Liveness is ensured when the *bounded fair links* [Yin 2003] assumption holds. Nodes authenticate the messages they send to other nodes; we assume that an adversary cannot break cryptographic techniques.

Replicas implement a state machine [Schneider 1990] to ensure consistency and deterministic replies to client requests; the state machine consists of a set of variables S encoding its state and a set of commands C operating on them. The execution of a command leads to an arbitrary number of state variables being read and/or modified and an output being provided to the environment. Our approach requires deterministic state machines: for the same sequence of inputs every non-faulty replica must produce the same sequence of outputs. Thereby, the replicas have to be in an identical state between processing the same two requests. The use of deterministic state machines guarantees that all non-faulty replicas executing a client request answer with identical replies.

We divide the state S into *objects*; an object O comprises a set of up to $|S|$ state variables ($0 < |O| \leq |S|$), sizes of different objects may vary. Altogether, objects cover the whole state ($\bigcup O_i = S$). For simplicity, we assume objects to be disjoint ($O_i \cap O_j = \emptyset$).

2.2 Architectural Properties

We assume that the BFT system architecture separates agreement from execution [Yin 2003] (SEP); our approach does not require agreement stage and execution stage to be located on different hosts. In particular, the following BFT systems[1] are suitable to integrate ODRC: PBFT [Castro 1999], BASE [Rodrigues 2001], CBASE [Kotla 2004], SEP [Yin 2003], TTCB [Correia 2004]), and VM-FIT [Reiser 2007]. Note that this list is not intended to be exhaustive.

To emphasize the generality of our approach, we use an abstract view of a BFT system as a composition of replicas (consisting of agreement stage and execution stage) and voters (responsible for identifying correct replies); voting is usually performed by the client. Replicas and voters may both be subject to Byzantine faults. The properties discussed in the following refer to non-faulty replicas and voters.

---

[1] Some of the systems listed do not explicitly separate agreement from execution. However, their basic concepts allow a separation of agreement stage and execution stage to the extent required for ODRC.

2.2.1 Replicas

A non-faulty replica must provide the following properties in order to be suitable for applying ODRC:

R1 **Total Request Ordering:** The agreement stage inputs an arbitrary sequence of client requests (that may differ between replicas) and outputs a stable totally-ordered sequence of requests (identical across all replicas).

R2 **Request Execution:** Each replica in the execution stage inputs the totally-ordered sequence of requests and outputs a set of replies that is identical to the output of all other non-faulty replicas.

R3 **Reply Cache:** Each replica caches replies in order to provide them to voters on demand.

$R1$ and $R2$ are standard techniques to ensure replica consistency. Note that we make no assumptions on how the total ordering of requests ($R1$) is achieved; that is, we treat the agreement stage as a black box. Therefore, ODRC can be implemented in systems using the standard BFT protocol [Castro 1999, Kotla 2004, Rodrigues 2001, Yin 2003], as well as in systems relying on trusted components [Chun 2007, Correia 2004, Reiser 2007].

In general, BFT systems provide the reply cache ($R3$) to be able to resend replies in case of network problems in order to guarantee liveness. Replicas usually store only the last reply sent to each client to limit cache size. In this paper, we assume that the reply cache is hosted by the agreement stage [Yin 2003].

2.2.2 Voters

A non-faulty voter must provide the following properties in order to be suitable for applying ODRC:

V1 **Reply Verification:** The result to a client request is accepted as soon as the voter receives $f + 1$ identical replies (or reply digests) from different replicas.

V2 **Incomplete-Voting Notification:** All non-faulty replicas eventually learn about an incomplete voting (i. e., the voter is not able to collect $f + 1$ identical replies within a certain period of time).

$V1$ is a standard property of BFT systems as $f + 1$ identical replies are the proof for correctness in the presence of at most f faulty replicas. $V2$ covers an essential mechanism to guarantee liveness. In systems using the standard BFT protocol [Castro 1999, Kotla 2007; 2004, Rodrigues 2001, Yin 2003], for example, it protects clients against a faulty primary not forwarding requests. In particular, on the expiration of a voting timeout, a client multicasts the affected request to all replicas, potentially triggering a view change.

2.3 Service-State Checkpointing

ODRC requires a BFT system to provide a mechanism to checkpoint the service state as well as a mechanism to restore the service state based on a checkpoint. In particular,

we assume the execution stage to provide an *object checkpoint function* that creates a snapshot of a state object. We also demand the execution stage to provide an *object update function* that updates an object based on a given checkpoint.

As most BFT systems [Castro 1999, Kotla 2007; 2004, Reiser 2007, Rodrigues 2001, Yin 2003] rely on periodic checkpoints to correct faulty replicas as well as to bring slow replicas up to date, both functions may be implemented using existing mechanisms. For example, all BASE-related systems [Kotla 2004, Rodrigues 2001, Yin 2003] implement a copy-on-write approach that only snapshots state objects modified since the last checkpoint. Furthermore, these systems also provide a function to selectively update state objects. Therefore, object checkpoint and object update functions that satisfy our requirements can be implemented without major implementation changes.

2.4 Request Analysis Function

We require requests to carry information about which objects they might read or modify during execution. To extract this information, we assume the existence of an application-specific *request analysis function (RAF)* that inputs a request *req* and outputs a set of objects:

$$Set_{<Object>}\ RAF(Request\ req)$$

For each request *req*, this function determines the maximal set of objects that might be accessed during execution; we assume each request to access at least one object. Note that the function can be implemented conservatively (i. e., it may return more objects than actually accessed) for services where a detailed analysis is too expensive. For the remainder of this paper, we state that a request *accesses* an object when the object is included in the set of objects returned by the request analysis function, independent of whether the request actually reads or modifies the object during execution.

In general, it may not be possible to specify a request-analysis function for arbitrary services as requests do not necessarily contain enough information to determine the state objects they access. However, most replicated BFT systems use similar functions to efficiently implement state-machine replication. For example, systems derived from BASE [Kotla 2004, Rodrigues 2001, Yin 2003] utilize information about state access of requests to determine state changes; CBASE [Kotla 2004] executes requests in parallel based, among other criteria, on the state objects they access.

3. Selective Request Execution

In this section, we present an extension to the common BFT state-machine–replication architecture that implements *selective request execution*. Instead of processing all requests on all replicas, a request is executed on only a subset of replicas, selected based on the objects accessed by the request. As a result, selective request execution reduces the load on a replica's execution stage. In this section, we assume the absence of faults; we drop this assumption in Section 4.

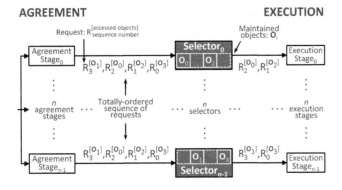

AGREEMENT   **EXECUTION**

Figure 1. Based on the objects to be accessed, selectors selectively execute agreed requests on local replicas.

```
1  void insert(Request req) {
2    Set<Object> O_req = RAF(req);
3    ack_unmain(O_req \ O_main, req);
4    if (O_req ∩ O_main == ∅) {
5      R_store.enqueue(req);
6    } else {
7      update_objects(O_req \ O_main, req);
8      R_exec.enqueue(req);
9  } }
10
11 Request next_request() {
12   return R_exec.dequeue();
13 }
```

Figure 2. Selector algorithm

3.1 State Distribution

Selective request execution requires all replicas to host all state objects. However, each replica is only responsible for keeping an assigned subset of objects up to date. In particular, objects are distributed across the system so that each object is assigned to $f + 1$ replicas; for the remainder of this paper, we refer to an object assigned to a replica as a *maintained object*. As a result, each object is maintained on $f + 1$ replicas and *unmaintained* on all others. We assume that a replica is able to determine its set of maintained objects based on a global assignment relation (e. g., for $f = 1$ and $n = 4$, a replica with ID $r \in [0, n-1]$ is assigned all objects with IDs o being $o \bmod 2 = r \bmod 2$; see Figure 1).

3.2 Selector

To apply selective request execution, we introduce a *selector* module between agreement stage and execution stage (see Figure 1). Each replica comprises its own selector that manages local request execution. Selectors of different replicas do not interact with each other, but rely on the same deterministic state machine (see Section 3.3) and operate on the same input; that is, the totally-ordered sequence of requests provided by the agreement stage ($R1$). As a result, all non-faulty selectors behave in a consistent manner. The selector provides the following lightweight interface:

```
void insert(Request req);
Request next_request();
void force_request(Request req);
```

Relying on the two functions insert and next_request, agreement stage and execution stage use the selector as a producer/consumer queue. The agreement stage submits requests to the selector by calling insert in the determined order. Independently, the execution stage fetches the next request to be executed by calling next_request, which blocks while there are no requests ready for processing. Please refer to Section 4.2 for a discussion of the use of force_request during fault handling.

3.3 Basic Algorithm

The main task of a selector is to determine whether to process a request on its associated replica. For each call to insert, the selector executes the algorithm shown in Figure 2 that relies on the following data structures:

- O_{main} the set of currently maintained objects.
- R_{exec} a FIFO queue containing all requests selected for execution. The selector uses this queue to *execute* requests by handing them over to the execution stage on calls to next_request (lines 11-12).
- R_{store} a FIFO queue containing all requests not selected for execution.

In general, a selector distinguishes between two categories of requests: First, requests selected for execution on the local replica are forwarded to the execution stage in the order defined by the agreement stage. Second, requests that are not selected for local execution are enqueued in R_{store} preserving their relative order. This approach allows the selector of a non-faulty replica to maintain the following invariant: At any time, the selector is able to bring each object on its local replica up to date. This is the case because either an object already is up to date, as all requests accessing the object have been executed on the local replica, or because the object subsequently can be updated by processing the state-modifying requests from R_{store} that access the object.

When the agreement stage calls insert, the selector executes the request analysis function (see Section 2.4) to extract the set of objects O_{req} the request req accesses during execution (line 2). If O_{req} only consists of unmaintained objects, the selector does not select req for execution (line 5). However, if there is *at least one maintained object* in O_{req}, req is selected for execution on the local replica (line 8). For the remainder of Section 3, we assume a request to access only maintained objects. In Section 4, we drop this assumption and also explain the function calls in lines 3 (see Section 4.2.2) and 7 (see Section 4.1).

With every request accessing a maintained object being processed (see Figure 1), local versions of maintained objects are always consistent with the current service state. As we assume that each request accesses at least one state object (see Section 2.4), the algorithm and our state distribution scheme (see Section 3.1) guarantee that, in the absence of faults, each request is processed on at least $f + 1$ replicas. In consequence, enough replies are provided to the voter to prove the result correct ($V1$). In case of faults, replies from additional replicas are needed to decide the vote, as further discussed in Section 4.2.1.

3.4 Checkpointing and Garbage Collection

Retaining requests that have not been selected for local execution in R_{store} allows a selector to update each local unmaintained object at any point in time. In order to limit the size of R_{store}, a selector basically uses the same mechanism as [Yin 2003]. It relies on periodic checkpoints that become stable as soon as $f + 1$ identical certificates from different replicas are available. When all state changes caused by a request are part of stable checkpoints, a selector is able to discard the request from R_{store}. In contrast to [Yin 2003], our approach involves *object checkpoints* covering only a single state object, instead of full checkpoints covering the whole replica state.

A selector i generates a checkpoint (using the object checkpoint function, see Section 2.3) for an object o for every kth *execution* of a request accessing o, with k being a system-wide constant (e. g., 100); besides object data, the checkpoint also includes digests of the replies to the k requests that led to the checkpoint (see Section 4.2.1). Next, the selector computes a digest d of the object checkpoint and multicasts $\langle CHECKPOINT, o, s, d \rangle_i$ to *all* selectors; s is the sequence number of the request that triggered the checkpoint creation. The selector assembles $f + 1$ identical checkpoint messages to a full object-checkpoint certificate that represents the proof for checkpoint correctness.

When the selector completes an object-checkpoint certificate, it discards older checkpoints and checkpoint certificates for the corresponding (maintained or unmaintained) object. A request with sequence number s is deleted from R_{store} as soon as checkpoint certificates (indicating sequence numbers of at least s) of all objects accessed by the request are available.

Using every kth access to an object to decide when to generate an object checkpoint guarantees that all non-faulty replicas assigned to the same object create a consistent checkpoint. As replicas execute all requests accessing a maintained object, they all create the checkpoint after the same request. Therefore, in the absence of faults, at least $f + 1$ checkpoint messages are available, enough to assemble a full object-checkpoint certificate. In case of faults, checkpoint messages from additional replicas assist in assembling a full object-checkpoint certificate (see Section 4.2.2).

```
1 void update_objects(Set<Object> O_update, Request req) {
2   Queue<Request> R_selected = ∅;
3   for(int i = R_store.get_index_latest(req); i ≥ 0; --i) {
4     Request r = R_store.get(i);
5     Set<Object> O_r = RAF(r);
6     if(O_r ∩ O_update ≠ ∅) {
7       R_selected.enqueue(r);
8       O_update = O_update ∪ O_r;
9   } }
10
11  for each Object o in O_update {
12    ObjectCheckpoint ocp = C_store.get(o);
13    if(ocp has not already been applied) {
14      update object using ocp;
15  } }
16
17  for(int i = (R_selected.size() - 1); i ≥ 0; --i) {
18    Request r = R_selected.get(i);
19    R_store.delete(r);
20    R_exec.enqueue(r);
21 } }
```

Figure 3. Algorithm for updating unmaintained objects.

4. On-Demand Replica Consistency

In this section, we drop the assumptions of all replicas being non-faulty and of requests accessing only maintained objects. As an immediate result of the latter, a selector needs to synchronize the state of its local replica with the current service state before executing a request that accesses unmaintained (and therefore possibly outdated) state objects. However, the selector does not perform a full state update. Instead, the selector ensures *on-demand replica consistency (ODRC)*. ODRC is "on demand" in two dimensions: Consistency is only ensured when a request to be executed actually demands it (time); furthermore, consistency is only ensured for the objects actually accessed by the request (space). In this paper, we use ODRC as a general term for applying selective request execution in conjunction with on-demand replica consistency.

4.1 Handling Cross-Border Requests

As a selector omits requests accessing only unmaintained objects, those objects may become outdated. Therefore, the selector has to ensure consistency of unmaintained objects prior to executing a request that accesses both maintained and unmaintained objects ("*cross-border request*"). The mechanism used by the selector to update unmaintained objects relies on a combination of object checkpoints and additional request execution. To ensure consistency of all objects accessed by a request, the selector calls update_objects (see Figure 2, line 7), which executes the two-step algorithm of Figure 3; the algorithm assumes a cache C_{store} holding the latest stable object checkpoints.

```
1 void force_request(Request req) {
2   if(req ∈ R_store) {
3     Set_<Object> O_req = RAF(req);
4     update_objects(O_req \ O_main, req);
5     R_exec.enqueue(req);
6     R_store.delete(req);
7 } }
```

Figure 4. Fault-handling function

```
1 void ack_unmain(Set_<Object> O_unmain, Request req) {
2   for each Object o in O_unmain {
3     if((++T_access[o] % k) == 0) {
4       attach req to a timer for o;
5       start the timer;
6   }
7 } }
```

Figure 5. Checkpoint monitoring function

In the first step, the selector determines, which requests to execute in order to update all unmaintained objects for a request req. Potential candidates are all requests from R_{store} that up to now were not selected for execution but contribute to updating the state of the unmaintained objects accessed by req. Starting with the latest request in R_{store} whose sequence number is smaller than the sequence number of req, the following operations are repeated for each request r in R_{store}. First, the set of objects O_r accessed by r is composed using the request analysis function (line 5). Second, if any object in O_r is a member in the set of objects to update O_{update}, r contributes to bringing them up to date and is therefore selected for execution (lines 6-7); furthermore, O_{update} is updated adding all objects contained in O_r (line 8), as these objects also have to be consistent when request r will be executed. Note that additional objects in O_r only have to be updated to the extent required by request r, they do not have to contain the current state of the object. In summary, this algorithm step goes back in time selecting all requests accessing objects to update. As these requests may require additional objects to be consistent, those objects are also updated to resemble their state at this point in time.

In the second step, the selector actually restores the replica state. First, it updates each unmaintained object in O_{update} using the checkpoint from C_{store} (lines 11-14). Next, the selector forwards all requests selected in the first step to the execution stage (lines 17-20); $R_{selected}$ is thereby traversed backwards to preserve request order.

Note that update_objects is the function that actually provides ODRC: when a request requires a set of objects to be consistent, the selector ensures consistency of exactly those objects, leaving other objects unaffected and possibly outdated. Therefore, executing the request req will produce the same output as if it were processed by a replica whose state is completely up to date.

4.2 Handling Faulty and Slow Replicas

During normal-case operation, a request is executed on $f+1$ replicas as voters only need $f+1$ identical replies to prove a response correct. However, in case of faulty or slow replicas, a voter might not be provided with enough replies to reach a decision. In this case, the voter sends an incomplete-voting notification to all replicas ($V2$).

4.2.1 Handling Incomplete Reply Voting

On the reception of an incomplete-voting notification, the agreement stage performs the standard fault-handling operations (e. g., a view change); in particular, the agreement stage resends the cached reply ($R3$) to the voter. If no reply is available for a request req, the agreement stage forces the selector to select req for local execution using force_request (see Figure 4). If req has not yet been processed (i. e., it is in R_{store}; lines 2, 6), the selector treats req like a cross-border request, updates unmaintained objects, and selects req for execution (lines 4-5). As each non-faulty replica eventually learns about an incomplete voting ($V2$), selectors of all non-faulty replicas will eventually select req for execution, providing additional replies to decide the vote.

A replica may receive an incomplete-voting notification for a request whose sequence number is smaller than the sequence number of the latest stable checkpoint of an accessed object. In this case, the selector sends the corresponding reply digest (which is included in the object checkpoint, see Section 3.4) to the voter (omitted in Figure 4). As the object checkpoint (and therefore the reply digest) is stable, there is at least one non-faulty replica that has actually executed the request and therefore resends the full reply on the reception of the incomplete-voting notification. As a result, other non-faulty replicas that have not executed the request may only return a reply digest.

Note that the agreement stage copes with most of the problems arising from faulty or slow replicas, hiding them from the selector. Agreement stages of different replicas decide independently whether to call force_request, based on their local reply cache. As voters continue request retransmissions while lacking $f+1$ identical replies, all non-faulty replicas that have originally omitted the execution of the request will eventually provide additional replies.

4.2.2 Handling Incomplete Checkpoint Voting

Faulty or slow replicas may (temporarily) prevent non-faulty replicas from assembling a full object-checkpoint certificate by providing faulty or no checkpoints. In this case, the mechanism in Figure 5 forces other non-faulty replicas to provide additional checkpoints when an object checkpoint does not become stable within a certain period of time. The mechanism relies on a table T_{access} containing an access counter for every unmaintained object and a set of timers.

The basic selector algorithm calls `ack_unmain` for every request req (see Figure 2, line 3); this function increments a counter in T_{access} for each unmaintained object o accessed by req (see Figure 5, lines 2-3). This way, a selector is able to determine the point in time at which the next checkpoint for an unmaintained object is due. A resulting counter value divisible by k indicates that all (non-faulty) replicas maintaining o will checkpoint this object after having executed req (see Section 3.4). In this case, the selector may expect a checkpoint for o to become stable within a certain (application-dependent) period of time. To monitor this, the selector attaches req to an object-specific timer and starts the timer (lines 4-5). The timer is stopped when the selector is able to assemble a full object-checkpoint certificate for o.

When the object timer expires, however, the selector calls `force_request` (see Section 4.2.1) for the request req. As a result, req is executed on the local replica; prior to that, the selector updates the unmaintained object o (see Figure 4, line 4), as o is accessed by req. Furthermore, processing req (i.e., the kth request accessing o since the last stable object checkpoint) triggers the creation and distribution of the next checkpoint for o. As all non-faulty replicas (that do not maintain o) behave in the same manner, each selector is eventually provided with enough additional object checkpoints for o to assemble a full object-checkpoint certificate.

4.3 Safety and Liveness

Introducing a selector between agreement stage and execution stage creates an additional potential point of failure, as a faulty selector may lead to a faulty replica, and vice versa. However, as selector interaction (besides the exchange of object checkpoints) is limited to the local replica (i.e., agreement stage and execution stage), selectors cannot be compromised by other replicas. With our approach treating the agreement stage as a black box, most mechanisms of the surrounding architecture ensuring safety (e.g., committing requests) and liveness (e.g., view changes) remain unaffected.

4.3.1 Safety

The safety properties of ODRC are primarily based on the safety properties of the underlying agreement protocol. As ODRC does not modify the agreement stage, correctness of the agreement protocol is preserved. In particular, it is guaranteed that, in the presence of at most f faults, the totally-ordered sequence of requests provided by the agreement stage is identical on all non-faulty replicas (R1).

In traditional BFT systems, this sequence I_A dictates the order in which all non-faulty replicas execute all requests. Applying ODRC, non-faulty replicas execute requests based on a sequence I_S that is provided by the selector. A correct selector transforms I_A into I_S ensuring the following two properties of I_S: first, requests whose access sets intersect in at least one object appear in I_S in the same relative order as in I_A; second, requests that access different state objects may be reordered. The latter is safe as, for a given

initial state, the result of executing those requests in any order places replicas in the same final state. A selector may delay a request accessing only unmaintained objects until the effects of the request are reflected in stable checkpoints for those objects. When a selector has obtained full checkpoint certificates for all objects accessed by a request, it is safe to remove the request from I_S, as at least $f + 1$ replicas have already executed the requests and agreed on the contents of the checkpoints.

A Byzantine selector may provide the selectors of different replicas with different checkpoints for the same object. However, as selectors only apply stable checkpoints, the safety of a non-faulty replica cannot be compromised by a Byzantine selector. Apart from exchanging checkpoints, selectors do not communicate with each other.

Voters (i.e., clients in traditional BFT systems) do not directly interact with ODRC selectors. However, a Byzantine voter may send an incomplete-voting notification for an arbitrary request r to the agreement stage of a non-faulty replica. If r has already been executed on the local replica, the agreement stage resends the cached reply for r (R3). Otherwise, the agreement stage forwards r to the local selector by calling `force_request` (see Section 4.2.1). As the selector ignores all requests that are not included in the totally-ordered sequence of requests provided by the agreement stage, a Byzantine voter is not able to force a selector into executing a request that has never been agreed on. Furthermore, a non-faulty selector will not execute the same request more than once (see Figure 4, lines 2 and 6).

4.3.2 Liveness

In the absence of faults, each request is executed on at least $f + 1$ replicas; that is, voters are able to collect enough identical replies to determine the correct result (V1). In case of network problems or faulty or slow replicas preventing a successful voting, voters eventually inform (e.g., via request retransmission) all selectors of non-faulty replicas about the lack of identical replies (V2). As a result, the affected request will finally be processed on all non-faulty replicas, allowing the client to make progress. Incomplete checkpoint voting is handled similarly, triggered by non-faulty selectors not able to assemble a stable object-checkpoint certificate.

4.4 Throughput Scalability

With all replicas executing all requests, traditional agreement-based BFT systems fail to provide throughput scalability: increasing the number of replicas n (while keeping f, the number of faults to tolerate, constant) does not increase system throughput. To improve throughput scalability, a number of BFT systems have been proposed where only a quorum of replicas handles and processes a request [Abd-El-Malek 2005, Cowling 2006]. However, in order to be safe, any two arbitrary quorums are required to overlap in at least one non-faulty replica. Therefore, quorum size increases with the total number of replicas n.

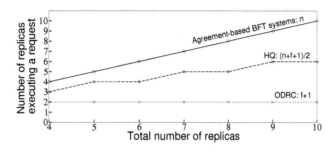

Figure 6. Overview of the number of replicas a request is actually executed on when more replicas are provided than necessary to tolerate $f = 1$ Byzantine fault.

With selectors operating on an agreed totally-ordered sequence of requests, in our approach, there is no need for subsets of replicas executing a request to overlap. Therefore, the minimal subset size of $f + 1$ replicas only depends on the number of faults to tolerate. As shown in Figure 6, for tolerating one Byzantine fault, a request is processed by two replicas, independent of n. In contrast, HQ [Cowling 2006], which requires the smallest quorums, processes a request on $\frac{n+f+1}{2}$ [Merideth 2008] replicas, not scaling as well.

Note that Figure 6 compares the average normal-case operation of a service requiring only few cross-border requests (see Section 8). To improve throughput scalability using ODRC, agreement-based systems need to solve the scalability bottleneck of quadratic costs for agreement [Castro 1999], for example, by using different clusters for agreement and execution [Clement 2009, Yin 2003]. This way, the number of execution replicas can be increased without degrading the performance of the agreement stage. However, our evaluation results show, that ODRC already enables a significant throughput increase for the minimal number of replicas.

5. Optimizations

This section presents optimizations that either directly increase performance of the ODRC algorithms presented in the previous two sections or contribute to the efficiency of the overall system. Furthermore, we describe approaches to configure selectors in order to customize system behavior.

5.1 Dynamic State Distribution

We expect application-specific heuristics to be used to assign objects to replicas; an optimal scheme distributes objects in a way that load is equally balanced across replicas. However, as object access patterns may be subject to change during the lifetime of a service, a static distribution scheme does not guarantee a permanent benefit. To compensate load imbalances, the selector i of an overloaded replica may therefore dynamically delegate the maintenance of an object to another selector j. Note that the fault handling mechanism (see Section 4.2) ensures that other selectors (including i) will step in to tolerate the fault in case j fails to maintain the object after the handover.

5.2 Optimized Checkpointing

We expect the execution stage to make use of copy-on-write and incremental cryptography to reduce the cost of producing object checkpoints [Castro 1999, Rodrigues 2001]. In addition, the following techniques can be applied to optimize checkpoint verification: first, the counter indicating the next checkpoint creation ignores read-only requests not modifying the state; as a result, state objects are only checkpointed when modified. Second, multiple objects may be verified together using a combined certificate.

5.3 Optimistic Updating of Unmaintained Objects

The selector presented in Section 3 updates an unmaintained object when the agreement stage inserts a committed request accessing the object. However, for agreement stages using the standard BFT three-phase protocol, the reception of a pre-prepare message in the first phase of the agreement protocol is already a good indication that a certain request is likely to be committed soon [Kotla 2007]. Therefore, an optimized selector may provide an additional function that allows the agreement stage to hand over a request on the reception of its corresponding pre-prepare message. Based on this information, the selector may optimistically start the updating process for the unmaintained objects accessed by the request in advance. As a result, less updating has to be done when the request is actually committed. In case an announced request is not committed (e. g., due to a faulty primary), this optimization would only lead to unnecessary object updates, it would not compromise safety.

5.4 Proactive Updating of Unmaintained Objects

Selectors may take advantage of periods of reduced workload to proactively update unmaintained objects. As a result, fewer changes have to be reproduced when requests actually demand consistency of unmaintained objects. In order to guarantee an upper bound for ODRC update procedures of unmaintained objects, one might also define a maximum number of outstanding modifying requests, forcing the selector to update an unmaintained object when a certain threshold is reached. In general, there is a tradeoff between reducing replica load and minimizing update duration.

5.5 Optimized Fault Handling

In general, a selector only processes requests exclusively accessing unmaintained objects on calls to `force_request`. Therefore, a crashed replica may result in bad performance, with voters on each request demanding replies from additional replicas due to incomplete voting. An optimized selector may compute statistics on forced requests; on an accumulation of forced requests accessing the same object, the selector may temporarily add the object to its maintained-objects set. This way, voters are provided with additional replies without having to explicitly demand them.

6. NFS Case Study

We have implemented a Byzantine fault-tolerant network file system (NFS) to show the practicality of ODRC. In this section, we discuss how to integrate NFS with ODRC and present an extensive evaluation.

6.1 Service Integration

NFS manages files and directories on the basis of *objects*; we define an ODRC state object to be a single NFS object. Internally, NFS uses *file handles* to uniquely identify an object; there is only one type of file handle, for both files and directories. Most NFS operations access only one object (e. g., SETATTR, GETATTR, READ, WRITE). However, some operations access two (e. g., CREATE, REMOVE, LOOKUP) or four (RENAME) objects and may therefore lead to cross-border requests.

In general, a request carries the file handle(s) of the object(s) its corresponding operation will read or modify, making it easy to implement a request analysis function. However, there is a set of operations (e. g., CREATE, REMOVE, RENAME) that identify some of the objects accessed by their *name*. Therefore, each selector also maintains a name-to-file-handle mapping for each object. Note that this mapping is only an inconvenience that could be obviated by refactoring the NFS service to either use only file handles or only names to identify objects.

Our prototype comprises parts of the CBASE-FS [Kotla 2004] prototype; CBASE-FS is an extension of the BASE implementation of a Byzantine fault-tolerant network file system [Rodrigues 2001] that supports concurrent execution of requests. In particular, our prototype reuses the CBASE-FS client implementation and the BASE conformance wrapper. Extending the CBASE approach with an ODRC selector module allows us to combine the performance gains offered by processing requests in parallel with the advantages of selective request execution.

As our file system is not the first Byzantine fault-tolerant NFS based on state-machine replication [Castro 1999, Kotla 2007; 2004, Rodrigues 2001, Yin 2003], we omit a discussion of replication-related problems (e. g., non-determinism) that are solved in the BASE abstraction layer. Like BASE, our file system provides NFSv2 [Sun Microsystems 1989].

6.2 Evaluation

We evaluate our file system using a cluster of dual-core hosts (2.4 GHz, 2 GB RAM) for the replicas and a cluster of quad-core hosts (2.4 GHz, 8 GB RAM) for the clients, all hosts are connected with switched 1 Gb/s Ethernet. All experiments use 32 NFS server daemons per replica and a block size of 4 kilobytes for both reads and writes[2]; the data is stored on disk.

---

[2] 4 kilobytes is the standard block size of CBASE-FS. Note that the maximum block size of NFSv2 is 8 kilobytes. Later versions allow a higher block size and are therefore capable of achieving a higher throughput.

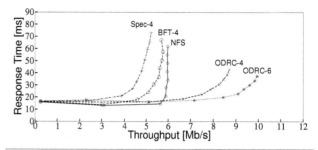

Figure 7. System throughput versus response time at the client for writes of 4 kilobytes with and without ODRC.

The baseline system (*BFT-4*) comprises four replicas and is therefore able to tolerate one Byzantine fault. Each replica hosts an agreement stage and an execution stage that run in separate processes and communicate via local sockets. BFT-4 replicas rely on the three-phase BFT protocol [Castro 1999] to reach agreement. The execution stage of a BFT-4 replica comprises a parallelizer module that allows concurrent execution of requests, as proposed in [Kotla 2004].

We also compare ODRC against a Zyzzyva [Kotla 2007]-like setting (*Spec-4*) that uses speculative execution: all Spec-4 replicas execute a state-modifying request after one protocol phase and send the reply back to the client. The client accepts the reply as soon as $3f + 1$ (read-only requests: $2f + 1$) replicas have provided matching responses. Note that Spec-4 does not implement the full Zyzzyva protocol, only its fast path, which is optimized for the absence of both faults and packet losses.

We evaluate ODRC using two different settings: *ODRC-4* extends the baseline system BFT-4 by introducing a selector component between the agreement stage and the execution stage of each replica; this setting allows us to evaluate the impact of ODRC on response time and throughput for the minimal number of replicas. *ODRC-6* extends ODRC-4 with two additional replicas that only comprise an execution stage but do not participate in the agreement protocol. Instead, they learn the total order of requests from the other replicas [Yin 2003]. ODRC-6 allows us to evaluate the impact of ODRC on scalability. In addition, we run the experiments on unreplicated NFS to get results for the unreplicated case, using the same configuration as for the BFT systems.

6.2.1 Normal-Case Operation

The following experiments evaluate the impact of ODRC on throughput and response time in the absence of faults.

Micro-Benchmark We use a micro-benchmark to evaluate the write throughput and response time of our file system with and without ODRC. In this experiment, we vary the number of clients that continuously write data in blocks of 4 kilobytes to separate files in a directory exported by the file system. In the ODRC test runs, the selector uniformly assigns files to replicas; that is, each ODRC-4 replica treats half (ODRC-6: a third) of all files as maintained objects.

The results of the benchmark (see Figure 7) show that for small workloads, the baseline system BFT-4 achieves slightly better response times than the ODRC variants. This is due to different voting conditions: an ODRC client is able to verify a reply as soon as both (i.e., $f + 1$) replicas processing the request have delivered the correct response. In contrast, a BFT-4 client only needs to wait for the replies of the two fastest non-faulty replicas. This way, the baseline system better compensates delays introduced by the agreement stage and other replication overhead that may vary between replicas. However, our results show that this weakness of ODRC becomes irrelevant for higher workloads.

The baseline system reaches a maximum throughput of 5.7 Mb/s for writes of 4 kilobytes. By selectively executing requests, ODRC-4 reduces the load on replicas and is therefore able to increase the overall throughput by 53%, using the same number of replicas as BFT-4. At the same time, the response time of ODRC-4 is about 30% lower than the response time of BFT-4. Our experiments show that these improvements in throughput and response time even outweigh the overhead of state-machine replication and Byzantine agreement allowing ODRC-4 to outperform unreplicated NFS in both categories. Relying on two additional execution replicas, ODRC-6 is able to increase the throughput to 9.9 Mb/s while further improving response time.

Note that using speculative execution (Spec-4), as for example implemented in Zyzzyva, does not offer any benefits over using traditional BFT in this experiment. The reason for that is that Byzantine agreement only contributes about one to four milliseconds to the response time observed by the clients (more than 14 milliseconds), most of the time is added by the application. Improving agreement through speculative execution therefore has little effect on the overall response time. In contrast, as clients are required to wait for $3f + 1$ matching replies instead of only $f + 1$, the slowest replica dictates the performance of Spec-4.

Macro-Benchmark We use the Postmark [Katcher 1997] benchmark to evaluate the performance of our Byzantine fault-tolerant file system in a realistic usage scenario. Postmark simulates the usage pattern of modern Internet services such as email or web-based commerce. We apply the same configuration as [Kotla 2004] and run a *read-mostly* experiment where the transaction phase of the benchmark favors reads over writes and a *write-mostly* experiment where reads are dominated by writes. We increase the number of clients (i.e., instances of the Postmark benchmark running in parallel) from five to fifty.

To measure the impact of object distribution on performance, we apply two different strategies to ODRC-4. The first strategy assigns file-system objects (i.e., files and directories) in a *round-robin* fashion to replicas. As this approach completely ignores dependencies between different objects, we consider it a worst-case strategy. The second object distribution strategy uses a simple *locality* heuristic: it assigns

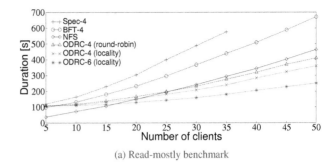

(a) Read-mostly benchmark

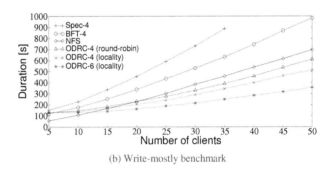

(b) Write-mostly benchmark

Figure 8. Results of the Postmark benchmark for two scenarios where (a) reads and (b) writes are favored during the transaction phase of the benchmark.

a file to the same replicas its parent directory is assigned to; directories are still assigned round-robin and may therefore be assigned to different replicas than their parent directories.

Figure 8 presents the results of the Postmark experiments. For both scenarios, it takes BFT-4 about 50% longer than unreplicated NFS to complete the benchmarks. Applying the round-robin distribution strategy allows ODRC-4 to run the benchmarks in 37% less time than BFT-4 for medium and high workloads. Cross-border requests represent about 11% of all requests for this strategy; assigning files to the same replicas as their parent directory removes almost all of them. Therefore, benchmark durations further decrease for ODRC-4 when using the locality strategy; compared to BFT-4, benchmarks complete in 44% (read-mostly) and 47% (write-mostly) less time. Note that these numbers are close to the theoretical optimum for ODRC-4 of 50% (i.e., a 100% increase in throughput). As a result, ODRC-4 is able to outperform unreplicated NFS by 19% and 25%, respectively.

Applying the locality strategy, ODRC-6 completes the benchmarks in about 60% less time than the baseline system; 67% is the optimum for ODRC-6. This confirms the good throughput scalability of the ODRC approach.

6.2.2 Introducing Faults

We now evaluate ODRC in the presence of faults. We distinguish between an *object fault* that leads to corrupted replies on all requests accessing the faulty object, and a *replica fault* that prevents a replica from providing replies at all.

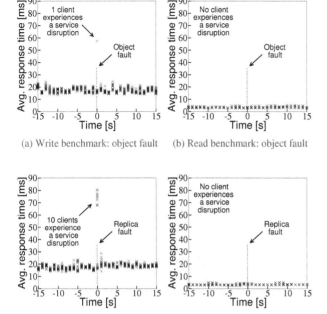

(a) Write benchmark: object fault (b) Read benchmark: object fault

(c) Write benchmark: replica fault (d) Read benchmark: replica fault

Figure 9. Impact of faults on the average response time at the client (1 sample/s) of ODRC-4 for a write-only and a read-only NFS micro-benchmark with twenty clients.

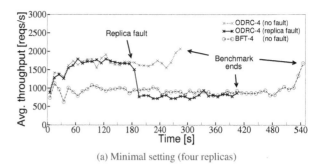

(a) Minimal setting (four replicas)

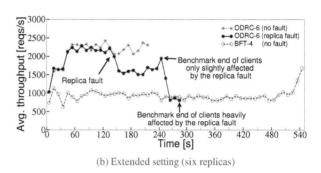

(b) Extended setting (six replicas)

Figure 10. Impact of a replica fault on the average system throughput of ODRC-4 and ODRC-6 (1 sample / 10 s) for the Postmark write-mostly benchmark with thirty clients.

Micro-Benchmarks This experiment evaluates the impact of faults on ODRC-4 running the NFS write-only micro-benchmark with twenty clients. For comparison, we also run a read-only micro-benchmark where clients read data in blocks of 4 kilobytes from separate files. In each case, we evaluate the worst-case scenario for an object checkpoint interval of $k = 100$ (see Section 3.4); that is, for the write-only benchmark, selectors on non-faulty replicas first have to replay the latest 100 write operations for the affected file(s) before being able to provide a reply to the pending request. For the read-only benchmark, no state modifications need to be replayed[3].

Figure 9a shows that when an object fault occurs during the write-only benchmark, ODRC-4 is not able to provide the client with enough replies to make progress for about one second (resulting in a peak in average response time). During this time, non-faulty replicas that do not maintain the affected file bring their local copies of the file up to date. When this procedure is complete, they process the pending request and provide additional replies. Please refer to Section 8 for a discussion of approaches to minimize the disruption. However, note that none of the clients accessing other files is penalized.

When we inject an object fault during the read-only benchmark, clients observe no notable response time increase (see Figure 9b). With no state modifications to replay, other replicas are prepared to process the pending request right away, tolerating the fault without disruption.

Figures 9c and 9d show that the failure of a replica has a similar impact on system performance than multiple concurrent object faults. In the write-only benchmark, half of clients experience a service disruption as their files are maintained objects on the faulty replica and therefore have to be updated on other replicas. With affected files remaining up to date on non-faulty replicas from then on, there are no disruptions on subsequent requests. However, the average response time increases after a replica fault, as there remain only three non-faulty replicas that process all requests.

Macro-Benchmark This experiment introduces a replica fault during the transaction phase of a write-mostly Postmark benchmark with thirty clients, for both ODRC-4 and ODRC-6. Figure 10 shows the average combined throughput of all clients for this scenario in comparison to BFT-4 and ODRC-4 runs without faults.

When the replica fault occurs, the throughput of ODRC-4 drops by about 50% due to files affected by the fault being restored on non-faulty replicas. Files are updated on the first access after the replica fault has occurred. In consequence, the impact of fault handling is not concentrated to a single point in time but distributed over a period of recovery, allowing ODRC-4 to keep throughput performance above

---

[3] Note that we use an optimization here: as stated in [Kotla 2004], the READ operation in NFS modifies the last-accessed-time attribute of a file (i.e., it is a state-modifying operation). However, when replaying multiple successive READ requests, only the latest one (i.e., the pending request) actually has to be processed in order to bring the last-accessed-time attribute up to date.

700 requests/s. With more and more objects being up to date, throughput steadily increases during this phase, eventually reaching the performance level of the baseline system. This behavior is in line with expectations, as in both cases, at this point, all non-faulty replicas process every request.

ODRC-6 comprises more replicas than actually required to tolerate the $f = 1$ Byzantine fault; that is, not all non-faulty replicas need to participate in the handling of a fault. To exploit this, we enable the fault handling for an object on only a set of $3f + 1 = 4$ selectors and disable it on the other two selectors; this is safe regardless of where a fault occurs. Figure 10b shows the benefit of this optimization. When we trigger the replica fault, the throughput of ODRC-6 only decreases by about 30% and does not drop to the BFT-4 level. Furthermore, some clients are able to complete the benchmark in almost the same time as in the absence of faults, as the files they access are maintained on replicas that only play a minor role in the handling of the replica fault; this explains the throughput drop at $t = 255$.

In general, tolerating a replica fault in ODRC-6 is not as costly as in ODRC-4. As objects are distributed across six (instead of four) replicas, a selector in ODRC-6 maintains only a third (instead of half) of all objects. As a result, when a replica fails, other selectors only need to update a third (instead of half) of the service state at most. Increasing the total number of replicas will enhance this advantage.

7. ZooKeeper Case Study

ZooKeeper [Hunt 2010] is a distributed coordination system that provides distributed applications with basic services like leader election, group membership, distributed synchronization, and configuration maintenance. ZooKeeper is widely in use at Yahoo for crucial tasks like failure recovery. Based on the original crash-tolerant implementation, we have built a Byzantine fault-tolerant version of the ZooKeeper service.

7.1 Service Integration

ZooKeeper stores information in a hierarchical namespace. Each tree node is able to store data, and to manage child nodes; we define an ODRC state object to be a single ZooKeeper node. A node is uniquely identified by its path. Each request carries the full path information of the node it will operate on as a string; the ODRC request analysis function uses this string to determine object access.

Our prototype comprises the application logic of the original ZooKeeper implementation. However, in order to make the service Byzantine fault-tolerant, we substitute the crash-tolerant protocol ZooKeeper uses to order requests with the standard three-phase BFT protocol. Furthermore, we introduce voting at the client, and add a small abstraction layer at the replica that ensures consistency of ZooKeeper node metadata (e. g., timestamps and node version counters). Please refer to [Clement 2009] for a discussion of the measures it takes to enforce deterministic ZooKeeper replicas in the context of Byzantine fault tolerance.

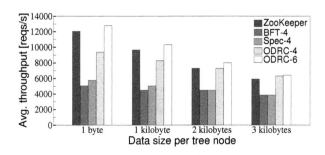

Figure 11. Realized throughput for repeatedly setting the data of ZooKeeper nodes using different data sizes between 1 byte and 3 kilobytes.

7.2 Evaluation

For our evaluation of ZooKeeper, we use the same cluster of machines as well as the same system settings as in Section 6.2. All systems are configured to tolerate one (Byzantine) fault; therefore, the setting for plain crash-tolerant ZooKeeper comprises three replicas.

7.2.1 Normal-Case Operation

We evaluate the write throughput of the different ZooKeeper variants with an experiment in which clients repeatedly write new data chunks of ZooKeeper-typical sizes to nodes; Figure 11 presents the results of this experiment. For one-byte writes, ODRC-4 achieves an 85% improvement over the baseline system BFT-4. As increasing the data size puts more load on the agreement stage, the benefit of ODRC-4 decreases to 63% for writes of 3 kilobytes. For the same reason, the improvement of ODRC-6 over ODRC-4 is higher for small data sizes. However, for large requests, ODRC-4 is able to outperform the crash-tolerant ZooKeeper implementation, using only the minimal number of replicas required for Byzantine fault tolerance. Note that in this experiment, again, speculation (Spec-4) does not provide a substantial performance gain due to the fact that clients have to wait for identical replies from all four replicas to make progress.

7.2.2 Introducing Faults

We repeat the write-only micro-benchmark experiments of Section 6.2.2 for ZooKeeper to examine the worst-case impact of an object fault and a replica fault on the response time of ODRC-4. Figure 12 shows that, in contrast to NFS, faults that occur during the write-only experiment do not lead to a service disruption that is noticeable at the ZooKeeper client. The reason for that lies in the fact that the data of a ZooKeeper node is always written (and also read) atomically and in its entirety; that is, each write operation replaces all the data stored at a node. In consequence, an ODRC selector is able to update the complete data of an unmaintained node by just replaying the latest write request and adjusting the node metadata to reflect the current version number. Exploiting this property of ZooKeeper, which is also provided

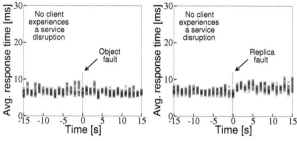

(a) Write benchmark: object fault   (b) Write benchmark: replica fault

Figure 12. Impact of faults on the average response time (1 sample/s) of ODRC-4 for a write-only ZooKeeper workload from twenty clients.

by other services (e. g., Chubby [Burrows 2006]), allows to significantly speed up fault handling. Of course, the response time also decreases for ZooKeeper after a replica fault (see Figure 12b), as the remaining replicas process all requests.

8. Discussion

Our evaluation shows that ODRC can increase the performance of a replicated BFT system for the default number of replicas by optimizing the resource utilization of execution stages. Thereby, cross-border requests are a key factor limiting ODRC performance gains, as they contradict the goal of executing every request on only a minimal subset of replicas. As the fraction of cross-border requests is not an intrinsic property of a service, but depends to a great extent on the service usage pattern and the object distribution scheme used in ODRC, the occurrence of cross-border requests can be contained by an effective object distribution strategy. For NFS, for example, it suffices to assign files and their parent directories to the same replicas (as done by the locality strategy, see Section 6.2.1) to reduce the fraction of cross-border requests to almost zero. Nevertheless, extensive use of cross-border requests may disqualify a service from ODRC benefits, if refactoring the service is not an option.

Like other BFT systems [Kotla 2007, Wood 2011], ODRC is optimized for fault-free operation. Our evaluation of NFS shows that when a fault occurs, duration of the service disruption depends on how outdated the local version of the affected object is on non-faulty replicas (see Section 6.2.2). In the worst case, the object has to be recreated before being able to make progress. Although this circumstance does not endanger liveness, in practice, additional measures may be taken to trade off some of the performance gains made possible by ODRC in return for a fault-handling speed up. In particular, this can be achieved by proactively updating unmaintained objects (see Section 5.4), limiting the extent to which the local copy of an object becomes outdated. However, as the ZooKeeper example shows, there are also services where such measures are not necessary, as handling a fault does not lead to a service disruption.

The evaluation results for NFS in Section 6.2.1 show that selective request execution improves response times for medium and high workloads. However, ODRC is not capable of reducing the minimal response time of a Byzantine fault-tolerant system, which is mainly dictated by the protocol used for Byzantine agreement. As ODRC uses the agreement stage as a black box and therefore does not rely on a specific protocol for request ordering, the approach can be combined with already available [Clement 2009, Correia 2004, Yin 2003] as well as future systems targeting to reduce the agreement overhead.

9. Related Work

Most related work in the fields of BFT state-machine replication has already been discussed in Section 2, in connection with our system model. In this context, we presented a list of BFT systems [Castro 1999, Correia 2004, Kotla 2004, Reiser 2007, Rodrigues 2001, Yin 2003] that may benefit from ODRC. Zyzzyva [Kotla 2007] and recent work from Guerraoui et al. [Guerraoui 2010] did not make the list as they improve performance by optimistically executing requests in an order that is speculated on. With the request order not being stable (which contradicts R1, see Section 2.2.1), replicas may become inconsistent and have to rely on the help of clients to converge. In contrast, the use of ODRC may lead to partly outdated but never inconsistent replicas. Furthermore, using a selector, a replica is able to update its state without the help of clients and only based on local knowledge.

Yin et al. [Yin 2003] proposed the separation of agreement and execution, which is essential to benefit from the throughput scalability of our approach (see Section 4.4). Clement et al. [Clement 2009] introduced a third stage to reduce authentication cost and optimize request ordering in cluster environments. Prophecy [Sen 2010] improves throughput and latency of BFT state-machine replication for read-centric workloads by introducing a trusted reply-checksum cache. We consider the latter two approaches orthogonal to our work.

Kotla et al. [Kotla 2004] (CBASE) proposed to increase the performance of a BFT system by executing independent requests in parallel. Two requests are independent, if replicas can process the requests in a different order without compromising correctness; this is true, for example, if the requests access different parts of the replica state. In CBASE, information about state access of requests is used by a parallelizer module located between agreement stage and execution stage that forwards concurrent requests to a set of worker threads. Still, all CBASE replicas process all requests. Like CBASE, ODRC uses information about the state access of requests to optimize system performance. However, ODRC replicas only process requests that have been selected for execution by the local selector module. Both approaches can be combined by forwarding the output of a selector to a parallelizer, as implemented in our prototype. Our evaluations for both NFS and ZooKeeper show that combining concurrent execution with selective execution (ODRC-4) allows a significant performance improvement over plain concurrent execution (BFT-4).

Processing full requests on only a subset of replicas was originally proposed by Pâris [Pâris 1986] in the context of a crash-tolerant file system that relies on quorums. The file system distinguishes between replicas that contain the complete file data and a version number ("copies") and replicas that only contain file version numbers but no data ("witnesses"). While copies execute the full request, witnesses only increment their local file version counter when processing a modifying request; still, witnesses participate in each operation. In case of faults, witnesses may be upgraded to copies. Liskov et al. implemented a related approach in the Harp file system [Liskov 1991] that makes active use of witnesses only during times of node failures or network partitions. Cheap Paxos [Lamport 2004] generalizes the idea to use lightweight auxiliary nodes for handling crashes of full-fledged replicas in order to reduce the resource requirements of a replicated service. Ladin et al. [Ladin 1992] have shown that by processing a request on only a subset of replicas and lazily updating the other replicas, a crash-tolerant replicated service can be built that outperforms an unreplicated service.

SPARE [Distler 2011] and ZZ [Wood 2011] are designed to minimize resource consumption in the context of Byzantine fault tolerance. Both systems rely on virtualization and use only $f + 1$ service replicas during normal-case operation. In case of faults or suspected faulty behavior, the systems quickly activate up to f additional replicas running in separate virtual machines. SPARE and ZZ focus on minimizing the resource footprint of a BFT system by reducing the number of service replicas; they were not designed to increase performance. ODRC also proposes an optimization that exploits the fact that $f + 1$ replies are enough to prove a response correct in the absence of faults. However, ODRC uses this property to improve resource utilization of execution stages for the default number of replicas (i. e., $3f + 1$ in traditional BFT systems), which allows to increase the upper throughput bound of the overall system.

State partitioning was previously applied to tolerate crashes and increase scalability in large-scale file systems and distributed data storage [Gribble 2000, MacCormick 2004, van Renesse 2004]. State partitioning was also used in the context of Byzantine fault tolerance: Farsite [Adya 2002] and OceanStore [Rhea 2003] are large-scale file systems that assign files to different groups of $3f + 1$ replicas, each separately executing a BFT protocol. As a result, the throughput of the overall system improves for an increased number of replica groups. However, within a replica group, all replicas execute all requests. Our approach increases throughput by applying state partitioning within a replica group, obviating the need for complex inter–replica-group protocols to handle cross-border requests. Therefore, ODRC already achieves a significant performance improvement for the default num-

ber of replicas. However, optimizing the performance of a single replica group, a possible use case of ODRC is to act as a building block for large-scale systems.

Malek et al. [Abd-El-Malek 2005] used quorums to build a BFT system for arbitrary services (Q/U). It requires a minimum of $5f + 1$ servers and a quorum size of $4f + 1$ to tolerate f Byzantine faults. HQ [Cowling 2006] implements a hybrid approach that combines quorum-based and agreement-based protocols to reduce the number of servers to $3f + 1$ and the minimal quorum size to $2f + 1$. Both systems make use of quorums to achieve fault scalability. As discussed in Section 4.4, ODRC offers better throughput scalability. In Q/U and HQ, concurrent object access may lead to inconsistent replica states, which requires replicas to revert previous state modifications. In contrast, replica states in our approach may partially become outdated, but never inconsistent.

10. Conclusion

A traditional agreement-based Byzantine fault-tolerant system executes all requests on all replicas to ensure consistency. As a result, the system usually provides more than the $f + 1$ identical replies to a request that are actually needed for a client to make progress in the absence of faults.

In this paper, we have shown that a selector module between agreement stage and execution stage is able to increase the performance of the overall BFT system by selectively executing each request on only a subset of $f + 1$ replicas, based on the state objects a request accesses. As this approach may lead to parts of the replica state being outdated, the selector ensures replica consistency by updating the state of objects on demand, that is, at the time and to the extent required by a request to be executed.

Our evaluation of Byzantine fault-tolerant variants of NFS and ZooKeeper shows that the use of on-demand replica consistency lowers the response time for medium and high workloads, and is able to increase the throughput of a Byzantine fault-tolerant service beyond the throughput of the corresponding non–fault-tolerant unreplicated service.

Acknowledgements

We would like to thank the anonymous reviewers for their comments and our shepherd, Liuba Shrira, for her guidance.

References

[Abd-El-Malek 2005] M. Abd-El-Malek, G. R. Ganger, G. R. Goodson, M. K. Reiter, and J. J. Wylie. Fault-scalable Byzantine fault-tolerant services. In *Proceedings of the 20th Symposium on Operating Systems Principles*, pages 59–74, 2005.

[Adya 2002] A. Adya, W. J. Bolosky, M. Castro, G. Cermak, R. Chaiken, J. R. Douceur, J. Howell, J. R. Lorch, M. Theimer, and R. P. Wattenhofer. FARSITE: Federated, available, and reliable storage for an incompletely trusted environment. In *Proceedings of the 5th Symposium on Operating Systems Design and Implementation*, pages 1–14, 2002.

[Burrows 2006] M. Burrows. The Chubby lock service for loosely-coupled distributed systems. In *Proceedings of the 7th Symposium on Operating Systems Design and Implementation*, pages 335–350, 2006.

[Castro 1999] M. Castro and B. Liskov. Practical Byzantine fault tolerance. In *Proceedings of the 3rd Symposium on Operating Systems Design and Implementation*, pages 173–186, 1999.

[Chun 2007] B.-G. Chun, P. Maniatis, S. Shenker, and J. Kubiatowicz. Attested append-only memory: making adversaries stick to their word. In *Proceedings of the 21st Symposium on Operating Systems Principles*, pages 189–204, 2007.

[Clement 2009] A. Clement, M. Kapritsos, S. Lee, Y. Wang, L. Alvisi, M. Dahlin, and T. Riche. UpRight cluster services. In *Proceedings of the 22nd Symposium on Operating Systems Principles*, pages 277–290, 2009.

[Correia 2004] M. Correia, N. F. Neves, and P. Veríssimo. How to tolerate half less one Byzantine nodes in practical distributed systems. In *Proceedings of the 23rd International Symposium on Reliable Distributed Systems*, pages 174–183, 2004.

[Cowling 2006] J. Cowling, D. Myers, B. Liskov, R. Rodrigues, and L. Shrira. HQ replication: A hybrid quorum protocol for Byzantine fault tolerance. In *Proceedings of the 7th Symposium on Operating Systems Design and Implementation*, pages 177–190, 2006.

[Distler 2011] T. Distler, R. Kapitza, I. Popov, H. P. Reiser, and W. Schröder-Preikschat. SPARE: Replicas on hold. In *Proceedings of the 18th Network and Distributed System Security Symposium*, 2011.

[Gribble 2000] S. D. Gribble, E. A. Brewer, J. M. Hellerstein, and D. Culler. Scalable, distributed data structures for Internet service construction. In *Proceedings of the 4th Symposium on Operating Systems Design and Implementation*, pages 319–332, 2000.

[Guerraoui 2010] R. Guerraoui, N. Knežević, V. Quéma, and M. Vukolić. The next 700 BFT protocols. In *Proceedings of the EuroSys 2010 Conference*, pages 363–376, 2010.

[Hendricks 2007] J. Hendricks, G. R. Ganger, and M. K. Reiter. Low-overhead Byzantine fault-tolerant storage. In *Proceedings of the 21st Symposium on Operating Systems Principles*, pages 73–86, 2007.

[Hunt 2010] P. Hunt, M. Konar, F. P. Junqueira, and B. Reed. ZooKeeper: Wait-free coordination for internet-scale systems. In *Proceedings of the 2010 USENIX Annual Technical Conference*, pages 145–158, 2010.

[Katcher 1997] J. Katcher. Postmark: A new file system benchmark. Technical Report 3022, Network Appliance Inc., 1997.

[Kotla 2007] R. Kotla, L. Alvisi, M. Dahlin, A. Clement, and E. Wong. Zyzzyva: speculative Byzantine fault tolerance. In *Proceedings of the 21st Symposium on Operating Systems Principles*, pages 45–58, 2007.

[Kotla 2004] R. Kotla and M. Dahlin. High throughput Byzantine fault tolerance. In *Proceedings of the 2004 Conference on Dependable Systems and Networks*, pages 575–584, 2004.

[Ladin 1992] R. Ladin, B. Liskov, L. Shrira, and S. Ghemawat. Providing high availability using lazy replication. *ACM Transactions on Computer Systems*, 10:360–391, 1992.

[Lamport 2004] L. Lamport and M. Massa. Cheap Paxos. In *Proceedings of the 2004 International Conference on Dependable Systems and Networks*, pages 307–314, 2004.

[Liskov 1991] B. Liskov, S. Ghemawat, R. Gruber, P. Johnson, L. Shrira, and M. Williams. Replication in the Harp file system. In *Proceedings of the 13th Symposium on Operating Systems Principles*, pages 226–238, 1991.

[MacCormick 2004] J. MacCormick, N. Murphy, M. Najork, C. A. Thekkath, and L. Zhou. Boxwood: abstractions as the foundation for storage infrastructure. In *Proceedings of the 6th Symposium on Operating Systems Design and Implementation*, pages 105–120, 2004.

[Merideth 2008] M. Merideth. Tradeoffs in Byzantine-fault-tolerant state-machine-replication protocol design. Technical Report CMU-ISR-08-110, Carnegie Mellon University, 2008.

[Pâris 1986] J.-F. Pâris. Voting with witnesses: A consistency scheme for replicated files. In *Proceedings of the 6th International Conference on Distributed Computer Systems*, pages 606–612, 1986.

[Reiser 2007] H. P. Reiser and R. Kapitza. Hypervisor-based efficient proactive recovery. In *Proceedings of the 26th International Symposium on Reliable Distributed Systems*, pages 83–92, 2007.

[Rhea 2003] S. Rhea, P. Eaton, D. Geels, H. Weatherspoon, B. Zhao, and J. Kubiatowicz. Pond: The OceanStore prototype. In *Proceedings of the 2nd Conference on File and Storage Technologies*, pages 1–14, 2003.

[Rodrigues 2001] R. Rodrigues, M. Castro, and B. Liskov. BASE: Using abstraction to improve fault tolerance. In *Proceedings of the 18th Symposium on Operating Systems Principles*, pages 15–28, 2001.

[Schneider 1990] F. B. Schneider. Implementing fault-tolerant services using the state machine approach: a tutorial. *ACM Computing Survey*, 22(4):299–319, 1990.

[Sen 2010] S. Sen, W. Lloyd, and M. J. Freedman. Prophecy: Using history for high-throughput fault tolerance. In *Proceedings of the 7th Symposium on Networked Systems Design and Implementation*, 2010.

[Sun Microsystems 1989] Sun Microsystems. NFS: Network file system protocol specification. Internet RFC 1094, 1989.

[van Renesse 2004] R. van Renesse and F. B. Schneider. Chain replication for supporting high throughput and availability. In *Proceedings of the 6th Symposium on Operating Systems Design and Implementation*, pages 91–104, 2004.

[Wester 2009] B. Wester, J. Cowling, E. B. Nightingale, P. M. Chen, J. Flinn, and B. Liskov. Tolerating latency in replicated state machines through client speculation. In *Proceedings of the 6th Symposium on Networked Systems Design and Implementation*, pages 245–260, 2009.

[Wood 2011] T. Wood, R. Singh, A. Venkataramani, P. Shenoy, and E. Cecchet. ZZ and the art of practical BFT execution. In *Proceedings of the EuroSys 2011 Conference*, 2011.

[Yin 2003] J. Yin, J.-P. Martin, A. Venkataramani, L. Alvisi, and M. Dahlin. Separating agreement from execution for Byzantine fault tolerant services. In *Proceedings of the 19th Symposium on Operating Systems Principles*, pages 253–267, 2003.

Efficient Middleware for Byzantine Fault Tolerant Database Replication

Rui Garcia

CITI / Departamento de Informática,
Faculdade de Ciências e Tecnologia,
Universidade Nova de Lisboa
Quinta da Torre, Caparica, Portugal
bomgarcia@gmail.com

Rodrigo Rodrigues

MPI-SWS
Kaiserslautern and Saarbrücken,
Germany
rodrigo@mpi-sws.org

Nuno Preguiça

CITI / Departamento de Informática,
Faculdade de Ciências e Tecnologia,
Universidade Nova de Lisboa
Quinta da Torre, Caparica, Portugal
nuno.preguica@di.fct.unl.pt

Abstract

Byzantine fault tolerance (BFT) enhances the reliability and availability of replicated systems subject to software bugs, malicious attacks, or other unexpected events. This paper presents Byzantium, a BFT database replication middleware that provides snapshot isolation semantics. It is the first BFT database system that allows for concurrent transaction execution without relying on a centralized component, which is essential for having both performance and robustness. Byzantium builds on an existing BFT library but extends it with a set of techniques for increasing concurrency in the execution of operations, for optimistically executing operations in a single replica, and for striping and load-balancing read operations across replicas. Experimental results show that our replication protocols introduce only a modest performance overhead for read-write dominated workloads and perform better than a non-replicated database system for read-only workloads.

Categories and Subject Descriptors D.4.5 [*Reliability*]: Fault-tolerance; H.2.4 [*Systems*]: Concurrency

General Terms Design, Performance, Reliability

Keywords Databases, Middleware, Byzantine Fault-tolerance.

1. Introduction

Database systems are a key component of the computer infrastructure of most organizations. It is thus crucial to ensure that database systems work correctly and continuously even in the presence of a variety of unexpected events. The key to ensuring high availability of database systems is to use replication. While many methods for database replication have been proposed [Cecchet 2008], most of these solutions only tolerate silent crashes of replicas, which occur when the system suffers hardware failures, power outages, etc.

While this approach suffices for many types of faults, it does not tolerate the effects of events such as software bugs or malicious attacks that can cause databases to fail in ways other than silently crashing. These types of events are of growing concern. Recent studies show that the majority of bugs reported for three commercial database systems would cause the system to fail in a non-crash manner [Gashi 2007, Vandiver 2007]; another study found that a significant fraction of concurrency bugs in MySQL led to subtle violations of database semantics [Fonseca 2010]. Intrusions have also been reported as being a problem: database systems have become a frequent target of attacks that can result in loss of data integrity or even permanent data loss [DISA 2004].

A promising approach for increasing the correctness and availability of systems in the face of these types of events is through Byzantine fault tolerant (BFT) replication. This class of replication protocols makes no assumptions about the behavior of faulty replicas (i.e., assumes a Byzantine fault model [Lamport 1982]), so it can tolerate arbitrary failures from a subset of its replicas (typically up to $\frac{1}{3}$ of the replicas).

However, the application of BFT techniques to database systems has been quite limited. Previous proposals either do not allow transactions to execute concurrently, which inherently limits the performance of the system [Garcia Molina 1986, Gashi 2007], or rely on a trusted coordinator node that chooses which requests to forward concurrently [Vandiver 2007]. In the latter case, the coordinator becomes a central point of failure: if the node crashes or is compromised the entire system becomes vulnerable.

In this paper we present Byzantium, a novel, middleware-based, Byzantine fault tolerant database replication solution. Byzantium improves on existing BFT replication for databases because it allows extensive concurrency with no centralized components (on whose correctness the integrity of the system depends), which is essential for achieving good performance and reliability.

Several design features in Byzantium are interesting in their own right, and several of them are useful beyond the scope of database replication. In particular, we highlight the following set of design choices and techniques that make the Byzantium design novel.

Snapshot isolation. Unlike previous BFT database systems, which provide 1-copy serializability [Bernstein 1986], Byzantium targets snapshot isolation semantics. We show how we can take advantage of these weaker semantics to increase concurrency, by restricting the use of a more expensive serialization protocol to a subset of the operations.

Middleware-based replication. This approach allows us to use existing database systems without modifying them, and even allows distinct implementations from different vendors to be used at different replicas. Such diversity is important in order to increase resilience against attacks triggered by software vulnerabilities or deterministic software bugs, but not having access to the database internals raises the bar for our protocols to achieve good performance.

Optimistic execution of groups of operations. Our system design proposes two alternative techniques for optimistically executing operations in a single replica. Each technique offers distinct advantages: one of them works transparently with any form of concurrency control, and the other offers better performance in the presence of Byzantine replicas, but requires extracting write-sets in databases that use locking. Furthermore, while optimistic execution takes advantage of transactional semantics, these techniques may be useful for replicating other types of systems, namely those that support some form of speculation [Nightingale 2005].

Striping with BFT replication. We also show how we can use BFT replication to improve the performance of operations that do not update the state of the system, by striping reads from different clients to different subsets of replicas, while maintaining the desired database semantics.

We implemented Byzantium and evaluated our prototype using variants of TPC-C. Our experimental results show that our replicated database has only a modest performance overhead for read-write dominated workloads and exhibits performance gains of up to 90% over executing transactions in a single replica for read-only workloads.

The remainder of this paper is organized as follows. Section 2 discusses related work. Section 3 introduces the system model and Section 4 details the system design and proposed algorithms. Section 5 argues the correctness of our solution. Section 6 discusses implementation issues and Section 7 presents an evaluation of our prototype. Section 8 concludes the paper.

2. Related work

We group the set of related proposals into two main areas: Byzantine fault tolerance and database replication.

Byzantine fault tolerance. Byzantine fault tolerant (BFT) replication has drawn a lot of attention in the last decade as a mechanism to mask malicious attacks and software errors. Over these years, there has been a series of proposals for replication protocols that tolerate Byzantine faults and claim to be efficient enough to be practical (e.g., PBFT [Castro 2002], or Zyzzyva [Kotla 2007]). Most such proposals have centered around building a *replicated state machine* [Schneider 1990]. These protocols allow the replication of a deterministic service that follows an RPC model: clients issue requests, servers maintain the state of the service, and given a certain request and current state, the servers deterministically issue a reply and move on to the subsequent state. These protocols have been used to provide Byzantine fault tolerance to different services, including distributed file systems and coordination services [Castro 2002, Clement 2009].

While these protocols could also be used to replicate a database server by executing individual database operations as operations of the replicated state machine, this would preclude operations from being executed concurrently, thus limiting performance. Furthermore, for database systems using lock-based concurrency control, an operation that blocked would prevent any further operations from being executed.

We build on these proposals since our system uses state machine BFT replication as one of its building blocks. However, we address these performance limitations by taking advantage of snapshot isolation semantics and optimistic execution, which allow us to craft our protocols in a way that avoids the use of state machine replication when possible, thus increasing concurrency.

Database replication. There exist several proposals, both from industry and academia, for replicating database systems in order to both increase their resilience to faults and improve their throughput (see [Cecchet 2008] for an overview of the topic).

The proposals that are more closely related to our work are the various proposals for middleware-based replication that provide snapshot isolation (SI) [Elnikety 2005, Lin 2005, Plattner 2004]. In particular, one of the two alternative protocols we present uses an approach that bears similarity to Ganymed [Plattner 2004], with all read-write transactions executing first in the same replica responsible for guaranteeing SI properties. The execution of read-only transactions is distributed among the other replicas. The other alternative protocol we present bears similarity to the approach proposed by Elnikety et. al. [Elnikety 2005], with read-write transactions executing first in a single replica (possibly using different replicas for concurrent transactions) and, at commit time, transactions being propagated to the other replicas that detect conflicts according to the SI properties. While Byzantium uses similar techniques, our solution differs in that it is designed with the assumption of a Byzantine fault model, instead of assuming that nodes fail only by crashing.

There have also been a few proposals for Byzantine fault tolerant database replication.

The initial protocols in this area were proposed by Garcia-Molina et al. [Garcia Molina 1986] and by Gashi et al. [Gashi 2007]. Their replication protocols serialize all requests and do not allow transactions to execute concurrently. Consequently, their protocols provide stronger semantics than ours, but serializing all requests inherently limits the performance of the system. We improve on these systems by using weaker semantics and novel protocols to obtain better performance.

HRDB [Vandiver 2007] provides BFT database replication using a trusted node to coordinate the replicas. The coordinator chooses which requests to forward concurrently, in a way that maximizes the amount of parallelism between concurrent requests. HRDB provides good performance, but requires trust in the coordinator, which can be problematic if replication is being used to tolerate attacks. Furthermore, the coordinator is a single point of failure: if the coordinator crashes, the availability of the system will be affected until it recovers. Finally, HRDB ensures 1-copy serializability, whereas our approach provides weaker (yet commonly used) semantics in order to achieve good performance.

In a prior workshop paper we described a preliminary design of the multi-master version [Preguiça 2008]. This paper improves on that work in several ways. First, we modified the design to accommodate changes that were necessary to obtain good performance (many of which were driven by observing the performance limitations of the original design in a real deployment). In particular, we have optimized the execution of read-only transactions to allow them to execute in a small subset of the replicas, and we changed the way we execute operations so that they are propagated to all replicas before commit time. Second, we propose a second version of the system, based on a single-master approach. This approach requires less support from the underlying database system. Finally, we present a complete implementation and an experimental evaluation of our prototype.

3. System model

We assume a Byzantine failure model where faulty nodes (client or servers) may behave arbitrarily, other than not being able to break the cryptographic techniques that are used. We assume at most f replicas are faulty out of a total of $n = 3f + 1$ replicas. In our current implementation we do not employ existing proactive recovery mechanisms [Castro 2002], which implies that we need to ensure that we have no more than f faulty nodes throughout the system lifetime. If we were to apply proactive recovery techniques, we would

be able to tolerate at most f faults during a window of vulnerability.

When the correctness conditions of the system are met, the safety property ensured by Byzantium is that the replicated database provides ACID semantics, with *snapshot isolation* (SI) level. In SI, a transaction logically executes in a database snapshot. A transaction can commit if it has no write-write conflict with any committed concurrent transaction. Otherwise, it must abort.

SI is an attractive level of isolation for several reasons: it allows for increased concurrency among transactions when compared to stronger properties such as 1-copy serializability, it is implemented by many commercial database systems, it provides identical results to 1-copy serializability for many typical workloads (including the most widely used database benchmarks, TPC-A, TPC-B, TPC-C, and TPC-W) [Elnikety 2006], and there exist techniques to transform a general application program so that its execution under SI is equivalent to strict serializability [Fekete 2005].

Our system guarantees these safety properties in an asynchronous distributed system: we assume nodes are connected by an unreliable network that may fail to deliver messages, corrupt them, delay them arbitrarily, or deliver them out of order. Many database replication protocols make stronger assumptions about the timely delivery of network messages, so they can use timeouts to detect replica faults. This assumption can be problematic because when it is not met safety violations may occur. To give a simple example, in a primary-backup scheme, two machines can believe erroneously that the other has failed. This could lead to the existence of two primaries that accept new updates independently, leading to state divergence. Furthermore, this assumption can be broken either by deliberate attacks (e.g., by flooding a correct node) or by other occurrences like longer than usual garbage collection cycles that lead to increases in message processing delays [Gribble 2001].

The fact that we are ensuring safety in an asynchronous network model implies that we need to assume some form of synchrony for liveness [Fischer 1985]. Thus our system only guarantees that clients can make progress during periods when the delay to deliver a message is bounded. This assumption is referred to as eventual synchrony, and is a common assumption for liveness in replicated systems that ensure safety despite asynchrony [Castro 2002].

Note that the goal of Byzantium is to ensure correctness and high availability of the system despite arbitrary faults, and not to defend against attacks that try to violate the confidentiality of database contents. There exist, however, extensions to BFT protocols to address the problem of confidentiality [Yin 2003] that we could leverage in our work.

3.1 Database and BFT protocol requirements

Our system employs two components that are used as black boxes, but are required to provide certain semantics.

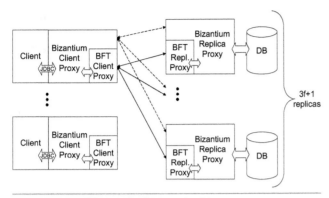

Figure 1. System Architecture.

The first component is a BFT state machine replication protocol. This protocol must implement linearizable semantics [Herlihy 1987], which is the case for most BFT state machine replication proposals [Castro 2002, Kotla 2007]. In Byzantium, we used a BFT system that requires $3f+1$ replicas, but as future work we would like to modify the implementation and evaluate the performance change when using a BFT system that only requires $2f+1$ execution replicas [Yin 2003]. This would bring advantages in terms of the aggregate machine load and the inter-replica communication costs, but also provide fewer opportunities for using striping to improve the performance of read-only transactions (which dominate several common workloads).

The other component is an off-the-shelf database system, for which we consider a standard model where the state is modified by applying transactions. A transaction is started by a BEGIN followed by a sequence of read or write operations, and ends with a COMMIT or ROLLBACK. When issuing a ROLLBACK, the transaction aborts and has no effect on the database. When issuing a COMMIT, if the commit succeeds, the effects of write operations are made permanent in the database. We require that the off-the-shelf database provides snapshot isolation semantics and supports savepoints (both of which are common in database systems).

While we do not place restrictions on the concurrency control mechanism the databases implement, one of the alternative designs we propose requires support for extracting write-sets in databases that use locking. As in other (non-BFT) replicated databases (e.g., [Elnikety 2006]) we extract write-sets from ongoing transactions using database triggers.

4. System design

In this section we present the design and algorithms used in the Byzantium system.

4.1 Architecture

Byzantium is built as a middleware system with the architecture depicted in Figure 1. The system is composed of a set of $n = 3f + 1$ server replicas and a finite number of clients.

Each replica is composed of a database system and the Byzantium replica proxy, which is linked with both a communication library and the replica-side BFT library. The communication library allows for the client to communicate directly with the replicas without going through the more expensive serialization provided by the BFT replication protocol. The communication library implements a light-weight retransmission protocol to ensure that messages are delivered in FIFO order, but, unlike BFT operations, these messages can reach different replicas in different orders. Each replica maintains a full copy of the data in an off-the-shelf database system; i.e., we use a shared-nothing architecture (to ensure fault isolation with Byzantine nodes) and we do not partition data. The replica proxy is responsible for handling client requests and controlling the execution of operations in the database system, guaranteeing that the operations execute with the desired semantics in all non-faulty replicas, and that, when the system quiesces, the state of the database in all non-faulty replicas is the same.

User applications run on client nodes and access our system using the JDBC interface. Thus, applications that access conventional database systems using a JDBC interface can use Byzantium without modification. The JDBC driver we built is responsible for implementing the client side of the Byzantium protocol. (And thus we refer to it as the Byzantium client proxy.) Some parts of the client side protocol consist of invoking operations that run through the BFT state machine replication protocol, and therefore this proxy is linked with the client side of the BFT replication library.

As mentioned, two of the components of the architecture, namely the database system and the BFT replication library, are used as black boxes. (In some cases, where write-set extraction is required, the approach may more accurately be termed "gray-box".)

Using an off-the-shelf, black-box database provides several advantages: it enables the use of third party databases that may not support replication and whose source code may not be accessible, it allows for upgrading the database server without having to update the replication code, and it allows for different replicas to run different implementations of the database server. The latter capability is important for ensuring a lower degree of fault correlation, in particular when these faults are caused by deterministic software bugs [Rodrigues 2001, Vandiver 2007]. Running distinct versions is facilitated by the fact that replicas use the JDBC interface to communicate with the database system. Thus we can easily swap between database servers that implement this interface. By default, we configured our prototype implementation to run an instance of PostgreSQL at each replica.

The other black-box component is the BFT replication library. Byzantium uses an implementation of the PBFT protocol [Castro 2002]. The PBFT library provides two main interfaces: on the client side, the library offers a "BFT_invoke" method that issues the request and returns the corresponding reply, implementing the client side of the protocol; on the server-side, the library executes the replica protocol for serializing all requests, and once a request is serialized it invokes a "BFT_execute" upcall, whereby the application-specific code executes the client request, updates the service state, and produces the corresponding reply. Note that these libraries only execute one request at a time, since they must assume that the state changes performed by a given request may affect the outcome of the next one. Since we use PBFT as a black box, we can easily switch this replication library with a different one, provided it offers the same guarantees (i.e., state machine replication with linearizable semantics) and has a similar programming interface.

4.2 Normal case operation

We start by describing the normal case in which clients and replicas are not Byzantine, and thus all nodes follow the protocols we present. We address the cases of malicious behavior in subsequent sections. The code executed by the client proxy is presented in Figure 2 and the code executed by the replica proxy is presented in Figure 3. For simplicity, the code omits some details such as error handling, message signing and optimizations. This code is used to implement two versions of the system, termed single-master and multi-master. The difference between the two versions, as far as the code in Figure 3 goes, lies only in the selection of replicas to be contacted, as explained later.

At a high-level, our approach is to only force a total order among the operations for which doing so is required to ensure that all transactions execute against the same snapshot at all replicas, i.e., the BEGIN and COMMIT/ROLLBACK operations. For these, we rely on the PBFT protocol to enforce a total order at all replicas despite Byzantine faults. The remaining operations (reads and writes) can be executed more efficiently by propagating them using the unreliable multicast mechanism, and executing them in a single replica, concurrently with other operations. The main problem then becomes how to deal with the case when that replica is Byzantine and returns wrong results. This is handled at commit time by validating the read and write results at all replicas.

In more detail, the application program starts a transaction by executing a BEGIN call on the database interface (*function db_begin*, Figure 2, line 1). The client starts by generating a unique identifier for the transaction and selecting one replica to speculatively execute the transaction – we call this replica the master replica for the transaction. (Note that it does not need to be the same primary replica that is used by the PBFT protocol.) Then, the client issues the corresponding PBFT operation that will serialize the transaction begin in all replicas by calling the *BFT_invoke(<BEGIN,...>)* method from the PBFT library. The execution of the protocol will eventually trigger the corresponding *BFT_execute* upcall (Figure 3, line 1) at all replicas. At that moment, a database transaction is started at the replica. Given the properties of the PBFT system, and since both BEGIN and COMMIT op-

erations execute serially as PBFT operations, the transaction is started in the same (equivalent) snapshot of the database in every correct replica.

After executing BEGIN, an application can execute a sequence of read and write operations (functions *db_read_op* and *db_write_op*, Figure 2, line 11 and line 24 respectively). In a write operation, the operation is multicast to all replicas (by calling *mcast*, which triggers the corresponding *receive* upcall in all replicas, shown in Figure 3, line 33). The operation is received by all replicas but only executed in the master and its result is returned to the client, who then returns the result to the application. Both the client proxy and the master replica keep a list of operations and their results.

Read operations can be executed in two different ways. If the transaction is known to be read-write (i.e., after the first write operation), the execution of a read operation is similar to the execution of a write operation. Otherwise, we perform an optimized read as described in Section 4.4.

The transaction concludes by executing a COMMIT operation (*function db_commit*, Figure 2, line 32). At commit time, it is necessary to (1) serialize all commit operations among themselves and relative to snapshots for beginning transactions, and (2) perform a series of validations, namely to confirm the outputs of the read and write operations that were executed in a single (potentially Byzantine) replica. To achieve this, the client issues the COMMIT PBFT operation that includes the digest of the operations that were issued and their results. This will trigger the *BFT_execute* upcall at all replicas (Figure 3, line 9), and the invocation of this upcall will occur in the same order relative to other BEGIN and COMMIT operations.

A transaction can also end with a ROLLBACK operation. A straightforward solution is to simply abort transaction execution in all replicas. We discuss the problems that arise with this approach when the master is Byzantine in Section 4.6.

4.3 Optimistic execution and recovery

As we explained, the read and write operations of a transaction execute optimistically in a single replica, and their results need to be subsequently validated at commit time. We developed two algorithms that differ in how the master replica is chosen and how operations execute (or not) optimistically in non-master replicas. While the single-master technique performs better for read-write dominated workloads, the multi-master version performs better when there is a large number of read-only transactions. Additionally, these techniques imply a trade-off between the support required from the database system and the performance in the presence of a Byzantine master. We discuss each of these techniques in turn.

Multi-master optimistic execution. In the multi-master version, each client can select a different replica as the master. This leads to more flexibility in terms of load balancing, and good resilience to a Byzantine master.

```
1   function db_begin() : txHandle
2       uid = generate new uid
3       (masterRep, readReps)= select(1 replica, f+1 replicas)
4       BFT_invoke(<BEGIN, uid ,(masterRep, readReps)>)
5       ops = new Map
6       readOnly = true
7       opCount = 0
8       trxHandle = new trxHandle(uid, masterRep, readReps, opCount, ops, readOnly)
9       return txHandle
10
11  function db_read_op(txHandle, op) : result
12      opNum = ++txHandle.opCount
13      mcast(<read(txHandle.uid, opNum, op)>)
14      recv(<readResult(txHandle.uid,opNum,HOp, res)>           // first result
15      txHandle.ops.put(opNum,<op,1,H(res),'read'>);
16      return res
17  background                                                  // additional results
18      recv(<readResult(txHandle.uid, opNum,HOp, res)>)
19      <_, count,HRes,_> = txHandle.ops.get(opNum)
20      if(HRes == H(res)) then
21          trxHandle.ops.put(opNum,<op,count+1,HRes,'read'>);
22      endif
23
24  function db_write_op(txHandle, op) : result
25      opNum = ++txHandle.opCount
26      mcast(<write(txHandle.uid, opNum, op)>)
27      recv(<writeResult(txHandle.uid,opNum,HOp, res)>)
28      txHandle.readOnly = false
29      txHandle.ops.put(opNum, <op,1,H(res),'write'>);
30      return res
31
32  function db_commit(txHandle)
33      concurrent
34          if(txHandle.readOnly) then
35              while LastReadConfirmed(txHandle)<txHandle.opCount
36                  wait
37              end while
38              return
39          endif
40      with
41          lastConfirmed=LastReadConfirmed(txHandle)
42          HOps = H(ListOps(txHandle))
43          HRes = H(ListRes(txHandle, lastConfirmed))
44          res = BFT_invoke(<COMMIT, txHandle.uid, lastConfirmed ,HOps,HRes>)
45          if(res == true) then
46              return
47          else
48              throw ByzantineExecutionException
49          endif
50      end concurrent
```

Figure 2. Byzantium client proxy code.

```
1   upcall for BFT_exec(<BEGIN, uid ,( masterRep , readReps )>)
2       DBTxHandle = db. begin ()
3       ops = new Map
4       readOnly = true
5       txSrvHandle = new txSrvHandle ( uid , DBTxHandle , masterRep , readReps ,
6                                       ops , readOnly )
7       openTxs . put ( uid , trxSrvHandle )
8
9   upcall for BFT_exec(<COMMIT, uid , lastConfirmed , cltHOps , cltHRes >) : boolean
10      txSrvHandle = openTxs . get ( uid )
11      openTxs . remove ( uid )
12      if ( NOT ThisIsMasterReplica ( txSrvHandle )) then
13          execOK = exec_and_verify ( txSrvHandle . DBTxHandle , lastConfirmed ,
14                                      txSrvHandle . ops , cltHOps , cltHRes )
15          if (NOT execOK ) then
16              DBTxHandle . rollback ()
17              return false
18          endif
19      endif
20      return DB_trx_handle . commit ()
21
22  upcall for recv(<read ( uid , opNum , op)>)
23      txSrvHandle = openTxs . get ( uid )
24      txSrvHandle . ops . put ( opNum , <op,->)
25      if (( txSrvHandle . readOnly AND ThisIsReadReplica ( txSrvHandle ))
26              OR ( NOT txSrvHandle . readOnly AND
27                          ThisIsMasterReplica ( txSrvHandle ))) then
28          result = txSrvHandle . DBTxHandle . exec (op)
29          txSrvHandle . ops . put ( opNum , <op, result >);
30          send_reply (<readResult ( uid , opNum ,H( op ), result )>)
31      endif
32
33  upcall for recv(<write ( uid , opNum , op)>)
34      txSrvHandle = openTxs . get ( uid )
35      txSrvHandle . ops . put ( opNum , <op,->)
36      if ( txSrvHandle . masterRep == THIS_REPLICA ) then
37          if (( txSrvHandle . readOnly AND NOT ThisIsReadReplica ( txSrvHandle )) then
38                  ExecReadPrefix ( trxHandle . ops )
39          endif
40          result = txSrvHandle . DBTxHandle . exec ( op)
41          txSrvHandle . ops . put ( opNum , <op, result >)
42          send_reply (<writeResult ( uid , opNum ,H( op ), result )>)
43      endif
44      txSrvHandle . readOnly = false
```

Figure 3. Byzantium replica proxy code.

In this version of the protocol, the master is selected in the beginning of the transaction either at random or following a more sophisticated load-balancing scheme [Elnikety 2007]. Subsequent reads and writes are then performed optimistically at the master replica, which, as we pointed out, may be Byzantine and return incorrect results. Therefore, when the transaction attempts to commit there are two validation steps that need to be performed: ensuring that the results the client observed were correct, and that the transaction can commit according to SI.

For the former, all non-master replicas have to execute the remaining transaction operations and verify that the returned results match the results previously output by the master (Figure 3 line 12). (For now we assume that these operations were received by all replicas, and in Section 4.5.2 we explain how to handle the case where replicas do not receive them, either due to message loss or to a Byzantine client.) Since the transaction executes in the same snapshot in every replica (due to the fact that both BEGIN and COMMIT operations are serialized by PBFT), if the master and client were correct, all other correct replicas should obtain the same results. In case the results do not match, either the client or the master was Byzantine and we rollback the transaction.

Additionally, all replicas including the master must guarantee that SI properties hold for the committing transaction. The way this is done depends on the concurrency control mechanism of the underlying database system.

For database systems with optimistic concurrency control, guaranteeing SI is immediate. As a transaction executes in the same snapshot and commits in the same order relative to other commits in all replicas, the validation performed by the database system suffices to guarantee that a transaction commit succeeds in all replicas or in none (when a conflicting transaction has previously committed). With these database systems, all operations can be executed optimistically in all replicas, reducing the work needed at commit.

However, most database systems rely on locks for concurrency control, including the main system we used to test our prototype, PostgreSQL. In such systems, before executing a write operation, the transaction must obtain a lock on rows to be written. When a replica is acting as master for some ongoing transaction, it will obtain locks on rows it changes. This can block the local execution of a committing transaction that has written in the same rows, and ultimately lead to a deadlock. We address this problem by temporarily undoing all operations of the ongoing transaction. After executing the committing transaction, if the commit succeeds, the ongoing transaction is rolled back due to a conflict. Otherwise, we replay the undone operations and the ongoing transaction execution may proceed.

To achieve this, we rely on the widely available transaction *savepoint* mechanism. Savepoints enable rolling back all operations executed inside a running transaction after the savepoint is established. Thus, when the BEGIN operation executes, a savepoint is created in the initial database snapshot. Later, when we need to undo the operations that were executed in the transaction but still use the same database snapshot, we just need to rollback the transaction to the savepoint that was previously created.

To know which local transaction would block a committing transaction, we use approaches similar to non-BFT replicated databases with SI semantics [Elnikety 2006] - we further discuss this issue in Section 6.

Single-master optimistic execution. Unlike in the multimaster protocol, where we have to extract write-sets in databases the use locking, in the single-master case we can avoid this by optimistically executing transactions in the same single node. This requires that clients and replicas agree on the single master that should be executing reads and writes in the course of transactions. This is achieved by augmenting the service state maintained by the PBFT service with this information, and augmenting the service interface with special operations to enable changing the current master. The properties of PBFT will then ensure that all clients and replicas agree on which replica is acting as a master, and this can be communicated to the client as part of the output of the BEGIN PBFT operation.

Given this agreement, validating the SI properties can be done in a straightforward manner just by executing read and write operations of each transaction when it commits. In this case, as transactions commit serially in all replicas, if the transaction can commit in the master, it will be able to commit in all non-master replicas independently of the concurrency control mechanism that is used.

This scheme can be extended to allow transactions to also execute speculatively in non-master replicas before commit time. In databases using optimistic concurrency control, operations can be broadcast to all replicas, which speculatively execute them as they are received. However, for this optimization to work in lock-based database systems, we must guarantee that the same transactions obtain the same set of locks in all replicas – otherwise, some committing transaction would not be able to obtain the needed locks. Guaranteeing this without controlling the internals of the database system requires guaranteeing that operations are issued in some given order to the database system. To ensure this, HRDB [Vandiver 2007] proposes the use of commit barriers controlled by a centralized coordinator for this purpose. In our system we implement a similar idea of using information provided by the master to coordinate other replicas, guaranteeing that all operations of a transaction but the last one can execute speculatively in non-master replicas before commit-time. However, unlike HRDB, if the master is suspected to be faulty, another replica is selected to act as primary.

In particular, our approach leverages the fact that, for transactions executing speculatively at non-master replicas, if an operation completes in the master, it can execute in other replicas without blocking. This follows from the fact that any other operation that would require the same locks would have blocked in the master. Thus, when a non-master replica receives an operation op_n from t_1, it knows that it can execute operation op_{n-1} from transaction t_1 (because op_{n-1} has not blocked in the master). However, this condition is not sufficient to avoid blocking when a transaction commits – e.g. suppose an operation op of t_2 executes at the master after the master committed t_1. If a non-master replica executes op before running some operation of t_1 that requires the same locks as op, t_1 would be unable to commit. To address this problem, we must guarantee that if an operation was executed after the commit of t_n at the master replica, that same operation will execute in all non-master replicas after committing t_n (this was known as the *Transaction-ordering rule* in HRDB). To guarantee this, the message with the result of a write operation, op_n, includes the number of the last committed transaction, t_m, at the master (serialized by the PBFT protocol). The message that propagates a read or write operation op_{n+1}, includes the value received in the result of the last write operation. Therefore, we can enforce the necessary order by imposing that operation op_n executes at non-master replicas only after that replica has executed the commit of t_m.

4.4 Read-only optimizations

When transactions begin, we assume they are read-only until the first write operation. While the transaction is flagged as read-only, we employ the following optimization to improve the performance of read-only transactions (and of read-only prefixes of other transactions). Read requests are executed in $f+1$ replicas (chosen randomly when the transaction begins) and the result from the first reply is returned to the client, while the remaining replies are collected in the background (Figure 2, lines 17-22). When the f additional replies that are received in the background match the first reply, the result of the read is considered to be confirmed without the need for executing the operation in any additional replica.

At commit time, if all returned values were correctly validated by $f+1$ replicas, the client immediately returns the commit successfully. In the case that some reads were not yet validated, the commit procedure falls back to the original, unoptimized version, which is run in parallel with trying to conclude the optimized validation. In the normal case, when the $f+1$ replicas reply at similar speeds, operations are confirmed by the optimized protocol. When a write operation occurs, the transaction is promoted to read-write, and starts executing the normal protocol. However, the confirmed read operations executed prior to the first write will not be included in the final commit-time validation.

This scheme enables a form of striping and load-balancing, since read-only transactions only execute their read operations in $f+1$ of the $3f+1$ replicas. For providing efficient load balancing, the selection of the replicas that execute the reads (Figure 2, line 3) could follow one of the various existing proposals (e.g., [Elnikety 2007]). In our prototype, the $f+1$ replicas are selected randomly with the constraint that, in the multi-master version, the $f+1$ replicas always include the master. This has the advantage that, when a transaction is upgraded to read-write, the master replica has already executed all previous operations and is ready to proceed with the execution of the subsequent operations. In the single-master version, the $f+1$ replicas selected to execute read operations do not include the master (as in Ganymed [Plattner 2004]). The rationale for this approach is to reduce the load in the master node – otherwise, the master node would have to execute all transactions. The downside of this approach is

that when a transaction is upgraded to read-write, the master node needs to execute all operations whose results are not known to be guaranteed, if any.

4.5 Tolerating Byzantine faults

So far we have mostly assumed that nodes follow the protocol. In this section we explain how the system handles Byzantine behavior, starting with the assumption that only replicas may exhibit Byzantine behavior, and later addressing the case of Byzantine clients.

4.5.1 Tolerating a faulty master

A faulty master can return erroneous results or fail to return any results to the clients. The first situation is addressed by having all replicas verify at commit time the correctness of results returned by the master. If the results of executing the operations in the transaction do not match the results that the client observed (and whose digests are passed as an argument to the PBFT COMMIT operation), the replicas will rollback the transaction and the client will throw an exception signaling Byzantine behavior. This guarantees that correct replicas will only commit transactions for which the master has returned correct results for every operation.

Addressing the case where a master fails to reply to an operation requires different solutions, depending on whether a single-master or a multi-master approach is used.

Multi-master In the multi-master version, if the master fails to reply to an operation, the client selects a new master to replace the previous one and starts by re-executing all previously executed transaction operations in the new master. If the obtained results do not match, the client rollbacks the transaction by executing a ROLLBACK operation and throws an exception signaling Byzantine behavior. If the results match, the client proceeds by executing the next operation in the new master.

Because of this mechanism, there may be situations where the master is unaware that it did not execute the entire transaction (e.g., if the client switched to a new master due to temporary unreachability of the original master). To handle this, at commit time, a replica that believes itself to be the master of a transaction must still verify that the sequence of operations sent by the client is the same as the sequence that it has itself executed. Thus, if the master that was replaced is active, it will find out that additional operations exist and will execute them.

In subsequent transactions, a client will not select as master a replica that it suspects of exhibiting Byzantine behavior.

Single-master In this scheme, if the master fails to reply to an operation, the client will forward the request to all replicas. Each replica will try to forward the request to the master on behalf of the client. If the master replies, the replica will forward the reply to the client. Otherwise, the replica will suspect the master, and request a master change.

A replica starts a master change by submitting a PBFT *master change* operation (as a consequence of a suspicion of Byzantine behavior). When $f + 1$ *master change* operations from different replicas are executed concerning the same master, a new master is automatically elected – in our prototype, replicas are numbered, and the next master replica is selected in a round-robin fashion. In this case, all ongoing transactions are marked for rollback. When executing the next operation, the client will be informed that the transaction will rollback.

Aside from this situation, a replica will also request a master change when, during the execution of the commit, the results observed by the client do not match the local results. This raises the possibility of Byzantine clients using this mechanism to cause false positives and trigger constant master changes. The next section discusses this and other avenues that Byzantine clients may use to cause the system to malfunction.

4.5.2 Tolerating Byzantine clients

The system also needs to handle Byzantine clients that might try to cause the replicated system to deviate from the intended semantics. Note that we are not trying to prevent a malicious client with legitimate access from writing incorrect data, or deleting entries in the database. Such attacks can only be limited by enforcing security/access control policies and maintaining database snapshots that can be used for data recovery [King 2005]. What we are trying to prevent are violations of database semantics or service availability due to clients that do not follow the protocol we described.

Since PBFT already ensures that Byzantine clients cannot affect the system other than by invoking operations through the service interface, our system only needs to address violations of the remaining protocols that are used.

An obvious check that replicas need to perform is whether they are receiving a valid sequence of operations from each client. These are simple checks, such as verifying that a BEGIN is always followed by a COMMIT/ROLLBACK and that the unique identifiers that are sent are valid.

There are, however, more subtle deviations that could be exploited by Byzantine clients. One avenue of attack follows from that fact that during a transaction operations are multicast to all replicas, and, at commit time, the client propagates a digest of operations and results to all replicas, but not the operations themselves. A Byzantine client could exploit this behavior by sending different sets of operations to different replicas. (A similar possibility is that some of the messages containing operations are lost and do not reach some of the replicas by commit time.) The consequence would be that at commit time, only those replicas that had a sequence that matched the digests would commit the transaction, leading to divergent database states at different replicas.

To address this problem, while avoiding a new round of messages among replicas during correct executions, we leverage a PBFT protocol mechanism that enables replicas to

agree on non-deterministic choices [Rodrigues 2001]. This feature was originally used for replicas to agree on things such as the current clock value. In this mechanism, the primary proposes a value for the non-deterministic choices and replicas that disagree with that value can reject it. If there is no set of $2f + 1$ replicas that accept that choice then the operation is not executed, a primary change will take place, and the new primary can then propose a different value.

We use this mechanism to allow replicas to vote on whether they have the sequence of operations that match the digests sent by the client. If $2f + 1$ replicas agree on the fact that they hold all the operations, the PBFT operation will proceed with transaction commitment. Correct replicas that were not in this set and do not have the correct sequence of operations must obtain it from other replicas. If the primary believes it does not have the sequence of operations and $2f + 1$ replicas agree on this fact (if the client sent incorrect digests, for instance) then the PBFT operation proceeds with all replicas rolling back the transaction. Otherwise, if there is no agreement among any set of $2f + 1$ nodes, the PBFT protocol automatically initiates a primary change and the new primary will repeat the process. In parallel, correct replicas that have the right sequence of operations will multicast them to all replicas. This ensures liveness, since either a correct replica has the set of operations and will eventually propagate them to all replicas, or the correct replicas will eventually agree on the fact that they do not have access to the operations and the transaction will rollback.

Another possible consequence of Byzantine clients is that the master could be forced to discard its previously executed sequence of operations. This would be the case if the set of values sent in the commit operation is accepted by $2f + 1$ replicas but these values do not correspond to the values sent to the master before committing. In this case we need to allow the master to undo the executed operations and execute the new sequence in the original snapshot. To this end, we set up a savepoint when the transaction starts. Later, if the master finds that the Byzantine client had sent other replicas a different set of values that match the commit digests, then the transaction is rolled back to the savepoint that was previously created before executing the new sequence of operations. This ensures that all replicas, including the master, execute the same operations on the same database snapshot, guaranteeing the correct behavior of our system.

Another possible point of exploitation arises if a Byzantine client send an incorrect digest for the results, leading all replicas but the master to rollback the transaction. To address this case, the master checks the received digest and rollbacks the transaction if a Byzantine behavior is detected.

Finally, to avoid the aforementioned problem of clients constantly changing the master due to false accusations in the single master approach, we can deploy a mechanism by which replicas suspect a client that causes too many master changes, and that client is forced to ask for a receipt, signed by the master replica, of the operation results that were returned to it before being able to cause a new master change. The operation to request a receipt can be handled just like read and write operations.

4.6 Handling rolled back transactions

When a transaction ends with a ROLLBACK operation, a possible approach is to simply rollback the transaction in all replicas without verifying if previously returned results were correct (e.g., this solution is adopted in [Vandiver 2007]). In our system, this could be easily implemented by executing a PBFT operation that rollbacks the transaction in each replica.

This approach does not lead to any inconsistency as the replicas are not modified. However, in case of a faulty master, the application might have received an erroneous result, leading to the decision to rollback the transaction. For example, consider a transaction trying to reserve a seat in a flight that has seats available. When the transaction queries the database for available seats, a Byzantine master might incorrectly return that none is available. As a consequence, the application program may end the transaction with a ROLL-BACK. If no verification of the results that were returned was performed, the client would have made a decision to rollback the transaction based on an incorrect database state.

To detect this, we decided to include an option to force the system to verify the correctness of the returned results even when a transaction ends with a ROLLBACK operation. When this option is activated, the execution of a rollback becomes similar to the execution of a commit (with the obvious difference that the transaction always rollbacks). If the verification fails, the ROLLBACK operation raises an exception. Note that a correct program should include the code to catch all exceptions raised by a database operation and take appropriate action depending on the content of the exception.

5. Correctness

In this section we sketch a proof that our design meets safety and liveness conditions.

The safety part of our correctness conditions states that transactions that are committed on the replicated database observe SI semantics. This follows from the linearizable semantics of PBFT [Castro 2002], and the fact that all BEGIN and COMMIT operations are serialized by PBFT and thus execute in the same total order at all non-faulty replicas. This, coupled with the fact that the output of the commit operation only depends on the sequence of begin and commit operations that happened previously, which is the same at all non-faulty replicas, implies that the output of commits will be the same as the output of the local commit at each non-faulty replica. Note that the output of commit is independent of the other operations that are not serialized through PBFT

because commit operations carry as argument the sequence of values that were read and written by the transaction.

Given this point, the proof that the system obeys SI semantics follows from the fact that each non-faulty replica applies the begin and commit operations on their local database that provides SI semantics and forwards the reply from the database, and that the commit validates all the outputs that the client received and applies all the updates that the client issued during the transaction.

Read-only transactions are a special case since the commit does not require invoking a PBFT operation, but these also conform to SI semantics, since when these transactions begin they run a PBFT begin operation, hence establishing a position for this transaction in the serial order of committed transactions at all non-faulty replicas, as stated above. Since the values that are read are confirmed by $f + 1$ replicas, there is at least one non-faulty replica in that set that will return a value that is correct according to the SI semantics and the total order set by the PBFT begin operation.

For liveness, we need to ensure that operations that are initiated by the client are eventually executed. This requires the same assumptions that are required for liveness of the PBFT protocol, which is that after some unknown point in the execution of the program messages message delays do not grow superlinearly [Castro 2002].

Given this assumption, and due to PBFT's liveness condition, we can guarantee that the BEGIN, COMMIT, and ROLL-BACK operations are eventually executed. Furthermore, operations that do not go through the PBFT protocol are simple RPCs which are eventually executed under the same set of assumptions. The execution of these operations does not block, by algorithm construction, and thus we can guarantee that all client operations eventually get executed.

6. Implementation

We implemented a prototype of Byzantium in Java. We developed a Java-based PBFT implementation, and our proxies use the JDBC interface to interact with the application and the underlying database system. We also built a communication library providing FIFO semantics and message authentication using Message Authentication Codes (MACs).

We use several techniques proposed in other middleware database replication systems [Cecchet 2008, Elnikety 2006, Rodrigues 2001, Vandiver 2007] in our system. We make non-deterministic database operations (e.g. select) deterministic by rewriting the operation and/or overwriting the non-deterministic components of each reply.

We implemented two mechanisms to avoid deadlocking in our multi-master version: one that relies on the extraction of write-sets using database triggers, and another that relies on the analysis of SQL code. These mechanisms work by maintaining the write-sets of on-going transactions and, before executing a remote operation, verify if it would conflict with local transaction. Sometimes it is not possible to verify this from the write code – in such situations, it is possible to obtain the additional needed information by relying on the SELECT ... FOR UPDATE NOWAIT SQL statement. Our experiments have shown that neither approach offers a clear performance advantage over the other.

Since Byzantium was designed to work with any database that supports the JDBC interface and provides Snapshot Isolation, we tried it with two different database implementations: PostgreSQL and HyperSQL[1], a Java-based database system that implements SI. In HyperSQL, the single-master version worked without changes, but for the multi-master version it was necessary to develop a workaround to address an unimplemented method of the JDBC interface. The performance of HyperSQL with the TPC-C benchmark configuration we used was much lower than that of PostgreSQL. Thus, in our evaluation, we only report the performance with PostgreSQL.

7. Evaluation

In this section, we evaluate the performance of our prototype. Our tests were performed with $f = 1$, leading to $n = 3f+1 = 4$ replicas. Studies show that this configuration is sufficient to mask almost all reported database bugs [Gashi 2007]. The tests were run on a cluster of machines, each one with a single-core 2.6 GHz AMD Opeteron 252 processor, 4 GB of memory, a 146 GB Ultra320 SCSI disk and 1 Gigabit ethernet ports. The machines were connected by a Nortel Ethernet Routing Switch 5520. The machines were running the Linux operating system, kernel version 2.6.30. The database used was PostgreSQL version 8.3.4, in synchronous commit mode to guarantee reliability. The JVM we used was Sun VM version 1.6.0_12.

The evaluation used an open-source implementation of TPC-C[2]. We made slight modifications to the benchmark, namely to include a warm-up phase before starting the performance measurements, and to allow clients to execute on different machines. An important point is that our benchmark does not use database batches, which makes it more demanding for the communication protocols and leads to worse performance when compared to the solution using batches. Our database configuration included 10 warehouses and an inter-transaction arrival time of 200 ms. The experiments show the average of 5 runs, with the error bars showing the lowest and highest value.

The goal of our experiments is to evaluate the overhead of providing BFT replication and the efficiency of the two versions of our protocols. For this, we compare the Byzantium multi-master, *Byz-N*, and single master, *Byz-1*, versions to both a non-replicated *proxy*-based solution and a *full BFT* replicated system where all operations are serialized by executing each operation as a PBFT operation. The *proxy* solution uses a proxy that relays the connections from all clients

[1] http://hsqldb.org/

[2] http://sourceforge.net/projects/benchmarksql/

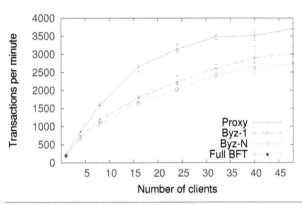

Figure 4. Performance on standard TPC-C workload (with no batches). Note that Full BFT has only a single data point, as expected, since concurrency leads to deadlock.

to a single database server using JDBC connections. The solution is multi-threaded, with each client being handled concurrently by a dedicated thread. This reflects the performance of a non-replicated database server, while only adding the overhead of implementing a Java-based middleware solution. We use the *proxy* and *full BFT* solutions as comparison points, because they represent the best and worst case of what an implementation using our code base is expected to achieve, with the former incurring no BFT overhead and the latter incurring the overhead of running PBFT for every operation.

In our experiments, we have used the mechanism that avoids deadlocks without using database triggers. We also disabled the mechanism to verify the correctness of the execution of rolled back transactions.

7.1 TPC-C standard: read-write dominated workload

Figure 4 presents the performance results obtained with the standard TPC-C workload, consisting of 92% read-write transactions and 8% read-only transactions. The results show that the performance of our versions is between 20% to 35% lower than the *proxy* solution.

There are two main reasons for this overhead. First, the workload consists mostly of read-write transactions, for which both read and write operations must execute at all replicas. (In fail-stop replication, part of this cost is avoided as it suffices to read from a single replica.) Thus, this workload introduces overhead associated with the replication protocols that is not compensated by any form of load-balancing. This overhead could have been minimized if transactions had long prefixes of read-only operations, as explained in section 4.4. However, TPC-C read-write transactions have very small prefixes of read-only operations – e.g., the new order transaction, which may include over 50 operations, has a prefix of two read operations before the first write operation.

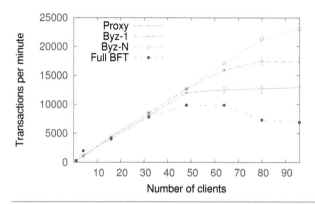

Figure 5. Performance on read-only workload, based on TPC-C transactions with no batches.

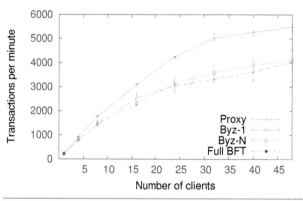

Figure 6. Performance on mixed workload, based on TPC-C transactions with no batches. Full BFT has only a single data point, as expected, since concurrency leads to deadlock.

When compared with the *full BFT* solution, our solution performs 5% better for a single client. For a larger number of clients, the *full BFT* solution always blocked. This was expected, as we did not include any mechanism to prevent a PBFT operation from blocking in the database when trying to acquire a lock that is taken (as we have in *Byz-N*).

When comparing our two versions, the difference tends to be about 10% and slightly increasing with the number of clients, with the single master version performing better. The reason for this lies in the optimization mechanism of the single-master version, which speculatively executes operations in all replicas, thus minimizing the execution time for the commit operation.

Two additional results are important. First, the induced rollback ratio due to concurrency problems increases with the number of clients, but it is similar for both *Byzantium* versions and the *proxy* solution (with a difference of less than 3% in all studied scenarios).

The second result worth mentioning is the time to execute a transaction. In this case, the results vary depending on the transaction type. Read-only transactions run up to $1.3\times$ faster in *Byzantium* versions than in *proxy* – we will discuss the reasons for this in more detail in the next section. Read-write transactions run slower in *Byzantium* than in *proxy* by a factor of 0.7 or better. This is due to the additional stages introduced by the replication algorithm.

7.2 Read-only workload

Figure 5 presents the performance results obtained with a modified TPC-C workload consisting of only read-only transactions, with 50% for each type of read-only transaction (check inventory level and check order status).

The results show that *Byz-N* improves on the performance of *proxy* by up to 10% when the number of clients is smaller than 32, and this improvement increases up to 90% with 96 clients. The main reason for this is related to the load of replicas. In the multi-master version, since operations of read-only transactions tend to execute only in $f + 1$

replicas, the load of each replica is half of the load when running *proxy*. Thus, by load balancing, our solution is able to achieve close to optimal throughput, almost doubling the result of *proxy*.

The performance of *Byz-1* is better than *proxy* and worse than *Byz-N*. In the case of the single master protocol, the load of replicas other than the master is at most $\frac{2}{3}$ of the load when running *proxy*.

As expected, the results show that the *full BFT* performance is the worst of the four setups, and the differences increase rapidly with the number of clients. This is due to having to use the PBFT protocol for executing each operation, and demonstrates the need for minimizing the use of this protocol, as proposed by our solution.

7.3 Mixed workload

Figure 6 presents the performance results obtained with a modified TPC-C workload consisting of 50% read transactions (with 25% for both check inventory level and check order status) and 50% write transactions (with 27% for new order, 21% for payment and 2% for delivery, keeping the original ratio among write transactions). The results show that the performance of our versions is between 20% and 30% lower than the *proxy* solution with up to 48 clients. This represents a slightly lower overhead when compared with the write-dominated workload. In this case, the results for *Byz-N* and *Byz-1* are very similar, as a result of the improved performance of the multi-master version for read-only transactions.

7.4 Byzantine behavior

Next, we evaluate the performance of Byzantium in the presence of Byzantine faults caused by database bugs. To this end, we have changed the code of the Byzantium replica proxy to simulate incorrect results from the database. The implementation is very simple: for each data item read, it randomly changes the returned value with a pre-defined

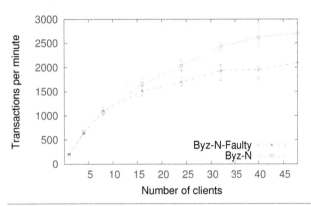

Figure 7. Performance on the presence of a single Byzantine server, using TPC-C standard workload .

probability. Although this scenario represents a rather benign case of Byzantine behavior, it is one that is probably more likely to happen in practice, with a replica sending the same incorrect messages to every other node. In the scenario of a malicious replica, the performance penalty could be higher, particularly in the single-master case.

We have measured the throughput of the system with the standard TPC-C benchmark, a single incorrect replica and an error probability of 10% in the period after the server starts exhibiting Byzantine behavior. We only present the results for our multi-master design – the results with a single master show a similar pattern.

Figure 7 presents the results, with *Byz-N-Faulty* representing the throughput with a Byzantine replica. The lower throughput was expected for two reasons. First, the load is divided among a smaller number of replicas. Second, in all PBFT protocol steps, the messages from the Byzantine replica cannot be used for obtaining the required conditions, thus delaying the execution of the protocol.

A more interesting and surprising result is related to the number of rolled back transactions, which increased to about 5% of all executed transactions. This result seemed to suggest that our mechanism to change the master replica worked very slowly. However, after analyzing the experiments, we discovered that a large number of transactions with incorrect results were rolled back by the benchmark code and not due to the detection of Byzantine behavior by our algorithms. The reason for this is that the transaction code often uses the results from a previous operation as a parameter in the subsequent operations. When the previous returned result is incorrect, the subsequent operation fails and the benchmark ends up rolling back the transaction. These results also show the importance of the mechanism to verify the execution of rolled back transactions introduced in Section 4.6.

8. Conclusions

We presented Byzantium, the first proposal for middleware-based BFT replication of database systems that allows for concurrent execution of database operations and does not rely on centralized components. Byzantium shows that it is possible to obtain the strong assurances that derive from using a Byzantine fault model, while paying only a modest penalty in terms of performance overhead. We showed how to minimize the use of the expensive BFT operations using two different techniques for optimistic execution, and how to optimize the execution of read-only transactions.

We evaluated Byzantium and our results show that replication introduces only a modest performance overhead for read-write dominated workloads and we perform up to 90% better than a non-replicated database system for read-only workloads. Our single-master version performs better in read-write dominated workloads while the multi-master version performs better with a large number of read-only transactions.

In the future, we intend to deploy different database systems in different replicas, and to explore the use of other BFT protocols as their implementations become available.

Acknowledgments

We would like to thank the anonymous reviewers and our shepherd, Ozalp Babaoglu, for the time they dedicated to providing valuable feedback on earlier versions of this paper. This work was partially supported by CITI and FCT/MCTES project # PTDC/EIA/74325/2006.

References

[Bernstein 1986] Philip A. Bernstein, Vassos Hadzilacos, and Nathan Goodman. *Concurrency Control and Recovery in Database Systems*. Addison-Wesley Longman Publishing Co., Inc., Boston, MA, USA, 1986. ISBN 0-201-10715-5.

[Castro 2002] Miguel Castro and Barbara Liskov. Practical Byzantine Fault Tolerance and Proactive Recovery. *ACM Trans. Comput. Syst.*, 20:398–461, November 2002.

[Cecchet 2008] Emmanuel Cecchet, George Candea, and Anastasia Ailamaki. Middleware-based Database Replication: The Gaps Between Theory and Practice. In *Proceedings of the 2008 ACM SIGMOD International Conference on Management of Data*, SIGMOD '08, pages 739–752. ACM, 2008.

[Clement 2009] Allen Clement, Manos Kapritsos, Sangmin Lee, Yang Wang, Lorenzo Alvisi, Mike Dahlin, and Taylor Riche. Upright Cluster Services. In *Proceedings of the ACM SIGOPS 22nd Symposium on Operating Systems Principles*, SOSP '09, pages 277–290. ACM, 2009.

[DISA 2004] Defense Information Systems Agency DISA. Database security technical implementation guide - version 7, release 1. White paper available at databasesecurity.com, October 2004.

[Elnikety 2006] Sameh Elnikety, Steven Dropsho, and Fernando Pedone. Tashkent: Uniting Durability With Transaction Ordering for High-performance Scalable Database Replication. In *Proceedings of the 1st ACM SIGOPS/EuroSys European Conference on Computer Systems 2006*, EuroSys '06, pages 117–130. ACM, 2006.

[Elnikety 2007] Sameh Elnikety, Steven Dropsho, and Willy Zwaenepoel. Tashkent+: Memory-aware Load Balancing and Update Filtering in Replicated Databases. In *Proceedings of the 2nd ACM SIGOPS/EuroSys European Conference on Computer Systems 2007*, EuroSys '07, pages 399–412, New York, NY, USA, 2007. ACM.

[Elnikety 2005] Sameh Elnikety, Willy Zwaenepoel, and Fernando Pedone. Database Replication Using Generalized Snapshot Isolation. In *Proceedings of the 24th IEEE Symposium on Reliable Distributed Systems*, pages 73–84. IEEE Computer Society, 2005.

[Fekete 2005] Alan Fekete, Dimitrios Liarokapis, Elizabeth O'Neil, Patrick O'Neil, and Dennis Shasha. Making Snapshot Isolation Serializable. *ACM Trans. Database Syst.*, 30:492–528, June 2005.

[Fischer 1985] Michael J. Fischer, Nancy A. Lynch, and Michael S. Paterson. Impossibility of Distributed Consensus with One Faulty Process. *J. ACM*, 32:374–382, April 1985.

[Fonseca 2010] Pedro Fonseca, Cheng Li, Vishal Singhal, and Rodrigo Rodrigues. A Study of the Internal and External Effects of Concurrency Bugs. In *Proceedings of the 40th IEEE/IFIP International Conference on Dependable Systems and Networks*, DSN 2010. IEEE, July 2010.

[Garcia Molina 1986] Hector Garcia Molina, Frank Pittelli, and Susan Davidson. Applications of byzantine agreement in database systems. *ACM Trans. Database Syst.*, 11:27–47, March 1986.

[Gashi 2007] Ilir Gashi, Peter Popov, and Lorenzo Strigini. Fault Tolerance via Diversity for Off-the-Shelf Products: A Study with SQL Database Servers. *IEEE Trans. Dependable Secur. Comput.*, 4:280–294, October 2007.

[Gribble 2001] Steven D. Gribble. Robustness in Complex Systems. In *Proceedings of the Eighth Workshop on Hot Topics in Operating Systems*, pages 21–26. IEEE Computer Society, 2001.

[Herlihy 1987] M. P. Herlihy and J. M. Wing. Axioms for Concurrent Objects. In *Proceedings of the 14th ACM SIGACT-SIGPLAN Symposium on Principles of Programming Languages*, POPL '87, pages 13–26. ACM, 1987.

[King 2005] Samuel T. King and Peter M. Chen. Backtracking Intrusions. *ACM Trans. Comput. Syst.*, 23:51–76, February 2005.

[Kotla 2007] Ramakrishna Kotla, Lorenzo Alvisi, Mike Dahlin, Allen Clement, and Edmund Wong. Zyzzyva: Speculative Byzantine Fault Tolerance. In *Proceedings of twenty-first ACM SIGOPS Symposium on Operating Systems Principles*, SOSP '07, pages 45–58. ACM, 2007.

[Lamport 1982] Leslie Lamport, Robert Shostak, and Marshall Pease. The Byzantine Generals Problem. *ACM Trans. Program. Lang. Syst.*, 4:382–401, July 1982.

[Lin 2005] Yi Lin, Kem Bettina, Marta Patiño Martínez, and Ricardo Jiménez-Peris. Middleware Based Data Replication Providing Snapshot Isolation. In *Proceedings of the 2005 ACM SIGMOD International Conference on Management of Data*, SIGMOD '05, pages 419–430. ACM, 2005.

[Nightingale 2005] Edmund B. Nightingale, Peter M. Chen, and Jason Flinn. Speculative Execution in a Distributed File System. In *Proceedings of the twentieth ACM Symposium on Operating Systems Principles*, SOSP '05, pages 191–205. ACM, 2005.

[Plattner 2004] Christian Plattner and Gustavo Alonso. Ganymed: Scalable Replication for Transactional Web Applications. In *Proceedings of the 5th ACM/IFIP/USENIX International Conference on Middleware*, Middleware '04, pages 155–174. Springer-Verlag New York, Inc., 2004.

[Preguiça 2008] Nuno Preguiça, Rodrigo Rodrigues, Cristóvão Honorato, and João Lourenço. Byzantium: Byzantine-fault-tolerant Database Replication Providing Snapshot Isolation. In *Proceedings of the Fourth conference on Hot Topics in System Dependability*, HotDep'08, pages 9–9. USENIX Association, 2008.

[Rodrigues 2001] Rodrigo Rodrigues, Miguel Castro, and Barbara Liskov. BASE: Using Abstraction to Improve Fault Tolerance. In *Proceedings of the eighteenth ACM Symposium on Operating Systems Principles*, SOSP '01, pages 15–28. ACM, 2001.

[Schneider 1990] Fred B. Schneider. Implementing Fault-tolerant Services Using the State Machine Approach: A Tutorial. *ACM Comput. Surv.*, 22:299–319, December 1990.

[Vandiver 2007] Ben Vandiver, Hari Balakrishnan, Barbara Liskov, and Sam Madden. Tolerating Byzantine Faults in Transaction Processing Systems Using Commit Barrier Scheduling. In *Proceedings of twenty-first ACM SIGOPS Symposium on Operating Systems Principles*, SOSP '07, pages 59–72. ACM, 2007.

[Yin 2003] Jian Yin, Jean-Philippe Martin, Arun Venkataramani, Lorenzo Alvisi, and Mike Dahlin. Separating Agreement From Execution for Byzantine Fault Tolerant Services. In *Proceedings of the nineteenth ACM Symposium on Operating Systems Principles*, SOSP '03, pages 253–267. ACM, 2003.

ZZ and the Art of Practical BFT Execution

Timothy Wood, Rahul Singh, Arun Venkataramani,
Prashant Shenoy, and Emmanuel Cecchet

University of Massachusetts Amherst
{twood,rahul,arun,shenoy,cecchet}@cs.umass.edu

Abstract

The high replication cost of Byzantine fault-tolerance (BFT) methods has been a major barrier to their widespread adoption in commercial distributed applications. We present ZZ, a new approach that reduces the replication cost of BFT services from $2f + 1$ to practically $f + 1$. The key insight in ZZ is to use $f + 1$ execution replicas in the normal case and to activate additional replicas only upon failures. In data centers where multiple applications share a physical server, ZZ reduces the aggregate number of execution replicas running in the data center, improving throughput and response times. ZZ relies on virtualization—a technology already employed in modern data centers—for fast replica activation upon failures, and enables newly activated replicas to immediately begin processing requests by fetching state on-demand. A prototype implementation of ZZ using the BASE library and Xen shows that, when compared to a system with $2f + 1$ replicas, our approach yields lower response times and up to 33% higher throughput in a prototype data center with four BFT web applications. We also show that ZZ can handle simultaneous failures and achieve sub-second recovery.

Categories and Subject Descriptors D.4.5 [*Operating Systems*]: Reliability—Fault-tolerance

General Terms Reliability, Design, Experimentation

Keywords Byzantine Fault Tolerance, Virtualization, Data Centers

1. Introduction

Today's enterprises rely on data centers to run their critical business applications. As users have become increasingly dependent on online services, malfunctions have become highly problematic, resulting in financial losses, negative publicity, or frustrated users. Consequently, maintaining high availability of critical services is a pressing need as well as a challenge in modern data centers.

Byzantine fault tolerance (BFT) is a powerful replication approach for constructing highly-available services that can tolerate arbitrary (Byzantine) faults. This approach requires replicas to agree upon the order of incoming requests and process them in the agreed upon order. Despite numerous efforts to improve the performance or fault scalability of BFT systems [Abd-El-Malek 2005, Castro 1999, Cowling 2006, Guerraoui 2010, Kotla 2007, Vandiver 2007], existing approaches remain expensive, requiring at least $2f + 1$ replicas to execute each request in order to tolerate f faults [Kotla 2007, Yin 2003]. This high replication cost has been a significant barrier to their adoption—to the best of our knowledge, no commercial data center application uses BFT techniques today, despite the wealth of research in this area.

Many recent efforts have focused on optimizing the agreement protocol used by BFT replicas [Cowling 2006, Kotla 2007]; consequently, today's state-of-the-art protocols can scale to a throughput of 80,000 requests/s and incur overheads of less than 10 μs per request for reaching *agreement* [Kotla 2007]. In contrast, request *execution* overheads for typical applications such as web servers and databases [Vandiver 2007] can be in the order of milliseconds or tens of milliseconds—three orders of magnitude higher than the agreement cost. Since request executions dominate the total cost of processing requests in BFT services, the hardware (server) capacity needed for request executions will far exceed that for running the agreement protocol. Hence, we argue that the total cost of a BFT service can be truly reduced only when the total overhead of request executions, rather than the cost to reach agreement, is somehow reduced.

In this paper we present ZZ, a new approach that reduces the cost of replication as well as that of request executions in BFT systems. Our approach enables general BFT services to be constructed with a replication cost close to $f + 1$, halving the $2f + 1$ or higher cost incurred by state-of-the-art approaches [Yin 2003]. ZZ targets shared hosting data center environments where replicas from multiple applications can share a physical server. The key insight in ZZ[1] is to run only $f + 1$ execution replicas per application in the graceful case

EuroSys'11, April 10–13, 2011, Salzburg, Austria.
Copyright © 2011 ACM 978-1-4503-0634-8/11/04... $10.00

---

[1] Denotes sleeping replicas; from the sleeping connotation of the term "zz.."

where there are no faults, and to activate additional sleeping replicas only upon failures. By multiplexing fewer replicas onto a given set of shared servers, our approach is able to provide more server capacity to each replica, and thereby achieve higher throughput and lower response times for request executions. In the worst case where all applications experience simultaneous faults, our approach requires an additional f replicas per application, matching the overhead of the $2f + 1$ approach. However, in the common case where only a *subset* of the data center applications are experiencing faults, our approach requires fewer replicas in total, yielding response time and throughput benefits. Like [Yin 2003], our system still requires $3f + 1$ agreement replicas; however, we argue that the overhead imposed by agreement replicas is small, allowing such replicas from multiple applications to be densely packed onto physical servers.

The ability to quickly activate additional replicas upon fault detection is central to our ZZ approach. While any mechanism that enables fast replica activation can be employed in ZZ, in this paper, we rely upon virtualization—a technique already employed in modern data centers—for on-demand replica activation.

The following are our contributions. We propose a practical solution to reduce the cost of BFT to nearly $f + 1$ execution replicas and define formal bounds on ZZ's replication cost. As reducing the execution cost in ZZ comes at the expense of potentially allowing faulty nodes to increase response times, we analyze and bound this response time inflation and show that in realistic scenarios malicious applications cannot significantly reduce performance. We also implement a prototype of ZZ by enhancing the BASE library and combining it with the Xen virtual machine and the ZFS file system. ZZ leverages virtualization for fast replica activation and optimizes the recovery protocol to allow newly-activated replicas to immediately begin processing requests through an amortized state transfer strategy. We evaluate our prototype using a BFT web server and ZZ-based NFS file server. Experimental results demonstrate that in a prototype data center running four BFT web servers, ZZ's use of only $f + 1$ execution replicas in the fault-free case yields response time and throughput improvements of up to 66%, and still enables rapid recovery after simultaneous failures occur. Overall, our evaluation emphasizes the importance of minimizing the execution cost of real BFT services and demonstrates how ZZ provides strong fault tolerance guarantees at significantly lower cost compared to existing systems.

2. State-of-the-art vs. the Art of ZZ

In this section, we compare ZZ to state-of-the-art approaches and describe how we reduce the execution cost to $f + 1$.

2.1 From 3f+1 to 2f+1

In the traditional PBFT approach [Castro 1999], during graceful execution a client sends a request Q to the replicas. The $3f + 1$ (or more) replicas agree upon the sequence

	PBFT'99	SEP'03	Zyzzyva'07	ZZ
Agreement replicas	$3f+1$	$3f+1$	$3f+1$	$3f+1$
Execution replicas	$3f+1$	$2f+1$	$2f+1$	$(1+r)f+1$
Agreement MACs/req per replica	$2+\frac{8f+1}{b}$	$2+\frac{12f+3}{b}$	$2+\frac{3f}{b}$	$2+\frac{10f+3}{b}$
Minimum work/req (for large b)	$(3f+1)\cdot$ $(E+2\mu)$	$(2f+1)E+$ $(3f+1)2\mu$	$(2f+1)E+$ $(3f+1)2\mu$	$(f+1)E+$ $(3f+1)2\mu$
Maximum throughput (if $E \gg \mu$)	$\frac{1}{(3f+1)E}$	$\frac{1}{(2f+1)E}$	$\frac{1}{(2f+1)E}$	$\frac{1}{(f+1)E}$

Table 1. ZZ versus existing BFT approaches. Here, f is the number of allowed faults, b is the batch size, E is execution cost, μ is the cost of a MAC operation, and $r \ll 1$ is a variable formally defined in §4.3.3. All numbers are for periods when there are no faults and the network is well-behaved.

number corresponding to Q, execute it in that order, and send responses back to the client. When the client receives $f + 1$ valid and matching responses from different replicas, it knows that at least one correct replica executed Q in the correct order. Figure 1(a) illustrates how the principle of separating agreement from execution can reduce the number of execution replicas required to tolerate up to f faults from $3f + 1$ to $2f + 1$. In this separation approach [Yin 2003], the client sends Q to a primary in the agreement cluster consisting of $3f + 1$ lightweight machines that agree upon the sequence number i corresponding to Q and send $[Q, i]$ to the execution cluster consisting of $2f + 1$ replicas that store and process application state. When the agreement cluster receives $f + 1$ matching responses from the execution cluster, it forwards the response to the client knowing that at least one correct execution replica executed Q in the correct order. For simplicity of exposition, we have omitted cryptographic operations above.

2.2 Circumventing 2f+1

The $2f + 1$ replication cost is believed necessary [Abd-El-Malek 2005, Cowling 2006, Kotla 2007] for BFT systems. However, more than a decade ago, Castro and Liskov concluded their original paper on PBFT [Castro 1999] saying "it is possible to reduce the number of copies of the state to $f + 1$ but the details remain to be worked out". In this paper, we work out those details.

Table 1 compares the replication cost and performance characteristics of several BFT State Machine Replication (BFT-SMR) approaches to ZZ. Quorum based approaches [Abd-El-Malek 2005, Cowling 2006] lead to a similar comparison. All listed numbers are for gracious execution, i.e., when there are no faults and the network is well-behaved. Note that all approaches require at least $3f + 1$ replicas in order to tolerate up to f independent Byzantine failures, consistent with classical results that place a lower bound of $3f + 1$ replicas for a safe Byzantine consensus protocol that is live under weak synchrony assumptions [Dwork 1988].

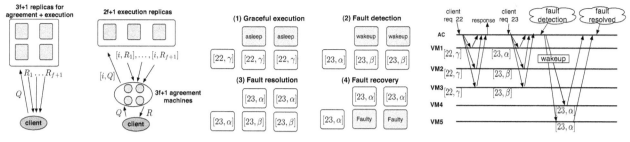

Figure 1. (a) The PBFT approach versus the separation of agreement from execution. **(b-c)** Various scenarios in the ZZ system for $f = 2$ faults. Request 22 results in matching responses γ, but the mismatch in request 23 initiates new virtual machine replicas on demand.

In contrast to common practice, we do not measure replication cost in terms of the total number of physical machines as we assume a virtualized environment that is common in many data centers today. Virtualization allows resources to be allocated to a replica at a granularity finer than an entire physical machine. Virtualization itself is useful in multiplexed environments, where a data center owner hosts many services simultaneously for better management of limited available resources. Note that virtualization helps all BFT approaches, not just ZZ, in multiplexed environments.

Cost: Our position is that execution, not agreement, is the dominant provisioning cost for most realistic data center services that can benefit from the high assurance provided by BFT. To put this in perspective, consider that state-of-the-art BFT approaches such as Zyzzyva show a peak throughput of over 80K requests/second for a toy application consisting of *null* requests, which is almost three orders of magnitude more than the achievable throughput for a database service on comparable hardware [Vandiver 2007]. Thus in realistic systems, the primary cost is that of hardware performing application execution, not agreement. ZZ nearly halves the data center provisioning cost by reducing the number of replicas actively executing requests (Table 1 row 2); however, this benefit can only be realized when BFT is used in a data center running multiple applications so that sleeping replicas can be distributed across a pool of servers.

Throughput: ZZ can achieve a higher peak throughput compared to state-of-the-art approaches when execution dominates request processing cost and resources are constrained. For a fair comparison, assume that all approaches are provisioned with the same total amount of resources. Then, the peak throughput of each approach is bounded by the minimum of its best-case execution throughput and its best-case agreement throughput (row 4). Agreement throughput is primarily limited by the overhead μ of a MAC operation and can be improved significantly through batching. However, batching is immaterial to the overall throughput when execution is the bottleneck (row 5).

The comparison above is for performance during periods when there are no faults and the network is well-behaved. In adverse conditions, the throughput and latency of all approaches can degrade significantly and a thorough comparison is nontrivial and difficult to characterize concisely [Clement 2009, Singh 2008].

When failures occur, ZZ incurs a higher latency to execute some requests until its failure recovery protocol is complete. Our experiments suggest that this additional overhead is modest and is small compared to typical WAN delays. In a world where failures are the uncommon case, ZZ offers valuable savings in replication cost or, equivalently, improvement in throughput under limited resources.

ZZ is not a new "BFT protocol" as that term is typically used to refer to the agreement protocol; instead, ZZ is an execution approach that can be interfaced with existing BFT-SMR agreement protocols. Our prototype uses the BASE implementation of the PBFT protocol as it was the most mature and readily available BFT implementation at the time of writing. The choice was also motivated by our premise that we do not seek to optimize agreement throughput, but to demonstrate the feasibility of ZZ's execution approach with a reasonable agreement protocol. Admittedly, it was easier to work out the details of augmenting PBFT with ZZ compared to more sophisticated agreement protocols.

3. ZZ design

3.1 System and Fault Model

We assume a Byzantine failure model where faulty replicas or clients may behave arbitrarily. There are two kinds of replicas: 1) agreement replicas that assign an order to client requests and 2) execution replicas that maintain application state and execute client requests. Replicas fail independently, and we assume an upper bound g on the number of faulty agreement replicas and a bound f on the number of faulty execution replicas in a given window of vulnerability. We initially assume an infinite window of vulnerability, and relax this assumption in Section 4.3.4. An adversary may coordinate the actions of faulty nodes in an arbitrary manner. However, the adversary can not subvert standard cryptographic assumptions about collision-resistant hashes, encryption, and digital signatures.

ZZ uses the state machine replication model to implement a BFT service. Replicas agree on an ordering of incoming requests and each execution replica executes all requests in the same order. Like all previous SMR based BFT systems, we assume that either the service is deterministic or the non-deterministic operations in the service can be transformed to deterministic ones via the agreement protocol [Castro 1999].

Our system ensures safety in an asynchronous network that can drop, delay, corrupt, or reorder messages. Liveness is guaranteed only during periods of synchrony when there is a finite but possibly unknown bound on message delivery time. The above system model and assumptions are similar to those assumed by many existing BFT systems [Castro 1999, Kotla 2007, Rodrigues 2001, Yin 2003].

Virtualization: ZZ assumes that replicas are run inside virtual machines. As a result, it is possible to run multiple replicas on a single physical server. To maintain the fault independence requirement, no more than one agreement replica and one execution replica of each service can be hosted on a single physical server.

ZZ assumes that the hypervisor may be Byzantine. Because of the placement assumption above, a malicious hypervisor is equivalent to a single fault in each service hosted on the physical machine. As before, we assume a bound f on the number of faulty hypervisors within a window of vulnerability. We note that even today sufficient hypervisor diversity (e.g., Xen, KVM, VMWare, Hyper-V) is available to justify this assumption.

3.2 ZZ Design Overview

ZZ reduces the replication cost of BFT from $2f+1$ to nearly $f + 1$ based on two simple insights. First, if a system is designed to be correct in an asynchronous environment, it must be correct even if some replicas are out of date. Second, during fault-free periods, a system designed to be correct despite f Byzantine faults must be unaffected if up to f replicas are turned off. ZZ leverages the second insight to turn off f replicas during fault-free periods requiring just $f+1$ replicas to actively execute requests. When faults occur, ZZ leverages the first insight and behaves exactly as if the f standby replicas were slow but correct replicas.

If the $f + 1$ active execution replicas return matching responses for an ordered request, at least one of these responses, and by implication all of the responses, must be correct. The problematic case is when the $f + 1$ responses do not match. In this case, ZZ starts up additional virtual machines hosting standby replicas. For example, when $f = 1$, upon detecting a fault, ZZ starts up a third replica that executes the most recent request. Since at most one replica can be faulty, the third response must match one of the other two responses, and ZZ returns this matching response to the client. Figure 1(b-c) illustrates the high-level control flow for $f = 2$. Request 22 is executed successfully generating the response γ, but request 23 results in a mismatch waking up the two standby VM replicas. The fault is resolved by comparing the outputs of all $2f + 1$ replicas, revealing α as the correct response.

The above design would be impractical without a quick replica wake-up mechanism. Virtualization provides this

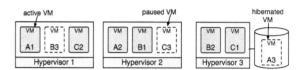

Figure 2. An example server setup with three $f = 1$ fault tolerant applications, A, B, and C; only execution replicas are shown.

mechanism by maintaining additional replicas in a "dormant" state. Figure 2 illustrates how ZZ can store additional replicas both in memory as prespawned but paused VMs and hibernated to disk. Paused VMs resume within milliseconds but consume memory resources. Hibernated replicas require only storage resources, but can incur greater recovery times.

3.3 Design Challenges

The high-level approach described above raises several further challenges. First, how does a restored replica obtain the necessary application state required to execute the current request? In traditional BFT systems, each replica maintains an independent copy of the entire application state. Periodically, all replicas create application checkpoints that can be used to bring up to speed any replicas which become out of date. However, a restored ZZ replica may not have any previous version of application state. It must be able to verify that the state it obtains is correct even though there may be only one correct execution replica (and f faulty ones), e.g., when $f = 1$, the third replica must be able to determine which of the two existing replicas possesses the correct state.

Second, transferring the entire application state can take an unacceptably long time. In existing BFT systems, a recovering replica may generate incorrect messages until it obtains a stable checkpoint. This inconsistent behavior during checkpoint transfer is treated like a fault and does not impede progress of request execution if there is a quorum of $f + 1$ correct execution replicas with a current copy of the application state. However, when a ZZ replica recovers, there may exist just one correct execution replica with a current copy of the application state. The traditional state transfer approach can stall request execution in ZZ until f recovering replicas have obtained a stable checkpoint.

Third, ZZ's replication cost must be robust to faulty replica or client behavior. A faulty client must not be able to trigger recovery of standby replicas. A compromised replica must not be able to trigger additional recoveries if there are at least $f + 1$ correct and active replicas. If these conditions are not met, the replication cost savings would vanish and system performance could be worse than a traditional BFT system using $2f + 1$ replicas.

4. ZZ Protocol

In this section we briefly describe the separated protocol from [Yin 2003], and present ZZ's modifications to support switching from $f+1$ to $2f+1$ execution replicas after faults are detected.

4.1 Graceful Execution

Client Request & Agreement: In Figure 3 step 1, a client c sends a request Q to the agreement cluster to submit an operation o with a timestamp t. The timestamps ensure exactly-once semantics for execution of client requests, and a faulty client's behavior does not affect other clients' requests.

Upon receiving a client request Q, the agreement replicas will execute the standard three phase BFT agreement protocol [Castro 1999] in order to assign a sequence number n to the request. When an agreement replica j learns of the sequence number n committed to Q, it sends a commit message C to all execution replicas (Fig. 3 step 2).

Execution: An execution replica i executes a request Q when it gathers a commit certificate $\{C_i\}, i \in A|2g+1$, i.e. a set of $2g+1$ valid and matching commit messages from the agreement cluster, and it has executed all other requests with a lower sequence number. Each execution node produces a reply R which it sends to the client and an execution report message ER sent to all agreement nodes (Fig. 3 steps 3-4).

In the normal case, the client receives a response certificate $\{R_i\}, i \in E|f+1$—matching reply messages from $f+1$ execution nodes. Since at most f execution replicas can be faulty, a client receiving a response certificate knows that the response is correct. If a client does not obtain matching replies, it resends its request to the agreement cluster. If an agreement node receives a retransmitted request for which it has received $f+1$ matching execution report messages, then it can send a reply affirmation, RA to the client (Fig. 3 step 5). If a client receives $g+1$ such messages containing a response digest, $\overline{R}$, matching one of the replies already received, then the client can accept that reply as valid. This "backup" solution is used by ZZ to prevent unnecessary wakeups where a partially faulty execution node may reply to the agreement cluster, but not to the client. If the agreement cluster cannot produce an affirmation for the client, then additional nodes must be started as described in subsequent sections.

4.2 Dealing with Faults

4.2.1 Checkpointing

Checkpoints are used so that newly started execution replicas can obtain a recent copy of the application state and so that replicas can periodically garbage collect their logs. The checkpoints are constructed at predetermined request sequence numbers, e.g., when it is exactly divisible by 1024.

With at least $2f+1$ execution replicas in other BFT-SMR systems, a recovering replica is guaranteed to get at least $f+1$ valid and matching checkpoint messages from other execution replicas, allowing checkpoint creation and validation to be done exclusively within the execution cluster [Yin 2003]. However, ZZ runs only $f+1$ execution replicas in the normal case, and thus a new replica may not be able to tell which of the checkpoints it is provided are correct.

To address this problem, ZZ's execution cluster must coordinate with the agreement cluster during checkpoint creation. ZZ execution nodes create checkpoints of applica-

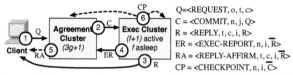

Figure 3. The normal agreement and execution protocol in ZZ proceeds through steps 1-4. Step 5 is needed only after a fault. Checkpoints (step 6) are created on a periodic basis. The notation $\langle \text{LABEL}, X \rangle$ denotes the message of type LABEL with parameters X. We indicate the digest of parameter Y as $\overline{Y}$.

tion state and their reply log, then assemble a proof of their checkpoint, *CP*, and send it to all of the execution *and* agreement nodes (Fig. 3 step 6). Informing the agreement nodes of the checkpoint digest allows them to assist recovering execution replicas in verifying the checkpoint data they obtain from potentially faulty nodes.

As in most BFT-SMR systems, ZZ uses a copy-on-write technique to preserve the state of each object when checkpoints must be made. All correct execution replicas will save a copy of each object as well as a cryptographic digest of the object's contents. Digests only need to be computed for objects modified since the last checkpoint; this can be done at the time of checkpoint creation, or proactively after an object is modified in order to decrease the checkpointing latency.

A checkpoint certificate $\{CP_i\}, i \in E|f+1$ is a set of $f+1$ *CP* messages with matching digests. When an execution replica receives a checkpoint certificate with a sequence number n, it considers the checkpoint stable and discards earlier checkpoints and request commit certificates with lower sequence numbers that it received from the agreement cluster. Likewise, the agreement nodes use these checkpoint certificates to determine when they can garbage collect messages in their communication logs with the execution cluster.

Newly awoken replicas in ZZ may not yet have the full application state at checkpoint time, but this does not prevent them from continuing to process requests, since out-of-date replicas are permitted to skip checkpoints. ZZ's window of vulnerability, defined in Section 4.3.4, assumes that additional faults will not occur until all execution replicas have obtained the full state and are able to create a complete, stable checkpoint.

4.2.2 Fault Detection

The agreement cluster is responsible for detecting faults in the execution cluster. Agreement nodes in ZZ are capable of detecting invalid execution or checkpoint messages; the fault detection and recovery steps for each of these are identical, so for brevity we focus on invalid or missing execution responses. In the normal case, an agreement replica j waits for an execution certificate, $\{ER_i\}, i \in E|f+1$, from the execution cluster. Replica j inserts this certificate into a local log ordered by the sequence number of requests. When j receives ER messages which do not match, or waits for longer than a predetermined timeout, j sends a *recovery request*, $W = \langle \text{RECOVER}, j, n \rangle_j$, to the f hypervisors controlling the standby execution replicas. When the hypervisor of

a sleeping execution replica receives a recovery certificate, $\{W_i\}, i \in A | g+1$, it wakes up the local execution replica.

4.2.3 Replica Recovery with Amortized State Transfer

When an execution replica k starts up, it must obtain the most recent checkpoint of the entire application state from existing replicas and verify that it is correct. Unfortunately, checkpoint transfer and verification can take an unacceptably long time. Worse, unlike previous BFT systems that can leverage incremental cryptography schemes to transfer only the objects modified since the last checkpoint, a recovering ZZ replica has no previous checkpoints.

How does replica k begin to execute requests without any application state? Instead of performing an expensive transfer of the entire state upfront, a recovering ZZ replica fetches and verifies the state necessary to execute each request on demand. Replica k first fetches a log of committed requests since the last checkpoint from the agreement cluster and a checkpoint certificate $\{CP_i\}, i \in A | g+1$ from $g+1$ agreement replicas. This checkpoint certificate includes digests for each state object, allowing the replica to verify that any state object it obtains has come from a correct replica.

After obtaining the checkpoint certificate with object digests, replica k begins to execute in order the recently committed requests. Let Q be the first request that reads from or writes to some object p since the most recent checkpoint. To execute Q, replica k fetches p on demand from any execution replica that can provide an object consistent with p's digest that k learned from the certificate. Replica k continues executing requests in sequence number order fetching new objects on demand until it obtains a stable checkpoint.

The recovery time can be optimized by only replaying requests which cause writes to state. Note that since on-demand transfers only fetch objects touched by requests, they are not sufficient for k to obtain a stable checkpoint, so the replica must also fetch the remaining state in the background. Recovery is complete only when replica k has obtained a stable checkpoint, although it will be able to correctly respond to replicas as soon as it obtains the subset of the application state needed to process the request that triggered the fault.

4.3 System Properties

We formally define the performance, replication cost, safety and liveness properties of ZZ. Due to space constraints we defer complete proofs to the appendix and [Wood 2011].

4.3.1 Response Time Inflation

ZZ relies on timeouts to detect faults in execution replicas. This opens up a potential performance vulnerability. A low value of the timeout can trigger fault detection even when the delays are benign and needlessly start new replicas. On the other hand, a high value of the timeout can be exploited by faulty replicas to degrade performance as they can delay sending each response to the agreement cluster until just before the timeout. The former can take away ZZ's savings

in replication cost as it can end up running more than $f+1$ (and up to $2f+1$) replicas even during graceful periods. The latter hurts performance under faults. Note that safety is not violated in either case.

To address this problem, we suggest the following simple procedure for setting timeouts to limit response time inflation. Upon receiving the first response to a request committed to sequence number n, an agreement replica sets the timeout τ_n to Kt_1, where t_1 is the response time of the first response and K is a pre-configured variance bound. If the agreement replica does not receive f more matching responses within τ_n, then it triggers a fault and wakes up f additional replicas.

This procedure trivially bounds the response time inflation of requests to a factor of K, but we can further constrain the performance impact by considering the response time distribution as follows. Given p, the probability of a replica being faulty,

THEOREM 1. *Faulty replicas can inflate average response time by a factor of:* $max \left(1, \sum_{0 \le m \le f} P(m)I(m) \right)$ *where:*

$$P(m) = \binom{f}{m} p^m (1-p)^{f-m}$$
$$I(0) = 1, \; else:$$
$$I(m) = max \left(1, \frac{K \cdot E[MIN_{f+1-m}]}{E[MAX_{f+1}]} \right)$$

$P(m)$ represents the probability of m simultaneous failures and $I(m)$ is the response time inflation that m faulty nodes can inflict. To get the total impact of response time inflation, we must sum this product for all possible values of m. $E[MIN_{f+1-m}]$ is the expected minimum response time for a set of $f+1-m$ replicas and $E[MAX_{f+1}]$ is the expected maximum response time of all $f+1$ replicas, assuming all response times are identically distributed as some distribution Ψ. The top term in $I(m)$ follows from the rule defined above: a faulty node can increase response time by at most K compared to the fastest correct replica (i.e. the replica with the minimum response time out of $f+1-m$ nodes). The bottom term is the non-faulty case where response time is limited by the slowest of the $f+1$ replicas.

As an example, suppose $K=4$, $f=3$, and response times are exponentially distributed with $E[\Psi] = 2ms$. Then $E[MIN_{f+1-m}] = \frac{2}{3+1-m}ms$ and $E[MAX_{f+1}] = 4.2ms$. If $p = 0.1$, then $I = 1.0009$, i.e., average response time rises by only 0.09%. Only for $p > 0.48$ is the inflation greater than 10%. Note that proactive recovery can be used to ensure p remains small [Castro 2002] and that to achieve this worst case bound faulty nodes must be able to predict the earliest response time of correct replicas. In practice, correct execution replicas may sometimes violate the variance bound due to benign execution or network delays, causing a *false timeout*. These false timeouts can impact overall replication cost as described in section 4.3.3.

4.3.2 Waking Up and Shutting Down

Since waking up nodes to respond to faults is an expensive procedure, ZZ distinguishes between "blocking" and "non-blocking" faults, and only triggers a wakeup event for blocking faults—those which cannot be resolved without a wakeup. Fortunately, blocking faults by definition are more widespread in their impact, and thus can always be traced back to a faulty execution node which can then be shutdown.

THEOREM 2. *If a wakeup occurs, ZZ will be able to terminate at least one faulty replica.*

This theorem is proved in the appendix, and is based on the following wake up rule.

Wakeup Rule: A wakeup happens if and only if a mismatch report is "blocking".

To understand the difference between blocking and non-blocking faults, consider the response matrix where position (i, j) indicates E_i's response as reported by agreement node A_j. Consider two examples where the client receives conflicting responses P and Q, and $f = g = 1$,

Non-blocking fault	Blocking fault
$A_1 A_2 A_3 A_4$	$A_1 A_2 A_3 A_4$
$E_1 : Q\ P\ P\ P$	$E_1 : Q\ P\ P\ P$
$E_2 : Q\ P\ P\ P$	$E_2 : Q\ Q\ Q\ P$

In the first scenario, it is impossible to distinguish whether only A_1 is faulty or if an execution replica and A_1 is faulty; however, $g + 1$ agreement nodes can provide a reply affirmation that P is the correct response. In the second case, there is no way to tell whether Q or P is the correct response, so a wakeup is required. Once this replica is started, ZZ will be able to determine which replicas were faulty so it can terminate them and reduce the number of active replicas back to only $f + 1$. To do this, ZZ employs the following rules:

Shutdown Rule: If any replicas can be convicted as faulty, shut down f replicas starting with all convictably faulty replicas, and followed by additional replicas that were just woken up if needed. If no replicas can be convicted, delay the shutdown procedure until all replicas produce a stable checkpoint. Then shut down any f of the original replicas.

Note that in most cases a blocking fault will allow ZZ to convict at least one faulty replica causing an immediate shutdown; however, in certain cases where multiple faulty execution and agreemnt nodes collude, it may not be possible to determine which nodes are faulty. ZZ prefers to shut down replicas immediately after a wakeup because this prevents malicious nodes from reducing ZZ's performance benefits that rely on running only $f + 1$ active replicas in the normal case. We define how ZZ is able to convict faulty replicas in the appendix, and describe how these rules ensure ZZ's safety and liveness properties in Section 4.3.4.

4.3.3 Overall Replication Cost

The expected replication cost of ZZ varies from $f + 1$ to $2f + 1$ depending on the probability of replicas being faulty p, and the likelihood of false timeouts, Π_1.

THEOREM 3. *The expected replication cost of ZZ is less than $(1 + r)f + 1$, where $r = 1 - (1 - p)^{f+1} + (1 - p)^{f+1}\Pi_1$.*

These two factors influence the replication cost because additional nodes are started only if 1) a replica is truly faulty (which happens with probability $1 - (1 - p)^{f+1}$), or 2) there are no faults, but a correct slow replica causes a false timeout (which happens with probability $(1 - p)^{f+1}\Pi_1$). In either case, the replication cost is increased by f, resulting in the theorem. The value of p can be reduced by proactive recovery, and Π_1 is dependent on the value of K. Adjusting K results in a tradeoff between the replication cost and the response time inflation bound. Note that in practice the replication cost may be even lower than this because ZZ will quickly shutdown nodes after the fault has been resolved.

4.3.4 Safety and Liveness Properties

We state the safety and liveness properties ensured by ZZ and outline the proofs.

Safety: ZZ ensures the safety property that if a correct client obtains either a response certificate or an affirmation certificate for a response $\langle \text{REPLY}, t, c, j, R \rangle_j$, then (1) the client issued a request $\langle \text{REQUEST}, o, t, c \rangle_c$ earlier; (2) all correct replicas agree on the sequence number n of that request and on the order of all requests with sequence numbers in $[1, n]$; (3) the value of the reply R is the reply that a single correct replica would have produced.

The first claim follows from the fact that the agreement cluster generates valid commit certificates only for valid client requests and the second follows from the safety of the agreement protocol that ensures that no two requests are assigned the same sequence number [Castro 2002]. To show the third claim, we must prove that if a client receives a reply in ZZ, that the reply must be the one produced by a correct execution replica. We consider two cases, depending on whether a wakeup is required to process request n. If a wakeup is required, we must show that after the fault is resolved, the system produces equivalent behavior as if no wakeup had been needed.

In the first scenario a client receives either $f + 1$ matching execution reports or a single execution report that matches $g + 1$ reply affirmations from the agreement cluster without requiring wakeups. The client is assured that reply R is the correct response because matching responses must have been produced by $f + 1$ different execution nodes; since at least one of those nodes is correct, the reply must be correct.

In the second case a client does not immediately receive matching execution reports or reply affirmations. We show that the client will retransmit its requests, and correct agreement replicas will wakeup an additional f execution replicas using the following Lemma in the appendix:

Lemma 1: A client will receive an affirmation certificate unless a mismatch is blocking.

This lemma allows us to guarantee that if clients are not able to directly obtain an affirmation certificate, then a

blocking mismatch must have occurred, which in turn will cause an additional f replicas to be started.

The new replicas will be able to obtain a correct snapshot since that only requires a single correct execution replica with application state. The new replicas will now be equivalent to any non-faulty replicas at the start of the last checkpoint epoch, and will correctly replay any requests up to n. Of the $2f + 1$ replicas now active, at most f can be faulty, leaving $f + 1$ responses made by correct replicas. These responses must be identical and will be used to produce the execution report needed by the client for a valid response.

This guarantees that after a *wakeup* the system behavior is equivalent to that of a correct replica. However, we still must show that once f replicas are *shutdown* that the system will continue to function properly even though newly awoken replicas may not yet have the full application state. In order to maintain the safety property, at least one of the "old" replicas with full state must remain active and not be shutdown. We ensure this with the following lemma proved in the appendix:

Lemma 2: If a mismatch is blocking, then (a) at least one faulty replica can be shutdown and (b) the system will be able to make a stable checkpoint.

If the blocking mismatch results in at least one convictably faulty replica, then ZZ will shut down that node and up to f new replicas. This direclty satisfies (a), and part (b) is guaranteed because this procedure will leave at least one correct original node active that will be able to provide the full application state and create a stable checkpoint.

If no nodes can be convicted as faulty, then ZZ will not shut down any nodes until a new stable checkpoint is created, fulfilling requirement (b). Once the checkpoint is made, ZZ will shut down f of the original replicas; since unconvictable faults can only occur if more than one node is faulty, this will eliminate at least one faulty node.

At this point, ZZ has ensured that a correct response has been returned to the client and that at least one active execution replica contains the full application state. This is equivalent to a correctly operating system, ensuring ZZ's safety properties.

Liveness: ZZ ensures the liveness property that if a correct client sends a request R with a timestamp exceeding previous requests and repeatedly retransmits the request, then it will eventually receive a response certificate or an affirmation certificate for R. We need eventual synchrony to show this liveness property. If the client repeatedly retransmits R, then the agreement cluster will eventually commit a sequence number to R. A correct execution replica will receive a commit certificate, i.e., messages from at least $g + 1$ agreement replicas assigning a common sequence number n to R, by the following property of the agreement protocol: Any request that commits locally at a correct agreement replica will eventually commit locally at, at least, $g + 1$ correct replicas. These properties are guaranteed by the agreement protocol used by ZZ [Castro 2002].

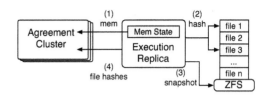

Figure 4. For each checkpoint an execution replica (1) sends any modified memory state, (2) creates hashes for any modified disk files, (3) creates a ZFS snapshot, and (4) returns the list of hashes to agreement nodes.

The existence of at least one correct execution replica ensures that the client gets at least one valid (but yet uncertified) response for R. The agreement cluster in ZZ ensures that it either obtains an execution certificate for R or wakes up the full $2f + 1$ execution replicas. In either case, the agreement cluster will eventually obtain an execution certificate, ensuring that the client eventually obtains an affirmation certificate for R.

Window of Vulnerability: ZZ's current implementation assumes a window of vulnerability equal to the time needed to detect a fault, wakeup additional replicas, and finish the state transfer so that at least one new replica can create a stable checkpoint. Since at least one faulty replica is detected and eliminated after each failure the system periodically "refreshes" itself. However, since there might only be one correct replica with the full application state, this replica cannot become faulty until it finishes transferring the full state to another replica. This is a similar requirement to other BFT-SMR systems which generally rely on proactive recovery (where a replica must be restarted and recover the full application state) in order to reduce the window of vulnerability from infinity.

5. ZZ Implementation

We implemented ZZ by enhancing the 2007 version of BASE [Rodrigues 2001] so as to 1) use virtual machines to run replicas, 2) incorporate ZZ's checkpointing, fault detection, rapid recovery and fault-mode execution mechanisms, and 3) use file system snapshots to assist checkpointing.

5.1 Replica Control Daemon

We have implemented a ZZ replica control daemon that runs on each physical machine and is responsible for managing replicas after faults occur. The control daemon, which runs in Xen's Domain-0, uses the certificate scheme described in Section 4.2.2 to ensure that it only starts or stops replicas when enough non-faulty replicas agree that it should do so.

Inactive replicas are maintained either in a paused state, where they have no CPU cost but incur a small memory overhead on the system, or hibernated to disk, which utilizes no resources other than disk space. To optimize the wakeup latency of replicas hibernating on disk, ZZ uses a paged-out restore technique that exploits the fact that hibernating replicas initially have no useful application state in memory, and thus can be created with a bare minimum allocation

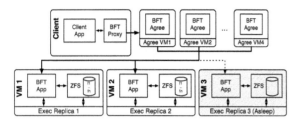

Figure 5: Experimental setup for a basic ZZ BFT service.

	Graceful performance				After failure		
	h_1	h_2	h_3	h_4	h_1	h_2	h_3
BASE	1234	1234	1234	1234	1234	1234	1234
SEP_{Agree}	1234	1234	1234	1234	1234	1234	1234
SEP_{Exec}	134	124	123	234	134	124	123
ZZ_{Agree}	1234	1234	1234	1234	1234	1234	1234
ZZ_{Exec}	12	12	34	34	123	124	34
ZZ_{Sleep}	3	4	1	2			1

Table 2. Placement of the 4 web servers' virtual machines (denoted 1 to 4) on the 4 data center hosts (h_1 to h_4) under graceful performance and on the 3 remaining hosts after h_4 failure.

of 128MB of RAM (which reduces their disk footprint and load times). After being restored, their memory allocation is increased to the desired level. Although the VM will immediately have access to its expanded memory allocation, there may be an application dependent period of reduced performance if data needs to be paged in.

5.2 Exploiting File System Snapshots

Checkpointing in ZZ relies on the existing mechanisms in the BASE library to save the protocol state of the agreement nodes and any memory state used by the application on the execution nodes. In addition, ZZ supports using the snapshot mechanism provided by modern journaled file systems [ZFS 2004] to simplify checkpointing disk state. Creating disk snapshots is efficient because copy-on-write techniques prevent the need for duplicate disk blocks to be created, and the snapshot overhead is independent of the disk state of the application. ZZ uses ZFS for snapshot support, and works with both the native Solaris and user-space Linux ZFS implementations.

ZZ includes meta-information about the disk state in the checkpoint so that the recovery nodes can validate the disk snapshots created by other execution nodes. To do so, execution replicas create a cryptographic hash for each file in the disk snapshot and send it to the agreement cluster as part of the checkpoint certificate as shown in Figure 4. Hashes are computed only for those files that have been modified since the previous epoch; hashes from the previous epoch are reused for unmodified files to save computation overheads.

Tracking Disk State Changes: The BASE library requires all state, either objects in memory or files on disk, to be registered with the library. In ZZ we have simplified the tracking of disk state so that it can be handled transparently without modifications to the application. We define functions bft_fopen() and bft_fwrite() which replace the ordinary fopen() and fwrite() calls in an application. The bft_fwrite() function invokes the modify() call of the BASE library which must be issued whenever a state object is being edited. This ensures that any files which are modified during an epoch will be rehashed during checkpoint creation.

For the initial execution replicas, the bft_fopen() call is identical to fopen(). However, for the additional replicas which are spawned after a fault, the bft_fopen call is used to retrieve a file from the disk snapshots and copy it to the replica's own disk on demand. When a recovering replica first tries to open a file, it calls bft_fopen(foo), but the replica

will not yet have a local copy of the file. The recovery replica fetches a copy of the file from any replica and verifies it against the hash contained in the most recent checkpoint. If the hashes do not match, the recovery replica requests the file from a different replica, until a matching copy is found and copied to its own disk.

6. Experimental Evaluation

6.1 Experiment Setup

Our experimental data-center setup uses a cluster of 2.12 GHz 64-bit dual-core Dell servers, each with 4GB RAM. Each machine runs a Xen v3.1 hypervisor and Xen virtual machines. Both domain-0 (the controller domain in Xen) as well as the individual VMs run the CentOS 5.1 Linux distribution with the 2.6.18 Linux kernel and the user space ZFS filesystem. All machines are interconnected over gigabit ethernet. Figure 5 shows the setup for agreement and execution replicas of a generic BFT app for $g = f = 1$; multiple such applications are assumed to be run in a BFT data center.

6.1.1 Throughput

Our experiments involve three fault-tolerant server applications: a Web Server, an NFS server, and a toy client-server microbenchmark.

Fault-tolerant Web Server: We have implemented a BFT-aware HTTP 1.0 Web server that mimics a dynamic web site with server side scripting. The request execution time is configurable to simulate more complex request processing. We generate web workloads using *httperf* clients which contact a local BFT web proxy that submits the requests to the agreement nodes.

Fault-tolerant NFS: BASE provides an NFS client relay and a BFT wrapper for the standard NFS server. We have extended this to support ZZ's on demand state transfer which allows a recovery replica to obtain file system state from ZFS snapshots as it processes each request.

Client-Server Microbenchmark: We utilize the simple client-server application from the BASE library to measure ZZ's performance for *null* requests and to study its recovery costs under different application state scenarios.

Our experiments compare three systems: ZZ, BASE, and Separated (SEP). BASE is the standard BFT library described in [Rodrigues 2001]. SEP is our extension of BASE which separates the agreement and execution replicas, and

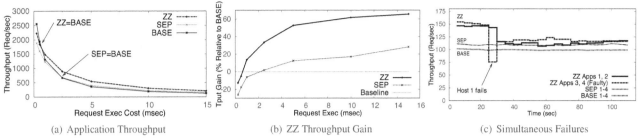

| (a) Application Throughput | (b) ZZ Throughput Gain | (c) Simultaneous Failures |

Figure 6. (a-b) When resources are constrained, ZZ significantly increases system throughput by using fewer replicas. **(c)** Under simultaneous failures of several applications, ZZ quickly recovers and still maintains good throughput.

requires $3f + 1$ agreement and $2f + 1$ execution replicas similar to [Yin 2003]. ZZ also requires $3f + 1$ agreement replicas, but extends SEP to use only $f + 1$ active execution replicas, with an additional f sleeping replicas. While more recent agreement protocols provide higher performance than BASE, our evaluation focuses on cases where execution is at least an order of magnitude more expensive than agreement; we believe our conclusions are consistent with what would be found with more optimized agreement protocols.

6.2 Graceful Performance

We study the graceful performance of ZZ by emulating a shared hosting environment running four independent web apps on four machines. Table 2 shows the placement of agreement and execution replicas on the four hosts. As the agreement and execution clusters can independently handle faults, each host can have at most one replica of each type per application.

We first analyze the impact of request execution cost under ZZ, SEP, and BASE, which require $f + 1$, $2f + 1$, and $3f + 1$ execution replicas per web server respectively. Figure 6(a) compares the throughput of each system as the execution cost per web request is adjusted. When execution cost averages $100 \ \mu s$, BASE performs the best since the agreement overhead dominates the cost of processing each request and our implementation of separation incurs additional cost for the agreement replicas. However, for execution costs exceeding 0.75 ms, the execution replicas become the system bottleneck. As shown in Figure 6(b), ZZ begins to outperform BASE at this point, and performs increasingly better compared to both BASE and SEP as execution cost rises. SEP surpasses BASE for request costs over 2ms, but cannot obtain the throughput of ZZ since it requires $2f + 1$ replicas instead of only $f + 1$. ZZ provides as much as a 66% increase in application throughput relative to BASE for requests with large execution costs.

6.2.1 Latency

We further characterize the performance of ZZ in graceful operation by examining the relation between throughput and response time for different request types. Figure 7 shows the relation between throughput and response time for increasingly CPU intensive requests. For null requests or very low loads, Figure 7(a), BASE beats SEP and ZZ because it has

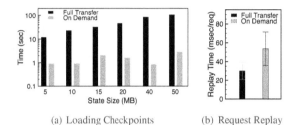

| (a) Loading Checkpoints | (b) Request Replay |

Figure 8. (a) The cost of full state transfer increases with state size. (b) On Demand incurs overhead when replaying requests since state objects must be verified.

less agreement overhead. At 1ms, ZZ's use of fewer execution replicas enables it to increase the maximum throughput by 25% over both SEP and BASE. When the execution cost reaches 10ms, SEP outperforms BASE since it uses $2f + 1$ instead of $3f + 1$ replicas. ZZ provides a 50% improvement over SEP, showing the benefit of further reducing to $f + 1$.

6.3 Simultaneous Failures

When several applications are multiplexed on a single physical host, a faulty node can impact all its running applications. In this experiment, we simulate a malicious hypervisor on one of the four hosts that causes multiple applications to experience faults simultaneously. Host h_4 in Table 2 is set as a faulty machine and is configured to cause faults on all of its replicas 20 seconds into the experiment as shown in Figure 6(c). For ZZ, the failure of h_4 directly impacts web servers 3 and 4 which have active execution replicas there. The replica for server 2 is a sleeping replica, so its corruption has no effect on the system. The failure also brings down one agreement replica for each of the web servers, however they are able to mask these failures since $2f + 1$ correct agreement replicas remain on other nodes.

ZZ recognizes the corrupt execution replicas when it detects disagreement on the request output of each service. It responds by waking up the sleeping replicas on hosts h_1 and h_2. After a short recovery period (further analyzed in the next section), ZZ's performance is similar to that of SEP with three active execution replicas competing for resources on h_1 and h_2. Even though h_3 only has two active VMs and uses less resources with ZZ, applications 3 and 4 have to wait for responses from h_1 and h_2 to make progress. Both ZZ and SEP maintain a higher throughput than BASE that runs all applications on all hosts.

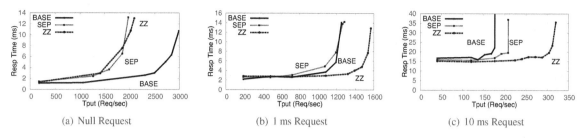

| (a) Null Request | (b) 1 ms Request | (c) 10 ms Request |

Figure 7: For high execution costs, ZZ achieves both higher throughput and lower response times.

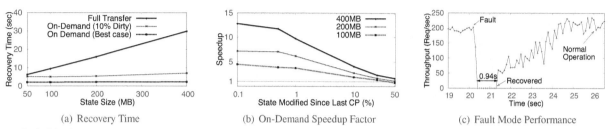

| (a) Recovery Time | (b) On-Demand Speedup Factor | (c) Fault Mode Performance |

Figure 9. (a-b) The worst case recovery time depends on the amount of state updated between the last checkpoint and the fault. **(c)** The recovery period lasts for less than a second. At first, requests see higher latency since state must be fetched on-demand.

6.4 Recovery Cost

The following experiments study the cost of recovering replicas in more detail using both microbenchmarks and our fault tolerant NFS server. We study the recovery cost, which we define as the delay from when the agreement cluster detects a fault until the client receives the correct response.

6.4.1 NFS Recovery Costs

We investigate the NFS server recovery cost for a workload that creates 200 files of equal size before encountering a fault. We vary the size of the files to adjust the total state maintained by the application, which also impacts the number of requests which need to be replayed after the fault.

ZZ uses an on-demand transfer scheme for delivering application state to newly recovered replicas. Figure 8(a) shows the time for processing the checkpoints when using full transfer or ZZ's on-demand approach (note the log scale). The full state transfer approach performs very poorly since the BFT NFS wrapper must both retrieve the full contents of each file and perform RPC calls to write out all of the files to the actual NFS server. When transferring the full checkpoint, recovery time increases exponentially and state sizes greater than a mere 20 megabytes can take longer than 60 seconds, after which point NFS requests typically will time out. In contrast, the on-demand approach has a constant overhead with an average of 1.4 seconds. This emphasizes the importance of using the on-demand transfer for realistic applications where it is necessary to make some progress in order to prevent application timeouts.

We report the average time per request replayed and the standard deviation for each scheme in Figure 8(b). ZZ's on demand system experiences a higher replay cost due to the added overhead of fetching and verifying state on-demand; it also has a higher variance since the first access to a file incurs more overhead than subsequent calls. While ZZ's replay time is larger, the total recovery time is much smaller when using on-demand transfer.

6.4.2 Obtaining State On-Demand

This experiment uses a BFT client-server microbenchmark which processes requests with negligible execution cost to study the recovery cost after faults are caused in applications with different state sizes.

In the best case, a fault occurs immediately after a checkpoint and new replicas only need to load and resume from the last save, taking a constant time of about 2s regardless of state size (Figure 9(a)). Otherwise, the cost of on-demand recovery varies depending on the amount of application state that was modified since the last checkpoint. The "10% Dirty" line shows the recovery cost when 10% of the application's state needs to be fetched during replay. In that case, ZZ's recovery time varies from 5.2s to 7.1s for states of 50 and 400MB, respectively. This remains much faster than the Full Transfer technique which requires over 30s to transfer and verify 400MB of state.

The tradeoff between amount of dirty state and recovery speed is further studied in Figure 9(b). Even when 10% of application state is modified between each checkpoint, on-demand transfers speed up recovery by at least five times. Only when more than 50% of state is dirty does it becomes more expensive to replay than to perform a full transfer. Fortunately, we have measured the additional cost of ZFS checkpoints at 0.03s, making it practical to checkpoint every few seconds, during which time most applications will only modify a small fraction of their total application state.

Next we examine the impact of on-demand recovery on throughput and latency. The client sends a series of requests involving random accesses to 100kB state objects and a fault is injected after 20.2s (Figure 9(c)). The faulty request experiences a sub-second recovery period, after which the ap-

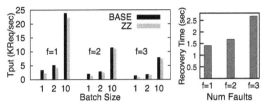

(a) Null Request Performance   (b) Recovery Time

Figure 10. Recovery time increases for larger f from message overhead and increased ZFS operations.

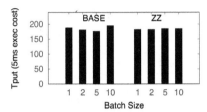

Figure 11. Throughput of ZZ and BASE with different batch sizes for a 5ms request execution time.

plication can handle new requests. The mean request latency prior to the fault is 5ms with very little variation. The latency of requests after the fault has a bimodal distribution depending on whether the request accesses a file that has already been fetched or one which needs to be fetched and verified. The long requests, which include state verification and transfer, take an average of 20ms. As the recovery replica rebuilds its local state, the throughput rises since the proportion of slow requests decreases. After 26s, the full application state has been loaded by the recovery replica, and the throughput prior to the fault is once again maintained.

6.4.3 Impact of Multiple Faults

We examine how ZZ's graceful performance and recovery time changes as we adjust f, the number of faults supported by the system when *null* requests are used requiring no execution cost. Figure 10(a) shows that ZZ's graceful mode performance scales similarly to BASE as the number of faults increases. This is expected because the number of cryptographic and network operations rises similarly in each.

We next examine the recovery latency of the client-server microbenchmark for up to three faults. We inject a fault to f of the active execution replicas and measure the recovery time for f new replicas to handle the faulty request. Figure 10(b) shows how the recovery time increases slightly due to increased message passing and because each ZFS system needs to export snapshots to a larger number of recovering replicas. We believe the overhead can be attributed to our use of the user-space ZFS code that is less optimized than the Solaris kernel module implementation, and messaging overhead which could be decreased with hardware multicast.

6.5 Trade-offs and Discussion

6.5.1 Agreement Protocol Performance

Various agreement protocol optimizations exist such as request batching, but these may have less effect when request execution costs are non-trivial. While Figure 10(a) shows a large benefit of batching for null requests, Figure 11 depicts a similar experiment with a request execution time of 5ms. We observe that batching improvements become insignificant with non-trivial execution costs. This demonstrates the importance of reducing execution costs, not just agreement overhead, for real applications.

6.5.2 Maintaining Spare VMs

In our previous experiments recovery VMs were kept in a paused state which provides a very fast recovery but consumes memory. Applications that have less stringent latency requirements can keep their recovery VMs hibernated on disk, removing the memory pressure on the system.

With a naive approach, maintaining VMs hibernated to disk can increase recovery latency by a factor proportional to their amount of RAM. This is because restoring a hibernated VM involves loading the VM's full memory contents from disk. The table below shows how our paged-out restore technique reduces the startup time for a VM with a 2GB memory allocation from over 40 seconds to less than 6 seconds.

Operation	Time (sec)
Xen Restore (2GB image)	44.0
Paged-out Restore (128MB→2GB)	5.88
Unpause VM	0.29
ZFS Clone	0.60

ZZ utilizes ZFS to simplify checkpoint creation at low cost. The ZFS clone operation is used during recovery to make snapshots from the previous checkpoint available to the recovery VMs. This can be done in parallel with initializing the recovery VMs, and incurs only minimal latency.

6.5.3 Limitations & Potential for Optimization

ZZ obtains lower hardware costs during graceful performance at the expense of increased delay when a fault occurs. We believe this is a valuable trade-off, and that obtaining state on demand will mitigate the recovery cost in many scenarios. However, there are some applications which contain too large a state or modify it too often for this to be reasonable. Recovering replicas in ZZ must obtain all the state objects relevant for replaying every request that occurred since the last checkpoint. Thus the recovery cost directly depends on the amount of state actively used during each checkpoint period. If the request replay procedure involves reads to a large number of state objects, then those objects will need to be transferred and verified before recovery can finish. One approach to reduce this cost is for ZZ's sleeping replicas to periodically wakeup and obtain the most recent application state. This approach is used in SPARE to reduce the amount of checkpoint data that must be obtained when recovering an out of date replica [Distler 2011b]. This form of proactive recovery can be easily employed by ZZ to prefetch state objects that only change across long time scales.

In addition to obtaining the relevant state, the recovering ZZ replicas must perform the computation required to

replay every request which involved a write since the last checkpoint. This cost again will depend on the rate at which checkpoints can be made and the execution cost of the write requests. For example, processing the writes in the BFT NFS application studied in Section 6.4.1 are relatively expensive, taking about 30ms per request replayed. If more than a few hundred such requests need to be processed, the recovery time could grow into the tens of seconds. We believe this cost could be reduced by having recovering replicas only replay a subset of the write requests which occurred since the last checkpoint. For example, there may be multiple writes to the same state object; only the most recent such request needs to be replayed. This would reduce both the number of requests that need to be replayed, and possibly the amount of state that needs to be immediately obtained from the last checkpoint. In fact, the replay process could be further optimized by only replaying the missed requests that are "dependencies" for processing the request which produced a fault (or any subsequent requests)—e.g. only those requests that modify a state object that is read by a later request.

7. Related Work

[Lamport 1982] introduced the problem of Byzantine agreement. Lamport also introduced the state machine replication approach [Lamport 1978] that relies on consensus to establish an order on requests. Consensus in the presence of asynchrony and faults has seen almost three decades of research. [Dwork 1988] established a lower bound of $3f + 1$ replicas for Byzantine agreement given partial synchrony, i.e., an unknown but fixed upper bound on message delivery time. The classic FLP [Fischer 1985] result showed that no agreement protocol is guaranteed to terminate with even one (benignly) faulty node in an asynchronous environment. Viewstamped replication [Oki 1988] and Paxos [Lamport 1998] describe an asynchronous state machine replication approach that is safe despite crash failures.

Early BFT systems [Kihlstrom 1998, Reiter 1995] incurred a prohibitively high overhead and relied on failure detectors to exclude faulty replicas. However, accurate failure detectors are not achievable under asynchrony, thus these systems effectively relied on synchrony for safety. Castro and Liskov's PBFT [Castro 1999] introduced a BFT SMR-based system that relied on synchrony only for liveness. The view change protocol at the core of PBFT shares similarities with viewstamped replication [Oki 1988] or Paxos [Lamport 1998] but incurs a replication cost of at least $3f + 1$ for safety. PBFT showed that the latency and throughput overhead of BFT can be low enough to be practical. The FARSITE system [Adya 2002] reduces the replication cost of a BFT file-system to $f + 1$; in comparison, ZZ has similar goals, but is able to provide the same cost reduction for any application which can be represented by a more general SMR system. ZZ draws inspiration from Cheap Paxos [Lamport 2004], which advocated the use of cheaper auxiliary nodes used only to handle crash failures of main nodes. Our contribution

is extending the idea to Byzantine faults and demonstrating its practicality through system design and implementation.

Virtualization has been used in several BFT systems recently since it provides a clean way to isolate services. The SPARE system also uses virtualized replicas to reduce the cost of BFT execution to $f + 1$ replicas [Distler 2011b]. SPARE exploits virtualization to reduce the cost of proactive recovery, allowing it to periodically wakeup passive replicas so they can efficiently catch up to the latest application state. ZZ amortizes recovery cost by obtaining only the necessary state on demand, but its sleeping replicas do not proactively obtain state like those in SPARE. Unlike ZZ, SPARE relies on a trusted component to order requests and does not fully explore the potential for response time inflation from malicious replicas. The cost of BFT execution has also been attacked in [Distler 2011a] where only $f + 1$ replicas process each request when there are no faults. The idea of "reactive recovery", where faulty replicas are replaced after fault detection, was used in [Sousa 2007], which employed virtualization to provide isolation between different types of replicas. In ZZ, reactive recovery is not an optional optimization, but a requirement since in order to make progress it must instantiate new replicas after faults are detected.

The Remus system uses virtualization to provide black-box crash fault tolerance using a standby VM replica [Cully 2008]. ZZ seeks to provide stronger Byzantine fault tolerance guarantees at a similar replication cost, although ZZ, like all BFT systems, requires application support for the BFT protocol. Terra is a virtual machine platform for trusted computing that employs a trusted hypervisor [Garfinkel 2003]; ZZ allows hypervisors to be Byzantine faulty.

8. Conclusions

In this paper, we presented ZZ, a new execution approach that can be interfaced with existing BFT-SMR agreement protocols to reduce the replication cost from $2f + 1$ to practically $f + 1$. Our key insight was to use $f + 1$ execution replicas in the normal case and to activate additional VM replicas only upon failures. We implemented ZZ using the BASE library and the Xen virtual machine and evaluated it on a prototype data center that emulates a shared hosting environment. The key results from our evaluation are as follows. (1) In a prototype data center with four BFT web servers, ZZ lowers response times and improves throughput by up to 66% and 33% in the fault-free case, when compared to systems using $3f + 1$ and $2f + 1$ replicas, respectively. (2) In the presence of multiple application failures, after a short recovery period, ZZ performs as well or better than $2f + 1$ replication and still outperforms BASE's $3f + 1$ replication. (3) The use of paused virtual machine replicas and on-demand state fetching allows ZZ to achieve sub-second recovery times. (4) We find that batching in the agreement nodes, which significantly improves the performance of null execution requests, yields no perceptible improvements for realistic applications with non-trivial request execution costs.

Overall our results demonstrate that in shared data centers hosting multiple applications with substantial request execution costs, ZZ can be a practical and cost-effective approach for providing BFT.

Acknowledgements: This work was supported in part by NSF grants CNS-0916972, CNS-0910671, CNS-0834243, and CNS-0720616. We also thank our shepherd, Jacob Lorch, for his many suggestions on improving the paper.

References

[Abd-El-Malek 2005] Michael Abd-El-Malek, Gregory R. Ganger, Garth R. Goodson, Michael K. Reiter, and Jay J. Wylie. Fault-scalable Byzantine Fault-Tolerant Services. *SIGOPS Oper. Syst. Rev.*, 39(5):59–74, 2005. ISSN 0163-5980.

[Adya 2002] Atul Adya, William J. Bolosky, Miguel Castro, Gerald Cermak, Ronnie Chaiken, John R. Douceur, Jon Howell, Jacob R. Lorch, Marvin Theimer, and Roger P. Wattenhofer. FARSITE: Federated, Available, and Reliable Storage for an Incompletely Trusted Environment. In *Proc. of the 5th Symposium on Operating Systems Design and Implementation (OSDI)*, 2002.

[Castro 1999] M. Castro and B. Liskov. Practical Byzantine Fault Tolerance. In *Proceedings of the Third Symposium on Operating Systems Design and Implementation*, February 1999.

[Castro 2002] Miguel Castro and Barbara Liskov. Practical Byzantine Fault Tolerance and Proactive Recovery. *ACM Transactions on Computer Systems (TOCS)*, 20(4), November 2002.

[Clement 2009] A. Clement, M. Marchetti, E. Wong, L. Alvisi, and M. Dahlin. Making Byzantine Fault Tolerant Systems Tolerate Byzantine Faults. In *6th USENIX Symposium on Networked Systems Design and Implementation (NSDI)*, April 2009.

[Cowling 2006] James Cowling, Daniel Myers, Barbara Liskov, Rodrigo Rodrigues, and Liuba Shrira. HQ Replication: A Hybrid Quorum Protocol for Byzantine Fault Tolerance. In *Proceedings of the Seventh Symposium on Operating Systems Design and Implementations (OSDI)*, Seattle, Washington, November 2006.

[Cully 2008] Brendan Cully, Geoffrey Lefebvre, Dutch Meyer, Mike Feeley, Norm Hutchinson, and Andrew Warfield. Remus: High Availability via Asynchronous Virtual Machine Replication. In *NSDI*, 2008.

[Distler 2011a] Tobias Distler and Rüdiger Kapitza. Increasing Performance in Byzantine Fault-Tolerant Systems with On-Demand Replica Consistency. In European Chapter of ACM SIGOPS, editor, *Proceedings of the EuroSys 2011 Conference (EuroSys '11)*, 2011.

[Distler 2011b] Tobias Distler, Rüdiger Kapitza, Ivan Popov, Hans P. Reiser, and Wolfgang Schröder-Preikschat. SPARE: Replicas on Hold. In Internet Society (ISOC), editor, *Proceedings of the 18th Network and Distributed System Security Symposium (NDSS '11)*, 2011.

[Dwork 1988] Cynthia Dwork, Nancy Lynch, and Larry Stockmeyer. Consensus in the Presence of Partial Synchrony. *Journal of the ACM*, 35(2), 1988.

[Fischer 1985] Michael J. Fischer, Nancy A. Lynch, and Michael S. Paterson. Impossibility of Distributed Consensus with One Faulty Process. *J. ACM*, 32(2):374–382, 1985. ISSN 0004-5411.

[Garfinkel 2003] Tal Garfinkel, Ben Pfaff, Jim Chow, Mendel Rosenblum, and Dan Boneh. Terra: a Virtual Machine-based Platform for Trusted Computing. In *SOSP '03: Proceedings of the nineteenth ACM symposium on Operating systems principles*, pages 193–206, New York, NY, USA, 2003. ACM Press.

[Guerraoui 2010] Rachid Guerraoui, Nikola Knežević, Vivien Quéma, and Marko Vukolić. The Next 700 BFT Protocols. In *EuroSys '10: Proceedings of the 5th European conference on Computer systems*, pages 363–376, New York, NY, USA, 2010. ACM. ISBN 978-1-60558-577-2.

[Kihlstrom 1998] Kim Potter Kihlstrom, L. E. Moser, and P. M. Melliar-Smith. The SecureRing Protocols for Securing Group Communication. In *HICSS '98: Proceedings of the Thirty-First Annual Hawaii International Conference on System Sciences-Volume 3*, Washington, DC, USA, 1998.

[Kotla 2007] Ramakrishna Kotla, Lorenzo Alvisi, Mike Dahlin, Allen Clement, and Edmund Wong. Zyzzyva: Speculative Byzantine Fault Tolerance. In *SOSP '07: Proceedings of twenty-first ACM SIGOPS Symposium on Operating Systems Principles*, New York, NY, USA, 2007. ACM.

[Lamport 1998] L. Lamport. Part Time Parliament. *ACM Transactions on Computer Systems*, 16(2), May 1998.

[Lamport 1982] L. Lamport, R. Shostack, and M. Pease. The Byzantine Generals Problem. *ACM Transactions on Programming Languages and Systems*, 4(3):382–401, 1982.

[Lamport 1978] Leslie Lamport. Time, Clocks, and the Ordering of Events in a Distributed System. *Commun. ACM*, 21(7), 1978.

[Lamport 2004] Leslie Lamport and Mike Massa. Cheap Paxos. In *DSN '04: Proceedings of the 2004 International Conference on Dependable Systems and Networks*, page 307, Washington, DC, USA, 2004. IEEE Computer Society. ISBN 0-7695-2052-9.

[Oki 1988] Brian M. Oki and Barbara H. Liskov. Viewstamped Replication: a General Primary Copy. In *PODC '88: Proceedings of the seventh annual ACM Symposium on Principles of distributed computing*, New York, NY, USA, 1988. ACM.

[Reiter 1995] Michael K. Reiter. The Rampart Toolkit for Building High-integrity Services. In *Selected Papers from the International Workshop on Theory and Practice in Distributed Systems*, London, UK, 1995. Springer-Verlag.

[Rodrigues 2001] Rodrigo Rodrigues, Miguel Castro, and Barbara Liskov. BASE: Using Abstraction to Improve Fault Tolerance. In *Proceedings of the eighteenth ACM symposium on Operating systems principles*, New York, NY, USA, 2001.

[Singh 2008] Atul Singh, Tathagata Das, Petros Maniatis, Peter Druschel, and Timothy Roscoe. BFT Protocols Under Fire. In *NSDI '08: Proceedings of the Usenix Symposium on Networked System Design and Implementation*, 2008.

[Sousa 2007] Paulo Sousa, Alysson N. Bessani, Miguel Correia, Nuno F. Neves, and Paulo Verissimo. Resilient Intrusion Tolerance Through Proactive and Reactive Recovery. In *Proceedings of the 13th Pacific Rim International Symposium on Dependable Computing*, Washington, DC, USA, 2007.

[Vandiver 2007] Ben Vandiver, Hari Balakrishnan, Barbara Liskov, and Sam Madden. Tolerating Byzantine Faults in Database Systems Using Commit Barrier Scheduling. In *Proceedings of the 21st ACM Symposium on Operating Systems Principles (SOSP)*, Stevenson, Washington, USA, October 2007.

[Wood 2011] Timothy Wood, Rahul Singh, Arun Venkataramani, Prashant Shenoy, and Emmanuel Cecchet. ZZ and the Art of Practical BFT. Technical report, University of Massachusetts Amherst, Feb. 2011.

[Yin 2003] J. Yin, J.P. Martin, A. Venkataramani, L. Alvisi, and M. Dahlin. Separating Agreement from Execution for Byzantine Fault Tolerant Services. In *Proceedings of the 19th ACM Symposium on Operating Systems Principles*, October 2003.

[ZFS 2004] ZFS. The Last Word in File Systems. http://www.sun.com/2004-0914/feature/, 2004.

9. Appendix

We defer proofs for Theorems 1 and 3 to a technical report [Wood 2011], and prove Theorem 2 here since it is the basis for ZZ's safety and liveness properties.

THEOREM 2. *If a wakeup occurs, ZZ will be able to terminate at least one faulty replica.*

To ensure this theorem, ZZ uses the following rule:

Wakeup Rule: A wakeup happens if and only if a mismatch report is "blocking".

A mismatch occurs when an agreement replica receives execution replies which are not identical. Suppose that for a particular request there are $g + c$ agreement replicas which experience a mismatch. Consider the *mismatch matrix* of size $(f + 1) * (g + c)$ where entry i, j corresponds to the reply by execution replica E_i as reported by agreement node A_j. Let the execution mismatch of a row be defined as the smallest number of entries that need to be changed in order to make all $g + c$ entries in that row identical. Let the smallest such execution mismatch across all $f + 1$ rows be m. The mismatch is considered blocking if $m < c$.

To understand the difference, consider two examples where the client receives conflicting responses P and Q, and $f = g = 1$,

Full Matrix	Mismatch Matrix	
$A_1 A_2 A_3 A_4$	$A_3 A_4$	
$E_1 : Q\,Q\,P\,Q$	$E_1 : P\,Q$	$c = 1$
$E_2 : Q\,Q\,Q\,P$	$E_2 : Q\,P$	$m = 1$

In this scenario, the mismatch matrix has size $g + c = 2$. Since $g = 1$, $c = 1$. Both rows require one entry to be changed in order to create a match, so the minimum mismatch is $m = 1$. Since $m = c$, this is not a blocking fault. The client will be able to receive a reply affirmation that Q is correct from A_1 and A_2. Note that if an additional node were woken up, it would not be possible to tell which execution node was faulty since it is impossible to tell if A_3 or A_4 is also faulty.

Full Matrix	Mismatch Matrix	
$A_1 A_2 A_3 A_4$	$A_2 A_3$	
$E_1 : Q\,Q\,Q\,P$	$E_1 : Q\,Q$	$c = 1$
$E_2 : Q\,P\,P\,P$	$E_2 : P\,P$	$m = 0$

The second scenario illustrates a blocking mismatch. In this case, the *rows* in the mismatch matrix require no changes, thus $m = 0$. Since $m < c$ we have a blocking fault. This makes sense because there is no way to tell whether Q or P is the correct response, so a wakeup is required. To ensure Theorem 2, we also must guarantee:

Lemma 1: A client will receive an affirmation certificate unless a mismatch is blocking.

Since $g + c$ agreement replicas report a mismatch, the number of replicas that have matching replies is $2g + 1 - c$. For a client to get an affirmation certificate, we must show that there are at least $g + 1$ *correct* agreement replicas with matching replies.

As all rows have an execution mismatch of at least m, at least m of the agreement replicas out of the $g + c$ reporting a mismatch must be faulty. To see why this is true, consider that some row in the matrix must correspond to a correct execution replica. If there are no faulty agreement replicas, then all entries in that row should be identical. However, recall that m is the minimum number of entries in any row which would need to be changed to make the whole row identical. This means that even the row for the correct execution replica requires at least m entries to be changed to match. Thus at least m agreement replicas in the mismatch matrix must be lying about the correct replica's response.

Since m agreement replicas that are part of the mismatch matrix are faulty, that means that at most $g - m$ of the remaining $2g + 1 - c$ replicas can be faulty. Therefore, $2g + 1 - c - (g - m) = g + 1 - c + m$ are correct. If the fault is categorized as non-blocking, then $m >= c$, giving us at least $g + 1$ correct agreement replicas with matching replies. These nodes will be able to provide an execution affirmation to the client without requiring a wakeup, proving the lemma.

Lemma 2: If a mismatch is blocking, then (a) at least one faulty replica can be shutdown and (b) the system will still be able to make a stable checkpoint.

The lemma is based on the shutdown rule defined in Section 4.3.2, and depends on whether any replicas can be convicted as faulty. To convict a faulty execution replica we need at least $g + 1$ agreement replicas to concur that the execution node produced an incorrect answer; such an execution replica is *convictably faulty*. Consider the complete $(2f + 1) * (3g + 1)$ response matrix obtained by the replica control daemon after all recovery replicas have produced a reply to the faulty request. Since at most g agreement replicas (columns) and f execution replicas (rows) can be faulty, there must be an $(f + 1) * (2g + 1)$ sub-matrix which contains identical, correct responses. The other entries in the matrix may contain some wrong response. The replica control daemon determines if an execution replica is faulty by looking at its row and checking if there are at least $g + 1$ entries with an incorrect response. Such an execution replica must be faulty because $g + 1$ agreement nodes report it gave an invalid reply, and at least one of those nodes must be correct.

If at least one replica can be convicted as faulty, then the shutdown rule trivially proves part (a) of the lemma. Part (b) must also hold because only new replicas or convictably faulty nodes will be terminated, leaving at least one old replica able to create a stable checkpoint.

If no replicas can be convicted due to collusion between the agreement and execution clusters, then ZZ will not shut down any nodes until the $f + 1$ correct execution replicas create a stable checkpoint, fulfilling part (b). After the checkpoint is made, f of the original replicas will be shutdown; this must include at least one faulty replica since multiple faulty replicas most collude to prevent conviction, satisfying part (a). This guarantees the lemma and proves Theorem 2.

Energy Management in Mobile Devices with the Cinder Operating System

Arjun Roy, Stephen M. Rumble, Ryan Stutsman, Philip Levis, David Mazières, Nickolai Zeldovich†

Stanford University and MIT CSAIL†

Abstract

We argue that controlling energy allocation is an increasingly useful and important feature for operating systems, especially on mobile devices. We present two new low-level abstractions in the Cinder operating system, *reserves* and *taps*, which store and distribute energy for application use. We identify three key properties of control – *isolation*, *delegation*, and *subdivision* – and show how using these abstractions can achieve them. We also show how the architecture of the HiStar information-flow control kernel lends itself well to energy control. We prototype and evaluate Cinder on a popular smartphone, the Android G1.

Categories and Subject Descriptors D.4.7 [*Operating Systems*]: Organization and Design

General Terms Design

Keywords energy, mobile phones, power management

1. Introduction

In the past decade, mobile phones have emerged as a dominant computing platform for end users. These very personal computers depend heavily on graphical user interfaces, always-on connectivity, and long battery life, yet in essence run operating systems originally designed for workstations (Mac OS X/Mach) or time-sharing systems (Linux/Unix).

Historically, operating systems have had poor energy management and accounting. This is not surprising, as their APIs standardized before energy was an issue. For example, the first commodity laptop with performance similar to a desktop, the Compaq SLT/286 [Com 1988], was released just one year before the C API POSIX standard. The resulting energy management limitations of POSIX have prompted a large body of research, ranging from CPU

scheduling [Flautner 2002] to accounting [Zeng 2003] to offloading networking. Despite this work, current systems still provide little, if any, application control or feedback: users have some simple high-level sliders or toggles.

This limited control and visibility of energy is especially problematic for mobile phones, where energy and power define system lifetime. In the past decade, phones have evolved from low-function proprietary applications to robust multiprogrammed systems with applications from thousands of sources. Apple announced that as of April 2010 their App Store houses 185,000 apps [App 2010] for the iPhone with more than 4 billion application downloads. This shift away from single-vendor software to complex application platforms means that the phone's software must provide effective mechanisms to manage and control energy as a resource. Such control will be even more important as the danger grows from buggy or poorly designed applications to potentially malicious ones.

In the past year, mobile phone operating systems began providing better support for understanding system energy use. Android, for example, added a UI that estimates application energy consumption with system call and event instrumentation, such as processor scheduling and packet counts. This is a step forward, helping users understand the mysteries of mobile device lifetime. However, while Android provides improved *visibility* into system power use, it does not provide *control*. Outside of manually configuring applications and periodically checking battery use, today's systems cannot do something as simple as controlling email polling to ensure a full day of device use.

This paper presents Cinder, a new operating system designed for mobile phones and other energy-constrained computing devices. Cinder extends the HiStar secure kernel [Zeldovich 2006] to provide new abstractions for controlling and accounting for energy: reserves and taps. *Reserves* are a mechanism for resource delegation, providing fine-grained accounting and acting as an allotment from which applications draw resources. Where reserves describe a quantity of a resource, *taps* place rate limits on resources flowing between reserves. By connecting reserves to one another, taps allow resources to flow to applications. Taps and reserves compose together to allow applications to express their intentions, enabling policy enforcement by the operating system.

Cinder estimates energy consumption using standard device-level accounting and modeling [Zeng 2002]. HiStar's explicit information flow control allows Cinder to track which parties are responsible for resource use, even across interprocess communication calls serviced in other address spaces. Without needing any additional state or support code, Cinder can accurately amortize costs across principals, such as the energy cost of turning on the radio to multiple applications that simultaneously need Internet access.

While Cinder runs on a variety of hardware platforms (AMD64, i386, ARM), the most notable is the HTC Dream, a.k.a. the Android G1. To the best of our knowledge, other than extensions to Linux, Cinder is the first research operating system that runs on a mobile phone. The reason for such a first is simple: the closed nature of phone platforms makes porting an operating system exceedingly difficult.

This paper makes three research contributions. First, it proposes reserves and taps as new operating system mechanisms for managing and controlling energy consumption. Second, it evaluates the effectiveness and power of these mechanisms in a variety of realistic and complex application scenarios running on a real mobile phone. Third, it describes experiences in writing a mobile phone operating system, outlining the challenges and impediments faced when conducting systems research on the dominant end-user computing platform of this decade.

2. A Case for Energy Control

This section motivates the need for low-level, fine-grained energy control in a mobile device operating system. It starts by reviewing some of the prior work on energy visibility and the few examples of coarse energy control. Using several application examples as motivation, it describes three mechanisms an OS needs to provide for energy: isolation, delegation, and subdivision. The next section describes reserves and taps, abstractions which provide these mechanisms at a fine granularity.

2.1 Prior Work on Visibility and Control

Managing energy requires accurately measuring its consumption. A great deal of prior work has examined this problem for mobile systems, including ECOSystem [Zeng 2002], Currentcy [Zeng 2003], PowerScope [Flinn 1999b], and PowerBooter [Zhang 2010]. These systems use a model of the power draw of hardware components based on hardware states. For example, an 802.11b card draws only slightly more power while transmitting than receiving, whereas a CPU's power draw increases with utilization. Current mobile phone energy accounting systems, such as Android's, use this approach. Cinder also does as well; Section 4 provides the details.

Early systems like ECOSystem [Zeng 2002] proposed mechanisms by which a user could control per-application energy expenditure. ECOSystem, in particular, introduced an abstraction called Currentcy, which gives an application the ability to spend a certain amount of energy, up to a fixed cap. This flat hierarchy of energy principals – applications – is reasonable for simple large applications. Mobile applications and systems today, however, are far more complex and involve multiple principals. For example, web browsers run active code as well as possibly untrusted plugins, network daemons control access to the cellular data network, and peripherals have complex energy profiles.

2.2 Isolation, Delegation, and Subdivision

We believe that for applications to effectively control energy, an operating system must provide three energy management mechanisms: *isolation*, *delegation*, and *subdivision*. We motivate these mechanisms through application examples that we follow through the rest of the paper.

The first mechanism is isolation. Isolation is a fundamental part of an operating system. Memory and inter-process communication (IPC) isolation provide security, while CPU and disk space isolation ensure that processes cannot starve others. Isolating energy consumption is similarly important. An application should not be permitted to consume inordinate amounts of energy, nor should it be able to deprive other applications. Consider two processes in a system, each with some share of system energy. To improve system reliability and simplify system design, the operating system should isolate each process' share from the other's. If one process forks additional processes, these children must not be able to consume the energy of the other.

The second mechanism is delegation. Delegation allows a principal to loan any of its available energy and power to another principal. After delegation, either the resource donor or the recipient can freely consume the delegated resources. Furthermore, if there are multiple donors delegating to this recipient, the resources are pooled for use by the recipient. Resource delegation is an important enabler of inter-application cooperation. For example, the Cinder *netd* networking stack transfers energy into a common radio activation pool when an application cannot afford the high initial expense of powering up the radio. By delegating their energy to the radio, multiple processes can contribute to expensive operations; this may not only improve quality of service, but even reduce energy consumption.

The third mechanism is subdivision. Subdivision allows applications to partition their available energy. Combined with isolation, subdivision allows an application to give another principal a partial share of its energy, while being assured that sure that the rest will remain for its own use. For example, modern web browsers commonly run plugins, some of which may even be untrusted. If a browser is granted a finite amount of power, it might want to protect itself from buggy or poorly written plugins that could waste CPU energy. Subdivision lets the browser give full control over a fraction of its energy allotment to plugins. Isolation further ensures that each plugin component does not consume more than its share.

2.3 Prior Systems

Prior systems like ECOSystem [Zeng 2002, 2003] only partially support isolation and subdivision: child processes share the resources of their parent. This is sufficient when applications are static entities, but not when they spawn new processes and invoke complex services. The web browser demonstrates the problem: it has no way to prevent its plugins from consuming its own resources once they are spawned. Cinder's subdivision lends naturally to familiar and standard abstractions such as process trees, resource containers, and quotas.

Furthermore, prior systems do not permit delegation, which is akin to priority inheritance. For always-on systems which have small variations in power draw, such as the laptops for which they were designed, this is not a serious limitation. On mobile phones, however, which have almost two orders of magnitude difference in active and sleep power, the cost of powering up peripherals, such as the wireless data interface, can be significant. Delegation provides a means to facilitate application cooperation.

3. Design

Cinder is based on HiStar [Zeldovich 2006], a secure operating system built upon information flow control. Cinder adds two new fundamental kernel object types: *reserves* and *taps*. This section gives a brief overview of HiStar and key features related to resource management, describes reserves and taps, gives examples of how they can be used, and details how they are secured.

3.1 HiStar

HiStar is composed of six first-class kernel objects, all protected by a security *label*. Its segments, threads, address spaces, and devices are similar to those of conventional kernels. *Containers* enable hierarchical control over deallocation of kernel objects – objects must be referenced by a container or face garbage collection. *Gates* provide protected control transfer of a thread from one address space to a named offset in another; they are the basis for all IPC.

3.2 Reserves

A reserve describes a right to use a given quantity of a resource, such as energy. When an application consumes a resource the Cinder kernel reduces the values in the corresponding reserve. The kernel prevents threads from performing actions for which their reserves do not have sufficient resources. Reserves, like all other kernel objects, are protected by a security label (§3.5) that controls which threads can observe, use, and manipulate it.

All threads draw from one or more energy reserves. Cinder's CPU scheduler is energy-aware and allows a thread to run only when at least one of its energy reserves is not empty. Threads that have depleted their energy reserves cannot run. Tying energy reserves to the scheduler prevents new spending, which is sufficient to throttle energy consumption.

Reserves allow threads to delegate and subdivide resources. As a simple example, an application granted 1000 mJ of energy can subdivide its reserve into an 800 mJ and a 200 mJ reserve, allowing another thread to connect to the 200 mJ reserve. However, threads rarely manage energy in such concrete quantities, preferring instead to use taps (§3.3). A thread can also perform a reserve-to-reserve transfer provided it is permitted to modify both reserves.

Reserves also provide accounting by tracking application resource consumption. Applications may access this accounting information in order to provide energy-aware features. Finally, reserves can be deleted directly or indirectly when some ancestor of their container is deleted, just as a file can be deleted either directly or indirectly when a directory containing it is deleted in a Unix system.

3.3 Taps

A tap transfers a fixed quantity of resources between two reserves per unit time, which controls the maximum rate at which a resource can be consumed. For example, an application reserve may be connected to the system battery via a tap supplying 1 mJ/s (1 mW).

Taps aid in subdividing resources between applications since partitioning fixed quantities is impractical for most policies. A user may want her phone to last at least 5 hours if she is surfing the web; the amount of energy the browser should receive is relative to the length of time it is used. Providing resources as a rate naturally addresses this.

Another approach, which Cinder does not take, would be to implement transfer rates between reserves through threads that explicitly move resources and enforce rate-limiting as well as accounting. Given five applications, each to be limited to consume an average of 1 W, the system could create five application reserves and threads, with each thread transferring while tracking and limiting energy into each of these applications' reserves. However, this fine-grained control would cause a proliferation of these special-purpose threads, adding overhead and decreasing energy efficiency.

Taps are made up of four pieces of state: a rate, a source reserve, a sink reserve, and a security label containing the privileges necessary to transfer the resources between the source and sink (§3.5). Conceptually, it is an efficient, special-purpose thread whose only job is to transfer energy between reserves. In practice, transfers are executed in batch periodically to minimize scheduling and context-switch overheads.

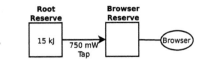

Figure 1. A 15 kJ battery, or *root reserve*, connected to a reserve via a tap. The battery is protected from being misused by the web browser. The web browser draws energy from an isolated reserve which is fed by a 750 mW tap.

3.4 Resource Consumption Graph

Reserves and taps form a directed graph of resource consumption rights. The root of the graph is a reserve representing the system battery; all other reserves are a subdivision of this root reserve. Figure 1 shows a simple example of a web browser whose consumption is rate limited using a tap. The tap guarantees that even if the browser is aggressively using energy the battery will last at least 5 hours (15,000 J at 0.750 J/s is about 5.6 hours).

3.5 Access Control & Security

Any thread can create and share reserves or taps to subdivide and delegate its resources. This ability introduces a problem of fine-grained access control. To solve this, reserves and taps are protected by a security label, like all other kernel objects. The label describes the privileges needed to observe, modify, and use the reserve or tap.

Using resources from a reserve requires both observe and modify privileges: observe because failed consumption indicates the reserve level (zero) and modify for when consumption succeeds. Since a tap actively moves resources between a source and sink reserve, it needs privileges to observe and modify both reserve levels; to aid with this, taps can have privileges embedded in them.

4. Cinder on the HTC Dream

Controlling energy requires measuring or estimating its consumption. This section describes Cinder's implementation and its energy model. The Cinder kernel runs on AMD64, i386, and ARM architectures. All source code is freely available under open-source licenses. Our principal experimental platform is the HTC Dream (Google G1), a modern smartphone based on the Qualcomm MSM7201A chipset.

4.1 Energy accounting

Energy accounting on the HTC Dream is difficult due to the closed nature of its hardware. It has a two-processor design, as shown in Figure 2. The operating system and applications run on an ARM11 processor. A secure, closed ARM9 coprocessor manages the most energy hungry, dynamic, and informative components (e.g. GPS, radio, and battery sensors). The ARM9, for example, exposes the battery level as an integer from 0 to 100.

Recent work on processors has shown that fine-grained performance counters can enable accurate energy estimates

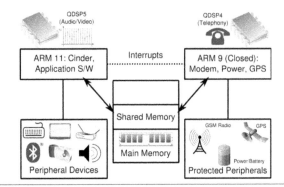

Figure 2. The two ARM cores in the MSM7201A chipset. Cinder runs on the ARM11, whereas the ARM9 controls access to sensitive hardware including the radio and GPS. The two communicate via shared memory and interrupt lines.

within a few percent [Economou 2006; Snowdon 2009]. Without access to such state in the HTC Dream, however, Cinder relies on the simpler well-tested technique of building a model from offline-measurements of device power states in a controlled setting [Flinn 1999b; Fonseca 2008; Zeng 2002]. Phones today use this approach, and so Cinder has equivalent accuracy to commodity systems.

4.2 Power Model

Our energy model uses device states and their duration to estimate energy consumption. We measured the Dream's energy consumption during various states and operations. All measurements were taken using an Agilent Technologies E3644A, a DC power supply with a current sense resistor that can be sampled remotely via an RS-232 interface. We sampled both voltage and current approximately every 200 ms, and aggregated our results from this data.

While idling in Cinder, the Dream uses about 699 mW and another 555 mW when the backlight is on. Spinning the CPU increases consumption by 137 mW. Memory-intensive instruction streams increase CPU power draw by 13% over a simple arithmetic loop. However, the HTC Dream does not have hardware support to estimate what percentage of instructions are memory accesses. The ARM processor also lacks a floating point unit, leaving us with only integer, control flow, and memory instructions. For these reasons, our CPU model currently does not take instruction mix into account and assumes the worst case power draw (all memory intensive operations).

4.3 Peripheral Power

The baseline cost of activating the radio is exceptionally high: small isolated transfers are about 1000 times more expensive, per byte, than large transfers. Figure 3 demonstrates the cost of activating the radio and sending UDP packets to an echo server that returns the same contents. Results demonstrate that the overhead involved dominates the total

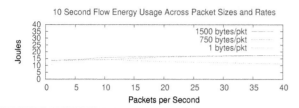

Figure 3. Radio data path power consumption for 10 second flows across six different packet rates and three packet sizes. Short flows are dominated by the 9.5 J baseline cost shown in Figure 4. For this simple static test, data rate has only a small effect on the total energy consumption. The average cost is 14.3 J (minimum: 10.5, maximum: 17.6).

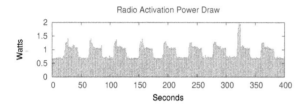

Figure 4. Cost of transitioning from the lowest radio power state to active. One UDP packet is transmitted approximately every 40 seconds to enable the radio. The device fully sleeps after 20 seconds, but the average plateau consumes an additional 9.5 J of energy over baseline (minimum 8.8 J, maximum 11.9 J). Power consumption for a stationary device can often be predicted with reasonable accuracy, but outliers, such as the penultimate transition, occur unpredictably.

power cost for flows lasting less than 10 seconds in duration, regardless of the bitrate.

Figure 4 shows this activation cost. An application powers up the radio by sending a single 1-byte UDP packet. The secure ARM9 automatically returns to a low power mode after 20 seconds of inactivity. Because the ARM9 is closed, Cinder cannot change this inactivity timeout.

With this workload, it costs 9.5 joules to send a single byte! One lesson from this is that coordinating applications to amortize energy start-up costs could greatly improve energy efficiency. In §5.5 we demonstrate how Cinder can use reserves and taps for exactly this purpose.

4.4 Mobility & Power Model Improvements

Cinder's aim is to leverage advances in energy accounting (see §8.2) to allow users and applications to provision and manage their limited budgets. Accurate energy accounting is an orthogonal and active area of research. Cinder is adaptable and can take advantage of new accounting techniques or information exposed by device manufacturers.

```
// Create a reserve
object_id_t res_id;
res_id = reserve_create(container_id, res_label);
objref res = OBJREF(container_id, res_id);

// Create a tap and connect it between
// the battery and the new reserve
object_id_t tap_id;
tap_id = tap_create(container_id, root_reserve,
                    res, tap_label);
objref tap = OBJREF(container_id, tap_id);
// Limit the child to 1 mW
tap_set_rate(tap, TAP_TYPE_CONST, 1);

if (fork() == 0) {
    // child process: switch to new reserve before exec
    self_set_active_reserve(res);
    execv(args[0], args);
}
```

Figure 5. energywrap excerpt without error handling.

5. Applications

To gain experience with Cinder's abstractions, we developed applications using reserves and taps. This section describes these applications, including a command-line utility that augments existing applications with energy policies, an energy constrained web browser that further isolates itself from its browser plugins, and a task manager application that limits energy consumption of background applications.

5.1 energywrap

Taking advantage of the composability of Cinder's resource graph, the energywrap utility allows any application to be sandboxed even if it is buggy or malicious. energywrap takes a rate limit and a path to an application binary. The utility creates a new reserve and attaches it to the reserve in which energywrap started by a tap with the rate given as input. After forking, energywrap begins drawing resources from the newly allocated reserve rather than the original reserve of the parent process and executes the specified program. This allows even energy-unaware applications to be augmented with energy policies.

The sandboxing policy provided by energywrap is implemented in about 100 lines of C++. An excerpt is shown in Figure 5. HiStar provides a wrap utility designed to isolate applications with respect to privileges and storage resources. Coupling this utility with energywrap allows any application or user to provide a virtualized environment to any thread or application. Section 6.1 evaluates the effectiveness of energy sandboxing and isolation.

energywrap has proved useful in implementing policies while designing and testing Cinder, particularly for legacy applications that have no notion of reserves or taps. Since energywrap runs an arbitrary executable, it is possible to use energywrap to wrap itself or shell scripts, which may invoke energywrap with other scripts or applications. This

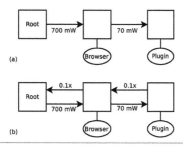

Figure 6. (a) A web browser configured to run for at least 6 hours on a 15 kJ battery. The web browser further ensures that its plugin cannot use more than 10% of its energy. (b) Adding 0.1x backward proportional taps promotes sharing of excess energy unused by the browser and plugin.

allows a wide class of ad hoc policies to be scripted using standard shell scripting or on-the-fly at the command line.

5.2 Fine-grained Control

Mobile browsers now support plugins like Adobe Flash [Fla 2009], and we can expect more plugins and extensions to follow. On a device where resources are precious, it is important to have tight control over these plugins.

In Cinder, an application may be given some fixed rate or quota of energy using reserves and taps. A web browser may, for example, want to also run a plugin while ensuring that it cannot starve other plugins or even the browser itself. Shown in Figure 6a, the browser can allocate a separate reserve for the plugin and connect it to its own energy via a low rate tap.

Often a single plugin (e.g. Flash) may be handling a number of pages or requests all in a single process. To scale the energy given to the plugin with the number of pages it is handling, the browser can add a tap per page. When a particular page is no longer being handled (e.g. the user navigates away) the taps associated with that page can be automatically garbage collected, effectively revoking those power sources.

Cinder includes a simple graphical web browser based on links2 that runs in Xorg or standalone against the framebuffer. It is augmented with an extension running in a separate process, whose energy usage is subdivided and isolated from the browser. The browser can send requests to the extension process (for ad blocking, etc.), and if the extension is unresponsive due to lack of energy the browser can display the unaugmented page.

5.2.1 Reclaiming Unused Resources

Consider a problem common to many applications: a web browser would like to allow a plugin to consume resources quickly while making sure it shares unused resources. The plugin may fully utilize peripherals and drive the device at peak power, requiring a reserve fed with a high rate tap. This raises a problem: if the plugin draws less than its tap rate, the reserve will slowly fill with energy that no other application can use.

To solve this problem, an application can use a proportional tap. These taps transfer a fraction of their source reserve's resource per unit time, rather than a fixed quantity. Figure 6b shows the fix to the browser; the plugin reserve on the right is limited to a maximum average power draw of 70 mW. The backwards proportional tap means the plugin reserve can store up to 10 s of this power (700 mJ) for bursty operations. Once the reserve reaches 700 mJ, the backwards proportional tap drains the reserve as quickly as the forward constant tap fills it. Similarly, the browser's reserve can accumulate up to 7000 mJ while being forced to share unused energy with other applications.

5.2.2 Hoarding and Resource Decay

Backward proportional taps alone are insufficient for preventing malicious applications from hoarding. Threads can sidestep taxation by creating a new reserve with no proportional taps and periodically transferring resources to it. The application could, over time, accumulate energy equal to the battery and starve the rest of the system.

To prevent this, Cinder could provide a `reserve_clone()` rather than `reserve_create()`. This call would take a reserve that an application has access to and create a new reserve taking care to duplicate any backward proportional taps that the application does not have the permission to remove. Additionally, Cinder would need to disallow system calls that transfer resources from a fast-draining reserve to a more slow-draining reserve unless the caller has proper permission (that is, the permission to remove all the backward taps from the source reserve that do not have a corresponding backward tap at the target reserve).

These constraints eliminate hoarding, but complicate applications that are not malicious. Therefore, in practice, Cinder prevents hoarding by imposing a global, long-term decay of resources across all reserves; every reserve has an implicit proportional backward tap to the battery.

By default, Cinder is configured to leak 50% of reserve resources after a period of 10 minutes. This long (but short compared to the period between battery recharges) half-life allows applications to accumulate and store energy for significant periods, and permits the system to make large-scale long-term hoarding impossible. ESX Server [Waldspurger 2002] successfully uses a similar "idle memory tax" to mitigate hoarding of unused memory between virtual machines.

Further experience with these abstractions is needed to understand whether the trade-offs associated with the more fundamental solution for hoarding are worth making.

5.3 Energy-Aware Applications

Using Cinder, developers can gain fine-grained control of resources within their applications, providing a better experience to end users. This includes adaptive policies for programs where partial or degraded results are still useful, and

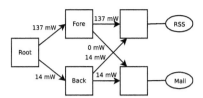

Figure 7. RSS is running in the foreground so the task manager has set its tap to give it additional power. Mail is running in the background, and can only draw energy from the background reserve. This ensures that actual battery consumption matches the user's expectation that the visible application is responsible for most energy consumption.

offer a compromise between battery life and user experience. For example, smart applications may scale the quality of streaming video or reduce texture quality in a game when available energy is low, since the user can still watch a video or play a game when insufficient resources are available to run at full fidelity.

As a concrete example, we have implemented an energy-aware network picture gallery. The application has a separate thread for downloading images, using an energy reserve distinct from the main thread. The rate the application consumes energy from this reserve depends on the frequency of image requests and the requested image sizes. The application checks the levels in the reserve periodically. A drop in the reserve level indicates that the downloader is consuming energy too quickly and will be throttled if it cannot curb consumption. In this case, the downloader only requests partial data from the remote interlaced PNG images, which yields a lower quality image in exchange for reduced data transfer over the network (and lower consumption by the device). Section 6.2 evaluates the effectiveness of these adaptations.

5.4 Background Applications

Background applications complicate resource management. Despite being invisible to the user, an application may be using resources. This discontinuity between reality and user perception makes the user suspicious of foreground applications they have used frequently, which may not be responsible. Cinder provides not only a means to understand which applications are using resources, but also a means to manage those resources to meet user expectations. Since the user naturally suspects foreground applications of using energy, he can easily manage his use of those applications. Cinder's job, then, is to manage background applications to prevent them from interfering with the user's natural intuition.

Figure 7 shows how Cinder accomplishes this. Each application has a reserve from which it draws energy. Each such application's reserve is then connected to two other reserves via taps. The first is the foreground reserve, which is connected to the battery via a high rate tap. The second is a low rate reserve connected to the battery via a low rate tap. An application's tap to the background reserve always

allows energy to flow; however, the foreground tap is set to a rate of 0 while the application is running in the background, and is set to a high value when the application is running in the foreground. The task manager is the creator of the tap connecting the application to the foreground reserve and, by default, is the only thread privileged to modify the parameters on the tap. Since programs are confined to low power while in the background, the user's expectations are respected. Section 6.3 evaluates this configuration.

5.5 Cooperative Network Stack

Some of the most energy-hungry devices on a mobile platform have complex, non-linear power models (e.g. the data path and the GPS). Careful control over how applications use such devices can result in energy savings. Section 4 shows that the radio has a high initial cost and a much smaller amortized price for bulk transfers. This power profile is a problem for some periodic background applications like email checkers, RSS feed downloaders, weather widgets, and time synchronization daemons. Cinder's network stack, *netd*, improves energy efficiency for this typical class of applications through using two mechanisms: precise resource accounting across process boundaries and flexible sharing and resource transfer control.

5.5.1 Accurate Accounting via Gates

To accurately track which threads cause resource consumption, Cinder uses HiStar's gates, which form the basis of inter-process communication. A gate is a named entry point in an address space, typically corresponding to a daemon or system service available over IPC. Unlike traditional IPC, in which a thread in a client process sends a message to a thread in a server process, here the calling thread itself enters the server's address space.

Since Cinder tracks resource consumption by the active reserve of a thread, the caller of a system-wide service, like *netd*, is billed for resource consumption it causes, even while executing in the other address space. Other systems, such as Linux, would need some form of message tracking during inter-process communication in order to heuristically bill the principals for resource consumption, whereas Cinder provides accurate accounting naturally. Section 7.1 details the complications that arose in reproducing Cinder atop Linux.

5.5.2 Encouraging Cooperation

To facilitate sharing, *netd* contains a reserve where threads cooperatively save up energy for a radio power up event. For each thread that makes a network system call, if the sum of its own reserve and *netd*'s reserve are not sufficient for the power on, the call blocks, contributes the energy acquired by its taps to the *netd* reserve, and sleeps to accumulate more. When there is sufficient energy to turn the radio on and perform the transmissions requested by the waiting threads, Cinder debits the reserve and permits the threads to proceed. The *netd* reserve is not subject to the system global half-

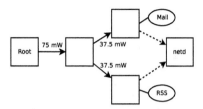

Figure 8. The mail checker and RSS feed downloader are constrained to use up to 37.5 mW apiece. When making network requests, netd explicitly transfers energy from their reserves into its own reserve. Once the requesting application's reserve, combined with the netd reserve, has enough energy, the radio will turn on. This simple policy helps synchronize applications' network access, reducing active radio time and saving energy.

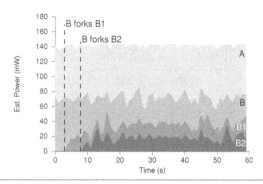

Figure 9. Stacked graph of Cinder's CPU energy accounting estimates during isolated process execution. Process A's energy consumption is isolated from other processes' energy use despite B's periodic spawning of child processes (B1 and B2). The sum of the estimated power of the individual processes closely matches the measured true power consumption of the CPU of about 139 mW during this experiment.

life, as the process is trusted not to hoard energy and, by construction, only stores enough energy to activate the radio before being expended.

Cinder estimates the cost of radio access by tracking when network transmit and receive events occur. For instance, if the radio has been idle for 20 seconds or more, threads wishing to use the network must contribute enough energy to turn the radio on and maintain the active power state until it idles again (§4). Once the radio is on, back-to-back actions are cheaper than ones with more delay between them because they extend the active period (delay the next idle period) less significantly.

For example, if the radio has been active for one second, it will automatically idle again 19 seconds later, so transmitting now only extends the active period by 1 second. However, if the radio is active but no packets have been sent or received for 15 seconds, transmitting now will extend the active period by an additional 15 seconds – the same action becomes much more expensive.

This leaves the problem of how to charge for incoming packets since energy has already been spent to receive them. To facilitate this, threads can debit their own reserves up to or into debt even if the cost can only be determined after-the-fact. This allows user space accounting; for example, in this case the receiving thread under the control of *netd*'s *send* gate debits its own reserve when packets are delivered to it.

Section 6.4 evaluates the effectiveness of *netd* in aiding cooperation between applications to increase the responsiveness of services while retaining their energy budget.

6. Evaluation

Using the applications described in §5, we evaluate whether Cinder meets the requirements described in Sections 1 and 2: can it control energy, provide visibility into the energy of a running system, and provide subdivision, delegation, as well as isolation? Furthermore, we evaluate whether Cinder can facilitate dynamic energy-aware applications and improve a system's energy efficiency by managing complex devices with non-linear power consumption.

All experiments exception the image viewer of §6.2 use Cinder running on an HTC Dream. The image viewer evaluation was performed on a Lenovo T60p laptop. To measure power draw, we connect the Dream to the Agilent E3644A DC power supply. To monitor reserve energy levels we use the Dream's serial port output.

6.1 Isolation, Subdivision, and Delegation: Buggy and Malicious Applications

We first show how a simple use case – protecting the system from a buggy or malicious energy hog – requires isolation, subdivision, and delegation. Figure 9 shows a stacked plot of Cinder's energy accounting estimates of two processes, A and B. In this experiment, the system is configured to evenly subdivide and delegate enough power to fully utilize the CPU between the two processes (about 68 mW to each process since running the CPU costs 137 mW).

Process B spawns a new child process at about 5 seconds (B1) and again at about 10 seconds (B2). Without reserves and taps, these additional processes would cause A to receive a smaller share of the CPU. Here, however, Process A is isolated from these forks and still consumes about 50% of the CPU (and power share).

This experiment highlights the fine-grained nature of Cinder's control: not only is A isolated from B, but B is also able to protect itself from its own children, B1 and B2. Rather than have its children draw from B's own reserve, B creates two new reserves subdividing and delegating its power to each using two taps. Each of the taps has one-quarter the power of B's tap, such that after spawning both they are using half of B's power. Figure 9 shows that both A and B's policies are composed and enforced in the expected way.

6.2 Subdivision and Delegation: Image Viewer

To demonstrate the practicality of energy-aware applications in Cinder, we used our image viewer described in §5.3,

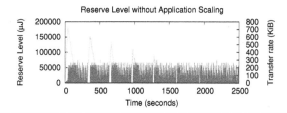

Figure 10. The same image viewer application as in §5.3, but without dynamic scaling of image quality. The line represents energy in the downloader's reserve while the bars represent the amount of data downloaded per image.

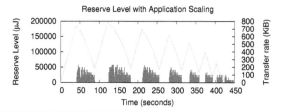

Figure 11. Image viewer with energy-aware scaling of image quality enabled. As energy becomes scarce, quality is lowered and less data is downloaded per image. The experiment takes less than one-fifth the time to complete within the energy budget versus the non-adaptive viewer due to adaptation to reduced available energy.

tested with and without energy-aware image scaling. The tests mimic a user loading a page of images, pausing to view the images, and then requesting more. We tracked the energy reserve levels, the amount of data transferred over the network interface, the download time for each batch of images, and the average bytes transferred per image over time. Each image was of similar size (∼2.7 MiB) and each batch contained the same number of images. Pausing between batches allowed the energy reserve for the downloader thread to fill at a constant rate. The first pause lasted for 40 seconds, with each successive pause being 5 seconds shorter than the previous pause, so a smaller amount of energy built up in the reserve after each batch was downloaded.

When image download sizes are not scaled back as in Figure 10, the amount of data transferred stays constant per batch. With each successive batch, the amount of energy in the reserve at the start of the batch decreases since the user pauses more briefly after downloading. Thus the reserve runs out soon after the start of each batch in this case, with the image transfers stalling until enough energy is available for the thread to continue, causing a long run time.

When image requests use energy-aware scaling as in Figure 11, the quality of images and bytes transferred for each image drops as the energy level dips, and the transfer time per image decreases with the smaller image data. Over the course of the test, the level of energy present in the reserve dropped below the threshold, but never to zero. The images

downloaded 5 times more quickly than the viewer which does not scale the images.

6.3 Delegation and Subdivision: Background Applications

Section 5.4 presented a configuration where system power is subdivided into a highly powered task manager reserve and a low powered background reserve. These reserves delegate their energy to applications running in the foreground and background, respectively, allowing background applications to continue to make slow forward progress, but keeping foreground applications responsive. This experiment uses a configuration identical to Figure 7.

Figure 12a shows two processes spinning on the CPU, initially in the background. The background tap provides the two of them 14 mW, enough to keep the 137 mW CPU at 10% utilization. At about 10 seconds, the task manager selects Process A as the foreground process, granting it enough energy to fully utilize the CPU (137 mW). Process B continues to run according to its background power share of 14 mW. At the 20 second mark, the task manager retires Process A to the background by setting its foreground tap rate to 0 mW. At 30 seconds, the task manager gives Process B access to the foreground resources and, similarly, returns it to the background at 40 seconds.

Figure 12b highlights the need for Cinder to prevent large-scale hoarding. The configuration is the same except the foreground tap gives 300 mW of power. Because 300 mW is greater than the CPU cost of 137 mW, applications in the foreground can accumulate excess energy. The two processes move in and out of the foreground as before, but this accumulated energy changes their behavior. When B is moved to the foreground, A still has plenty of energy, and so competes with B for the CPU such that each receives a 50% share. After A exhausts its energy, it returns to its original 14 mW. Shortly thereafter, B moves to the background as well. But just as A did, B accumulated energy during its time in the foreground and so is able to use ∼90% of the CPU until it exhausts its reserve.

The system-wide half-life both caps the total energy hoarding possible during foreground operation and returns applications to the natural background power over a 10 minute period. This allows a process to perform an elevated amount of work briefly after returning to background status if it underutilized its resources while in the foreground.

6.4 Delegation: Cooperative Network Stack

We demonstrate the effectiveness of Cinder's modified *netd* (shown in Figure 8), comparing it to an energy-unrestricted network stack. In both experiments, an RSS feed downloader starts with a poll interval of 60 seconds. Fifteen seconds later, a mail fetcher daemon starts, also with a 60 second poll interval. Both applications are provided enough power to start the radio every 60 seconds, if they work in unison.

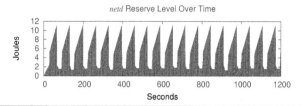

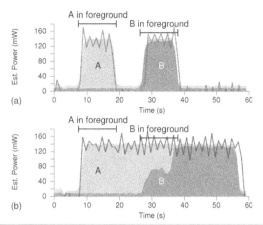

Figure 12. Stacked graph of Cinder's CPU energy accounting estimates as processes A and B spin on the CPU. Together, they are allowed 14 mW while in the background. The task manager runs A in the foreground in the 10 - 20 second interval and B in the foreground during the 30 - 40 second interval. (a) shows the results for the foreground tap providing the process with 137 mW (the precise cost of using the CPU at 100%). (b) shows the foreground tap providing the process with 300 mW. The dotted line shows actual power measurements compensated for baseline power draw with an idle CPU and averaged over 1 second intervals.

Figure 14. The level of the reserve into which the two background applications transfer their allotted joules. When the reserve reaches a level sufficient to pay for the cost of transitioning the radio to the active state, it is debited, the radio is turned on, and the processes proceed to use the network. Although Figure 4 showed an average 9.5 J cost to power up the radio, *netd* requires 125% of this level before turning the radio on, essentially mandating that applications have extra energy to transmit and receive subsequent packets. Therefore, the reserve does not empty to 0.

	Non-Coop	Coop	Improv
Total Time	1201s	1201s	N/A
Total Energy	1238J	1083J	12.5%
Active Time	949s	510s	46.3%
Active Energy	1064J	594J	44.2%

Table 1. Improvements in energy consumption and active radio time using cooperative resource sharing in Cinder. Energy use due to the radio is reduced, resulting in a 12.5% total system power reduction over the 20 minute experiment.

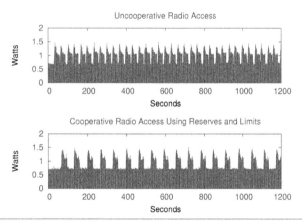

Figure 13. Two background applications, a pop3 mail and an RSS fetcher, each poll every sixty seconds. a) Since they are not coordinated, their use of the radio is staggered, resulting in increased power consumption. Each application uses the radio for at most a few seconds, but neither takes advantage of the other having brought the radio out of the low power idle state. b) The same mail and RSS background applications using reserves and limits to coordinate their access to the radio data path. Enough energy is allocated to each application to turn the radio on every two minutes. By pooling their resources, they are able to turn the radio on at most every sixty seconds.

Figure 13a shows the uncooperative applications wasting energy – running when the radio is idle and powering it up independently. Neither combines efforts to amortize costs.

In comparison, Figure 13b shows what happens with the modified *netd*. Each application still only receives enough energy to activate the radio every two minutes; however, when they initiate network operations, their threads block and contribute acquired energy to the *netd* reserve (Figure 14). Since the two threads combine their power in the *netd* reserve, every 60 seconds enough energy is saved to use the radio and both applications proceed simultaneously.

Using delegation, independent applications in Cinder automatically collaborate, improving quality of service. In this case, the improved quality of service is increasing the frequency of mail and news checks by a factor of two, using the same energy budget. Table 1 shows the energy savings of the modified *netd*. In total, 12.5% less energy is used in the same time interval for an equivalent amount of work. While significant, we stress that our baseline power consumption is artificially dominant, as Cinder does nothing to place the hardware into lower power states while idle in contrast to Linux. We expect Cinder to provide greater improvement on a mature mobile platform that makes full use of power saving features.

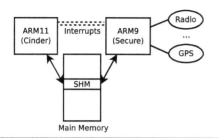

Figure 15. Cinder may only indirectly access many hardware features, such as the radio and GPS, by passing messages to a secure ARM9 coprocessor.

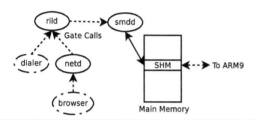

Figure 16. The user-level smdd daemon manages the shared memory interface on the ARM11 and exports interfaces to the radio, GPS, battery sensor, and so on via gate calls. Consumers of this interface, such as the radio daemon, *rild*, may also export their own gate calls. *netd*, for example, implements gates for libOS TCP/IP sockets. Gates are used by both user applications (*browser, dialer*) and OS daemons (*netd*, etc.).

7. Experience Developing on a Mobile Platform

We ported Cinder to the HTC Dream mobile phone. Because developing a kernel for a mobile phone platform is a nontrivial task that is rarely attempted, we describe our process here in detail.

To run Cinder on the HTC Dream, we first ported the kernel to the generic ARM architecture (2,380 additional lines of C and assembly). MSM7201A-specific kernel device support for timers, serial ports, framebuffer, interrupts, GPIO pins, and keypad required another 1,690 lines of C. Cinder implements the GSM/GPRS/EDGE radio functionality in userspace with Android driver ports.

Implementing radio functionality is particularly difficult, as it requires access to secure and undocumented hardware that is not directly accessible from the processor. For instance, the MSM7201A chipset includes two cores: the ARM11 runs application code (Cinder), while a secure ARM9 controls the radio and other sensitive features (Figure 15). Accessing these features requires communicating between the cores using a combination of shared memory and interrupt lines. We first mapped the shared memory segment into a privileged user-level process and ported the Android Linux kernel's shared memory device to userspace (Figure 16). This daemon, *smdd* (4,756 lines),

exports ARM9 services via gate calls to other consumers, including the radio interface library (RIL). The RIL generates and consumes messages between cores that initiate and respond to radio events, such as dialing a number or being notified of an incoming call.

In Android, the radio interface library consists of two parts: an open source generic interface library that provides common radio functions across different hardware platforms, and a device-specific, Android-centric shared object that interfaces with specific modem hardware (libril.so). Unfortunately, libril.so is closed-source and precompiled for Android: this makes it excessively difficult to incorporate into another operating system. Without hardware documentation or tremendous reverse engineering, using the radio requires running this shared object in Cinder. To do so, we wrote a compatibility shim layer to emulate both Android's "bionic" libc interface, as well as the various /dev devices it normally uses to talk to the ARM9 (1,302 lines of C). We rewrote the library's symbol table to link against our compatibility calls, rather than the binary-incompatible uClibc functions and syscalls that regular Cinder applications use. Finally, we wrote a port of the radio interface library frontend that provides gates to service radio requests from applications needing network access.

Cinder currently supports the radio data path (IP), and can send and receive SMS text messages. Cinder can also initiate and receive voice calls, but as it does not yet have a port of the audio library, calls are silent. In retrospect, since hardware documentation is unavailable, basing our solution on Android, rather than HiStar, would have been far simpler from a device support perspective. Crucially, however, our implementation atop Linux trades the simple and accurate IPC resource accounting needed in energy management for device drivers (§7.1). We felt that a cleaner slate justified the additional tedium as well as the reduced hardware support present in our prototype.

In summary, even trivial radio operation is quite complicated, requiring about 12,000 lines of userspace code along with the 263 KiB closed libril.so. In comparison, the entire Cinder kernel consists of about 27,000 lines of C for all four CPU architectures and all device drivers. The kernel is only 644 KiB – less than 2.5 times the size of libril.so.

7.1 Cinder-Linux versus Cinder-HiStar

Cinder was initially implemented on HiStar because several key behaviors of the platform are naturally expressible using HiStar's abstractions.

One such feature of Cinder is resource delegation between principals. Consider a common situation where a client process P requires work to be performed on its behalf by a daemon process D. A real world example is the radio interface layer daemon on the Android platform. Cinder must ensure P is charged for any work D performs on its behalf – or, equivalently, it must ensure that P provides the resources that D's code uses to run.

HiStar's abstractions achieve this behavior cleanly and simply. A process in HiStar is a container, containing an address space and one or more threads. IPC is performed through special gates defined by the process – a thread belonging to process P can enter a gate defined by process D, after which the thread has access to D's address space, though while under control of D's code text. When process P requires service from daemon D a thread, T, belonging to P enters D's address space via a gate. Cinder debits T for work it performs as usual even though it executes under the control of D's code, correctly billing consumption to P. This way, HiStar's IPC mechanism easily achieves the desired delegation behavior.

Linux, on the other hand, uses several different facilities to provide IPC, many of which are based upon message passing between processes. A few examples are pipes, Unix domain sockets, message queues, and semaphores. These forms of IPC occur without any resource sharing or attribution between processes. This subverts delegation since process P may elicit work by daemon D on its behalf without providing the resources for the work.

To compensate, Linux needs to verify that the calling process has provided adequate resources to perform the desired request. However, existing IPC mechanisms in Linux are not built with the goal of discovering the identity of the caller in mind. Consider a daemon D that reads requests from a named pipe in the filesystem. When D reads from the pipe, it only knows the writing process has permissions to access the pipe. In general, it cannot identify which process in the system made the request, and thus does not know which process to debit.

To mitigate this problem, Cinder-Linux needs a way for the daemon to determine the identity of the calling process. One possibility is to have a user level protocol in which a calling process P encodes both its identity and a description of how D can access resources that P has set aside for D within the request. For example, it could format a request as a triple: ⟨pid, reserve_id, request⟩. D accesses the reserve named in the request, and only performs work once it ensures the caller has provided sufficient resources in payment. Since a user level process can lie abouts it credentials, the protocol is not robust against malicious applications. A more robust mechanism would require new kernel IPC mechanisms.

Both Cinder-HiStar and Cinder-Linux must prevent resource misuse. In particular, D must not co-opt P's resources for performing unrelated tasks, and process P must provide resources for work performed by D on its behalf. Providing these guarantees on Linux requires either a fine grained permissions system or, alternatively, some form of information flow control or tracking (with which the daemon could determine which process sent a given request). In contrast, HiStar's existing information flow control mechanisms easily provide the necessary protection.

Linux has the benefit of being an established operating system with vast device driver support and the entire Android platform. As a result, it is easier to write real-world applications. Consequently, we have written an initial implementation of Cinder that runs on top of the Linux and the Android platform on the Dream. The basics of Cinder-Linux remain the same as Cinder-HiStar aside from resource attribution issues for IPC and fine-grained permissions. Most implementation of the Cinder abstractions are independent of the underlying operating system and similar on HiStar and Linux. Some differences in the implementation do exist, however. For example, Cinder-HiStar flows taps during scheduler timer interrupts, while Cinder-Linux uses a kernel thread. One area of future work is further testing the concepts and features of Cinder on the Cinder-Linux platform.

8. Related Work

We group related work into three categories: resource management, energy accounting, and energy efficiency.

8.1 Resource Management

Cinder's taps and reserves build on the abstraction of resource containers [Banga 1999]. Like resource containers, they provide a platform for attributing resource consumption to a specific principal. By separating resource management into rates and quantities, however, Cinder allows applications to delegate with reserves, yet reclaim unused resources. This separation also makes policy decisions much easier. Since resource containers serve both as limits and reservations, hierarchical composition either requires a single policy (limit or reserve) or ad hoc rules (a guaranteed CPU slot cannot be the child of a CPU usage limit).

Linux has recently incorporated "cgroups" [Menage 2008] into the mainline kernel, which are similar to resource containers, but group processes rather than threads. They are hierarchical and rely on "subsystem" modules that schedule particular resources (CPU time, CPU cores, memory).

ECOSystem [Zeng 2002, 2003] presents an abstraction for energy, "currentcy", which unifies a system's device power states. It represents logical tasks using a flat form of resource containers [Banga 1999] by grouping related processes in the same container. This flat approach makes it impossible for an application to delegate, as it must either share its container with a child or put it in a new container that competes for resources. Like ECOSystem, Cinder estimates energy consumption with a software-based model that ties runtime power states to power draw.

ECOSystem achieves pooling similar to Cinder's netd for devices with non-linear power consumption (disk and network access), using unique cost models for each device. Cinder simplifies construction of these policies using its fine-grained protection mechanism and reserves to provide the same result in userspace.

8.2 Measurement, Modeling, and Accounting

Accurately estimating a device's energy consumption is an ongoing area of research. Early systems, such as ECOSystem [Zeng 2002], use a simple linear combination of device states. Most modern phone operating systems, such as Symbian and OS X, follow this approach.

PowerScope improves CPU energy accuracy by correlating instrumented traces of basic blocks with program execution [Flinn 1999b]. A more recent system, Koala, explores how modern architectures can have counter-intuitive energy/performance tradeoffs, presenting a model based on performance counters and other state [Snowdon 2009]. A Koala-enabled system can use these estimates to specify a range of policies, including minimizing energy, maximizing performance, and minimizing the energy-delay product. The Mantis system achieves similar measurement accuracy to Koala using CPU performance counters [Economou 2006].

Quanto [Fonseca 2008] extends the TinyOS operating system to support fine-grained energy accounting across activities. Using a custom measurement circuit, Quanto generates an energy model of a device and its peripherals using a linear regression of power measurements. By monitoring the power state of each peripheral and dynamically tracking which activity is active, Quanto can give precise breakdowns of where a device is spending energy.

PowerBooter and PowerTutor [Zhang 2010] explore the generation of detailed power models for a full-featured smartphone (the HTC Magic) providing application power consumption estimation and feedback for tuning.

Cinder complements this work on modeling and accounting. Improved hardware support to determine where energy is going would make its accounting and resource control more accurate. On top of these models, Cinder provides a pair of abstractions that allow applications to flexibly and easily enforce a range of policies.

8.3 Energy Efficiency

There is rich prior work on improving the energy efficiency of individual components, such as CPU voltage and frequency scaling [Flautner 2002; Govil 1995], spinning down disks [Douglis 1995; Helmbold 1996], or carefully selecting memory pages [Lebeck 2000]. Phone operating systems today tend to depend on much simpler, but still effective optimization schemes than in the research literature, such as hard timeouts for turning off devices. The exact models or mechanisms used for energy efficiency are orthogonal to Cinder: they allow applications to complete more work within a given power budget. The image viewer described in §5.3 is an example of an energy-adaptive application, as is typical in the Odyssey system [Flinn 1999a].

9. Future Work

We believe that the reserve and tap abstractions may be fruitfully applied to other resource allocation problems beyond energy consumption. For instance, the high cost of mobile data plans makes network bits a precious resource. Applications should not be able to run up a user's bill due to expensive data tariffs, just as they should not be able to run down the battery unexpectedly. Since data plans are frequently offered in terms of megabyte quotas, Cinder's mechanisms could be repurposed to limit application network access by replacing the logical battery with a pool of network bytes. Similarly, reserves could also be used to enforce SMS text message quotas.

Using the HTC Dream's limited battery level information Cinder could adapt its energy model based on past component and application usage, dynamically refining its costs. Though Cinder can facilitate this, and we have made some adjustments to test this, evaluating the complex and dynamic system this would yield will require additional research.

10. Conclusion

Cinder is an operating system for modern mobile devices. It uses techniques similar to existing systems to model device energy use, while going beyond the capabilities of current operating systems by providing an IPC system that fundamentally accounts for resource usage on behalf of principals. It extends this accounting to add subdivision and delegation, using its reserve and tap abstractions. We have described and applied this system to a variety of applications demonstrating, in particular, their ability to partition applications to energy bounds even with complex policies. Additionally, we showed Cinder facilitates policies which enable efficient use of expensive peripherals despite non-linear power models.

Acknowledgments

We thank John Ousterhout, the anonymous reviewers, and our shepherd, Liuba Shrira, for their feedback. This work was supported by generous gifts from DoCoMo Capital, the National Science Foundation under grants #0831163, #0846014, and #0832820 POMI (Programmable Open Mobile Internet) 2020 Expedition Grant, the King Abdullah University of Science and Technology (KAUST), Microsoft Research, T-Mobile, NSF Cybertrust award CNS-0716806, and an NSERC Post Graduate Scholarship. This research was performed under an appointment to the U.S. Department of Homeland Security (DHS) Scholarship and Fellowship Program, administered by the Oak Ridge Institute for Science and Education (ORISE) through an interagency agreement between the U.S. Department of Energy (DOE) and DHS. ORISE is managed by Oak Ridge Associated Universities (ORAU) under DOE contract number DE-AC05-06OR23100. All opinions expressed in this paper are the authors' and do not necessarily reflect the policies and views of DHS, DOE, or ORAU/ORISE.

References

[Com 1988] THE EXECUTIVE COMPUTER; Compaq Finally Makes a Laptop. http://www.nytimes.com/1988/10/23/business/the-executive-computer-compaq-finally-makes-a-laptop.html, 1988.

[Fla 2009] Adobe and HTC Bring Flash Platform to Android, June 2009. http://www.adobe.com/aboutadobe/pressroom/pressreleases/pdfs/200906/062409AdobeandHTC.pdf.

[App 2010] Apple Previews iPhone OS 4, April 2010. http://www.apple.com/pr/library/2010/04/08iphoneos.html.

[Banga 1999] Gaurav Banga, Peter Druschel, and Jeffrey C. Mogul. Resource containers: a new facility for resource management in server systems. In *Proceedings of the 3rd Symposium on Operating Systems Design and Implementation*, pages 45–58, New Orleans, LA, 1999.

[Douglis 1995] Fred Douglis, P. Krishnan, and Brian N. Bershad. Adaptive Disk Spin-down Policies for Mobile Computers. In *Proceedings of the 2nd Symposium on Mobile and Location-Independent Computing*, pages 121–137, 1995.

[Economou 2006] Dimitris Economou, Suzanne Rivoire, and Christos Kozyrakis. Full-system power analysis and modeling for server environments. In *Proceedings of the 2nd Workshop on Modeling, Benchmarking and Simulation*, Boston, MA, 2006.

[Flautner 2002] Krisztian Flautner and Trevor Mudge. Vertigo: automatic performance-setting for linux. In *Proceedings of the 5th Symposium on Operating Systems Design and Implementation*, pages 105–116, Boston, MA, 2002.

[Flinn 1999a] Jason Flinn and M. Satyanarayanan. Energy-aware adaptation for mobile applications. In *Proceedings of the 17th ACM Symposium on Operating Systems Principles*, pages 48–63, Charleston, SC, 1999.

[Flinn 1999b] Jason Flinn and M. Satyanarayanan. PowerScope: A Tool for Profiling the Energy Usage of Mobile Applications. In *Proceedings of the 2nd IEEE Workshop on Mobile Computer Systems and Applications*, New Orleans, LA, 1999.

[Fonseca 2008] Rodrigo Fonseca, Prabal Dutta, Philip Levis, and Ion Stoica. Quanto: Tracking Energy in Networked Embedded Systems. In *Proceedings of the 8th Symposium on Operating Systems Design and Implementation*, pages 323–338, 2008.

[Govil 1995] Kinshuk Govil, Edwin Chan, and Hal Wasserman. Comparing algorithm for dynamic speed-setting of a low-power CPU. In *Proceedings of the 1st Conference on Mobile Computing and Networking*, pages 13–25, Berkeley, CA, 1995.

[Helmbold 1996] David P. Helmbold, Darrell D. E. Long, and Bruce Sherrod. A dynamic disk spin-down technique for mobile computing. In *Proceedings of the 2nd Conference on Mobile Computing and Networking*, pages 130–142, Rye, NY, 1996.

[Lebeck 2000] Alvin R. Lebeck, Xiaobo Fan, Heng Zeng, and Carla Ellis. Power aware page allocation. In *Proceedings of the 9th International Conference on Architectural Support for Programming Languages and Operating Systems*, pages 105–116, Cambridge, MA, 2000.

[Menage 2008] Paul Menage. cgroups, October 2008. http://git.kernel.org/?p=linux/kernel/git/torvalds/linux-2.6.git;a=blob;f=Documentation/cgroups/cgroups.txt;hb=b851ee7921fabdd7dfc96ffc4e9609f5062bd12.

[Snowdon 2009] David C. Snowdon, Etienne Le Sueur, Stefan M. Petters, and Gernot Heiser. Koala: a platform for OS-level power management. In *Proceedings of the 4th ACM European Conference on Computer Systems*, pages 289–302, Nuremberg, Germany, 2009.

[Waldspurger 2002] Carl A. Waldspurger. Memory resource management in VMware ESX server. *SIGOPS Oper. Syst. Rev.*, 36: 181–194, December 2002. ISSN 0163-5980.

[Zeldovich 2006] Nickolai Zeldovich, Silas Boyd-Wickizer, Eddie Kohler, and David Mazières. Making information flow explicit in HiStar. In *Proceedings of the 7th Symposium on Operating Systems Design and Implementation*, pages 263–278, Seattle, WA, 2006.

[Zeng 2002] Heng Zeng, Carla S. Ellis, Alvin R. Lebeck, and Amin Vahdat. ECOSystem: managing energy as a first class operating system resource. In *Proceedings of the 10th International Conference on Architectural Support for Programming Languages and Operating Systems*, pages 123–132, San Jose, CA, 2002.

[Zeng 2003] Heng Zeng, Carla S. Ellis, Alvin R. Lebeck, and Amin Vahdat. Currentcy: A unifying abstraction for expressing energy management policies. In *Proceedings of the 2003 USENIX Annual Technical Conference*, pages 43–56, San Antonio, TX, 2003.

[Zhang 2010] Lide Zhang, Birjodh Tiwana, Zhiyun Qian, Zhaoguang Wang, Robert P. Dick, Zhuoqing Morley Mao, and Lei Yang. Accurate online power estimation and automatic battery behavior based power model generation for smartphones. In *Proceedings of the eighth IEEE/ACM/IFIP international conference on Hardware/software codesign and system synthesis*, CODES/ISSS '10, pages 105–114, New York, NY, USA, 2010. ACM. ISBN 978-1-60558-905-3.

Fine-Grained Power Modeling for Smartphones Using System Call Tracing

Abhinav Pathak

Purdue University

pathaka@purdue.edu

Y. Charlie Hu

Purdue University

ychu@purdue.edu

Ming Zhang

Microsoft Research

mzh@microsoft.com

Paramvir Bahl

Microsoft Research

bahl@microsoft.com

Yi-Min Wang

Microsoft Research

ymwang@microsoft.com

Abstract

Accurate, fine-grained online energy estimation and accounting of mobile devices such as smartphones is of critical importance to understanding and debugging the energy consumption of mobile applications. We observe that state-of-the-art, utilization-based power modeling correlates the (actual) utilization of a hardware component with its power state, and hence is insufficient in capturing several power behavior not directly related to the component utilization in modern smartphones. Such behavior arise due to various low level power optimizations programmed in the device drivers. We propose a new, system-call-based power modeling approach which gracefully encompasses both utilization-based and non-utilization-based power behavior. We present the detailed design of such a power modeling scheme and its implementation on Android and Windows Mobile. Our experimental results using a diverse set of applications confirm that the new model significantly improves the fine-grained as well as whole-application energy consumption accuracy. We further demonstrate fine-grained energy accounting enabled by such a fined-grained power model, via a manually implemented *eprof*, the energy counterpart of the classic *gprof* tool, for profiling application energy drain.

Categories and Subject Descriptors   D.4.8 [Operating Systems]: Performance–Modeling and Prediction.
General Terms   Design, Experimentation, Measurement.
Keywords   Smartphones, Mobile, Energy.

1.   Introduction

Mobile devices such as smartphones provide significant convenience and capability to the users. A recent market analysis [Com] shows that the smartphone market is the fastest growing segment of the mobile phone market; in 2010 over 45.5 million people in the United States owned smartphones. Despite the incredible market penetration of smartphones, their utility has been and will remain severely limited by their battery life. As such, understanding the power consumption of applications running on mobile devices has attracted much research effort. Early research [Flinn 1999a;b, Mahesri 2005] has focused on power measurement, i.e., measuring the power consumption of the mobile device during the execution of an application using a power meter, with the goal of understanding energy consumption by individual applications. These studies directly rely on the availability of power meters and do not develop a power estimation model for use in the "wild" without a power meter.

More recent efforts have focused on developing online power models for mobile devices. Typically, during the training phase, a power consumption model is developed by running sample applications, and correlating certain application behavior, or *triggers*, with specific power states or power state transitions, of individual components or the entire system, measured using an external power meter. The generated power model during this training phase can then be used online, without any measurement from a power meter, for estimating the energy consumption in running any application. Thus such an online power model enables application developers to develop energy profiling tools to profile and consequently optimize the energy consumption of mobile applications, without the expensive power meters, much like how performance profiling enabled by *gprof* [Graham 1982] has facilitated performance optimization in the past several decades.

There are two desirable features for such an online power model: (1) It should incur low overhead in logging the triggers, so that the energy estimation based on the model can be performed online; (2) It should enable accurate fine-grained energy accounting, e.g., on a per-subroutine basis, and in the presence of multiple threads and processes. This is because as in performance profiling and optimization, the natural granularity in profiling and optimizing the energy consumption of an application is at the subroutine level.

The large body of work on power modeling on smartphones [Shye 2009, Zhang 2010], and more generally for desktops [Zeng 2002] and servers [Fan 2007, Kansal 2010], are based on the fundamental yet intuitive assumption that the (actual) utilization of a hardware component (e.g., disk, NIC) corresponds to a certain power state and the change of utilization is what triggers the power state change of that component. Consequently, their design all use the notion of utilization of a hardware component as the "trigger" in modeling power states and state transitions. The usage statistics of each hardware component (e.g. disk) are typically provided by the OS, for example, /proc on Linux.

In this paper, we make a key observation that the fundamental assumption behind utilization-based power modeling does not hold in several scenarios on smartphones, due to the increasingly sophisticated power optimization in the device drivers and OS-level power management.

- Several components (e.g., NIC, drive, and GPS) have tail power states.

- System calls that do not imply utilization (e.g., file open, close, socket close) can change power states.

- Several components (e.g. GPS, camera) do not have quantitative utilization.

An immediate consequence of these non-utilization-based power behavior of smartphone components is that while a utilization-based model may still achieve reasonable accuracies in estimating the energy consumption of whole applications due to the cancellation of per-interval estimation errors in both directions [Shye 2009], they suffer poor accuracy in fine-grained energy estimation. For example, they are incapable of modeling the energy consumption due to lingering tail power states which can last up to several seconds, far beyond the completion of the triggering subroutine. Consequently, such models cannot be used to develop profiling tools that support accurate energy accounting on a per-subroutine basis.

In this paper, we propose a new, *system-call-based power model* that overcomes the above limitations of utilization-based power modeling. Our design is motivated by the following observations. First, system calls provide the only means via which applications gain access to the hardware (I/O) components. As such their names along with the parameters give clear indication of the components and the level of utilization being requested. Hence, they already encompass the triggers used in utilization-based power modeling. Second, such a model can capture all the power behavior of I/O system calls that do not imply component workload or utilization. Finally, a system call can be naturally related back to the calling subroutine and the hosting thread and process. Together, the above observations suggest system-call-based power modeling can achieve accurate fine-gained energy estimation, and enable fine-gained energy accounting, e.g., on a per-subroutine, per-thread, and per-process basis.

The design of our system-call-based power modeling scheme consists of two major components. First, it uses Finite State Machines (FSM) to model the power states and state transitions of each component as well as the whole smartphone. Some states have constant power consumption; they capture non-utilization-based power behavior. Other states leverage a linear regression (LR) model to capture the power consumption due to system calls that generate workload. Second, it uses a carefully designed testing application suite which leverages the domain knowledge of system calls (i.e., their semantics and causal invocation ordering) to systematically uncover the FSM transition rules. We have implemented the system-call-based power modeling scheme on two smartphone OSes and validated that it significantly improves fine-grained power estimation as well as whole-application energy consumption estimation for a diverse set of applications.

In summary, this paper makes following contributions.

- We make the observation that the fundamental assumption behind utilization-based power modeling, that the power state of a component is correlated with its utilization, often does not hold on modern day smartphones, due to the increasingly sophisticated power optimizations in the device drivers. Consequently, utilization-based power models can suffer poor accuracy in fine-grained power estimation.

- We propose a new power modeling approach based on tracing system calls of the applications, which gracefully captures both utilization-based and non-utilization-based power behavior of I/O components, and hence can achieve accurate fine-grained energy estimation.

- We present the detailed design and implementation of our new approach on two smartphone operating systems, Windows Mobile and Android.

- We present experimental results that demonstrate the accuracy of our new modeling scheme using a diverse set applications. When estimating the energy consumption for 50ms time intervals, the 80th percentile intervals across the applications report an error under 10%, but vary between 16% and 52% across the applications under the utilization-based modeling. Further, the whole-application energy estimation error varies in the range of 0.2% to 3.6% under our scheme, compared to between 0.4% to 20.3% under utilization-based modeling.

- We further demonstrate fine-grained energy accounting enabled by such a fined-grained power model, via a manually implemented *eprof*, the energy counterpart of the classic *gprof* tool, for profiling application energy consumption on a per-subroutine basis.

2. Background: Power Management in Smartphones

Modern day smartphones come with a wide variety of hardware *components* embedded in them. Typical components include CPU, memory, Secure Digital card (sdcard for short), WiFi NIC, cellular, bluetooth, GPS, camera (may be multiple), accelerometer, digital compass, LCD, touch sensors, microphone, and speakers. It is common for smartphone applications to utilize several components simultaneously to offer richer user experience. Unlike on desktop and server machines, the power consumed by each I/O component is often comparable to or higher than that by the CPU on smartphones.

Each component can be in several operating modes, each draining a different amount of power. We call such different modes different *power states* for that component. We note that in our work and in all previous work, power modeling is concerned with estimating the power consumption of the whole phone as the components switch between operating modes, not when they are turned off by the user or the OS power manager, i.e., optimizations of their sleep-wakeup cycles. The later cases can be easily incorporated into energy accounting. In principle, the power state of the component should simply correspond to the throughput of its work done, i.e., the actual utilization of the component. For example, we observed that for a fixed channel condition, the WiFi NICs of the smartphones studied in this paper transmit at 5.5 Mbps at a lower power state, but may switch to a higher power state in order to transmit at 11 Mbps. We denote this assumption as the *utilization-power-state correlation assumption*.

However, as the device drivers in modern day smartphone operating systems incorporate more and more power management "smarts", the above simple assumption on the tight correlation between the utilization of a component and its power state often does not hold. In particular, the power state of a component could potentially depend on non-utilization-based factors. These include external conditions (e.g., signal strength for WiFi and Cellular) and semantics of system calls such as initiating a component or terminating the usage of a component (e.g., closing a socket). Using these factors as input, several power saving optimizations can be programmed in a component's device driver which decides the component's power transition rules. For example, the device drivers for the wireless NICs on the smartphones we have studied adjust the transmission power when the signal strength changes. In principle, all the information about power states and transition rules of a component can be uncovered from reading the device drivers of the component.

However, most of the device drivers for smartphones, including Android handsets, are proprietary.

3. State-of-the-Art: Utilization-based Modeling

The large body of work on power modeling for desktops, servers, and more recently for smartphones are based on the *utilization-power-state correlation assumption* stated in Section 2. Consequently, their designs all use the notion of utilization of a hardware component as the "trigger" in modeling power states and state transitions. We call this class of modeling as *utilization-based modeling*.

Early utilization-based models focused on estimating the power consumption of individual components, using the corresponding performance counters, e.g., CPU [Bellosa 2000, Snowdon 2009, Stanley-Marbell 2001, Tiwari 1996], memory [Rawson 2004], disk [Zedlewski 2003], and of the entire system [Bircher 2007, Flinn 1999b]. These models do not relate the power consumption of the system with the applications. More recent utilization-based power models tried to estimate the power consumption of applications running on desktops [Zeng 2002], sensor nodes [Shnayder 2004], virtual machines [Kansal 2010], data center servers [Fan 2007], and most recently, mobile devices [Shye 2009, Zhang 2010].

There are two key ingredients in constructing a power estimation model in a utilization-based power modeling scheme. First, during the training phase, it collects the utilization of individual hardware components, typically via OS-provided utilization statistics (e.g. usage statistics in /proc in Linux in [Shye 2009, Zhang 2010] for power modeling in smartphones) while running some sample applications, and measures the corresponding power consumption of those components, e.g., using power meters. Second, it typically develops an LR model to correlate the sampled trigger values, i.e., the utilization, and the measured power consumption at that sampling moment (e.g. [Shye 2009, Zhang 2010]). Once the model is constructed, it can be used to perform online estimation of power consumption, by continuously collecting the utilization of the components and feeding them into the model.

4. New Challenges

Utilization-based power modeling, however, cannot handle the following non-utilization-based power behavior of smartphone components.

Several components have tail power states. Components such as NICs, sdcard and GPS on many smartphones exhibit the so-called "tail" power state phenomenon; they stay at high power state for a period of time after active I/O activities. Figure 1(a) shows after a read (same is true for write) system call of 10 bytes which sent the sdcard to high power state and lasted about 10 milliseconds, the sdcard stays in high power state for 5 seconds, on the HTC Touch Pro phone (touch for short) running Windows Mobile 6.1 (WM6 for

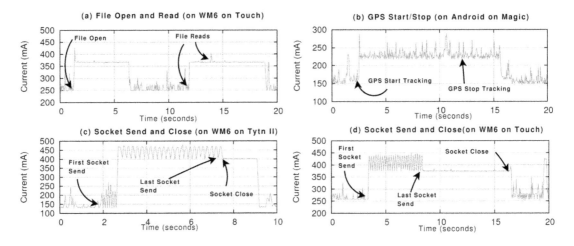

Figure 1: System call based power transitions: Top row: Disk on touch , GPS on magic , bottom row: network on touch and tytn2.

short). Similar tail power state lasts for 3 seconds on the HTC Tytn 2 (tytn2 for short) phone running WM6. On all the smartphones (WM6 on touch and tytn2, and Android on HTC Magic (magic for short)) we have tested, the NIC continues to be in high power state for a few seconds after an active send/recv is completed. Figure 1(c) shows the tail power state lasts about 1.7 second on the tytn2 phone. Clearly the utilization of the components (NIC or sdcard) is zero during the tail power states. This breaks the fundamental assumption behind utilization-based power models that the usage of a components determines its power state.

System calls that do not imply utilization can change power states. Many systems calls that do not imply high utilization of components can send the components to high or low power states. This could be due to power optimizations programmed in device drivers. For example, on windows mobile, file open, close, create, delete system calls trigger a power level change in the sdcard device driver; after these calls the component remains in a high power state for a few seconds. Figure 1(a) gives the example of file open on the touch phone. Similarly, a socket close system call immediately ends the high tail power state of the NIC on windows mobile touch smartphones, shown in Figure 1(d). Such low level power optimizations are typically done in the device drivers and they also break the utilization-power-state correlation assumption in utilization-based power models.

Several components do not have quantitative utilization. Several "exotic" components on smartphones, such as GPS and camera, do not have a notion of quantitative utilization, that is parallel to the amount of data sent or received by a NIC. These components typically are turned on or off by system calls. For example, the "requestLocationUpdate" call in Android sets GPS in a high power state, shown in Figure 1(b). The "opencamerahardware" system call in Android sets camera in a high power state. In principle, one can add a counter (e.g. to /proc) to record the binary state of

these components to facilitate utilization-based power modeling of these components. However, since utilization-based modeling fundamentally assumes periodic sampling of performance counters, it can suffer delay in reading the change of power states of such components.

An immediate consequence of these limitations is that utilization-based modeling can suffer poor accuracy in fine-grained energy estimation, i.e., for small intervals such as the duration of subroutines in an application, for two reasons. First, utilization-based modeling relies on periodic sampling of the usage counters (e.g. reading /proc). The time interval at which the sampling is performed can be too course-grained compared to the duration of subroutines, or the sampling can become costly if done at a very fine granularity. Second, more importantly, a tail power state triggered by a system call in a subroutine can last till long after that subroutine has returned. In addition, using a diverse set of applications we show in Section 7 that the error in per-interval power estimation can add up and lead to poor accuracy in whole-application energy consumption estimation for many applications.

5. System-Call-Based Power Modeling

5.1 Key Idea

In this paper, we propose *system-call-based power modeling* which overcomes the above limitations of utilization-based power modeling. Our new approach is based on the following five observations:

- System calls provide the only means via which applications gain access to the hardware (I/O) components. As such their names along with the parameters give clear indication of which components and what level of utilization are being requested. Hence, they already encompass the triggers used in utilization-based power modeling. In fact, the utilization statistics used in utilization-based power modeling such as for network and disk activities in

/proc in Linux are exactly updated based on the parameters in selected system calls such as read/write. To model CPU power consumption, we log context switch events in the kernel.

- Using system calls as triggers to power state transitions naturally solves the second limitation of utilization-based power modeling. Those system calls that do not imply utilization but trigger power state changes can be identified during the training phase and incorporated in the power model.

- Similarly, the invocation of system calls that turn on and off "exotic" components immediately triggers power state change for those components, which avoids the delay due to periodic sampling of performance counters in utilization-based modeling.

- Using system calls as triggers naturally suggests using a Finite State Machine (FSM) to model the state transitions. The states in an FSM can be easily annotated with the timing and workload of recent events to accurately model state transitions due to non-utilization-based power behavior (e.g. tail states) and accumulated-utilization-based power behavior (e.g., transmitting enough packets causes the WiFi NIC to go to a higher power state to increase the bitrate.)

- System calls can be easily related back to the calling subroutine, the hosting thread and process. This, combined with the above observations that they capture non-utilization-based power behavior, enables fine-grained energy accounting, e.g., on a per-subroutine, per-thread, or per-process basis.

In the following, we first analyze the power states and state transitions involved in a single system call to motivate the Finite State Machine (FSM) implementation of system-call-based power modeling. To systematically uncover the FSM of power states for a given phone, we develop the CTester application suite, based on the domain knowledge of system calls for each OS, to automate the construction of the FSM in three incremental steps. Step 1 uncovers the FSM for individual system calls for a component. Steps 2 and 3 then exercise superposition of different power states for one and multiple components, respectively.

5.2 Modeling Single System Call Power Consumption

The actual mechanism of system-call-based modeling is motivated by the typical sequence of power states and state transitions involved in a single system call.

5.2.1 Power Consumption Behavior of System Calls

The first column of Figure 2 plots the actual power levels of a smartphone, measured using a power meter, from the start of a typical system call and during the corresponding power states and state transitions resulted. The three rows are for a single disk read, a network send with a few packets, and

a network send with many packets, respectively. A system call that specifies a certain amount of workload, e.g., sending X bytes to the NIC, can send the component into one of several possible base power states. For example, on tytn2, touch and magic, sending a few packets within a short period sends the NIC to a lower power base state, while sending many packets, via one or multiple back-to-back send system calls, sends the NIC to a high power base state. The workload specified by a system call eventually is carried out by the component via a sequence of atomic I/O operations, e.g., a sequence of packet transmissions by the NIC, each corresponding to an instantaneous burst of power consumption from the current base power state.

5.2.2 Abstraction of Power States

Since a system call is often turned into a sequence of low-level hardware component operations, e.g., packet transmissions, the key to developing a system-call-based power model which only uses system call information as input is to devise power states that abstract away the low-level power behavior, e.g., in between individual packet transmissions, yet still capture the power consumption of that component due to the system calls.

We denote the power state spanning the duration covering the sequence of the component-level I/O operations due to a system call a *productive power state* of that component. Such a productive power state is characterized by the burst power of individual I/O operations, and the duration of the sequence of I/O operations. Each component can potentially have multiple such productive states. Since the duration of a productive state depends on the workload specified in the input parameter of the system call, and the power bursts due to individual packet transmissions cannot be exactly captured (drivers are closed-source), we develop an LR-based power model for that *productive state* which correlates the input workload (e.g., bytes to be sent) with the total duration of the power burst for that system call. We note this usage of LR is different from in utilization-based power modeling (e.g. [Shye 2009]) where LR is used to estimate the power drain of the whole phone.

Next, we observe after a productive power state, i.e., when there are no more I/O operations, a component can stay at the base power of the productive power state for a period of time. This is previously known for cellular NICs [Balasubramanian 2009], but we discovered it can also hold true for disks and GPS on smartphones. Since the component is not doing useful work, we call this state the *tail power state*. A tail power state is characterized by its base power and its duration.

Continuing on our example in Figure 2, the three figures in the middle column depict the abstract productive and tail power states and the transitions that model the measured power states and transitions in the first column. The above abstraction of two (types of) power states and state transitions triggered by system calls is general in that it cap-

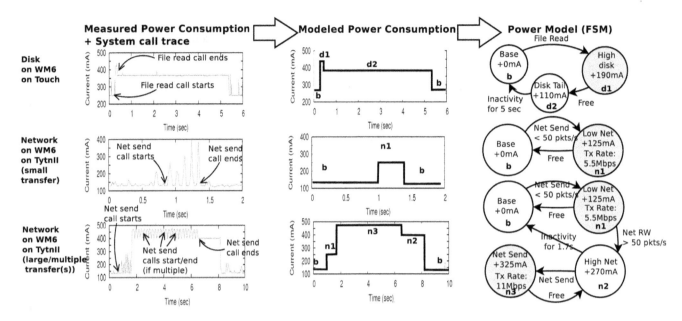

Figure 2: Modeling power states following a system call.

tures that different I/O system calls can have different duration/power in either state, or may experience only one of the two states. Finally, we observe that even consecutive burst powers during a productive power state can fluctuate, as observed in other work [Bellosa 2000] as well. We suspect this is largely due to the sampling nature of power measurement by the power meter. We calculate the average of all the burst powers measured during the training phase and use it as the burst power of that productive state.

5.2.3 System-Call-based Power Modeling using FSMs

The above abstraction of power states, which captures the power behavior of individual I/O system calls, naturally suggests a Finite State Machine (FSM)-based implementation to capture the transitions between the power states. Formally, in our FSM-based power model, each state represents a power state of a component, or of the set of all components when extended to model the power of the entire smartphone as discussed below. Each state is annotated with a (power, timeout duration) tuple, and the timing and workload of recent events of the component. The transitions between states capture the conditions that trigger state changes. There are three types of conditions: a timeout activity (e.g., the duration of the tail power state is over), a new system call, or a combination of timeout and past history of device utilization. As an example for the third condition above, a new send system call of fewer than 50 packets only takes the NIC to a low productive power state on WM6 on tytn2. A subsequent system call that together with previous send system call generates more than 50 packets per second will cause the NIC to enter a higher productive state followed by a tail state. To capture this history information of device utilization, at each state in the FSM, we need to keep information of all

system calls in the recent past. We find in our experiments with two operating systems and three smartphone handsets that storing system call information up to the past 60 seconds while staying in a power state is sufficient to capture all state transition conditions from that state due to recent device utilization.

Continuing our example in Figure 2, the last column shows the three FSMs that model the power states and state transitions of the three system calls, respectively. Unless otherwise stated, all power measures in FSMs in the rest of paper refer to the additional power consumption on top of an idle phone, i.e., with no application activities. For simplicity, we represent power consumed at an instant using the current drawn by the phone in milliAmperes. The actual power consumed would be the current drawn multiplied by 3.7V, the voltage supply of the battery. Similarly, energy is reported in uAH (micro Ampere Hours), and the actual energy would be the uAH value multiplied by 3.7V. These metrics are used since smartphone batteries are rated using these metrics and hence is easy to correlate.

5.3 Modeling Multiple System Calls

Once the FSMs for individual system calls are generated for a component, we systematically integrate them to model the power consumption when there are multiple concurrent system calls to the same component. Concurrent system calls here are defined as where the second system call is invoked before the component is out of the productive or tail state due to the first system call. Concurrent system calls can be issued from the same process or from concurrently running processes. For example, when multiple processes are running, a system call causes the context switch from the calling process (which is now blocked) to another process, which then issues another system call before the first one is completed.

We first observe that since a component can only be in a small number of possible power states, taking the union of all power states discovered by modeling individual system calls is sufficient to discover all possible power states for that component. We next focus on modeling state transitions.

There are two possible timings in which two concurrent system calls can arrive. In the first case, a subsequent system call arrives after the previous one is out of its productive power state (it could still be in the subsequent tail power state). In this case, the resulting power behavior when the second system call arrives can be modeled by simply superimposing the FSM of the second system call on top of the first, i.e., taking the maximum of the power states due to them for the overlapping time period.

In the second case, the second system call arrives while the first is still in its productive state. Since a productive state models the power consumption when the component is performing actual I/O operations, if the second system call is a workload-based system call, e.g., read/write, the effect on the component power behavior will be as if the first system call was invoked with the total workload of both system calls. If the second call is an initialization-based system call, e.g., open, the component will first come out of the productive state of the previous read/write, enters the tail state of that read/write, and then starts the FSM due to the open system call. From this time on, the FSM of the open call will be superimposed on the tail state of the previous call.

5.4 Modeling Multiple Components

After we generate the FSM models for the individual components, we develop one FSM model for the entire smartphone, by driving all components simultaneously. One may expect the total power consumption of multiple components to be a simple summation of those of individual components when active in isolation. Our experiments show that this was indeed the case on Android, but not on Windows Mobile. In particular, the tail states of different components interfere with each other on Windows Mobile. The basic idea behind combining FSMs of different smartphone components is to try out all possible combinations of the sets of conditions, each set for driving one component into all possible states of that component, and measure the corresponding total power consumption. This process is automated via the CTester application suite described later.

Complexity. While the above approach can result in a combinatorial number of power states and testing runs, in practice it remains practical. First, the major components we tested (CPU, disk, network, GPS, camera) typically have only up to three power states each (one or two productive states and one tail state), and hence even the total number of combinations is still under 20 for the three major compo-

nents (CPU, disk, network). Second, since energy modeling is a one-time procedure per smartphone per OS, it is acceptable for this procedure to take some time.

Completeness of the methodology. We expect the methodology above can capture all the possible states and transition conditions in practice. All the power states are likely to be accounted for as each component can have a finite number of power states and our CTester application suite exercises all combinations of the states across components. Furthermore, we run the CTester application suite multiple times to ensure the the derived FSMs are consistent. This repeatability serves as an assurance of the completeness of the FSM power model. Finally, we validate the derived FSM power model by performing fine-grained energy estimation against actual measurement from a power meter, using a diverse set of applications (Section 7.3).

We note the above completeness issue also exists in utilization-based modeling. Since our methodology for constructing the FSM-based power model offers a systematic way of searching for all possible states and state transitions, it is more likely to result in a more complete model than utilization-based modeling, which is typically based on trying random sample applications (e.g. [Shye 2009]).

5.5 The CTester Applications

To support the above methodology for uncovering the power states and state transitions of various components, we design a testing benchmark suite, called CTester, which includes an application for each component carefully designed to exercise all the relevant system calls, and a wrapper application that invokes individual applications at predetermined timing to create scenarios of concurrent system calls on the same or multiple components. During a CTester application run, we measure the power dissipated through an external power meter which reports fine-grained power consumption of the smartphone. The difference from the base power when the phone is in idle state is the extra power due to running the application.

Creating the benchmark applications requires the knowledge of all the possible system calls and their ordering (e.g. a read call cannot proceed an open system call) through which applications or the kernel can access a component. The first step is to classify all I/O system calls into two categories: initialization-based which includes calls that initialize or uninitialize a component such as file open, close, create, delete, and workload-based which includes calls that generate the actual workload to the component, such as file read and write. Next, for each component, we develop a testing application to exercise the relevant system calls by interleaving initialization and workload system calls. Figure 3 shows the generic structure of a CTester application for one component which consists of a sequence of initialization (e.g. file open/close) and workload (e.g. send/recv) system calls, which respects their correct usage ordering, i.e., a workload-

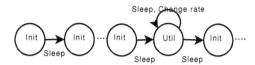

Figure 3: Structure of a CTester program.

based call has to follow an open system call. We introduce sufficient interval (on the order of several seconds) between the system calls.

The range of input parameters to the consecutive workload-based calls is decided based on the component's throughput, with the goal of uncovering all productive states and the threshold on the workload that triggers the transition from a low productive power state to a high productive power state if there are multiple. For example, the WiFi NIC of some smartphones has two productive states and the switching happens when there are more than 50pkts/sec transmitted or received. To search for such transitions efficiently, we increase the input parameters exponentially and perform binary search between two consecutive parameters if they lead to different productive states. In principle, the search process should continue until reaching the limits of the argument space. In practice, the thresholds for power transitions usually occur at small parameter values, e.g. 50 packets/sec for the WiFi NICs. Hence the search finishes very quickly, i.e., after observing that there is no more power state change after a few iterated doubling of the input parameter values.

6. Implementation

We have implemented system call tracing on Windows Mobile 6.5 running Windows CE 5.2 kernel and Android 2.2 running Linux Kernel 2.6.34. We note such system call tracing is previously available in desktop OSes [etw, str] but not in smartphone OSes.

6.1 Tracing System Calls in Windows Mobile 6

WM6 provides a logging mechanism called CeLog [cel], which only tracks all the events related to CPU (context switch and thread suspend/resume), memory (virtual and heap page fault), TLB, and interrupt. CElog is implemented by instrumenting the corresponding functions in the kernel source code. In principle, we could extend CElog to log other system calls needed for our power modeling, such as file system, GUI, and network interface, by instrumenting other parts of the kernel. However, this requires kernel source code and is tedious.

Instead, we leveraged the mechanism used by WM6 to implement system calls, a process called *thunking*. A system call is made to a special invalid address, which will result in a prefetch-abort trap, and the trap handler recognizes which system call was being made based on the invalid address. To log the needed additional system calls, we reroute the system calls in `coredll.dll` and `ws2.dll` (for networking) to enable logging. Essentially, we replace the thunks (e.g. `xxx_send`) with our own functions (e.g. `celog_send`),

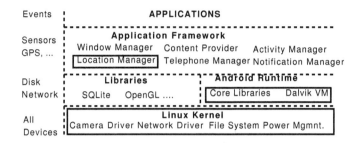

Figure 4: System call instrumentation in Android.

which are simply wrappers that log the system calls before calling the thunk exported by `coredll.dll` and `ws2.dll`. The new functions are implemented as a part of a library that is statically linked with `coredll.dll` and `ws2.dll` during compilation. System calls to GPS and various sensors are logged in a similar way.

6.2 Tracing System Calls in Android

Figure 4 shows the Android framework. It consists of an application framework which is built on top of a few C/C++ libraries and Android runtime. The runtime consists of core java libraries and Dalvik Virtual Machine which are optimized for mobile platforms. The entire Android stack sits on top of the standard Linux kernel. Applications on Android can be either written using Android SDK in java which will run in isolation in Dalvik VM, or written in native C/C++ compiled using the arm compiler which will run outside Dalvik, or written in both using Google's Native Development Kit (NDK). NDK assists compute-intensive applications written in java on Android SDK, to run part of their code (usually the compute-intensive part) outside Dalvik to improve their performance.

Ideally we just need to log all the events at the kernel layer. However, since Android uses many daemon processes that run on Dalvik to coalesce requests from different application processes, it would be difficult to attribute the cause of a system call back to the specific caller, which is needed to perform accurate energy accounting, e.g., on a per-process basis. For this reason, we log events not only at the kernel level, but also the framework level and the VM level (the three boxes in Figure 4). In particular, we use SystemTap [sys] to log system calls and CPU scheduling events in the `sched.switch` function in the kernel, use the logging library that comes with Android to additionally log different sensors, GPS, accelerometer, camera accesses at the framework level, and log disk/network at the DalvikVM level. We changed the default timer granularity of 10 ms to 1ms.

6.3 Flashing Customized Kernel Images

We built a WM6 and Android image with our modifications using platform builder [pb] and shared WM kernel source code [cec] for WM6 and cynogenmod tools [cyn] for Android. The customized images can directly work in corre-

Table 1: Mobile handsets used throughout the paper.

Name	Handset HTC-	CPU (MHz)	OS (kernel)	Base power
magic	Magic	528	Android 2.0 (Linux 2.6.34)	160mA
touch	Touch	528	WM6.1 (CE5.2)	250mA
tytn2	Tytn II	400	WM6.1 (CE5.2)	130mA

sponding emulators running on a desktop PC. We flashed commodity smartphones with customized kernel images.

7. Evaluation

We now present the FSM power models using our system-call-based power modeling approach for Android and WM running on three smartphones and validate their accuracy in energy estimation on a set of diverse applications.

7.1 Hardware Platform

Table 1 lists the hardware, the OS, and the base power of three smartphones used throughout this paper. We use out-of-box settings for LCD brightness and other system parameters. We use PowerMonitor [pow] to measure the power consumed by the smartphone. The smartphone battery is bypassed; the power terminals of the phone are connected to the power supply from the power monitor and the phone is powered through the monitor. PowerMonitor samples current draw once every 200 microseconds.

7.2 FSM Models Constructed

We start with the FSMs constructed for individual components of the three smartphones, followed by the FSM models for the entire smartphones.

7.2.1 FSMs for Individual Components

Figure 5 depicts the FSMs for CPU, sdcard and WiFi NIC for the three smartphones. We also modeled LCD, GPS, and camera, but omit them from the FSMs for clarify. The state transitions are labeled by the system calls or history-based conditions. The productive states in the FSMs are color shaded while the tail/base states are not shaded.

For CPU, we observe a simple FSM for all three smartphones. A process (thread) scheduled to run on the processor consumes a constant current. When the null loop executes on the CPU (thread 0x0 (idle) in process NK.EXE on WM6 and process pid 0 on Android), the system goes to the base state. The productive state for magic, tytn2 and touch consumes +100mA, +130mA and +200mA, respectively.

For sdcard, we observe a simple FSM for Android on magic where disk read/write system calls trigger the state change to a productive state with an average power consumption of +105mA. Other disk system calls do not have any noticeable effect on power consumption. The duration of the productive state is determined by the amount of data written or read and the sdcard throughput. On WM6, we find that system calls file open, close, create, delete, read

and write transit the sdcard to a high power state which consumes +125mA on tytn2 and +190mA on touch. The duration of read/write system calls is determined similarly for Android. A productive state is followed by a tail state which consumes +75mA and +110mA and lasts for 3 and 6 seconds for tytn2 and touch, respectively, before going back to the base state. If another system call is observed in tail state, it jumps into productive state and the above cycle repeats.

For wireless NIC, we observe a few power optimizations implemented by the device drivers. For all the handsets, there were two productive states and one tail state. We see that state transitions occur only on network send/recv system calls in all the handsets, except touch where a socket close call also triggers a transition. For all handsets, we find that transmitting (or receiving) packets in the base state causes a state transition to a low productive state (called "low net"). If the transmission rate increases beyond a threshold (12 packets in the past 1 second), the component transits to another base state (called "high net"), which is the same as the tail state. Any further workload-based system calls, i.e., send/recv, cause the NIC to transit to the second productive state. When the sending rate drops below 12 packets in the past 1.5 seconds, the NIC transits to the base state in magic. When there is no network activity for 1.7 seconds, similar transition occurs on tytn2. On touch, a socket close call or 12 seconds of no activity, whichever occurs first, causes the NIC to transit to the base state.

The signal strength of the wireless network affects the power consumption of the productive states. The signal strength can vary from -20dBm to -90dBm (best to worst), and increases the productive state power consumption by a maximum of 20% from best to worst (nearly linearly). Using simple network API calls, the model periodically samples the network signal strength and corrects the power consumption accordingly.

We also developed FSMs for LCD, GPS and camera. The later two are turned on and off using system calls. For example, GPS and camera on Android can be accessed by applications on the Android framework through requestLocationUpdates and OpenCameraHardware system calls. One interesting observation about the FSM for GPS is that after the GPS is turned off, a tail state of +70mA follows for 3 seconds on Android (Figure 1(b)).

7.2.2 FSMs for the Entire Smartphone

Figure 6 plots the FSMs for entire phones for all three phones. For clarity, we only include CPU, sdcard and WiFi NIC, and omit GPS and camera. The names of the states are simply composed from the state names in the FSMs of individual components in Figure 5 (separated by ","). The composite states that exhibit similar characteristics have been merged; their names contain names before merging separated by a "/". As in Figure 5, the power consumed in each state in Figure 6 represents the additional current (in mA) consumed in that state above the base current. Specifically,

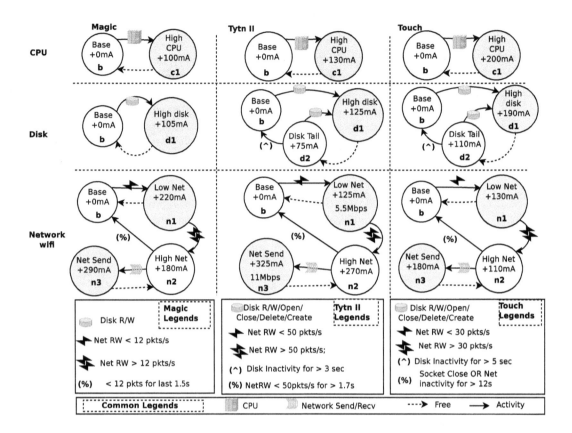

Figure 5: FSM for CPU, Disk and WiFi NIC for three smartphones. The circles in shades are productive states.

we run the cTester benchmark programs to drive different components to reach the desired states (e.g. CPU busy and WiFi NIC in tail) and measure the current consumed by the entire smartphone above the base level.

Android on magic has the simplest FSM. It has a base state, one productive state for each component (CPU, sdcard, network), and a network tail state. When multiple components are active, the total power consumption equals the summation of those of individual components when active alone. For WM6 on the other two handsets, there are two tail states, one of sdcard and the other of the network. We find that these states interact with each other; they do not add up as in case of Android. For example, an application consuming high CPU consumes +200mA (+130mA) irrespective of whether the transition came from the tail state of disk (which consumes +110mA (+75mA)) or from the base state (at 0mA) on the touch (tytn2) handset. A similar interaction exists between network and CPU on the touch handset. Another feature we observe is that when the tail states of sdcard and network overlap, the power draw does not equal the sum. For example, when sdcard and network are in their tail states, the power consumed on tytn2 will be the maximum of the power of the two states (in this case +270mA which is that of network). When the network tail state expires, the system falls back to the disk tail state (+75 mA), which lasts for 3 seconds from the last disk activity.

7.3 Energy Consumption Estimation

We now compare the accuracy of fine-grained and whole-application energy consumption estimation under our system-call-based modeling (FSM model for short) and under utilization-based modeling. For utilization-based modeling, we implemented the LR model for smartphones as described in [Shye 2009], by considering four components: LCD, CPU, sdcard, and NIC. We do not compare applications that use GPS or camera, since they are not modeled in [Shye 2009]. LR is developed by training with random application runs (as in [Shye 2009]).

For both OSes, we created /proc like utilization entries using our system call tracing, since our tracing captures non-utilization-based calls and hence forms a superset of the /proc-like logging utilities. However, the real benefit of using the same /proc under FSM and LR is that it allows us to run each application once, and perform offline comparison of their energy estimation accuracy for that run. This is important as most of the smartphone applications are interactive; the run time (and energy consumption) in different runs of the same application can easily fluctuate by over 10%. Further, if we ignore the extra overhead of periodically sampling /proc in LR for now, we can isolate and study the impact of /proc sampling frequency on LR's accuracy. We compare the overhead of FSM and LR in Section 7.4.

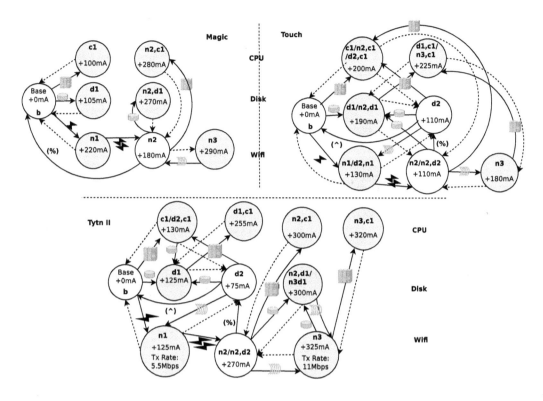

Figure 6: FSMs for entire smartphones when considering only CPU, Disk and WiFi NIC. The legends are the same as in Figure 5.

Table 2: Applications used throughout the paper.

App.	Description
Windows Mobile (on tytn2)	
game	First person shooter game (graphical)
chess	Mobile optimized version of chess
diskB	Open a file, sleep for 10 sec, write to file for 5 sec
netB	Connect, sleep for 10 sec, write over network for 5 sec
ie.cnn	User browsing mobile version of CNN on IE
pviewer	Slideshow of photos from SD card
docConv	A document converter (from .abc to .def)
virusScan	Performs an antivirus scan on SDcard
youtube	User watching a youtube video of 150 sec
puploader	Upload 8 photos to desktop using NIC from SDcard
Android (on magic)	
csort	sort program in native C
dropbear	ssh server process in native C, user listing all files
maps	google maps using GPS and network
facebook	facebook logging process
youtube	User watching a youtube video of 41 sec

Table 2 describes the applications used for both WM6 and Android, running on one handset each. We selected the tytn2 handset for evaluating WM6 applications since it has a more complex FSM when compared to touch, and the magic handset for Android applications. We selected a mixture of applications that utilize either a single, two, or more components at a time. For Android, we also selected a mixture of applications that are either written in native C (csort and dropbear), or in java based on Android SDK and

Google NDK (youtube, facebook), and included ones which use GPS (maps).

7.3.1 Fine-grained Energy Estimation

We first measure the error in fine-grained energy estimation. For each application, we split its execution time into 50ms intervals and calculate the absolute error of per-interval energy estimation under FSM and LR. Figure 7 shows the CDF of these errors for eight applications on WM6 and Android. We omitted simpler applications from these figures. They have comparable or better accuracy. Figure 7(a) shows that for all of the applications, the 80th percentile energy estimation error under FSM is less than 10%, and is less than 5% if we increase the time interval to 1s (not shown).

In contrast, Figure 7(b) shows that LR with 1s intervals (LR/1s) performs far worse than FSM. The error at the 80th percentile varies between 16% and 52%. A close look at the ie.cnn and youtube applications on WM6 shows why LR incurs much larger error. The ie.cnn application is CPU-intensive, with minimal network activity (about 20 packets of data). For each 1s interval, LR attributes a uniform power value for all 50ms intervals based on the average CPU utilization in that 1s, as shown in Figure 8(d). For youtube, when streaming a video from the youtube server, packet arrival is spread over the entire time duration of the video. There are several periods lasting between a few hundred ms to 1s during which there is no network activity (no network receive). During a specific 1s time period, if there is no network activity, LR attributes 0 power for the NIC. But in re-

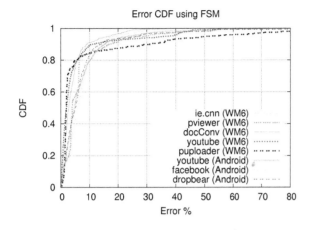

Error CDF using FSM

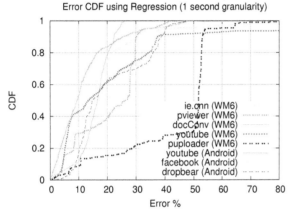

Error CDF using Regression (1 second granularity)

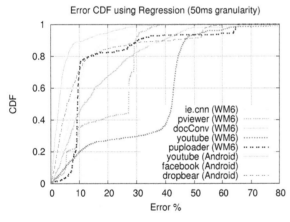

Error CDF using Regression (50ms granularity)

Figure 7: CDF of absolute error percentage between measured and estimated energy consumption per 50 ms for applications using (a) FSM, (b) LR with 1s granularity, (c) LR with 50ms granularity.

ality, due to recent high network activity (> 12 packets in past 1.5 second), the NIC is in tail power state consuming +180mA. As a result, all 20 50ms time intervals in this second have high error, as shown in Figure 8(f).

7.3.2 Impact of Fine-grained Sampling in LR

An intuitive way to improve LR's estimation accuracy is to improve its frequency in sampling utilization counters. However, high sampling frequency can lead to very high over-

head, as we show in Section 7.4. Ignoring the overhead for now, we increase the sampling frequency from once per second to once every 50ms, and the CDF of the resulting energy estimation error of LR (LR/50ms) is shown in Figure 7(c). We see that compared to LR/1s, the accuracy (a) increases for ie.cnn, docConv, puploader, and dropbear, (b) is mostly unaffected for pviewer, and (c) degrades for youtube (on WM6 and Android) and facebook.

Figure 8(g)-(i) plot the energy estimation under LR/50ms for the same three applications. Compared to LR/1s, LR/50ms performs quit well for the CPU-intensive ie.cnn since the more frequent sampling captures the fine-grained CPU utilization. However, since LR can not capture the disk tail states exhibited in running pviewer, like LR/1s (Figure 8(e)), LR/50ms (Figure 8(h)) also incurs about 30-40% error during the tail states. LR/50ms gives worse accuracy for applications in category (c) because of the network tail phenomenon. As explained earlier, packet arrival to NIC is spread throughout the time period of video. When we increase the sampling frequency of LR, LR only reports NIC utilization for the 50ms durations when there is actually a network receive, and zero NIC utilization in all other 50ms intervals. Relatively few 50ms intervals contain a recv call. As a result, though more precisely capturing the timing of network activities, LR's accuracy actually degrades. Puploader and dropbear are unaffected by this phenomenon as their destination servers are nearby (1ms RTT) and hence the gap between consecutive packets is only a few ms.

7.3.3 Whole-application Energy Estimation

Finally, we also compare the whole-application energy estimation accuracy under FSM and LR. Figure 9 plots the percentage error for all applications on both OSes under FSM and under LR/1s (using 1 second time granularity for training and estimation as in [Shye 2009]). We see that the estimation error under FSM varies between 0.2% and 3.6%. We note that if we just use the FSM for each component and add up the estimated energy consumption, the estimation error would be 22%, 45%, 11% and 36% for docConv, virusScan, youtube and puploader, respectively, on WM6 (not shown). In contrast, LR performs well for the applications that do not have dominant tail power patterns, but the estimation error varies between 3.5% to 20.3% for the rest, including diskB, pviewer, youtube, facebook, and dropbear. These results show system-call-based power modeling is not only far more accurate in fine-grained energy estimation, but can also significantly improve the accuracy of whole-application energy estimation.

7.4 Logging Overhead

Since /proc comes from logging system calls (though a subset of those in FSM), in principle both FSM and LR incur system call tracing overhead. In addition, LR incurs the overhead of periodic sampling of /proc. Below, we conser-

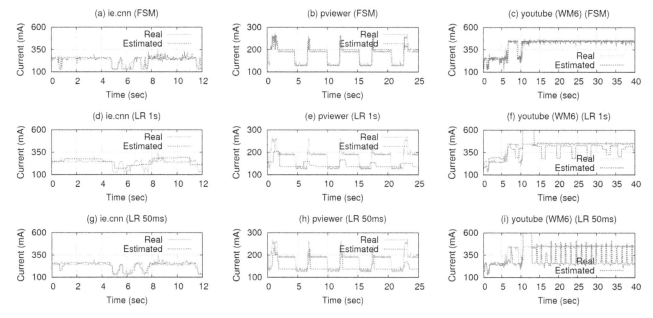

Figure 8: Power Profile: Measured vs estimated energy consumption over time on WM6 (tytn2 handset) under FSM, LR/1s, and LR/50ms.

vatively compare the system call tracing overhead in FSM with only the periodic sampling of /proc in LR.

We measure the system call logging overhead of FSM by comparing the execution time of the applications when logging is turned on versus off. Since re-execution of an interactive application (such as youtube) can experience significant variability in execution time due to external factors such as variability in server response time, network delay, and authentication schemes, we focus on applications that are not affected by external factors in measuring the logging overhead. The logging overhead under FSM varies between 5.4%-9.8% on Android and 1.1%-8.9% on WM6.

For LR, note the overhead of sampling /proc is proportional to the sampling frequency and independent of the frequency of system calls triggered by the application. We measured the overhead of reading /proc (CPU and IO (per-process network utilization is not available in /proc)) stats values for all processes from a user-level C application on Android. The overhead ranges between 7.5%-10.0% when sampling once per second, 15.2%-17.5% once every 500ms, and 35.3%-52.5% once every 200ms. The results show that due to high overhead, it is not even practical to obtain fine grained utilization information using /proc.

8. Applications: Energy Profiler

We give a proof-of-concept demonstration of an important application of fine-grained energy estimation enabled by our system-call-based power modeling, by showing a manually implemented *eprof*, the energy counterpart of the classic *gprof* tool [Graham 1982], for profiling application energy consumption.

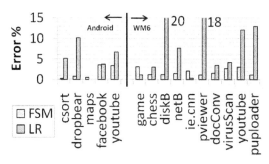

Figure 9: Absolute error in end-to-end energy consumption estimation.

Building on top of our FSM power model, there are two remaining challenges in the design of *eprof*. First, as in gprof, we need to construct the call graph and perform bookkeeping of each invocation of each subroutine, i.e., execution counts and energy consumption. Currently, we manually annotate every subroutine in the source code to log the entry and exit points. This also allows us to trace each system call to the caller subroutine.

The second challenge is per-subroutine energy accounting. A major difference between per-subroutine (or per-process, per-thread) execution time accounting in gprof and energy accounting in eprof comes from the tail energy consumption by many components. Figure 10 shows an example where a network system call in subroutine F1 sends the NIC to a tail state which lasts beyond the completion of both F1 and subroutine F2. The energy consumption during F2 was an accumulative effect of both CPU activities of F2 and the tail state of the NIC due to F1. A complete solution to the above problem of fine-grained energy accounting is beyond the scope of this paper. In our proof-of-concept implemen-

Table 3: Eprof output for three applications: The columns of the table are: % time and % energy spent in self, cumulative time and energy spent by all the function descendants, actual time spent in a function itself, number of times a function is called, time and energy spent in self per call, time and energy spent non-cumulative per call, and name of the function.

%		cum.		self		#	self		total		name
time	energy	s	uAH	s	uAH	calls	ms/call	nAH/call	ms/call	nAH/call	
						chess					
30.4	30.4	34.4	1243.0	21.6	779.9	61931	0.4	12.6	0.56	20.1	Is_Move_Legal
19.7	19.7	22.9	825.9	14.0	505.0	73665	0.2	6.9	0.3	11.2	Is_Move_Valid
14.9	14.9	75.1	2713.0	10.6	382.1	60416	0.2	6.3	1.2	44.9	CheckHumanMove
12.0	12.0	8.6	308.8	8.6	308.8	919	9.3	336.0	9.3	336.0	CountScore_Human
5.9	5.9	4.2	150.4	4.2	150.4	433	9.6	347.3	9.6	347.3	CountScore
						docConv					
42.0	36.6	34.0	1063.0	24.9	762.9	77	323.4	9907.3	441.9	13805.6	ExtractTextFromFile
28.3	25.9	45.8	1443.6	16.8	539.6	148133	0.1	3.6	0.3	9.8	CheckToken
19.5	28.0	55.1	1951.9	11.5	583.6	1	11533.7	583569.2	55120.9	1951933.5	ExtractText
7.0	6.3	59.3	2082.6	4.2	130.6	1	4186.6	130639.5	59307.5	2082573.0	main
3.2	3.2	1.88	65.9	1.9	65.9	77	24.5	855.8	24.5	855.8	getpagecontent
						puploader					
50.2	67.9	2.0	167.9	2.0	167.9	8	245.8	20987.6	245.8	20987.6	sendfile
18.0	15.0	0.7	37.0	0.7	37.0	8	88.2	4628.0	88.2	4628.0	readfile
17.5	9.0	0.7	22.2	0.7	22.1	8	85.9	2769.7	85.9	2769.7	calchash
5.7	3.3	0.2	8.1	0.2	8.1	1	224.8	8118.7	224.8	8118.7	GUI_Form
5.2	3.0	0.2	7.4	0.2	7.4	1	203.4	7345.7	203.4	7345.7	initnet

tation, we take the following simple approach: we always break up the power consumed in a power state into power consumed per component, by assigning a continuing power state (e.g. tail state) the same power when occurring alone. For example in Figure 10, F1 and F2 will be charged P1 and (P2-P1) during T2 to T3, respectively. We leave more complicated cases where multiple subroutines (threads or processes) in the past intervals are responsible for the total power as future work.

Table 3 shows the gprof-style time and energy breakdown for top 6 functions (sorted by runtime) in three applications (chess, docConv and puploader on WM6 on tytn2). We observe that the time and energy percentages are about the same for all the functions in chess (only CPU), but differ in docConv (CPU, sdcard). The most interesting application is puploader, where three functions, calchash, readfile and sendfile, use CPU, disk and network respectively. Sendfile consumes the most time (50.2%) and energy (67.9%). Calchash, though consuming 17.5% of the time, spends only 9.0% of the energy. If we did not apply fine-grained energy accounting as discussed above, sendfile and calchash would have been reported to consume 57.5% and 13.0% of energy.

9. Related Work

System call tracing has been exploited to develop useful tools on desktop and server machines, in particular, for accounting resource utilization [Barham 2004], building debuggers/replay tools [Guo 2008], automatic fault detection and diagnosis [Yuan 2006], and energy measurement and accounting [Zeng 2002]. It has also been used on mobile

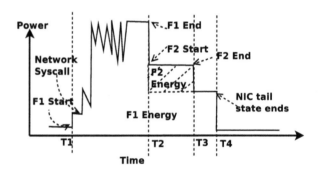

Figure 10: Eprof: Function level breakdown.

devices for malware detection [Bose 2008], and on sensor nodes for energy measurement [Shnayder 2004].

Our system-call-based power modeling is related to [Shnayder 2004, Zeng 2002], which also exploit events as opposed to utilization rates of components, but for desktop machines and sensor nodes. Further, these work do not use system calls to capture non-utilization-based power behavior. In contrast, we discover non-utilization-based power behavior on smartphones and use system call tracing to capture both utilization-based and non-utilization-based power behavior to achieve accurate fine-grained energy estimation.

Finally, RevNIC [Chipounov 2010] aims to reverse-engineer the exact behavior of device drivers, which is a harder task than reverse-engineering device drivers' power optimization behavior. Our approach works well in practice as fundamentally the power behavior of the drivers are correlated with the intention and consequence of system calls that trigger them, and there are only a small number of power states per component in practice.

10. Conclusion

In this paper, we have presented the design and implementation of a system-call-based power modeling approach which gracefully captures both utilization-based and non-utilization-based power behavior. Our experimental results on Android and Windows Mobile using a diverse set of applications show that the new model drastically improves the accuracy of fine-grained energy estimation as well as whole-application energy estimation. We further presented a proof-of-concept demonstration of *eprof*, the energy-counterpart of *gprof*, for optimizing the energy usage of application programs. We are developing a full-fledged *eprof* and plan to release the modified Android image and tools to the public. Our power modeling study also exposed significant diversity of power behavior of different OSes and smartphone handsets. As future work, we plan to develop detailed classification of power behavior of different OSes and handsets and for different applications.

Acknowledgments

We thank the program committee and reviewers for their helpful comments, and especially our shepherd, M. Satyanarayanan, whose detailed feedback significantly improved the paper and its presentation.

References

[Com] Android phones steal market share. URL http://bmighty.informationweek.com/mobile/showArticle.jhtml?articleID=224201881.

[cel] Celog event tracking. URL http://msdn.microsoft.com/en-us/library/aa462467.aspx.

[cyn] Cyanogenmod: Android community rom based on froyo. URL http://www.cyanogenmod.com/.

[etw] Event tracing for windows (etw). URL http://msdn.microsoft.com/en-us/library/ms751538.aspx.

[pb] Microsoft platform builder. URL http://msdn.microsoft.com/en-us/library/ms938344.aspx.

[pow] Monsoon power monitor. URL http://www.msoon.com/LabEquipment/PowerMonitor/.

[str] Strace. URL http://linux.die.net/man/1/strace.

[sys] System tap. URL http://sourceware.org/systemtap/.

[cec] Windows embedded ce shared source. URL http://msdn.microsoft.com/en-us/windowsembedded/ce/dd567722.aspx.

[Balasubramanian 2009] Niranjan Balasubramanian, Aruna Balasubramanian, and Arun Venkataramani. Energy consumption in mobile phones: a measurement study and implications for network applications. In *Proc of IMC*, 2009.

[Barham 2004] Paul Barham, Austin Donnelly, Rebecca Isaacs, and Richard Mortier. Using magpie for request extraction and workload modelling. In *Proc. of OSDI*, 2004.

[Bellosa 2000] F. Bellosa. The benefits of event: driven energy accounting in power-sensitive systems. In *Proc. ACM SIGOPS European workshop*, 2000.

[Bircher 2007] W.L. Bircher and L.K. John. Complete system power estimation: A trickle-down approach based on performance events. In *Proc. of ISPASS*, 2007.

[Bose 2008] Abhijit Bose, Xin Hu, Kang G. Shin, and Taejoon Park. Behavioral detection of malware on mobile handsets. In *Proc. of MobiSys*, 2008.

[Chipounov 2010] Vitaly Chipounov and George Candea. Reverse Engineering of Binary Device Drivers with RevNIC. In *Proc. of EuroSys*, 2010.

[Fan 2007] X. Fan, W.D. Weber, and L.A. Barroso. Power provisioning for a warehouse-sized computer. In *Proc. of ISCA*, 2007.

[Flinn 1999a] Jason Flinn and M. Satyanarayanan. Energy-aware adaptation for mobile applications. In *Proc. of SOSP*, 1999.

[Flinn 1999b] Jason Flinn and M. Satyanarayanan. Powerscope: A tool for profiling the energy usage of mobile applications. In *Proc. of WMCSA*, 1999.

[Graham 1982] S. L. Graham, P. B. Kessler, and M. K. McKusick. gprof: A call graph execution profiler. In *Proc. of ACM PLDI*, 1982.

[Guo 2008] Zhenyu Guo, Xi Wang, Jian Tang, Xuezheng Liu, Zhilei Xu, Ming Wu, M. Frans Kaashoek, and Zheng Zhang. R2: An application-level kernel for record and replay. In *OSDI*, pages 193–208, 2008.

[Kansal 2010] A. Kansal, F. Zhao, J. Liu, N. Kothari, and A.A. Bhattacharya. Virtual machine power metering and provisioning. In *Proc. of SOCC*, 2010.

[Mahesri 2005] A. Mahesri and V. Vardhan. Power consumption breakdown on a modern laptop. *Proc. of PACS*, 2005.

[Rawson 2004] F. Rawson. MEMPOWER: A simple memory power analysis tool set. *IBM Austin Research Laboratory*, 2004.

[Shnayder 2004] Victor Shnayder, Mark Hempstead, Bor rong Chen, Geoff Werner Allen, and Matt Welsh. Simulating the power consumption of large-scale sensor network applications. In *Proc. of Sensys*, 2004.

[Shye 2009] A. Shye, B. Scholbrock, and G. Memik. Into the wild: studying real user activity patterns to guide power optimizations for mobile architectures. In *Proc. of MICRO*, 2009.

[Snowdon 2009] David C. Snowdon, Etienne Le Sueur, Stefan M. Petters, and Gernot Heiser. Koala: a platform for os-level power management. In *Proc. of EuroSys*, 2009.

[Stanley-Marbell 2001] P. Stanley-Marbell and M. Hsiao. Fast, flexible, cycle-accurate energy estimation. In *Proc. of ISLPED*, 2001.

[Tiwari 1996] V. Tiwari, S. Malik, A. Wolfe, and M. Tien-Chien Lee. Instruction level power analysis and optimization of software. *The Journal of VLSI Signal Processing*, 13(2), 1996.

[Yuan 2006] C Yuan, N Lao, J Wen, J Li, Z Zhang, Y Wang, and W Ma. Automated known problem diagnosis with event traces. In *EuroSys*, 2006.

[Zedlewski 2003] J. Zedlewski, S. Sobti, N. Garg, F. Zheng, A. Krishnamurthy, and R. Wang. Modeling hard-disk power consumption. In *Proc. of FAST*. USENIX Association, 2003.

[Zeng 2002] Heng Zeng, Carla S. Ellis, Alvin R. Lebeck, and Amin Vahdat. Ecosystem: Managing energy as a first class operating system resource. In *Proc. of ASPLOS*, 2002.

[Zhang 2010] L. Zhang, B. Tiwana, Z. Qian, Z. Wang, R.P. Dick, Z.M. Mao, and L. Yang. Accurate Online Power Estimation and Automatic Battery Behavior Based Power Model Generation for Smartphones. In *Proc. of CODES+ISSS*, 2010.

Sierra: Practical Power-proportionality for Data Center Storage

Eno Thereska

Microsoft Research

etheres@microsoft.com

Austin Donnelly

Microsoft Research

austind@microsoft.com

Dushyanth Narayanan

Microsoft Research

dnarayan@microsoft.com

Abstract

Online services hosted in data centers show significant diurnal variation in load levels. Thus, there is significant potential for saving power by powering down excess servers during the troughs. However, while techniques like VM migration can consolidate computational load, storage state has always been the elephant in the room preventing this powering down. Migrating storage is not a practical way to consolidate I/O load.

This paper presents Sierra, a power-proportional distributed storage subsystem for data centers. Sierra allows powering down of a large fraction of servers during troughs without migrating data and without imposing extra capacity requirements. It addresses the challenges of maintaining read and write availability, no performance degradation, consistency, and fault tolerance for general I/O workloads through a set of techniques including power-aware layout, a distributed virtual log, recovery and migration techniques, and predictive gear scheduling. Replaying live traces from a large, real service (Hotmail) on a cluster shows power savings of 23%. Savings of 40–50% are possible with more complex optimizations.

Categories and Subject Descriptors C.2.4 [*Computer-communication networks*]: Distributed Systems

General Terms Design, Performance, Reliability

Keywords Data center, energy, power-proportionality

1. Introduction

Server power consumption is a major problem for small and large data centers, since it adds substantially to an organization's power bills and carbon footprint. Barroso and Hölzle have argued for *power proportionality*: the power used should be proportional to the dynamic system load,

even though the system is provisioned for the peak load [Barroso 2007]. Such a system could exploit diurnal variations in load, such as the "Pacific Ocean trough", seen in many user-facing services [Hamilton 2008].

Achieving power proportionality in hardware is hard, despite some support such as dynamic voltage scaling for CPUs. Servers continue to draw significant power even when idle, which can be up to 50% of the peak power consumption [Fan 2007]. A feasible alternative for large data centers is to turn off entire servers during troughs and consolidate load on the remaining servers. Virtualization techniques can transparently migrate computational state and network connections [Clark 2005]. However, storage presents a challenge. It is not feasible to migrate petabytes of storage several times a day for load consolidation (it takes too long).

This paper describes Sierra, a power-proportional distributed storage system. Sierra handles general workloads with read and writes (in contrast to recent work e.g., [Amur 2010] that only addresses read-only workloads) and Sierra supports in-place overwrites of data (this differs from append-only stores such as GFS [Ghemawat 2003]). Sierra exploits the redundancy (replication) in such systems to put the system in a lower "gear" during troughs by turning servers off. There are several challenges in doing this. To maintain read availability, at least one replica of each object must be on an active server always. This must be done without increasing the replication level and hence the system capacity requirements. For write availability, updates must succeed even when many servers are turned off. Durability, consistency, and fault-tolerance guarantees must be maintained on all data including the updates in low gear. Performance must not degrade: there should always be enough servers on to handle the load at any given time, and the load should be balanced across them. Powering servers down must not increase recovery time for transient or permanent server failures.

To our knowledge, Sierra is the first distributed storage system that meets all the above challenges and also provides power-proportionality. Our specific contributions are as follows. A new *power-aware layout* allows a significant fraction of the servers to be powered down without losing availability, load balancing, or fault tolerance. A novel use of a *distributed virtual log* allows updates to objects when some replicas are turned off. Such updates have the same

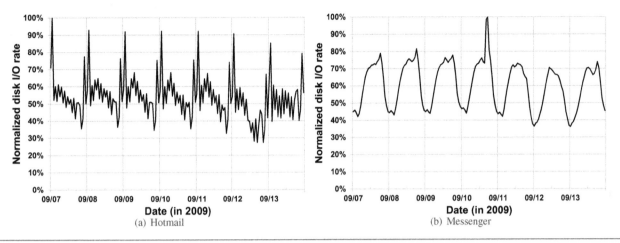

(a) Hotmail   (b) Messenger

Figure 1. One week of I/O load for two large online services

consistency, durability, and replication for fault tolerance as updates made when no servers are powered down. An *evaluation* of a full prototype running on a server cluster, using I/O traces from Hotmail, shows that significant power savings can be achieved with little performance impact.

2. Motivation, goals and non-goals

Sierra is based on the idea of *gear-shifting*: turning servers off and on to track diurnal variations in load. The savings achievable from this method depend on the average utilization of the system relative to the peak, and the predictability of load peaks and troughs. Figure 1(a) and 1(b) show the aggregated I/O load over time for two large online services, for 1 week. The load is aggregated over tens of thousands of servers for Hotmail (Windows Live Mail), and thousands of servers for Windows Live Messenger. The graphs show the disk bandwidth usage aggregated across all the back-end storage in the service at one-hour intervals and normalized to the peak value for each service (Section 4 will give more information about the services). We observe clear periodic patterns with significant peak-to-trough ratios.

Figure 2 shows the CDF of time spent at different utilization levels, again relative to the peak load. Substantial periods of time are spent with utilization much lower than the peak. The average utilization for Hotmail is 42% of its peak and that of Messenger is 60%. Thus, there is substantial scope for power savings through power proportionality. As an example, 50% savings in power for a 50,000 server data center could generate savings of $6 million/year assuming a power bill of $1 million/month [Hamilton 2009].

The primary scenario for Sierra, and the focus of this paper, is clusters of commodity servers where computation and storage are co-located on the same nodes. However, Sierra is also applicable to dedicated storage clusters. It is also relevant to SANs (storage area networks) but would require changing the internal behavior of the SAN. It is also

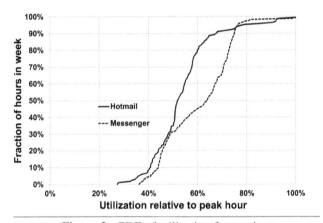

Figure 2. CDF of utilization for services

possible to use Sierra to spin down disks rather than turn off servers, e.g., if all server CPU resources are required.

This paper does not address migration and consolidation of computational and network load, since others have addressed them successfully (e.g., see [Chen 2008, Clark 2005]). In deployments where computation and storage are co-located, the selection of servers to turn off would need to be co-ordinated, for example between the VM migration layer and Sierra, to ensure that compute and I/O load are consolidated onto the same physical servers.

This paper does not address "trough-filling": eliminating troughs by filling them with useful work. Trough-filling by service consolidation was not effective for the online services we examined for two reasons. First, even when all users within a service are consolidated, we still see troughs (the Hotmail load graph captures all Hotmail users globally). Geo-distribution of these services will only lead to larger peak-to-trough ratios. Second, when consolidating across services, some are much larger and dominate (e.g., Hotmail has an order of magnitude more servers than Messenger). Filling troughs by selling the idle resources is an alternative for large cloud providers. Users' tasks could be run during troughs whenever the users' bid price (described by [Amazon 2010, Stokely 2009]) exceeds the cost of powering the servers. This approach is complementary to Sierra.

3. Design and implementation

Sierra adopts existing best practices to provide scalable replicated distributed storage. This "baseline" architecture for Sierra is described in Section 3.1. It provides scalability, load balancing, high availability, fault tolerance, and strong read/write consistency. The challenges lie in powering down significant numbers of servers at low load without sacrificing the above properties.

The first challenge is *layout*: the assignment of data replicas to servers. The commonly used "naïve random" approach is simple, and provides high availability, good load balancing, and parallelism when rebuilding data after a failure. However, it does not allow significant power savings. Perhaps surprisingly, it is possible to lay out data such that significant numbers of servers can be turned off, while still maintaining good availability, load balancing and rebuild parallelism. Section 3.2 describes such a layout, used by Sierra. The layout is also optimal in that it provides the maximum possible power savings for any desired level of availability. Further, this is achieved without increasing the replication level of objects or over-provisioning the system.

The second challenge is to maintain read and write availability at the original levels. This must be done in low gear (while a significant number of servers are off), during gear transitions (servers are being powered up or down), and on failures. Section 3.3 describes the distributed virtual log (DVL) used by Sierra to keep the system available for writes. The DVL builds on the concept of write off-loading for RAID-based storage [Narayanan 2008]. We have extended it to support a distributed storage system, which introduces new challenges of network efficiency, availability, consistency, and fault tolerance. By solving these challenges we were able to considerably simplify the rest of the system and yet provide write availability and consistency at all times.

The third challenge is to predict the number of servers required at any time to sustain the system load, so that performance is not degraded. This is done by the *gear scheduler* component of Sierra, described in Section 3.5.

3.1 Baseline architecture

The baseline design of Sierra (Figure 3) is not novel, and resembles existing scalable distributed storage systems (e.g., see [Azu 2009, Abd-El-Malek 2005, Ghemawat 2003, Saito 2004]); we describe it here for context.

Sierra provides read/write access to objects in units of *chunks*. The chunk size is a system parameter: a typical value is 64 MiB. Each chunk is replicated on multiple *chunk servers*; a typical replication factor for data is 3. We use primary-secondary replication; in general, at any time, a chunk server will be the primary for some chunks and a

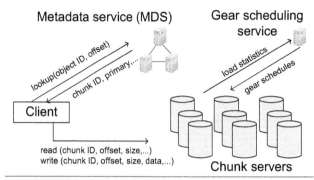

Figure 3. Baseline architecture for Sierra. We extend this architecture to support power-proportionality.

secondary for others. Clients can read or write arbitrary byte ranges within a chunk. All client read and write requests are sent to the primary, which determines request ordering and ensures read/write consistency. Write requests are sent by the primaries to all replicas in parallel and acknowledged to the client when all replicas are updated. Read requests are served by primaries directly from the local replica.

Sierra is intended to be a general-purpose storage system for use by a variety of applications. Hence, Sierra supports overwriting of existing chunk data; this differs from append-only stores such as GFS [Ghemawat 2003]. This allows Sierra to support a wider variety of workloads, but also introduces the challenge of supporting overwrites of chunks when one or more replicas are powered down.

A centralized metadata service (MDS) maps each object to its constituent chunks, and each chunk to its current primary. The MDS is an in-memory, deterministic state machine that could be replicated for high availability [Schneider 1990]. The MDS is not on the data path; clients cache metadata lookups from the MDS, and metadata is only updated on chunk creation or deletion and server failure, power-up, or power-down. The MDS grants leases to chunk primaries, which are periodically renewed through a heartbeat mechanism. If a lease expires it is reassigned by the MDS to a different replica of the chunk. Each such "view change" causes the MDS to increment a per-chunk *epoch*, which functions as a viewstamp [Oki 1988]. Requests from older epochs are rejected by chunk servers, ensuring that all replicas see a consistent ordering of client requests.

To reduce the metadata overheads, Sierra groups chunks into *chunk groups*. Leases and epochs are per chunk group rather than per chunk. All chunks in a chunk group are replicated on the same servers and have the same primary at any given time. Chunks are assigned randomly to chunk groups on creation. The number of chunk groups is a system parameter that trades off MDS load for fine-grained load balancing across servers. By default Sierra currently uses 64 chunk groups per chunk server. Load balancing in Sierra is done by uniform random assignment of chunks to chunk groups, chunk groups to servers, and primaries to replicas.

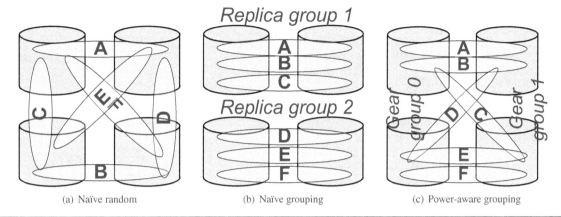

	Replica group 1				

(a) Naïve random   (b) Naïve grouping   (c) Power-aware grouping

Figure 4. Different layouts for 6 chunks (A-F), 4 servers, and 2-way replication. The *naïve random* layout replicates each chunk on 2 servers chosen at random. This layout is excellent for load balancing and rebuild parallelism but leads to little or no power savings. The *naïve grouping* layout limits the location of chunks to replica groups (two in this case). A chunk must not be replicated across replica groups. This layout leads to great power savings but has poor rebuild parallelism. The *power-aware grouping* layout relaxes the constraint placed by *naïve grouping* and introduces a reasonable tradeoff between load balancing, rebuild parallelism, and power savings.

Individual chunks may experience different diurnal access patterns and loads, however chunk groups and chunk servers see the same aggregated load pattern.

Currently the system uses primary/secondary replication with 1-phase commit. This guarantees that all client requests are applied in the same order on all replicas. However a failure of a primary during a write request can result in some but not all replicas applying the write. On such a failure, the client library corrects the divergence by re-sending the write to the new primary, which uses the DVL to ensure that the write succeeds. Hence, unlike optimistic concurrency control systems (e.g., [DeCandia 2007]) there is no long period of divergence. Alternatively two-phase commit or chain replication could be used to prevent temporary replica divergence at the cost of higher update latencies [van Renesse 2004]. Any of these techniques are reasonable and would work well with the power-saving aspects of Sierra.

3.2 Power-aware layout

We find that layout, the way in which chunks are assigned to chunk servers, makes a large difference in how power-proportional a system can be. The most commonly used layout today is a *naïve random* layout or a variant of it, where each chunk is replicated on r servers chosen at random. Figure 4(a) shows a simple example with 4 servers, 6 chunks, and 2-way replication.

Our goal is a layout that can maintain g available copies of each chunk (out of a total of r, the replication level) using only $\frac{g}{r}$ of the servers. If data is evenly distributed over servers, then this is the minimum number of servers required to host $\frac{g}{r}$ of the data. However, the layout must ensure that this fraction corresponds to exactly g replicas of each chunk. We define a layout as optimal for power proportionality if it

	Active servers	Rebuild	Load distr.
Naïve random	$N - (r - g)$	N	rC / N
Naïve grouping	$N \frac{g}{r}$	1	$C / \frac{N}{r}$
Power-aware	$N \frac{g}{r}$	$\frac{N}{r}$	$C / \frac{N}{r}$

Table 1. Active servers, rebuild parallelism, and load distribution for different layouts with N servers, C chunks, and r-way replication. g refers to the available copies of each chunk, or gear level ($1 \leq g \leq r$). A load distribution of rC / N means rC chunk replicas are uniformly distributed over N servers.

achieves this property for every integer g, $1 \leq g \leq r$. The value of g is now the *gear level* of the system at any time.

Unfortunately, the naïve random layout is far from optimal. In Figure 4(a)'s example, to keep 1 replica of every chunk available, we need 3 servers out of 4, rather than 2. As the number of servers and chunks increases, the minimum number of active servers required approaches $N - (r - g)$ rather than $N \frac{g}{r}$ where N is the total number of servers. An alternative approach is to put servers into *replica groups*, each of size r (Figure 4(b)). A chunk is then assigned to one replica group rather than to r independently chosen servers. This layout is optimally power-proportional, since $r - g$ servers in each replica group can be turned off in gear g. However, naïve grouping of servers suffers from a lack of *rebuild parallelism* because of the constraint on where replicas can reside. When a server suffers a permanent failure, intuitively there are fewer servers to rebuild *from* (in the illustrated example just 1 — this could lead to a read bottleneck) and fewer servers that one can rebuild *on*, which could lead to a write bottleneck.

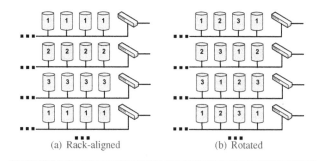

(a) Rack-aligned   (b) Rotated

Figure 5. Two ways of assigning gear groups.

Sierra generalizes naïve grouping to achieve both power savings and high rebuild parallelism with *power-aware grouping* (Figure 4(c)). Each server is assigned to exactly one of r *gear groups*, each of size $\frac{N}{r}$. Each chunk is replicated once in each gear group. Now any g out of r gear groups can serve g replicas of each chunk. If a server fails, then its data can be rebuilt in parallel on all remaining servers in its gear group. Thus, the rebuild parallelism is $\frac{N}{r}$ where N is the total number of servers.

Table 1 summarizes the three approaches. The small examples in Figure 4 do not show this, but it is important to note that in realistic deployments, all three layouts will balance load well by distributing a large number of chunk replicas over a much smaller number of servers. Compared to the naïve random layout, the two power-proportional layouts spread $\frac{1}{r}$ as many replicas over $\frac{1}{r}$ as many resources. However, this is still a very large number: we are aiming at systems with millions of chunks spread across 1000s of servers. The same load balancing is maintained at lower gears.

For a large data center, servers might be further grouped into *clusters*. Different clusters could store different data, and be independently organized using power-aware grouping. Different clusters could also have different replication factors or coding schemes. E.g., small hot objects might be replicated with $r > 3$, and large cold objects might be erasure-coded. With an m-of-n erasure code, power-aware grouping allows $\frac{g}{n}$ of the servers to serve all data with a redundancy of $g - m$, $(m \le g \le n)$.

Smaller clusters reduce the dependencies between servers and hence the overhead of waking up sleeping replicas when an active replica fails. However, the cluster should be large enough so that rebuild speeds are limited by network bandwidth rather than server performance; with current hardware we expect each cluster to contain 100–1000 servers.

Sierra also places chunk replicas in different fault domains, i.e., racks. There are two options here for power-aware layout: rack-aligned and rotated (Figure 5). Both options give the same fault tolerance, power proportionality, and load balancing. They differ in the selection of servers to be turned off leading to different tradeoffs on power and thermal balance. In the rack-aligned case, all servers in a rack are in the same gear group and hence always in the same power

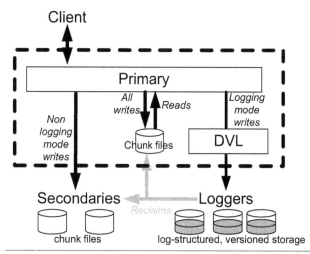

Figure 6. Data paths in logging and non-logging modes. The dotted box represents a single chunk server acting as a primary.

state. This permits entire racks to be turned off, allowing additional power savings by turning off rack-wide equipment such as switches. The rotated layout distributes the powered-up servers, and hence the thermal load, evenly across racks.

Our layout scheme does not add any additional complexity to data migration or adding and removing servers (or whole racks) to the system. As part of any data migration the constraint that the new location must be in the same gear group as the old location must be preserved.

3.3 Distributed virtual log

Each primary in Sierra is associated with one instance of a distributed virtual log or DVL (Figure 6). The DVL is used to reliably absorb updates to replicas that are on powered-down or failed servers. Hence, it is optimized for writes and for short-term storage. When one or more secondaries is unavailable (powered down or failed), a Sierra primary enters "logging mode". In this mode it sends writes to the primary replica and to the DVL but not to the secondaries. Data written to the DVL is replicated $r_L = r - 1$ times, guaranteeing a total of r replicas of all data. Along with the data, the DVL writes metadata including the chunk ID, the byte range written, a version number, and the primary's location. When all secondaries are available again, the primary enters "reclaim mode", where it scans the DVL and applies the deferred updates to the secondaries in the background. Once reclaimed, the data is deleted from the DVL.

No dedicated disk resources are assigned to a DVL; instead all DVLs share a common pool of *loggers*. Each logger uses a local on-disk log to store updates. Data from different DVLs is physically interleaved on the loggers but is kept logically separate. Write requests are load-balanced by the DVL across the set of available loggers. The DVL locally maintains a small in-memory map (recoverable from on-disk records at any time) that tracks the location of its data, and

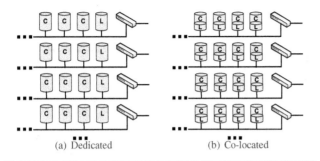

Figure 7. Two ways of configuring loggers (L) and chunk servers (C)

can also be queried by applications. Each DVL has a *logger view*, which is the set of all loggers available for use by it. This logger view is stored in the MDS, since it is small and infrequently updated.

The DVL provides the same semantics as a local on-disk log, even though successive writes and deletes can be sent to different loggers. These are:

- Write requests can be issued concurrently and are committed in FIFO order.
- Writes and deletions are durable when acknowledged.
- Reads always return the latest written data.
- Reads never return deleted data.

Write ordering is guaranteed by storing a monotonically increasing version (monotonic per-primary, not the whole cluster) with each write sent to the loggers. Correct deletion is guaranteed by writing a versioned, durable deletion marker. The deletion marker is deleted only when all older versions of the data have been deleted from all loggers.

Loggers can be run on dedicated servers or co-located with chunk servers as shown in Figure 7. The dedicated configuration minimizes contention between the chunk server workload and the logger workload, allowing the loggers to service mostly writes for which they are optimized. The dedicated configuration requires additional resources, but still provides significant total power savings. The provisioning method for both cases is described in Section 3.5.

3.3.1 Network-aware logging

Sierra minimizes the network overheads of logging and reclaim, especially the usage of scarce cross-rack (top-level switch) bandwidth. We use several optimizations to achieve this. First, the primary always serves reads from the local replica when the latest data is available there. This reduces disk and network load by avoiding demand reads on the DVLs, allowing the loggers to service mostly writes in logging mode and sequential scans in reclaim mode. A second network optimization is a direct data path from the loggers to the secondaries while reclaiming; for correctness, the control path always goes through the primary.

Third, the DVL uses a network-aware logger choice policy. In general, the DVL can send each update to any r_L

loggers in its logger view. However, these loggers must be in different fault domains for fault tolerance. Also, if loggers are co-located with chunk servers, the r_L loggers must be in different gear groups. This ensures that turning off chunk servers while gearing down does not make logged data unavailable. Subject to these constraints, the logger choice policy minimizes both total network bandwidth and cross-rack bandwidth usage. For every write request, the DVL for a chunk group primary G considers its loggers in this order:

1. Loggers on the same server as a replica of G,
2. Loggers in the same rack as a replica of G, and
3. Loggers in other racks.

Within each of these groups, loggers are sorted by disk load. The DVL then greedily chooses the first r_L loggers in this ordering that satisfy the constraint of being in different fault domains from each other and from the primary. Typically this will result in one logger in the same rack as each secondary. Reclaiming is done by each replica transferring data from the logger closest to it. In rare cases the DVL gets demand read requests; these are sent to the logger(s) closest to the primary and holding the required data.

With these optimizations, logging of updates uses no more network bandwidth than normal replication without a DVL. Reclaiming uses a modest amount of additional in-rack bandwidth, but no cross-rack bandwidth is used except for control messages. Furthermore, reclaiming is a background process that is run at low priority.

3.3.2 DVL recovery and migration

The primary for a chunk group always holds a lease at the MDS for that chunk group. If the primary fails to renew the lease, the MDS initiates a view change with a new epoch number. It grants the lease to another replica of the chunk group, which becomes the new primary. Before accepting requests, the new primary must instantiate the correct local state of its DVL, i.e., the metadata which identifies the location and version of all logged data. This is done by requesting the metadata from all L loggers in the DVL's logger view, concurrently. Each logger first completes all outstanding requests for that DVL, and then atomically updates the epoch and returns the in-memory metadata for that DVL. The logger then does not accept any further requests from the old epoch. When $L - r_L + 1$ loggers have responded, the DVL's state is complete and the DVL can begin operation.

When a server is powering down rather than failed, we use an optimization that avoids running DVL recovery. Chunk group primaries on a server about to power down proactively migrate themselves. First, each primary completes any outstanding requests on the chunk group, while rejecting new ones (these will be retried by the client at the new primary). Then, it serializes its DVL state and sends it as part of a lease cancellation message to the MDS. The MDS forwards this state to the new primary as part of the lease grant message.

When all primaries have been migrated, a server S that is powering down sends a final "standby" message to the MDS and to all primaries for which it is a secondary. The MDS then marks S as "powered down". S's peers then use the DVL for updates rather than attempting to update S. When S wakes up from standby it resumes sending heartbeats to the MDS. The MDS then re-balances load by sending S a list of chunk groups to acquire leases on. S then contacts the current primary of each chunk group to initiate the migration.

3.4 Fault tolerance and availability

This section describes novel aspects of fault tolerance in Sierra. The lease reassignment protocol maintains read availability after a transient chunk server failure; this is standard and we do not discuss it further. Write availability on failure is maintained by using the DVL to absorb writes; the DVL was described in Section 3.3. After a permanent chunk server failure, its data is re-replicated in parallel on to other servers in the same gear group (Section 3.2). Here we consider other failure modes and their implications for Sierra.

Transient failures in low gear: When the MDS detects failure of a chunk server S, it wakes up all servers in the cluster holding other replicas of S's chunks. This takes a few seconds (if waking from standby) or minutes (if powering up) which does not significantly increase the window of vulnerability for two additional failures. However, when the system is already in the lowest gear, failure of a single server can cause the last active replica of a chunk to become temporarily unavailable during this wakeup time. This can be avoided by setting the *minimum gear level* g_{min} to 2. g_{min} is a policy parameter that trades power savings for a higher probability of unavailability on failure with typical values of 1 (for more power savings), and 2 (for higher availability).

Permanent failures in low gear: When a transient failure of a server S is detected, the MDS wakes up all sleeping replicas of S's chunks within a few minutes. Since detection of permanent failure usually takes longer than this, all replicas are available to begin rebuild as soon as permanent failure is detected. In any case, the rebuild time is much larger than the wakeup time and dominates the total recovery time. Hence, there is no significant increase in the window of vulnerability to additional failures.

Transient and permanent logger failures: When a server fails any logger on it also fails. However, replicas of each record on a failed logger are still available on $r_L - 1$ other servers. One option to maintain this fault tolerance is to re-replicate the at-risk data (log records with fewer than r_L available replicas) on other loggers. However, since the loggers are only intended for short-term storage, this results in wasted work. Instead, primaries first try to reclaim at-risk data at high priority, waking up secondaries as necessary. If a logger fails permanently it is deleted from the logger view of all DVLs. Before a new logger is added to a DVL's logger view, all data on it is deleted since it might be stale. This is done efficiently by writing a single "delete all" record to the log head.

Failures due to power cycling: We are not aware of any reliable data on the effect of power cycling on failures. However, Sierra minimizes the power cycling of servers in two ways. First, cycling is done at a coarse time granularity aimed at changing gears a few times per day. Second, the gear scheduler (Section 3.5) rotates the selection of gear groups for power-down over several days. This evens out the amount of power cycling per server (and also the amount of idle time for maintenance activities such as scrubbing.)

Garbage collection: When data is reclaimed from the DVL it is marked as deleted. The DVL does so through appending a versioned deletion marker to the loggers. The data and deletion markers must eventually be garbage collected from the DVL. When data in the DVL is overwritten, the older version becomes stale and becomes a candidate for garbage collection as well. Garbage collection of deletion markers must wait until all stale versions of the deleted data are removed. If the stale versions are on a failed logger, garbage collection of those deletion markers is blocked until the logger becomes unavailable or is deleted from the logger view. Deletion markers are small (i.e., they do not take up much space in the DVL) and this is not a problem in general. Similarly, during DVL recovery, garbage collection of all deletion markers is blocked until all loggers in the logger view have responded.

Garbage collection is a background maintenance process (it has background I/O priority). Write and delete requests are never blocked as long as r_L fault-uncorrelated loggers are available. The DVL is a best-effort service and is not meant to be used as a long-term file system. Hence, it is simpler than a general log-structured file system. In the worst-case that all loggers are unavailable (e.g., because a huge data burst has filled their capacity) and a write arrives that must be logged (e.g., because it is an overwrite of data already in the DVL), the write must block until the garbage collection frees up space in the DVL. In practice, past data should be used to provision the loggers with enough capacity to absorb data until it is reclaimed and such unavailability should be a very rare case. Section 11 provides some illustrative numbers on our logger capacity requirements.

3.5 Gear scheduler

The gear scheduler predicts system load and schedules servers to power down or up accordingly. It is a centralized component that periodically aggregates load measurements from chunk servers, and computes gear schedules for the future. Our gear scheduler is simple, because our workloads exhibit predictable patterns. It predicts load for each hour of the present day based on historical averages from that same hour in previous days. It is important to note that not all workloads will exhibit predictable patterns. For example, load will be higher than predicted for flash crowds and lower than predicted for holidays. A hybrid scheduler based on historic averages and reacting to the current load is likely to be superior to ours for such cases. We opted to use a simple one because it worked well for us.

The load metric used by the gear scheduler is the overall I/O rate generated by all the clients of the storage system. Random-access I/O rate (measured in IOPS) and streaming I/O rate (measured in MiB/s) are considered separately. Reads and writes are also considered separately, and the write rate is multiplied by r to account for replication. The raw load metrics are measured every second at each primary, and periodically sent to the scheduler, which aggregates the load across the whole system. It then uses the peak values for each hour as the load metrics for that hour. We use the peak value since I/O load is often bursty and using the mean value can significantly degrade performance during bursts.

Given the measured performance per chunk server and the load, the scheduler computes the number of chunk servers required to sustain the system load in each hour:

$$N_{nonseq} = \frac{TotalIOPS_{read}}{ServerIOPS_{read}} + r \cdot \frac{TotalIOPS_{write}}{ServerIOPS_{write}}$$

$$N_{seq} = \frac{TotalMiB/s_{read}}{ServerMiB/s_{read}} + r \cdot \frac{TotalMiB/s_{write}}{ServerMiB/s_{write}}$$

$$N_{load} = \lceil \max(N_{nonseq}, N_{seq}) \rceil$$

This corresponds to a gear level $g = r\frac{N_{load}}{N}$; in general this will not be a whole number but have a fractional part. All servers in the lowest $\lfloor g \rfloor$ gear groups are left on, while all servers in the highest $r - \lceil g \rceil$ gear groups are turned off. This leaves up to 1 "fractional" gear group in which we turn off some servers (chosen at random) to achieve a total of N_{load} active servers. Alternatively the system can be configured for full rather than fractional gearing. In this mode, we power entire gear groups up and down, while always keeping at least N_{load} servers up. Hence, for full gearing the number of powered-up servers is $\lceil \frac{N\lceil g \rceil}{r} \rceil$ rather than $\lceil \frac{Ng}{r} \rceil$.

3.6 Implementation status

The evaluation in the following section is based on our Sierra prototype, which is implemented entirely at user level, with the MDS and every chunk server running as a user-level process. A client-side library exports object *read()*, *write()*, *delete()* and *create()* calls. The core Sierra implementation (chunk servers, MDS, client library) is 11 kLOC. The DVL is 7.6 kLOC. There is an additional 17 kLOC of support code (RPC libraries, etc.) NTFS is used as the local file system on each chunk server and its lines of code are not included in the above measurements.

Although the MDS is implemented as a deterministic state machine we have not currently implemented MDS replication; however, standard techniques exist for state machine replication and we are confident that the MDS could be replicated if required.

3.7 Summary of tradeoffs and limitations

We summarize the tradeoffs described in this section and the limitations of our approach. First, we had to redesign the data layout to allow for the possibility of servers to be off. The main tradeoff involved power savings on one hand and rebuild speed and load distribution on the other hand. Because of this tradeoff our method works best for large clusters (hundreds of servers) and the effects of the tradeoff become more pronounced for small clusters.

Second, we introduce a new service to the data center: a distributed virtual log (DVL). This service absorbs any writes that happen to a server that is powered down. Physically it is implemented as a short-term versioned store and it can reside on dedicated servers, or co-located with the existing chunk stores. The DVL would be simpler if the underlying chunk store file system were versioned, but ours (NTFS) is not. As such, the DVL adds some complexity in terms of lines of code and failure cases to the system. We hope it is still conceptually simple, since the notion of a "log" is well-understood and a distributed one is a rather natural extension for the data center environment.

Third, we described new failure scenarios that can be handled without introducing any new tradeoffs, and one failure scenario that introduces a tradeoff between power savings and availability. Specifically, when the system is already in the lowest gear, failure of a single server can cause the last active replica of a chunk to become temporarily unavailable during this wakeup time (a window of seconds to minutes). Depending on the workload characteristics and server failure profiles, some deployments might choose never to go to the lowest gear to avoid this tradeoff.

Fourth, the gear scheduler allows a spectrum of tradeoffs that can be explored (e.g., predictive vs. reactive gear selection policies, gearing timescales in the order of minutes and hours, and full vs. fractional gearing.) Several of these tradeoffs are shown in Figure 9(b). These are tradeoffs among power savings, complexity of workload characterization and ability of servers to rapidly switch on and off. We chose a simple default mode that works well for our workloads.

4. Evaluation

In this section we answer the following questions. First, does our *gear scheduler* work well? Second, what are the *power savings*? Third, what is the impact on *performance*? Finally, how does *rebuild* performance scale with the number of servers, for the Sierra layout. Our evaluation is driven by two sets of traces as well as microbenchmarks. We have obtained *load traces* from Hotmail (Windows Live Mail) and Windows Live Messenger. These are measurements of I/O load at a 1-hour granularity for 1 week, aggregated over thousands of servers as shown in Figure 1. The load traces are used to measure how well the gear scheduler works and expected power savings.

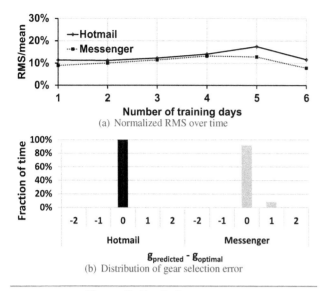

(a) Normalized RMS over time

(b) Distribution of gear selection error

Figure 8. Prediction accuracy

The load traces are not sufficiently fine-grained to measure performance on real hardware. For this, we have obtained a small set of *I/O traces* which contain I/O requests to e-mail messages stored on 8 Hotmail back-end servers over a 48-hour period, starting at midnight (PDT) on Monday, August 4, 2008. These I/O traces are from a different time period than the Hotmail load traces, but represent the same service. The I/O traces are taken at the block device level, i.e., below the main memory buffer cache but above the storage hardware (RAID). We map them onto our object-based file system as detailed in Section 4.3. The overall read/write ratio of these traces is 1.3 (i.e. slightly more reads). Approximately 276 GiB of data were written and 460 GiB of data were read from these servers during that time. The diurnal pattern for these traces is shown in Figure 10.

4.1 Gear scheduler

We applied the simple "hour of day" load prediction algorithm (see Section 3.5) to the Hotmail and Messenger load traces. The load metric used was mean bytes transferred per hour, since that is all we have available from this data. For each workload, we have 7 days of data. At the end of each day, we train the predictor using all the previous days' data, with equal weighting for all past days. We then test the predictor on the remaining days in the trace. The aim is to find how quickly, and to what value, the error converges.

We measure error taking the root-mean-square (RMS) of the difference between predicted and actual load. We normalize this by the mean load during the test period to give a scale-free measure of error. Figure 8(a) shows how the error changes over time. For both Hotmail and Messenger, the error is low (around 10% of the mean) even after a single day of training and does not change significantly afterward. Hence, even with this simple predictive method, the number

of active servers when using fractional gearing will always be within 10% of the ideal.

When using full gearing, we can also measure how often the predictor selects the wrong gear. For 3-way replication, the difference between the correct gear for the load, and the chosen gear based on predicted load, can range from -2 to +2, since gear values range from 1 to 3. Figure 8(b) shows how load prediction errors translate to gear selection errors in this case. The histograms show the frequency of occurrence of each value; perfect prediction would lead to a single bar centered on zero. With Hotmail we achieve this, i.e., the system always selects the correct gear, and for Messenger it selects the correct gear 90% of the time.

4.2 Power

Figure 9(a) shows the analytical estimation of power consumption of Sierra for the Hotmail and Messenger load traces as a fraction of the baseline system. The estimation is based on the number of servers expected to be turned off over the week. To separate the effects of prediction and power savings, here we use an "oracle" predictor that always predicts the correct gear, i.e., we correct the 10% misprediction that occurs in Messenger (an incorrect gear selection could either save more power — and have bad performance — or could save less power than a correct one). The loggers are assumed to be co-located with the servers. We are not allowed to reveal the exact number of servers in the baseline system, but it is all the Hotmail and Messenger servers (several thousands) as first described in Section 2.

We show the power savings for both fractional and full gearing, and for two different settings of the minimum gear value: $g_{min} = 1$ and $g_{min} = 2$. Setting g_{min} to 2 increases power consumption very slightly for Hotmail, and more for Messenger. This is because, for Hotmail, the system stays above gear 2 most of the time for load reasons, even when $g_{min} = 1$. We also see that fractional gearing, by being finer-grained, achieves significant power savings compared to full gearing, as expected from the analysis in Section 3.5.

The results so far used gearing based on mean load per hour, and with a maximum gear-shift frequency of once per hour, since this is the finest granularity obtainable from the load traces. However, from the Hotmail I/O traces we are able to measure the effect of using the peak load metric, and also of shifting gears more frequently. Figure 9(b) shows the analytical estimate of power consumption for the 2-day Hotmail I/O traces, using the peak load metric with the oracle predictor, full gearing and $g_{min} = 1$, and varying the timescale of prediction. At the 1-hour timescale, the power consumption is 72% of the baseline system compared to 69% predicted from the load traces. Higher power savings seem possible by gear-shifting frequently, e.g., once per minute; however this requires frequent power state cycling and also accurate fine-grained load prediction. It is not clear whether such fine-grained prediction is possible: the fine-grained traces only cover 48 hours which is insufficient

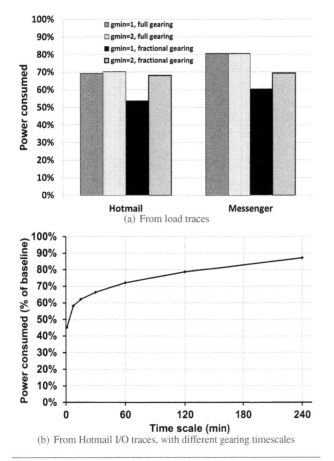

(a) From load traces

(b) From Hotmail I/O traces, with different gearing timescales

Figure 9. Power consumption

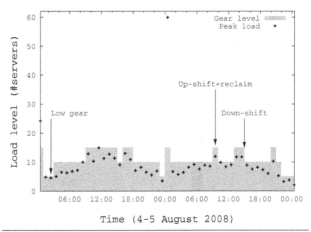

Figure 10. Gear schedule for Hotmail I/O trace

to evaluate this. All experiments in this paper use a 1-hour granularity as the 7-day load traces show good predictability at this granularity.

4.3 Performance on I/O traces

Our performance evaluation is based on replaying the Hotmail I/O traces on a Sierra system deployed on a small scale cluster. Our experimental testbed consists of 31 identical servers in 3 racks. Each rack has a Cisco Catalyst 3750E as a Top-of-Rack (ToR) switch providing 1 Gbps ports for the servers, and a 10 Gbps fiber uplink to a Cisco Nexus 5000. The testbed is assigned 10 servers in each rack, plus an extra server in one of the racks on which we run the MDS. Each server has two four-core 2.5 Ghz Intel Xeon processors, 16 GiB of RAM, a 1 TB system disk and a 1 TB disk that holds the Sierra chunk files and log files. Each server runs Windows Server 2008 Enterprise Edition, SP1.

Trace replay: We map the Hotmail I/O traces to Sierra objects using *virtual disks*. Each block device in the trace maps to a virtual disk, which corresponds to a unique object ID in Sierra. The virtual disk object is then stored in Sierra as a set of chunks corresponding to logical extents within the disk. Thus, the trace replay mechanism converts an access of *<block device, logical block number, size in blocks>*

to *<object ID, offset in bytes, size in bytes>*. The traces are then split by virtual disk and replayed from multiple client machines. Although both client and server machines have plentiful RAM, we do not use it for caching in our experiments, to match the traces available, which are taken below the main memory buffer cache. We also disable prefetching and write-back for the same reason.

Provisioning: For meaningful experimental results it is important to correctly provision the system for the workload, i.e., chose the correct number of chunk servers. Over-provisioning the system would increase the baseline system's power consumption unnecessarily, and thereby inflate the relative power savings of Sierra. Under-provisioning the system would also be meaningless because the system could not sustain the peak load even with all servers powered up. To calculate the number of chunk servers needed to support a workload, we use the methodology based on the peak load metric described in Section 3.5. Provisioning was based on measured single-server performance (see Section 4.4 for a description of the microbenchmarks used).

In addition to performance, we must also match the availability and capacity requirements of the workload. For availability, we replicate chunks on servers in three different fault domains, i.e., racks. For capacity, we are limited by the total storage capacity of our servers, which is smaller than the entirety of the virtual disks in the trace. However, it is sufficient to store the specific chunks that are accessed during any of our experimental runs, if we use a chunk size of 1 MiB. Hence, for each experiment, we pre-create the chunks that are accessed during the experiment, using a chunk size of 1 MiB. In practice a larger chunk size, e.g., 64 MiB is more common [Ghemawat 2003].

For the Hotmail I/O traces this methodology resulted in 15 chunk servers to meet the performance requirement (5 in each rack). The trace replay clients are load balanced on 9 of the remaining machines (3 in each rack). For all the performance experiments, we compared two configurations. The *Sierra* configuration had the above 5 chunk servers

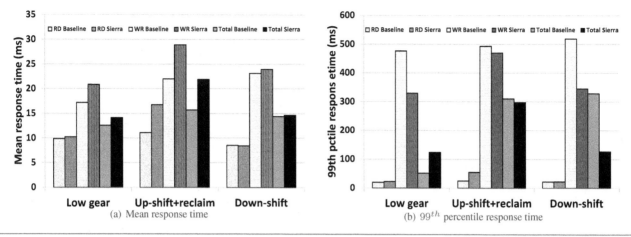

Figure 11. Performance comparison for Hotmail I/O trace. The "transition" experiment is for a downward gear shift.

and 1 dedicated logger per rack. The *Baseline* configuration was provisioned with the same total resources, i.e., 6 chunk servers per rack and no loggers. In the Sierra configuration, all 3 loggers are always left powered up to ensure that 3-way, cross-rack replication is always possible. Due to the small size of our testbed, the power cost of the dedicated loggers is not amortized well and their power overhead is 6% over that expected from Section 4.2 (which looked only at chunk server power savings).

We did not have access to a power measurement kit (i.e., the servers did not have a watt-meter attached) and furthermore the servers were in a remote datacenter and we did not have permission to turn them off. To emulate a server being off we instruct the RPC layer of that server to drop all received packets silently. This approach has the shortcoming that it does not capture the time it takes to power down and up a server, however we expect those times to be negligible compared to the time a server stays off (at least 1 hour).

Trace segment selection: For our performance experiments we selected three trace segments that represent the worst case for three different aspects of Sierra. First we chose the 1-hour period when the system was in the lowest gear, and saw the highest peak load of all such periods. The aim is to show the worst-case performance impact of putting the system in the lowest gear. Second, we chose the transition into highest gear when the amount of logged data in the DVL was the highest, and the 1-hour period following the transition. The aim here is to show the effect of an up shift (i.e., turning servers on and rebalancing load) as well as the performance impact of reclaiming logged data to the chunk servers. Finally, we chose the downward transition having the highest load in the minute immediately following the transition, and replayed the trace from the transition until the end of the minute. The aim here is to show the worst-case impact of turning servers off. Figure 10 shows the load metric for each hour of the Hotmail I/O trace, the gear chosen by the oracle predictor for each hour, and the periods corresponding to the three experiments[1]. All the experiments are based on full gearing, which gives the largest possible gear transitions; the lowest and highest gear levels are the same for both full and fractional gearing. In all experiments we pre-create the necessary system state, including logger state, by first replaying the relevant preceding portions of the trace.

Request response times: Figure 11 shows the mean and 99th percentile (note the different scales on the two y-axes) of performance of the baseline and Sierra configurations during the three experiments. We show the performance of read and write requests separately as well as the total performance. We make several observations. First, given that these are the worst three scenarios, the performance penalty for Sierra is small. Second, the low gear experiment shows that our provisioning and gear selection methodology is reasonable; the performance of Sierra with 2/3 of the chunk servers turned off is comparable to that of the baseline, i.e., all servers turned on. Third, the performance penalty is significantly smaller for the 99th percentile, showing that Sierra does not make the slow requests slower (Sierra writes are sometimes slightly faster than in the baseline system because they are sent to loggers which are optimized for writes.)

The performance penalty is slightly more pronounced for the up-shift/reclaim experiment. This is due to performance interference between the foreground workload and the reclaim process, which is faster than it needs to be. Table 2 shows this long-term reclaim rate required for this workload as well as the actual reclaim rate achieved during the up-shift/reclaim experiment. The long-term reclaim rate is obtained by measuring the number of unique bytes logged just before each up-shift in the 48-hour period, and summing

---

[1] The "peak" load in the Hotmail I/O traces happens just after midnight during a 2-hour period of background maintenance. Provisioning the system for the maintenance peak would only improve the I/O response time of background tasks. It would also artificially increase the baseline power consumption and hence improve Sierra's relative power savings. To avoid this effect, we exclude this maintenance period when provisioning the system, and keep the system in the highest gear during the window.

Total data logged in 48 hrs	166 GiB
Time in top gear	14 hrs
Required reclaim rate	3.37 MiB/s
Data reclaimed in 1 hour after up-shift	27 GiB
Reclaim rate achieved	7.7 MiB/s

Table 2. Reclaim statistics

# primaries migrated in down-shift	98
Total migration time	1.3 s
Total # of requests	11532
# of requests retrying	118
Average/worst retry penalty	9.8 / 74 ms
# primaries migrated in up-shift	116
Total migration time	2.4 s
Total # requests	18618
# number of requests retrying	152
Average/worst retry penalty	224 / 2168 ms

Table 3. Primary migration statistics

these values. This gives an upper bound on the amount of data that needs to be reclaimed for those 48 hours. Although all reclaim requests have background priority, we believe a better control mechanism [Douceur 1999] could reduce the rate to the long-term average. We note that, for this setup, provisioning the loggers with 30 GiB of capacity is sufficient to ensure all writes are absorbed without blocking.

Details on primary migration: Chunk group primaries are migrated during down-shifts to maintain availability and during up-shifts to rebalance load. After a primary migrates, the first request from each client for that chunk group will go to the old primary, fail, and be retried after re-fetching metadata from the MDS. Subsequently it is cached at the client and future requests do not pay the retry penalty.

Table 3 summarizes the migration statistics for the down-shift and the up-shift experiments. The total migration time is dominated by the time for the busiest primary to drain its queue (i.e., complete outstanding requests) before the new primary can begin accepting requests. This also determines the worst-case retry penalty. We could optimize this by waiting for primaries to have short queues before migrating them, rather than starting them at a fixed time dictated by the gear scheduler. However, given that so few requests are affected we have not prioritized this optimization.

Metadata state size: We measured the amount of metadata for the longest experiment, the up-shift one. The metadata service had about 100 MiB of in-memory state, corresponding to 4.6 million chunks in 320 chunk groups. This is a memory requirement of 23 bytes per chunk or 320 KiB per chunk group.

4.4 Microbenchmarks

The goal of this section is to measure, using microbenchmarks, the scalability of Sierra's read/write performance as well as the impact of layout on rebuild rates.

Single-server performance: This experiment establishes a baseline single-server performance. First, we measure single client streaming read and write bandwidth from a single server in MiB/s using 64 KiB reads and writes to a 2.4 GiB object (system chunk size is the default of 1 MiB). Second, we measure random-access read and write performance in IOPS (I/Os per second) by sending 100,000 I/Os to the server. The client keeps 64 requests outstanding at all times. Table 4 shows the results. Write have more variable performance than reads due to inherent properties of NTFS.

	Writes	Reads
Bandwidth (MiB/s)	82/**82**/82	82/**82**/82
IOPS	144/**179**/224	129/**137**/147

Table 4. Single-server performance. The min/**avg**/max metric is shown for 5 runs.

Multi-server performance: This experiment shows the peak performance of our system when multiple servers and clients are accessing data. Rack 1 is used for chunk servers. Rack 2 is used for the clients. The metadata service is placed on a machine in rack 1. 9 clients (each identical in setup to the single one above, but r is 3 in this case) make read and write requests to 9 chunk servers. Table 5 shows the results in terms of aggregate server performance. Variance is measured across clients. For all write experiments it is negligible. For the streaming read experiment the client performance was 37–41 MiB/s. For the random-access read experiment the client performance was 109–137 IOPS.

	Writes	Reads
Bandwidth (MiB/s)	96 *(246)*	348 *(738)*
IOPS	465 *(537)*	1152 *(1233)*

Table 5. Peak system performance. In brackets is the *ideal* performance as nine times the performance of a single server for reads, and a third of that for writes.

We observe that in all cases, write performance is about a third of read performance as expected because of 3-way replication. For the random-access workloads the servers' disks become the bottleneck. All numbers in those cases are close to the ideal. For streaming reads and writes, Sierra gets only around a third of the expected bandwidth. This is because of a well-known problem: lack of performance isolation between concurrent streams. A mechanism like Argon [Wachs 2007] would potentially be beneficial in our case but we have not explored it.

Rebuild speed: Sierra's power-aware layout provides not only power savings but also rebuild parallelism when a server fails permanently and its data must be rebuilt. To

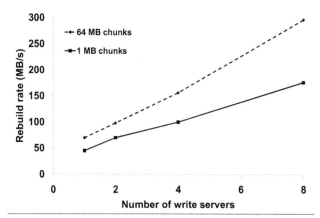

Figure 12. Rebuild rate as a function of cluster size

show the effect of this parallelism we ran a microbenchmark measuring rebuild times while varying the cluster size N, which determines the rebuild parallelism. In all cases we used 3-way replication, with data being read from $\frac{2N}{3}$ nodes and written in parallel to $\frac{N}{3}$ nodes. The write servers (being fewer) are the bottleneck rather than the read servers, and hence, Figure 12 shows the rebuild rate as a function of the number of write servers.

Rebuild scales with the number of servers, showing the importance of rebuild parallelism. Note that the single-server performance is lower than the raw disk bandwidth due to chunk creates (i.e., NTFS "create" system call) being the bottleneck. We show performance for a small 1 MiB chunk size (which was selected to accommodate the Hotmail I/O trace's capacity requirement on our testbed), and for a larger 64 MiB chunk size that amortizes better the overhead of the chunk creates. With 8 write servers and 64 MiB chunks, and no extra network optimizations, Sierra achieves a reasonable rebuild performance of 296 MiB/s (1 hour per TB rebuilt, or 2.3 Gbps of cross-rack network bandwidth usage).

5. Related Work

Section 3 described the (non power-proportional) storage systems that Sierra builds on [Azu 2009, Abd-El-Malek 2005, Ghemawat 2003, Saito 2004]. Here we contrast Sierra with related work on power savings.

The DVL component of Sierra builds on write off-loading [Narayanan 2008]. However, that work focused on traditional RAID-based storage on individual servers. Other previous work such as PARAID [Weddle 2007] (which first introduces the notion of gears in this context) and power-aware storage caching [Zhu 2005] are also based on enterprise storage. This previous work does not address the challenges of large data centers: providing power proportionality in while maintaining scalability, availability, consistency, and fault tolerance. Specifically, the Sierra DVL extends the previous work on write off-loading with network awareness, fast failure recovery, and asynchronous primary migration.

Work has been done on power-proportionality for read workloads [Amur 2010]. Sierra handles general workloads with read and writes. It is non-trivial to support writes while maintaining availability, fault tolerance and consistency. Achieving this through the design and implementation of the DVL is one of our key technical contributions. Some approaches rely on increasing the replication factor r, and hence the capacity requirements of the system, to achieve power proportionality [Amur 2010, Weddle 2007]. In addition to increasing capacity, these techniques come at the cost of write performance — more copies need to be updated on each write. This can result in more servers being provisioned for the baseline system, increasing its power consumption. Sierra achieves power-proportionality without increasing the replication factor of the original system.

Popular Data Concentration (PDC) exploits spatial locality by migrating hot data onto a small number of "hot" disks, and spinning down the cold disks [Pinheiro 2004]. However, when the cold data is accessed, there is a large latency in waiting for a disk to spin up. PDC also unbalances the load across disks, which means that the system must be over-provisioned for performance, which increases the baseline power consumption. Other work on analyzing effects of data layout targets availability [Yu 2007]. Sierra's layout builds on that body of work and shows surprising effects of layout on power-proportionality.

Current hardware-based approaches, such as CPU voltage scaling and multi-speed disks [Zhu 2005], do not offer a wide enough range of power consumption scaling. Sierra offers power proportionality through software, by turning off entire servers, without requiring specific power-proportionality support in hardware.

6. Conclusion and future work

Sierra is, to the best of our knowledge, the first power-proportional distributed storage system for general read and write workloads. It achieves power-proportionality in software while maintaining the consistency, fault-tolerance and availability of the baseline system. Analysis of load traces from two large online services show significant power savings from exploiting diurnal load patterns, and a performance evaluation on a small server cluster shows a modest overhead in I/O response time.

We have identified several directions for future work. First, work is needed to align methods for consolidating computational tasks (e.g., virtualization) with the I/O load consolidation that Sierra offers. Ideally the system should also preserve locality while shifting gears, i.e., the co-location of computation with the data it computes on. Second, more thinking is needed to achieve power-proportionality for optimistic concurrency control systems such as Amazon's Dynamo, which uses "sloppy quorums" [DeCandia 2007] and for Byzantine fault-tolerant systems.

7. Acknowledgments

We thank our shepherd Alistair Veitch, the anonymous reviewers, and Ant Rowstron, James Hamilton, Miguel Castro, Zhe Zhang and Paul Barham for their feedback. We thank Bruce Worthington and Swaroop Kavalanekar for the Hotmail I/O traces, Tom Harpel and Steve Lee for the Hotmail and Messenger load traces, Dennis Crain, Mac Manson and the MSR cluster folks for the hardware testbed.

References

[Azu 2009] Windows Azure Platform, October 2009. URL http://www.microsoft.com/azure.

[Abd-El-Malek 2005] Michael Abd-El-Malek, William V. Courtright II, Chuck Cranor, Gregory R. Ganger, James Hendricks, Andrew J. Klosterman, Michael Mesnier, Manish Prasad, Brandon Salmon, Raja R. Sambasivan, Shafeeq Sinnamohideen, John D. Strunk, Eno Thereska, Matthew Wachs, and Jay J. Wylie. Ursa Minor: versatile cluster-based storage. In *Proc. USENIX Conference on File and Storage Technologies (FAST)*, December 2005.

[Amazon 2010] Amazon. Amazon EC2 spot instances, April 2010. URL http://aws.amazon.com/ec2/spot-instances/.

[Amur 2010] Hrishikesh Amur, James Cipar, Varun Gupta, Michael Kozuch, Gregory Ganger, and Karsten Schwan. Robust and flexible power-proportional storage. In *Proc. Symposium on Cloud Computing (SOCC)*, Indianapolis, IN, June 2010.

[Barroso 2007] Luiz André Barroso and Urs Hölzle. The case for energy-proportional computing. *IEEE Computer*, 40(12):33–37, 2007.

[Chen 2008] Gong Chen, Wenbo He, Jie Liu, Suman Nath, Leonidas Rigas, Lin Xiao, and Feng Zhao. Energy-aware server provisioning and load dispatching for connection-intensive internet services. In *Proc. Symposium on Networked Systems Design and Implementation (NSDI)*, San Francisco, CA, 2008.

[Clark 2005] Christopher Clark, Keir Fraser, Steven Hand, Jacob Gorm Hansen, Eric Jul, Christian Limpach, Ian Pratt, and Andrew Warfield. Live migration of virtual machines. In *Proc. Symposium on Networked Systems Design and Implementation (NSDI)*, Boston, MA, May 2005.

[DeCandia 2007] Giuseppe DeCandia, Deniz Hastorun, Madan Jampani, Gunavardhan Kakulapati, Avinash Lakshman, Alex Pilchin, Swaminathan Sivasubramanian, Peter Vosshall, and Werner Vogels. Dynamo: Amazon's highly available key-value store. In *Proc. ACM Symposium on Operating Systems Principles (SOSP)*, October 2007.

[Douceur 1999] John R. Douceur and William J. Bolosky. Progress-based regulation of low-importance processes. In *Proc. ACM Symposium on Operating Systems Principles (SOSP)*, Kiawah Island, SC, December 1999.

[Fan 2007] Xiaobo Fan, Wolf-Dietrich Weber, and Luiz Andre Barroso. Power provisioning for a warehouse-sized computer. In *Proc. International Symposium on Computer architecture (ISCA)*, San Diego, CA, 2007.

[Ghemawat 2003] Sanjay Ghemawat, Howard Gobioff, and Shun-Tak Leung. The Google file system. In *Proc. ACM Symposium on Operating System Principles (SOSP)*, Lake George, NY, October 2003.

[Hamilton 2008] James Hamilton. Resource consumption shaping, 2008. URL http://perspectives.mvdirona.com/.

[Hamilton 2009] James Hamilton. Cost of power in large-scale data centers, 2009. URL http://perspectives.mvdirona.com/.

[Narayanan 2008] Dushyanth Narayanan, Austin Donnelly, and Antony Rowstron. Write off-loading: Practical power management for enterprise storage. In *Proc. USENIX Conference on File and Storage Technologies (FAST)*, San Jose, CA, February 2008.

[Oki 1988] Brian M. Oki and Barbara Liskov. Viewstamped replication: A general primary copy method to support highly-available distributed systems. In *Proc. Symposium on Principles of Distributed Computing (PODC)*, Toronto, Canada, August 1988.

[Pinheiro 2004] Eduardo Pinheiro and Ricardo Bianchini. Energy conservation techniques for disk array-based servers. In *Proc. Annual International Conference on Supercomputing (ICS'04)*, June 2004.

[Saito 2004] Yasushi Saito, Svend Frølund, Alistair Veitch, Arif Merchant, and Susan Spence. FAB: building distributed enterprise disk arrays from commodity components. In *Proc. International Conference on Architectural Support for Programming Languages and Operating Systems (ASPLOS)*, October 2004.

[Schneider 1990] Schneider. Implementing fault-tolerant services using the state machine approach: A tutorial. *CSURV: Computing Surveys*, 22, 1990.

[Stokely 2009] Murray Stokely, Jim Winget, Ed Keyes, Carrie Grimes, and Benjamin Yolken. Using a market economy to provision compute resources across planet-wide clusters. In *Proc. of the International Parallel and Distributed Processing Symposium (IPDPS)*, pages 1–8, Rome, Italy, May 2009.

[van Renesse 2004] Robbert van Renesse and Fred B. Schneider. Chain replication for supporting high throughput and availability. In *Proc. Symposium on Operating Systems Design and Implementation (OSDI)*, pages 91–104, 2004.

[Wachs 2007] Matthew Wachs, Michael Abd-El-Malek, Eno Thereska, and Gregory R. Ganger. Argon: performance insulation for shared storage servers. In *Proc. USENIX Conference on File and Storage Technologies (FAST)*, 2007.

[Weddle 2007] Charles Weddle, Mathew Oldham, Jin Qian, An-I Andy Wang, Peter Reiher, and Geoff Kuenning. PARAID: The gear-shifting power-aware RAID. In *Proc. USENIX Conference on File and Storage Technologies (FAST'07)*, February 2007.

[Yu 2007] Haifeng Yu and Phillip B. Gibbons. Optimal inter-object correlation when replicating for availability. In *Proc. Symposium on Principles of Distributed Computing (PODC)*, Portland, OR, August 2007.

[Zhu 2005] Q. Zhu, Z. Chen, L. Tan, Y. Zhou, K. Keeton, and J. Wilkes. Hibernator: Helping disk arrays sleep through the winter. In *Proc. ACM Symposium on Operating Systems Principles (SOSP)*, Brighton, United Kingdom, October 2005.

Parallel Symbolic Execution for Automated Real-World Software Testing

Stefan Bucur   Vlad Ureche   Cristian Zamfir   George Candea

School of Computer and Communication Sciences
École Polytechnique Fédérale de Lausanne (EPFL), Switzerland
{stefan.bucur,vlad.ureche,cristian.zamfir,george.candea}@epfl.ch

Abstract

This paper introduces Cloud9, a platform for automated testing of real-world software. Our main contribution is the scalable parallelization of symbolic execution on clusters of commodity hardware, to help cope with path explosion. Cloud9 provides a systematic interface for writing "symbolic tests" that concisely specify entire families of inputs and behaviors to be tested, thus improving testing productivity. Cloud9 can handle not only single-threaded programs but also multi-threaded and distributed systems. It includes a new symbolic environment model that is the first to support all major aspects of the POSIX interface, such as processes, threads, synchronization, networking, IPC, and file I/O. We show that Cloud9 can automatically test real systems, like memcached, Apache httpd, lighttpd, the Python interpreter, rsync, and curl. We show how Cloud9 can use existing test suites to generate new test cases that capture untested corner cases (e.g., network stream fragmentation). Cloud9 can also diagnose incomplete bug fixes by analyzing the difference between buggy paths before and after a patch.

Categories and Subject Descriptors D.2.5 [*Software Engineering*]: Testing and Debugging—Symbolic Execution

General Terms Reliability, Verification

1. Introduction

Software testing is resource-hungry, time-consuming, labor-intensive, and prone to human omission and error. Despite massive investments in quality assurance, serious code defects are routinely discovered after software has been released [RedHat], and fixing them at so late a stage carries substantial cost [McConnell 2004]. It is therefore imperative

EuroSys'11, April 10–13, 2011, Salzburg, Austria.

to overcome the human-related limitations of software testing by developing automated software testing techniques.

Existing automated techniques, like model checking and symbolic execution, are highly effective [Cadar 2008, Holzmann 2008], but their adoption in industrial general-purpose software testing has been limited. We blame this gap between research and practice on three challenges faced by automated testing: scalability, applicability, and usability.

First, path explosion—the fact that the number of paths through a program is roughly exponential in program size—severely limits the extent to which large software can be thoroughly tested. One must be content either with low coverage for large programs, or apply automated tools only to small programs. For example, we do not know of any symbolic execution engine that can thoroughly test systems with more than a few thousand lines of code (KLOC).

Second, real-world systems interact heavily with the environment (e.g., through system calls, library calls) and may communicate with other parties (e.g., over sockets, IPC, shared memory). For an automated testing tool to be used in practice, it must be capable of handling these interactions. Third, an important hurdle to adoption is that, in order to productively use most current tools, a user must become as versed in the underlying technology as the tool's developers.

Our goal in building Cloud9 is to address these challenges: we envision Cloud9 as a testing platform that bridges the gap between symbolic execution and the requirements of automated testing in the real world. As will be seen later, doing so requires solving a number of research problems.

Cloud9 helps cope with path explosion by parallelizing symbolic execution in a way that scales well on large clusters of cheap commodity hardware. Cloud9 scales linearly with the number of nodes in the system, thus enabling users to "throw hardware at the problem." Doing so without Cloud9 is hard, because single computers with enough CPU and memory to symbolically execute large systems either do not exist today or are prohibitively expensive.

We built into Cloud9 features we consider necessary for a practical testing platform: Besides single-threaded single-node systems, Cloud9 handles also multi-threaded and distributed software. It offers fine grain control over the behavior being tested, including the injection of faults and the scheduling of threads. Cloud9 embeds a new symbolic model of a program's environment that supports all major aspects of the POSIX interface, including processes, threads, synchronization, networking, IPC, and file I/O. Cloud9 provides an easy-to-use API for writing "symbolic tests"—developers can specify concisely families of inputs and environment behaviors for which to test the target software, without having to understand how symbolic execution works, which program inputs need to be marked symbolic, or how long the symbolic inputs should be. By encompassing entire families of behaviors, symbolic tests cover substantially more cases than "concrete" regular tests.

This paper makes three contributions: (1) the first cluster-based *parallel symbolic execution* engine that scales linearly with the number of nodes; (2) a *testing platform* for writing symbolic tests; and (3) a quasi-complete *symbolic POSIX model* that makes it possible to use symbolic execution on real-world systems. We built a Cloud9 prototype that runs on Amazon EC2, private clusters, and multicore machines.

In this paper we present Cloud9 and report on our experience using it for testing several parallel and distributed systems, describe some of the bugs found, and explain how we debugged flawed bug fixes. We describe the problem (§2), Cloud9's design (§3-§5), our prototype (§6), we evaluate the prototype (§7), survey related work (§8), and conclude (§9).

2. Problem Overview

Testing a program consists of exercising many different paths through it and checking whether they "do the right thing." In other words, testing is a way to produce partial evidence of correctness, and thus increase confidence in the tested software. Yet, due to the typically poor coverage one can get today, testing often turns into a mere hunt for bugs.

In practice, most software test harnesses consist of manually written tests that are run periodically; regression test suites provide an automated way of checking whether new bugs have entered the code [McConnell 2004]. Such suites tend to be tedious to write and maintain but, once in place, they can be reused and extended. In practice, the state of the art consists mostly of fuzzing, i.e., trying various inputs in the hope of finding bugs and improving test coverage.

In research, the state of the art consists of model checkers and automated test generators based on symbolic execution [Cadar 2008, Godefroid 2005]. Instead of running a program with regular concrete inputs (e.g., $x = 5$), symbolic execution consists of running a program with "symbolic" inputs that can take on all values allowed by the type (e.g., $x = \lambda$, where $\lambda \in \mathbb{N}$). Whenever a conditional branch is encountered that involves a predicate π that depends (directly or indirectly) on x, state and execution are forked into two alternatives: one following the then-branch (π) and another following the else-branch ($\neg\pi$). The two executions can now

be pursued independently. When a bug is found, test generators can compute concrete values for program inputs that take the program to the bug location. This approach is efficient because it analyzes code for entire classes of inputs at a time, thus avoiding the redundancy inherent in fuzzing.

The *first challenge* for such tools is path explosion, as mentioned earlier. One way to cope is to memoize the symbolic execution of sub-paths into test summaries that can be subsequently reused when the same sub-path is encountered again, as done in compositional test generation [Godefroid 2007]. Alternatively, it is possible to use various heuristics to prioritize the most interesting paths first, as done in KLEE [Cadar 2008]. Another approach is to execute symbolically only paths that are of interest to the test, as done in selective symbolic execution [Chipounov 2009].

We pursue a complementary approach—*parallel symbolic execution*—in which we symbolically execute a program in parallel on a cluster, thus harnessing the machines into a "distributed computer" whose aggregate CPU and memory surpass that of an individual machine. An alternative to a cluster-based approach would be to run a classic single-node symbolic execution engine on a Blue Gene-like supercomputer with vast shared memory and CPUs communicating over MPI. Supercomputers, however, are expensive, so we favor instead clusters of cheap commodity hardware.

One way to parallelize symbolic execution is by statically dividing up the task among nodes and having them run independently. However, when running on large programs, this approach leads to high workload imbalance among nodes, making the entire cluster proceed at the pace of the slowest node [Staats 2010]. If this node gets stuck, for instance, while symbolically executing a loop, the testing process may never terminate. Parallelizing symbolic execution on shared-nothing clusters in a way that scales well is difficult.

The *second challenge* is mediating between a program and its environment, i.e., symbolically executing a program that calls into libraries and the OS, or communicates with other systems, neither of which execute symbolically. One possible approach is to simply allow the call to go through into the "concrete" environment (e.g., to write a file) [Cadar 2006, Godefroid 2005]; unfortunately, this causes the environment to be altered for *all* forked executions being explored in parallel, thus introducing inconsistency. Another approach is to replace the real environment with a symbolic model, i.e., a piece of code linked with the target program that provides the illusion of interacting with a symbolically executing environment. For instance, KLEE uses a symbolic model of the file system [Cadar 2008]. Of course, real-world programs typically interact in richer ways than just file I/O: they fork processes, synchronize threads, etc.

We originally viewed the building of a complete environment model as an engineering task, but our "mere engineering" attempt failed: for any functionality that, in a normal execution, requires hardware support (such as enforcing isolation between address spaces), the core symbolic execution engine had to be modified. The research challenge therefore is to find the minimal set of engine primitives required to support a rich model of a program's environment.

The *third challenge* is using an automated test generator in the context of a development organization's quality assurance processes. To take full advantage of the automated exploration of paths, a testing tool must provide ways to control all aspects of the environment. For example, there needs to be a clean API for injecting failures at the boundary between programs and their environment, there must be a way to control thread schedules, and so on. There should be a way to programmatically orchestrate all environment-related events, but doing so should not require deep expertise in the technology behind the testing tools themselves.

The work presented here aims to address these three challenges. Cluster-based parallel symbolic execution (§3) provides the illusion of running a classic symbolic execution engine on top of a large, powerful computer. Without changing the exponential nature of the problem, parallel symbolic execution harnesses cluster resources to make it feasible to run automated testing on larger systems than what was possible until now. Our work complements and benefits all tools and approaches based on symbolic execution. We describe a way to accurately model the environment (§4) with sufficient completeness to test complex, real software, like the Apache web server and the Python interpreter. We present the APIs and primitives that we found necessary in developing a true testing platform (§5). We show how using these APIs enables, for instance, finding errors in bug patches by reproducing environment conditions which otherwise would have been hard or impossible to set up with regular test cases.

3. Scalable Parallel Symbolic Execution

In this section we present the design of the Cloud9 engine, focusing on the algorithmic aspects: after a conceptual overview (§3.1), we describe how Cloud9 operates at the worker level (§3.2) and then at the cluster level (§3.3).

3.1 Conceptual Overview

Classic Symbolic Execution Cloud9 employs symbolic execution, an automated testing technique that has recently shown a lot of promise [Cadar 2008, Godefroid 2005].

A symbolic execution engine (SEE) executes a program with unconstrained symbolic inputs. When a branch involves symbolic values, execution forks into two parallel executions (see §2), each with a corresponding clone of the program state. Symbolic values in the clones are *constrained* to make the branch condition evaluate to false (e.g., $\lambda \geq$ MAX) respectively true (e.g., $\lambda <$ MAX). Execution recursively splits into sub-executions at each subsequent branch, turning an otherwise linear execution into an *execution tree* (Fig. 1).

In this way, all execution paths in the program are explored. To ensure that only feasible paths are explored,

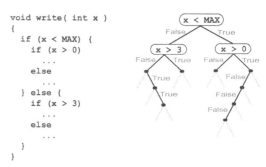

Figure 1: Symbolic execution produces an execution tree.

the SEE uses a constraint solver to check the satisfiability of each branch's predicate, and it only follows satisfiable branches. If a bug is encountered (e.g., a crash or a hang) along one of the paths, the solution to the constraints accumulated along that path yields the inputs that take the tested program to the bug—these inputs constitute a *test case*.

Parallel Symbolic Execution Since the size of the execution tree is exponential in the number of branches, and the complexity of constraints increases as the tree deepens, state-of-the-art SEEs can quickly bottleneck on CPU and memory even for programs with just a couple KLOC. We therefore build a parallel SEE that runs on a commodity cluster and enables "throwing hardware at the problem."

The key design goal is to enable individual cluster nodes to explore the execution tree independently of each other. One way of doing this is to statically split the execution tree and farm off subtrees to worker nodes. Alas, the contents and shape of the execution tree are not known until the tree is actually explored, and finding a balanced partition (i.e., one that will keep all workers busy) of an unexpanded execution tree is undecidable. Besides subtree size, the amount of memory and CPU required to explore a subtree is also undecidable, yet must be taken into account when partitioning the tree. Since the methods used so far in parallel model checkers [Barnat 2007, Holzmann 2008] rely on static partitioning of a finite state space, they cannot be directly applied to the present problem. Instead, Cloud9 partitions the execution tree *dynamically*, as the tree is being explored.

Dynamic Distributed Exploration Cloud9 consists of worker nodes and a load balancer (LB). Workers run independent SEEs, based on KLEE [Cadar 2008]. They explore portions of the execution tree and send statistics on their progress to the LB, which in turn instructs, whenever necessary, pairs of workers to balance each other's work load. Encoding and transfer of work is handled directly between workers, thus taking the load balancer off the critical path.

The goal is to dynamically partition the execution tree such that the parts are *disjoint* (to avoid redundant work) and together they *cover* the global execution tree (for exploration to be complete). We aim to minimize the number of work transfers and associated communication overhead. A fortuitous side effect of dynamic partitioning is the transparent handling of fluctuations in resource quality, availability, and cost, which are inherent to large clusters in cloud settings.

Cloud9 operates roughly as follows: The first component to come up is the load balancer. When the first worker node W_1 joins the Cloud9 cluster, it connects to the LB and receives a "seed" job to explore the entire execution tree. When the second worker W_2 joins and contacts the LB, it is instructed to balance W_1's load, which causes W_1 to break off some of its unexplored subtrees and send them to W_2 in the form of *jobs*. As new workers join, the LB has them balance the load of existing workers. The workers regularly send to the LB status updates on their load in terms of exploration jobs, along with current progress in terms of code coverage, encoded as a bit vector. Based on workers' load, the LB can issue job transfer requests to pairs of workers in the form ⟨ source worker, destination worker, # of jobs ⟩. The source node decides which particular jobs to transfer.

3.2 Worker-level Operation

A worker's visibility is limited to the subtree it is exploring locally. As W_i explores and reveals the content of its local subtree, it has no knowledge of what W_j's ($i \neq j$) subtree looks like. No element in the system—not even the load balancer—maintains a global execution tree. Disjointness and completeness of the exploration (see Fig. 2) are ensured by the load balancing algorithm.

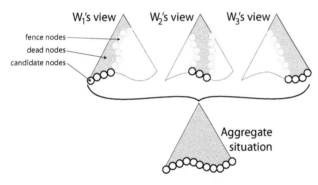

Figure 2: Dynamic partitioning of exploration in Cloud9.

As will be explained later, each worker has the root of the global execution tree. The tree portion explored thus far on a worker consists of three kinds of nodes: (1) internal nodes that have already been explored and are thus no longer of interest—we call them *dead* nodes; (2) *fence* nodes that demarcate the portion being explored, separating the domains of different workers; and (3) *candidate* nodes, which are nodes ready to be explored. A worker exclusively explores candidate nodes; it never expands fence or dead nodes.

Candidate nodes are leaves of the local tree, and they form the *exploration frontier*. The work transfer algorithm ensures that frontiers are disjoint between workers, thus ensuring that no worker duplicates the exploration done by another worker. At the same time, the union of all frontiers in the system corresponds to the frontier of the global execution tree. The goal of a worker W_i at every step is to choose the next candidate node to explore and, when a bug is encountered, to compute the inputs, thread schedule, and system call returns that would take the program to that bug.

The implementation of this conceptual model lends itself to many optimizations, some of which we cover in §6. Broadly speaking, judicious use of copy-on-write and a novel state-encoding technique ensure that actual program state is only maintained for candidate and fence nodes.

Worker-to-Worker Job Transfer When the global exploration frontier becomes poorly balanced across workers, the load balancer chooses a loaded worker W_s and a less loaded worker W_d and instructs them to balance load by sending n jobs from W_s to W_d. In the extreme, W_d is a new worker or one that is done exploring its subtree and has zero jobs left.

W_s chooses n of its candidate nodes and packages them up for transfer to W_d. Since a candidate node sent to another worker is now on the boundary between the work done by W_s and the work done by W_d, it becomes a fence node at the sender. This conversion prevents redundant work.

A job can be sent in at least two ways: (1) serialize the content of the chosen node and send it to W_d, or (2) send to W_d the path from the tree root to the node, and rely on W_d to "replay" that path and obtain the contents of the node. Choosing one vs. the other is a trade-off between time to encode/decode and network bandwidth: option (1) requires little work to decode, but consumes bandwidth (the state of a real program is typically at least several megabytes), while encoding a job as a path requires replay on W_d. We assume that large commodity clusters have abundant CPU but meager bisection bandwidth, so in Cloud9 we chose to encode jobs as the path from the root to the candidate node. As an optimization, we exploit common path prefixes: jobs are not encoded separately, but rather the corresponding paths are aggregated into a job tree and sent as such.

When the job tree arrives at W_d, it is imported into W_d's own subtree, and the leaves of the job tree become part of W_d's frontier (at the time of arrival, these nodes may lie "ahead" of W_d's frontier). W_d keeps the nodes in the incoming jobs as *virtual* nodes, as opposed to *materialized* nodes that reside in the local subtree, and replays paths only lazily. A materialized node is one that contains the corresponding program state, whereas a virtual node is an "empty shell" without corresponding program state. In the common case, the frontier of a worker's local subtree contains a mix of materialized and virtual nodes, as shown in the diagram above.

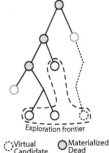

As mentioned earlier, a worker must choose at each step which candidate node to explore next—this choice is guided by a *strategy*. Since the set of candidate nodes now contains both materialized and virtual nodes, it is possible for the strategy to choose a virtual node as the next one to explore. When this happens, the corresponding path in the job tree is replayed (i.e., the symbolic execution engine executes that path); at the end of this replay, all nodes along the path are dead, except the leaf node, which has converted from virtual to materialized and is now ready to be explored. Note that, while exploring the chosen job path, each branch produces child program states; any such state that is not part of the path is marked as a fence node, because it represents a node that is being explored elsewhere, so W_d should not pursue it.

Summary A node N in W_i's subtree has two attributes, $N^{status} \in \{$materialized, virtual$\}$ and $N^{life} \in \{$candidate, fence, dead$\}$. A worker's frontier F_i is the set of all candidate nodes on worker W_i. The worker can only explore nodes in F_i, i.e., dead nodes are off-limits and so are fence nodes, except if a fence node needs to be explored during the replay of a job path. The union $\cup F_i$ equals the frontier of the global execution tree, ensuring that the aggregation of worker-level explorations is complete. The intersection $\cap F_i = \emptyset$, thus avoiding redundancy by ensuring that workers explore disjoint subtrees. Fig. 3 summarizes the life cycle of a node.

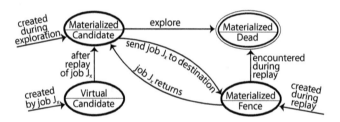

Figure 3: Transition diagram for nodes in a worker's subtree.

As suggested in Fig. 3, once a tree node is dead, it has reached a terminal state; therefore, a dead node's state can be safely discarded from memory. This enables workers to maintain program states only for candidate and fence nodes.

3.3 Cluster-level Operation

Load Balancing When jobs arrive at W_d, they are placed conceptually in a queue; the *length* of this queue is sent to the load balancer periodically. The LB ensures that the worker queue lengths stay within the same order of magnitude. The balancing algorithm takes as input the lengths l_i of each worker W_i's queue Q_i. It computes the average $\bar{l}$ and standard deviation σ of the l_i values and then classifies each W_i as underloaded ($l_i < max\{\bar{l} - \delta \cdot \sigma, 0\}$), overloaded ($l_i > \bar{l} + \delta \cdot \sigma$), or OK otherwise; δ is a constant factor. The W_i are then sorted according to their queue length l_i and placed in a list. LB then matches underloaded workers from the beginning of the list with overloaded workers from the end of the list. For each pair $\langle W_i, W_j \rangle$, with $l_i < l_j$, the load balancer sends a job transfer request to the workers to move $(l_j - l_i)/2$ candidate nodes from W_j to W_i.

Coordinating Worker-level Explorations Classic symbolic execution relies on heuristics to choose which state on the frontier to explore first, so as to efficiently reach the chosen test goal (code coverage, finding a particular type of bug, etc.). In a distributed setting, local heuristics must be coordinated across workers to achieve the global goal, while keeping communication overhead at a minimum. What we have described so far ensures that eventually all paths in the execution tree are explored, but it provides no aid in focusing on the paths desired by the global strategy. In this sense, what we described above is a *mechanism*, while the exploration strategies represent the *policies*.

Global strategies are implemented in Cloud9 using its interface for building *overlays* on the execution tree structure. We used this interface to implement distributed versions of all strategies that come with KLEE [Cadar 2008]; the interface is also available to Cloud9 users. Due to space limitations, we do not describe the strategy interface further, but provide below an example of how a global strategy is built.

A coverage-optimized strategy drives exploration so as to maximize coverage [Cadar 2008]. In Cloud9, coverage is represented as a bit vector, with one bit for every line of code; a set bit indicates that a line is covered. Every time a worker explores a program state, it sets the corresponding bits locally. The current version of the bit vector is piggybacked on the status updates sent to the load balancer. The LB maintains the current global coverage vector and, when it receives an updated coverage bit vector, ORs it into the current global coverage. The result is then sent back to the worker, which in turn ORs this global bit vector into its own, in order to enable its local exploration strategy to make choices consistent with the global goal. The coverage bit vector is an example of a Cloud9 overlay data structure.

4. The POSIX Environment Model

Symbolically executing real-world software is challenging not only because of path explosion but also because real-world systems interact with their environment in varied and complex ways. This section describes our experience building Cloud9's symbolic model of a POSIX environment, which supports most essential interfaces: threads, process management, sockets, pipes, polling, etc. We believe the described techniques are general enough to model other OSes and environments as well.

4.1 Environment Model Design

The goal of a symbolic model is to simulate the behavior of a real execution environment, while maintaining the necessary symbolic state behind the environment interface. The symbolic execution engine (SEE) can then seamlessly transition back and forth between the program and the environment.

While writing and maintaining a model can be laborious and prone to error [Chipounov 2011], there exist cases in which models provide distinct advantages. First, symbolic execution with a model can be substantially faster than without. For instance, in the Linux kernel, transferring a packet between two hosts exercises the entire TCP/IP networking stack and the associated driver code, amounting to over 30 KLOC. In contrast, Cloud9's POSIX model achieves the same functionality in about 1.5 KLOC. Requirements that complicate a real environment/OS implementation, such as performance and extensibility, can be ignored in a symbolic model. Second, when an interface is as stable as POSIX, investing the time to model it becomes worthwhile.

We designed a minimal yet general "symbolic system call" interface to the Cloud9 SEE, which provides the essential building blocks for thread context switching, address space isolation, memory sharing, and sleep operations. These are hard to provide solely through an external model. We give more details about symbolic system calls in §4.2.

In some cases, it is practical to have the host OS handle parts of the environment via *external calls*. These are implemented by concretizing the symbolic parameters of a system call before invoking it from symbolically executing code. Unlike [Cadar 2008; 2006, Godefroid 2005], Cloud9 allows external calls *only* for stateless or read-only system calls, such as reading a system configuration file from the /etc directory. This restriction ensures that external concrete calls do not clobber other symbolically executing paths.

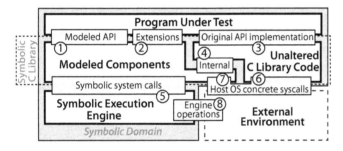

Figure 4: Architecture of the Cloud9 POSIX model.

Cloud9 builds upon the KLEE symbolic execution engine, and so it inherits from KLEE the mechanism for replacing parts of the C Library with model code; it also inherits the external calls mechanism. Cloud9 adds the symbolic system call interface and replaces parts of the C Library with the POSIX model. The resulting architecture is shown in Fig. 4.

Before symbolic execution starts, the Cloud9 system links the program under test with a special symbolic C Library. We built this library by replacing parts of the existing uClibc library in KLEE with the POSIX model code. Developers do not need to modify the code of to-be-tested programs in any way to make it run on Cloud9.

In the C Library, we replaced operations related to threads, processes, file descriptors, and network operations with their corresponding model ①, and augmented the API with Cloud9-specific extensions ②. A large portion of the C Library is reused, since it works out of the box ③ (e.g. memory and string operations). Finally, parts of the original

Primitive Name	Description
cloud9_make_shared	Share object across a CoW domain
cloud9_thread_create cloud9_thread_terminate	Create and destroy threads
cloud9_process_fork cloud9_process_terminate	Fork and terminate the current process
cloud9_get_context	Get the current context (pid and tid)
cloud9_thread_preempt	Preempt a thread
cloud9_thread_sleep	Thread sleep on waiting queue
cloud9_thread_notify	Wake threads from waiting queue
cloud9_get_wlist	Create a new waiting queue

Table 1: Cloud9 primitives used to build the POSIX model.

C Library itself use the modeled code ④ (e.g., Standard I/O stdio relies on the modeled POSIX file descriptors).

The modeled POSIX components interface with the SEE through symbolic system calls ⑤, listed in Table 1. Occasionally, the unmodified part of the C Library invokes external system calls ⑥, and the model code itself needs support from the host OS ⑦—in order to make sure the external calls do not interfere with the symbolic engine's own operations ⑧, such access is limited to read-only and/or stateless operations. This avoids problems like, for instance, allowing an external close() system call to close a network connection or log file that is actually used by the SEE itself.

4.2 Symbolic Engine Modifications

In order to support the POSIX interface, we augmented KLEE with two major features: multiple address spaces per state and support for scheduling threads and processes. This functionality is accessed by model code through the symbolic system call interface (Table 1). Additional models of non-POSIX environments can be built using this interface.

Address Spaces KLEE uses copy-on-write (CoW) to enable memory sharing between symbolic states. We extend this functionality in two ways. First, we enable multiple address spaces within a single execution state, corresponding to multiple processes encompassed in that state. Address spaces can thus be duplicated both across states (as in classic KLEE) and within a state, when cloud9_process_fork is invoked, e.g., as used by the POSIX model's fork().

Second, we organize the address spaces in an execution state as *CoW domains* that permit memory sharing between processes. A memory object can be marked as shared by calling cloud9_make_shared; it is then automatically mapped in the address spaces of the other processes within the CoW domain. Whenever a shared object is modified in one address space, the new version is automatically propagated to the other members of the CoW domain. The shared memory objects can then be used by the model as global memory for inter-process communication.

Multithreading and Scheduling Threads are created in the currently executing process by calling cloud9_thread_

create. Cloud9's POSIX threads (pthreads) model makes use of this primitive in its own `pthread_create()` routine.

Cloud9 implements a cooperative scheduler: An enabled thread runs uninterrupted (atomically), until either (a) the thread goes to sleep; (b) the thread is explicitly preempted by a `cloud9_thread_preempt` call; or (c) the thread is terminated via symbolic system calls for process/thread termination. Preemption occurs at explicit points in the model code, but it is straightforward to extend Cloud9 to automatically insert preemptions calls at instruction level (as would be necessary, for instance, when testing for race conditions).

When `cloud9_thread_sleep` is called, the SEE places the current thread on a specified waiting queue, and an enabled thread is selected for execution. Another thread may call `cloud9_thread_notify` on the waiting queue and wake up one or all of the queued threads.

Cloud9 can be configured to schedule the next thread deterministically, or to fork the execution state for each possible next thread. The latter case is useful when looking for concurrency bugs, but it can be a significant source of path explosion, so it should be disabled when not needed.

If no thread can be scheduled when the current thread goes to sleep, then a hang is detected, the execution state is terminated, and a corresponding test case is generated.

Note that parallelizing symbolic execution is orthogonal to providing the multithreading support described above. In the former case, the execution engine is instantiated on multiple machines and each instance expands a portion of the symbolic execution tree. In the latter case, multiple symbolic threads are multiplexed along the same execution path in the tree; execution is serial along each path.

4.3   POSIX Model Implementation

In this section, we describe the key design decisions involved in building the Cloud9 POSIX model, and we illustrate the use of the symbolic system call interface. This is of particular interest to readers who wish to build additional models on top of the Cloud9 symbolic system call interface.

The POSIX model uses shared memory structures to keep track of all system objects (processes, threads, sockets, etc.). The two most important data structures are stream buffers and block buffers, analogous to character and block device types in UNIX. Stream buffers model half-duplex communication channels: they are generic producer-consumer queues of bytes, with support for event notification to multiple listeners. Event notifications are used, for instance, by the polling component in the POSIX model. Block buffers are random-access, fixed-size buffers, whose operations do not block; they are used to implement symbolic files.

The symbolic execution engine maintains only basic information on running processes and threads: identifiers, running status, and parent–child information. However, the POSIX standard mandates additional information, such as open file descriptors and permission flags. This information is stored by the model in auxiliary data structures associated with the currently running threads and processes. The implementations of `fork()` and `pthread_create()` are in charge of initializing these auxiliary data structures and making the appropriate symbolic system calls.

Modeling synchronization routines is simplified by the cooperative scheduling policy: no locks are necessary, and all synchronization can be done using the sleep/notify symbolic system calls, together with reference counters. Fig. 5 illustrates the simplicity this engenders in the implementation of pthread mutex lock and unlock.

```
typedef struct {
  wlist_id_t wlist;
  char taken;
  unsigned int owner;
  unsigned int queued;
} mutex_data_t;

int pthread_mutex_lock(pthread_mutex_t *mutex) {
  mutex_data_t *mdata = ((mutex_data_t**)mutex);
  if (mdata->queued > 0 || mdata->taken) {
    mdata->queued++;
    cloud9_thread_sleep(mdata->wlist);
    mdata->queued--;
  }
  mdata->taken = 1;
  mdata->owner = pthread_self();
  return 0;
}

int pthread_mutex_unlock(pthread_mutex_t *mutex) {
  mutex_data_t *mdata = ((mutex_data_t**)mutex);
  if (!mdata->taken ||
      mdata->owner != pthread_self()) {
    errno = EPERM;
    return -1;
  }
  mdata->taken = 0;
  if (mdata->queued > 0)
    cloud9_thread_notify(mdata->wlist);
  return 0;
}
```

Figure 5: Example implementation of pthread mutex operations in Cloud9's POSIX environment model.

Cloud9 inherits most of the semantics of the file model from KLEE. In particular, one can either open a symbolic file (its contents comes from a symbolic block buffer), or a concrete file, in which case a concrete file descriptor is associated with the symbolic one, and all operations on the file are forwarded as external calls on the concrete descriptor.

In addition to file objects, the Cloud9 POSIX model adds support for networking and pipes. Currently, the TCP and UDP protocols are supported over IP and UNIX network types. Since no actual hardware is involved in the packet transmission, we can collapse the entire networking stack into a simple scheme based on two stream buffers (Fig. 6). The network is modeled as a single-IP network with multiple available ports—this configuration is sufficient to connect multiple processes to each other, in order to simulate and test

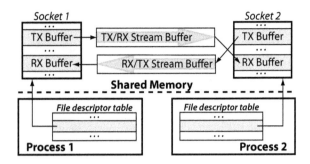

Figure 6: A TCP network connection is modeled in Cloud9 using TX and RX buffers implemented as stream buffers.

distributed systems. The model also supports pipes through the use of a single stream buffer, similar to sockets.

The Cloud9 POSIX model supports polling through the `select()` interface. All the software we tested can be configured to use `select()`, so it was not necessary to implement other polling mechanisms. The `select()` model relies on the event notification support offered by the stream buffers that are used in the implementation of blocking I/O objects (currently sockets and pipes).

The constraint solver used in Cloud9 operates on bit vectors; as a result, symbolic formulas refer to contiguous areas of memory. In order to reduce the constraint solving overhead, we aim to reduce the amount of intermixing of concrete and symbolic data in the same memory region. Thus, Cloud9's POSIX model segregates concrete from symbolic data by using static arrays for concrete data and linked lists (or other specialized structures) for symbolic data. We allocate into separate buffers potentially-symbolic data passed by the tested program through the POSIX interface.

In order to enable testing the systems presented in the evaluation section (§7), we had to add support for various other components: IPC routines, `mmap()` calls, time-related functions, etc. Even though laborious, this was mostly an engineering exercise, so we do not discuss it further.

5. Symbolic Test Suites

Software products and systems typically have large "handmade" test suites; writing and maintaining these suites requires substantial human effort. Cloud9 aims to reduce this burden while improving the quality of testing, by offering an easy way to write "symbolic test suites." First, a symbolic test case encompasses many similar concrete test cases into a single symbolic one—each symbolic test a developer writes is equivalent to many concrete ones. Second, a symbolic test case explores conditions that are hard to produce reliably in a concrete test case, such as the occurrence of faults, concurrency side effects, or network packet reordering, dropping and delay. Furthermore, symbolic test suites can easily cover unknown corner cases, as well as new, untested functionality. In this section, we present the API for symbolic tests and illustrate it with a use case.

Function Name	Description
cloud9_make_symbolic	Mark memory regions as symbolic
cloud9_fi_enable cloud9_fi_disable	Enable/disable the injection of faults
cloud9_set_max_heap	Set heap size for symbolic `malloc`
cloud9_set_scheduler	Set scheduler policy (e.g., round-robin)

Table 2: Cloud9 API for setting global behavior parameters.

Extended Ioctl Code	Description
SIO_SYMBOLIC	Turns this file or socket into a source of symbolic input
SIO_PKT_FRAGMENT	Enables packet fragmentation on this socket (must be a stream socket)
SIO_FAULT_INJ	Enables fault injection for operations on this descriptor

Table 3: Cloud9 extended `ioctl` codes to control environmental events on a per-file-descriptor basis.

5.1 Testing Platform API

The Cloud9 symbolic testing API (Tables 2 and 3) allows tests to programmatically control events in the environment of the program under test. A test suite needs to simply include a `cloud9.h` header file and make the requisite calls.

Symbolic Data and Streams The generality of a test case can be expanded by introducing bytes of symbolic data. This is done by calling `cloud9_make_symbolic`, a wrapper around `klee_make_symbolic`, with an argument that points to a memory region. `klee_make_symbolic` is a primitive provided by KLEE to mark data symbolic. In addition to wrapping this call, we added several new primitives to the testing API (Table 2). In Cloud9, symbolic data can be written/read to/from files, can be sent/received over the network, and can be passed via pipes. Furthermore, the `SIO_SYMBOLIC ioctl` code (Table 3) turns on/off the reception of symbolic bytes from individual files or sockets.

Network Conditions Delay, reordering, or dropping of packets causes a network data stream to be fragmented. Fragmentation can be turned on or off at the socket level using one of the Cloud9 `ioctl` extensions. §7 presents a case where symbolic fragmentation enabled Cloud9 to prove that a bug fix for the lighttpd web server was incomplete.

Fault Injection Calls in a POSIX system can return an error code when they fail. Most programs can tolerate such failed calls, but even high-quality production software misses some [Marinescu 2009]. Such error return codes are simulated by Cloud9 whenever fault injection is turned on.

Symbolic Scheduler Cloud9 provides multiple scheduling policies that can be controlled for purposes of testing on a per-code-region basis. Currently, Cloud9 supports a round-robin scheduler and two schedulers specialized for bug finding: a variant of the iterative context bounding scheduling algorithm [Musuvathi 2008] and an exhaustive exploration of all possible scheduling decisions.

5.2 Use Case

Consider a scenario in which we want to test the support for a new `X-NewExtension` HTTP header, just added to a web server. We show how to write tests for this new feature.

A symbolic test suite typically starts off as an augmentation of an existing test suite; in our scenario, we reuse the existing boilerplate setup code and write a symbolic test case that marks the extension header symbolic. Whenever the code that processes the header data is executed, Cloud9 forks at all the branches that depend on the header content. Similarly, the request payload can be marked symbolic to test the payload-processing part of the system:

```
char hData[10];
cloud9_make_symbolic(hData);
strcat(req, "X-NewExtension: ");
strcat(req, hData);
```

The web server may receive HTTP requests fragmented in a number of chunks, returned by individual invocations of the `read()` system call—the web server should run correctly regardless of the fragmentation pattern. To test different fragmentation patterns with Cloud9, one simply enables symbolic packet fragmentation on the client socket:

```
ioctl(ssock, SIO_PKT_FRAGMENT, RD);
```

To test how the web server handles failures in the environment, we can ask Cloud9 to selectively inject faults when the server reads or sends data on a socket by placing in the symbolic test suite calls of the form:

```
ioctl(ssock, SIO_FAULT_INJ, RD | WR);
```

Cloud9 can also enable/disable fault injection globally for all file descriptors within a certain region of the code using calls to `cloud9_fi_enable` and `cloud9_fi_disable`. For simulating low-memory conditions, Cloud9 provides a `cloud9_set_max_heap` primitive, which can be used to test the web server with different maximum heap sizes.

6. Cloud9 Prototype

We developed a Cloud9 prototype that runs on private clusters as well as cloud infrastructures like Amazon EC2 [Amazon] and Eucalyptus [Eucalyptus]. The prototype has 10 KLOC; the POSIX model accounts for half of the total code size. Cloud9 workers embed KLEE [Cadar 2008], a state-of-the-art single-node symbolic execution engine; the Cloud9 fabric converts a cluster of individual engines into one big parallel symbolic execution engine. This section presents selected implementation decisions underlying the prototype. More details are available at http://cloud9.epfl.ch.

Broken Replays   As discussed in §3.2, when a job is transferred from one worker to another, the replay done during materialization must successfully reconstruct the transferred state. Along the reconstruction path, the destination must execute the same instructions, obtain the same symbolic memory content, and get the same results during constraint solving as on the source worker. Failing to do so causes the replayed path to be *broken*: it either diverges, or terminates prematurely. In both cases, this means the state cannot be reconstructed, and this could affect exploration completeness.

The main challenge is that the underlying KLEE symbolic execution engine relies on a global memory allocator to service the tested program's `malloc()` calls. The allocator returns actual host memory addresses, and this is necessary for executing external system calls that access program state. Unfortunately, this also means that buffers are allocated at addresses whose values for a given state depend on the history of previous allocations in *other* states. Such cross-state interference leads to frequent broken replays.

We therefore replaced the KLEE allocator with a per-state deterministic memory allocator, which uses a per-state address counter that increases with every memory allocation. To preserve the correctness of external calls (that require real addresses), this allocator gives addresses in a range that is also mapped in the SEE address space using `mmap()`. Thus, before external calls are invoked, the memory content of the state is copied into the `mmap`-ed region.

Constraint Caches   KLEE implements a cache mechanism for constraint-solving results; this cache can significantly improve solver performance. In Cloud9, states are transferred between workers without the source worker's cache. While one might expect this to hurt performance significantly, in practice we found that the necessary portion of the cache is mostly reconstructed as a side effect of path replay, as the path constraints are re-sent to the local solver.

Custom Data Structures   We developed two custom data structures for handling symbolic execution trees: *Node pins* are a kind of smart pointer customized for trees. Standard smart pointers (e.g., the ones provided by Boost libraries) can introduce significant performance disruptions when used for linked data structures: chained destructors can introduce noticeable deallocation latency and may even overflow the stack and crash the system. The node pin allows trees to be treated akin to a "rubber band" data structure: as nodes get allocated, the rubber band is stretched, and some nodes act as pins to anchor the rubber band. When such a pin is removed, the nodes with no incoming references are freed up to the point where the rubber band reaches the pin next closest to the root. Tree nodes between two pins are freed all at once, avoiding the use of the stack for recursive destructor calls.

Another custom data structure is the *tree layer*. At first, Cloud9 used a single tree to represent the entire symbolic execution. As Cloud9 evolved, tree nodes acquired an increasing number of objects: program states, imported jobs, breakpoints, etc. This made tree searches inefficient, complicated synchronization, and generally impeded our development effort. We therefore adopted a layer-based structure similar to that used in CAD tools, where the actual tree is a superposition of simpler layers. When exploring the tree, one chooses the layer of interest; switching between layers can be done dynamically at virtually zero cost. Cloud9 currently uses separate layers for symbolic states, imported jobs, and several other sets of internal information.

7. Evaluation

There are several questions one must ask of a parallel symbolic execution platform, and we aim to answer them in this section: Can it be used on real-world software that interacts richly with its environment (§7.1)? Does it scale on commodity shared-nothing clusters (§7.2)? Is it an effective testing platform, and does it help developers gain confidence in their code (§7.3)? How do its different components contribute to overall efficiency (§7.4)?

For all our experiments, we used a heterogeneous cluster environment, with worker CPU frequencies between 2.3–2.6 GHz and with 4–6 GB of RAM available per core.

On each worker, the underlying KLEE engine used the best searchers from [Cadar 2008], namely an interleaving of random-path and coverage-optimized strategies. At each step, the engine alternately selects one of these heuristics to pick the next state to explore. Random-path traverses the execution tree starting from the root and randomly picks the next descendant node, until a candidate state is reached. The coverage-optimized strategy weighs the states according to an estimated distance to an uncovered line of code, and then randomly selects the next state according to these weights.

To quantify coverage, we report both line coverage and path coverage numbers. Line coverage measures the fraction of program statements executed during a test run, while path coverage reports how many execution paths were explored during a test. Path coverage is the more relevant metric when comparing thoroughness of testing tools. Nevertheless, we also evaluate properties related to line coverage, since this is still a de facto standard in software testing practice. Intuitively, and confirmed experimentally, path coverage in Cloud9 scales linearly with the number of workers.

7.1 Handling Real-World Software

Table 4 shows a selection of the systems we tested with Cloud9, covering several types of software. We confirmed that each system can be tested properly under our POSIX model. In the rest of this section, we focus our in-depth evaluation on several networked servers and tools, as they are frequently used in settings where reliability matters.

Due to its comprehensive POSIX model, Cloud9 can test many kinds of servers. One example is lighttpd, a web server used by numerous high-profile web sites, such as YouTube, Wikimedia, Meebo, and SourceForge. For lighttpd, Cloud9 proved that a certain bug fix was incorrect, and the bug could still manifest even after applying the patch (§7.3.4). Cloud9 also found a bug in curl, an Internet transfer application

System	Size (KLOC)	Type of Software
Apache httpd 2.2.16	226.4	Web servers
Lighttpd 1.4.28	39.5	
Ghttpd 1.4.4	0.6	
Memcached 1.4.5	8.3	Distributed object cache
Python 2.6.5	388.3	Language interpreter
Curl 7.21.1	65.9	Network utilities
Rsync 3.0.7	35.6	
Pbzip 2.1.1	3.6	Compression utility
Libevent 1.4.14	10.2	Event notification library
Coreutils 6.10	72.1	Suite of system utilities
Bandicoot 1.0	6.4	Lightweight DBMS

Table 4: Representative selection of testing targets that run on Cloud9. Size was measured using the `sloccount` utility.

that is part of most Linux distributions and other operating systems (§7.3.2). Cloud9 also found a hang bug in the UDP handling code of memcached, a distributed memory object cache system used by many Internet services, such as Flickr, Craigslist, Twitter, and Livejournal (§7.3.3).

In addition to the testing targets mentioned above, we also tested a benchmark consisting of a multi-threaded and multi-process producer-consumer simulation. The benchmark exercises the entire functionality of the POSIX model: threads, synchronization, processes, and networking.

We conclude that Cloud9 is practical and capable of testing a wide range of real-world software systems.

7.2 Scaling on Commodity Shared-Nothing Clusters

We evaluate Cloud9 using two metrics:

1. The time to reach a certain goal (e.g., an exhaustive path exploration, or a fixed coverage level)—we consider this an *external* metric, which measures the performance of the testing platform in terms of its end results.

2. The useful work performed during exploration, measured as the number of useful (non-replay) instructions executed symbolically. This is an *internal* metric that measures the efficiency of Cloud9's internal operation.

A cluster-based symbolic execution engine *scales* with the number of workers if these two metrics improve proportionally with the number of workers in the cluster.

Time Scalability We show that Cloud9 scales linearly by achieving the same testing goal proportionally faster as the number of workers increases. We consider two scenarios.

First, we measure how fast Cloud9 can exhaustively explore a fixed number of paths in the symbolic execution tree. For this, we use a symbolic test case that generates all the possible paths involved in receiving and processing two symbolic messages in the memcached server (§7.3 gives more details about the setup). Fig. 7 shows the time required to finish the test case with a variable number of workers: every doubling in the number of workers roughly halves the time to completion. With 48 workers, the time to complete is about 10 minutes; for 1 worker, exploration time exceeds our 10-hour limit on the experiment.

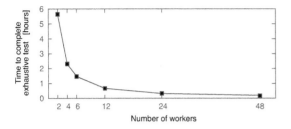

Figure 7: Cloud9 scalability in terms of the time it takes to exhaustively complete a symbolic test case for memcached.

Second, we measure the time it takes Cloud9 to reach a fixed coverage level for the `printf` UNIX utility. `printf` performs a lot of parsing of its input (format specifiers), which produces complex constraints when executed symbolically. Fig. 8 shows that the time to achieve a coverage target decreases proportionally with the number of added workers. The low 50% coverage level can be easily achieved even with a sequential SEE (1-worker Cloud9). However, higher coverage levels require more workers, if they are to be achieved in a reasonable amount of time; e.g., only a 48-worker Cloud9 is able to achieve 90% coverage. The anomaly at 4 workers for 50% coverage is due to high variance; when the number of workers is low, the average (5±4.7 minutes over 10 experiments) can be erratic due to the random choices in the random-path search strategy.

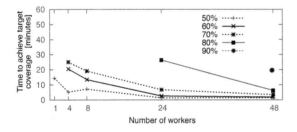

Figure 8: Cloud9 scalability in terms of the time it takes to obtain a target coverage level when testing `printf`.

Work Scalability We now consider the same scalability experiments from the perspective of useful work done by Cloud9: we measure both the total number of instructions (from the target program) executed during the exploration process, as well as normalize this value per worker. This measurement indicates whether the overheads associated with parallel symbolic execution impact the efficiency of exploration, or are negligible. Fig. 9 shows the results for memcached, confirming that Cloud9 scales linearly in terms of useful work done (top graph). The average useful work done by a worker (bottom graph) is relatively independent

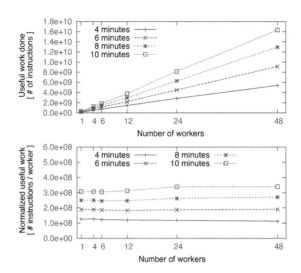

Figure 9: Cloud9 scalability in terms of useful work done for four different running times when testing memcached.

of the total number of workers in the cluster, so adding more workers improves proportionally Cloud9's results.

In Fig. 10 we show the results for `printf` and `test`, UNIX utilities that are an order of magnitude smaller than memcached. We find that the useful work done scales in a similar way to memcached, even though the three programs are quite different from each other (e.g., `printf` does mostly parsing and formatting, while memcached does mostly data structure manipulations and network I/O).

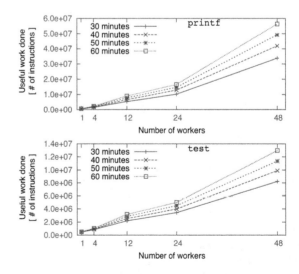

Figure 10: Cloud9's useful work on `printf` (top) and `test` (bottom) increases roughly linearly in the size of the cluster.

In conclusion, Cloud9 scales linearly with the number of workers, both in terms of the time to complete a symbolic testing task and in terms of reaching a target coverage level.

7.3 Effectiveness as a Testing Platform

In this section we present several case studies that illustrate how Cloud9 can explore and find new bugs, confirm/disprove that existing bugs have been correctly fixed, and regression-test a program after it has been modified. In the common case, Cloud9 users start with a concrete test case (e.g., from an existing test suite) and generalize it by making data symbolic and by controlling the environment.

7.3.1 Case Study #1: UNIX Utilities

KLEE is an excellent tool for testing command-line programs, in particular UNIX utilities. It does not tackle more complex systems, like the ones in Table 4, mainly due to path explosion (since KLEE is a single-node engine) and insufficient environment support. We cannot compare Cloud9 to KLEE on parallel and distributed systems, but we can compare on the Coreutils suite of UNIX utilities [Coreutils].

We run KLEE on each of the 96 utilities for 10 minutes, and then run a 12-worker Cloud9 on each utility for 10 minutes. Fig. 11 reports the average coverage increase obtained with Cloud9 over 7 trials, using KLEE's 7-trial average results as a baseline; the experiment totals $2 \times 7 \times 96 \times 10 = 13,440$ minutes > 9 days. The increase in coverage is measured as *additional* lines of code covered, expressed as a percentage of program size (i.e., we do not report it as a percentage of the baseline, which would be a higher number).

Cloud9 covers up to an additional 40% of the target programs, with an average of 13% additional code covered across all Coreutils. In general, improving coverage becomes exponentially harder as the base coverage increases, and this effect is visible in the results: a $12\times$ increase in hardware resources does not bring about a $12\times$ increase in coverage. Our results show that Cloud9 allows "throwing hardware" at the automated testing problem, picking up where KLEE left off. In three cases, Cloud9 achieved 100% coverage in 10 minutes on real-world code. This experiment does not aim to show that Cloud9 is a "better" symbolic execution engine than KLEE—after all, Cloud9 is based on KLEE—but rather that Cloud9-style parallelization can make existing symbolic execution engines more powerful.

The way we compute coverage is different from [Cadar 2008]—whereas KLEE was conceived as an automated *test generator*, Cloud9 is meant to *directly test* software. Thus, we measure the number of lines of code tested by Cloud9, whereas [Cadar 2008] reports numbers obtained by running the concrete test cases generated by KLEE. Our method yields more-conservative numbers because a test generated by KLEE at the end of an incomplete path (e.g., that terminated due to an environment failure) may execute further than the termination point when run concretely.

7.3.2 Case Study #2: Curl

Curl is a popular data transfer tool for multiple network protocols, including HTTP and FTP. When testing it, Cloud9

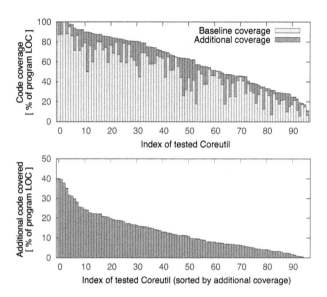

Figure 11: Cloud9 coverage improvements on the 96 Coreutils (1-worker Cloud9 vs. 12-worker Cloud9).

found a new bug which causes Curl to crash when given a URL regular expression of the form "`http://site.{one, two,three}.com{`". Cloud9 exposed a general problem in Curl's handling of the case when braces used for regular expression globbing are not matched properly. The bug was confirmed and fixed within 24 hours by the developers.

This problem had not been noticed before because the globbing functionality in Curl was shadowed by the same functionality in command-line interpreters (e.g., Bash). This case study illustrates a situation that occurs often in practice: when a piece of software is used in a way that has not been tried before, it is likely to fail due to latent bugs.

7.3.3 Case Study #3: Memcached

Memcached is a distributed memory object cache system, mainly used to speed up web application access to persistent data, typically residing in a database.

Memcached comes with an extensive test suite comprised of C and Perl code. Running it completely takes about 1 minute; it runs 6,472 different test cases and explores 83.66% of the code. While this is considered thorough by today's standards, two easy Cloud9 test cases further increased code coverage. Table 5 contains a summary of our results, presented in more details in the following paragraphs.

Symbolic Packets The memcached server accepts commands over the network. Based on memcached's C test suite, we wrote a test case that sends memcached a generic, symbolic binary command (i.e., command content is fully symbolic), followed by a second symbolic command. This test captures all operations that entail a pair of commands.

A 24-worker Cloud9 explored in less than 1 hour all 74,503 paths associated with this sequence of two symbolic packets, covering an additional 1.13% of the code relative to

Testing Method	Paths Covered	Isolated Coverage*	Cumulated Coverage**
Entire test suite	6,472	83.67%	—
Binary protocol test suite	27	46.79%	84.33% (+0.67%)
Symbolic packets	74,503	35.99%	84.79% (+1.13%)
Test suite + fault injection	312,465	47.82%	84.94% (+1.28%)

Table 5: Path and code coverage increase obtained by each symbolic testing technique on memcached. We show total coverage obtained with each testing method (*), as well as total coverage obtained by augmenting the original test suite with the indicated method (**); in parentheses, we show the increase over the entire test suite's coverage.

the original test suite. What we found most encouraging in this result is that such exhaustive tests constitute first steps toward using symbolic tests to *prove* properties of real-world programs, not just to look for bugs. Symbolic tests may provide an alternative to complex proof mechanisms that is more intuitive for developers and thus more practical.

Symbolic Fault Injection We also tested memcached with fault injection enabled, whereby we injected all feasible failures in memcached's calls to the C Standard Library. After 10 minutes of testing, a 24-worker Cloud9 explored 312,465 paths, adding 1.28% over the base test suite. The fact that *line* coverage increased by so little, despite having covered almost $50\times$ more paths, illustrates the weakness of line coverage as a metric for test quality—high line coverage should offer no high confidence in the tested code's quality.

For the fault injection experiment, we used a special strategy that sorts the execution states according to the number of faults recorded along their paths, and favors the states with fewer fault injection points. This led to a uniform injection of faults: we first injected one fault in every possible fault injection point along the original C test suite path, then injected pairs of faults, and so on. We believe this is a practical approach to using fault injection as part of regular testing.

Hang Detection We tested memcached with symbolic UDP packets, and Cloud9 discovered a hang condition in the packet parsing code: when a sequence of packet fragments of a certain size arrive at the server, memcached enters an infinite loop, which prevents it from serving any further UDP connections. This bug can seriously hurt the availability of infrastructures using memcached.

We discovered the bug by limiting the maximum number of instructions executed per path to 5×10^6. The paths without the bug terminated after executing $\sim 3 \times 10^5$ instructions; the other paths that hit the maximum pointed us to the bug.

7.3.4   Case Study #4: Lighttpd

The lighttpd web server is specifically engineered for high request throughput, and it is quite sensitive to the rate at

Fragmentation pattern (data sizes in bytes)	ver. 1.4.12 (pre-patch)	ver. 1.4.13 (post-patch)
1×28	OK	OK
$1 \times 26 + 1 \times 2$	crash + hang	OK
$2 + 5 + 1 + 5 + 2 \times 1 + 3 \times 2 + 5 + 2 \times 1$	crash + hang	crash + hang

Table 6: The behavior of different versions of lighttpd to three ways of fragmenting the HTTP request "GET /index.html HTTP/1.0CRLFCRLF" (string length 28).

which new data is read from a socket. Alas, the POSIX specification offers no guarantee on the number of bytes that can be read from a file descriptor at a time. lighttpd 1.4.12 has a bug in the command-processing code that causes the server to crash (and connected clients to hang indefinitely) depending on how the incoming stream of requests is fragmented.

We wrote a symbolic test case to exercise different *stream fragmentation* patterns and see how different lighttpd versions behave. We constructed a simple HTTP request, which was then sent over the network to lighttpd. We activated network packet fragmentation via the symbolic `ioctl()` API explained in §5. We confirmed that certain fragmentation patterns cause lighttpd to crash (prior to the bug fix). However, we also tested the server right after the fix and discovered that the bug fix was incomplete, as some fragmentation patterns still cause a crash and hang the client (Table 6).

This case study shows that Cloud9 can find bugs caused by specific interactions with the environment which are hard to test with a concrete test suite. It also shows how Cloud9 can be used to write effective regression test suites—had a stream-fragmentation symbolic test been run after the fix, the lighttpd developers would have promptly discovered the incompleteness of their fix.

7.3.5   Case Study #5: Bandicoot DBMS

Bandicoot is a lightweight DBMS that can be accessed over an HTTP interface. We exhaustively explored all paths handling the GET commands and found a bug in which Bandicoot reads from outside its allocated memory. The particular test we ran fortuitously did not result in a crash, as Bandicoot ended up reading from the libc memory allocator's metadata preceding the allocated block of memory. However, besides the read data being wrong, this bug could cause a crash depending on where the memory block was allocated.

To discover and diagnose this bug without Cloud9 is difficult. First, a concrete test case has little chance of triggering the bug. Second, searching for the bug with a sequential symbolic execution tool seems impractical: the exhaustive exploration took 9 hours with a 4-worker Cloud9 (and less than 1 hour with a 24-worker cluster).

7.3.6   Discussion

Cloud9 inherits KLEE's capabilities, being able to recognize memory errors and failed assertions. We did not add much in terms of bug detection, only two mechanisms for detecting hangs: check if all symbolic threads are sleeping (deadlock) and set a threshold for the maximum number of instructions executed per path (infinite loop or livelock). Even so, Cloud9 can find bugs beyond KLEE's abilities because the POSIX model allows Cloud9 to reach more paths and explore deeper portions of the tested program's code—this exposes additional potentially buggy situations. Cloud9 also has more total memory and CPU available, due to its distributed nature, so it can afford to explore more paths than KLEE. As we have shown above, it is feasible to offer proofs for certain program properties: despite the exponential nature of exhaustively exploring paths, one can build small but useful symbolic test cases that can be exhaustively executed.

7.4 Utility of Load Balancing

In this section we explore the utility of dynamic load balancing. Consider the example of exhaustively exploring paths with two symbolic packets in memcached, using 48 workers, but this time from a load balancing perspective. Fig. 12 shows that load balancing events occur frequently, with 3–6% of all states in the system being transferred between workers in almost every 10-second time interval.

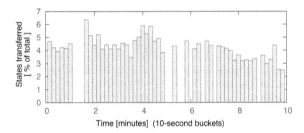

Figure 12: The fraction of total states (candidate nodes) transferred between workers during symbolic execution.

To illustrate the benefits of load balancing, we disable it at various moments in time and then analyze the evolution of total useful work done. Fig. 13 shows that the elimination of load balancing at any moment during the execution significantly affects the subsequent performance of exploration due to the ensuing imbalance. This demonstrates the necessity of taking a dynamic approach to parallel symbolic execution, instead of doing mere static partitioning of the execution tree.

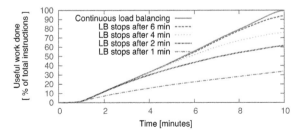

Figure 13: Instruction throughput of Cloud9 with load balancing disabled at various points during the exhaustive test.

8. Related Work

To our knowledge, parallel symbolic execution was first described in [Ciortea 2009]. Cloud9 builds upon those ideas.

Recently, [Staats 2010] described an extension to Java Pathfinder (JPF) that parallelizes symbolic execution by using parallel random searches on a static partition of the execution tree. JPF pre-computes a set of disjoint constraints that, when used as preconditions on a worker's exploration of the execution tree, steer each worker to explore a subset of paths disjoint from all other workers. In this approach, using constraints as preconditions imposes, at *every* branch in the program, a solving overhead relative to exploration without preconditions. The complexity of these preconditions increases with the number of workers, as the preconditions need to be more selective. Thus, per-worker solving overhead increases as more workers are added to the cluster. This limits scalability: the largest evaluated program had 447 lines of code and did not interact with its environment. Due to the *static* partitioning of the execution tree, total running time is determined by the worker with the largest subtree (as explained in §2). As a result, increasing the number of workers can even increase total test time instead of reducing it [Staats 2010]. Cloud9 mitigates these drawbacks.

Several sequential symbolic execution engines [Cadar 2008, Godefroid 2005; 2008, Majumdar 2007] have had great success in automated testing. These state-of-the-art tools exhaust available memory and CPU fairly quickly (as explained in §2). Cloud9 can help such tools scale beyond their current limits, making symbolic execution a viable testing methodology for a wider spectrum of software systems.

To our knowledge, we are the first to scalably parallelize symbolic execution to shared-nothing clusters. There has been work, however, on parallel model checking [Barnat 2007, Grumberg 2006, Kumar 2004, Lerda 1999, Stern 1997]. The SPIN model checker has been parallelized two-way for dual-core machines [Holzmann 2007]. Nevertheless, there are currently no model checkers that can scale to many loosely connected computers, mainly due to the overhead of coordinating the search across multiple machines and transferring explicit states. Cloud9 uses an encoding of states that is compact and enables better scaling.

Swarm verification [Holzmann 2008] generates series of parallel verification runs with user-defined bounds on time and memory. Parallelism is used to execute different search strategies independently. The degree of parallelism is limited to the number of distinct search strategies. Cloud9 is not limited in this way: due to the use of dynamic load balancing, Cloud9 affords arbitrary degrees of parallelism.

Korat [Misailovic 2007] is a parallel testing system that demonstrated $\sim 500\times$ speedup when using 1024 nodes. Korat is designed for cases when all program inputs—within a certain bound—can be generated offline. In contrast, Cloud9 handles arbitrarily complex inputs, even if enumeration is infeasible, and can handle system calls and thread schedules.

VeriSoft [Godefroid 1997] introduced stateless search for model checkers, in which a list of state transitions is used to reconstruct an execution state whenever needed. We extend this idea in the Cloud9 job model, which is in essence a form of stateless search. However, jobs are replayed from nodes on the frontier of the execution tree, instead of being replayed from the root. This is an important optimization: in practice, most programs have either a long linear path in the beginning (i.e., initialization code that does not depend on the symbolic input), or long linear executions between two consecutive state forks. Replaying from the root would represent a significant waste of resources for large programs.

Other techniques can be used to improve the scalability of symbolic execution. For example, compositional test generation [Boonstoppel 2008, Godefroid 2007] automatically computes summaries that condense the result of exploration inside commonly used functions. S^2E [Chipounov 2011] improves the scalability of symbolic execution by selectively executing symbolically only those parts of a system that are of interest to the tests. Cloud9 is complementary to these techniques and could be used to scale them further.

9. Conclusion

This paper presented Cloud9, an automated testing platform that employs parallelization to scale symbolic execution by harnessing the resources of commodity clusters. Cloud9 can automatically test real systems, such as memcached, that interact in complex ways with their environment. It includes a new symbolic environment that supports all major aspects of the POSIX interface, and provides a systematic interface to writing symbolic tests. Further information on Cloud9 can be found at http://cloud9.epfl.ch.

Acknowledgments

We are indebted to Ayrat Khalimov, Cristian Cadar, Daniel Dunbar, and our EPFL colleagues for their help on this project. We thank Michael Hohmuth, our shepherd, and the anonymous EuroSys reviewers for their guidance on improving our paper. We are grateful to Google for supporting our work through a Focused Research Award.

References

[Amazon] Amazon. Amazon EC2. http://aws.amazon.com/ec2.

[Barnat 2007] Jiri Barnat, Lubos Brim, and Petr Rockai. Scalable multi-core LTL model-checking. In *Intl. SPIN Workshop*, 2007.

[Boonstoppel 2008] Peter Boonstoppel, Cristian Cadar, and Dawson R. Engler. RWset: Attacking path explosion in constraint-based test generation. In *Intl. Conf. on Tools and Algorithms for the Construction and Analysis of Systems*, 2008.

[Cadar 2008] Cristian Cadar, Daniel Dunbar, and Dawson R. Engler. KLEE: Unassisted and automatic generation of high-coverage tests for complex systems programs. In *Symp. on Operating Systems Design and Implementation*, 2008.

[Cadar 2006] Cristian Cadar, Vijay Ganesh, Peter M. Pawlowski, David L. Dill, and Dawson R. Engler. EXE: Automatically generating inputs of death. In *Conf. on Computer and Communication Security*, 2006.

[Chipounov 2009] Vitaly Chipounov, Vlad Georgescu, Cristian Zamfir, and George Candea. Selective symbolic execution. In *Workshop on Hot Topics in Dependable Systems*, 2009.

[Chipounov 2011] Vitaly Chipounov, Volodymyr Kuznetsov, and George Candea. S2E: A platform for in-vivo multi-path analysis of software systems. In *Intl. Conf. on Architectural Support for Programming Languages and Operating Systems*, 2011.

[Ciortea 2009] Liviu Ciortea, Cristian Zamfir, Stefan Bucur, Vitaly Chipounov, and George Candea. Cloud9: A software testing service. In *Workshop on Large Scale Distributed Systems and Middleware*, 2009.

[Coreutils] Coreutils. Coreutils. http://www.gnu.org/software/coreutils/.

[Eucalyptus] Eucalyptus. Eucalyptus software. http://open.eucalyptus.com/, 2010.

[Godefroid 2005] P. Godefroid, N. Klarlund, and K. Sen. DART: Directed automated random testing. In *Conf. on Programming Language Design and Implementation*, 2005.

[Godefroid 1997] Patrice Godefroid. Model checking for programming languages using VeriSoft. In *Symp. on Principles of Programming Languages*, 1997.

[Godefroid 2007] Patrice Godefroid. Compositional dynamic test generation. In *Symp. on Principles of Programming Languages*, 2007. Extended abstract.

[Godefroid 2008] Patrice Godefroid, Michael Y. Levin, and David Molnar. Automated whitebox fuzz testing. In *Network and Distributed System Security Symp.*, 2008.

[Grumberg 2006] Orna Grumberg, Tamir Heyman, and Assaf Schuster. A work-efficient distributed algorithm for reachability analysis. *Formal Methods in System Design*, 29(2), 2006.

[Holzmann 2007] G. J. Holzmann and D. Bosnacki. Multi-core model checking with SPIN. In *Intl. Parallel and Distributed Processing Symp.*, 2007.

[Holzmann 2008] Gerard J. Holzmann, Rajeev Joshi, and Alex Groce. Tackling large verification problems with the Swarm tool. In *Intl. SPIN Workshop*, 2008.

[Kumar 2004] Rahul Kumar and Eric G. Mercer. Load balancing parallel explicit state model checking. In *Intl. Workshop on Parallel and Distributed Methods in Verification*, 2004.

[Lerda 1999] Flavio Lerda and Riccardo Sisto. Distributed-memory model checking with SPIN. In *Intl. SPIN Workshop*, 1999.

[Majumdar 2007] Rupak Majumdar and Koushik Sen. Hybrid concolic testing. In *Intl. Conf. on Software Engineering*, 2007.

[Marinescu 2009] Paul D. Marinescu and George Candea. LFI: A practical and general library-level fault injector. In *Intl. Conf. on Dependable Systems and Networks*, 2009.

[McConnell 2004] Steve McConnell. *Code Complete*. Microsoft Press, 2004.

[Misailovic 2007] Sasa Misailovic, Aleksandar Milicevic, Nemanja Petrovic, Sarfraz Khurshid, and Darko Marinov. Parallel test generation and execution with Korat. In *Symp. on the Foundations of Software Eng.*, 2007.

[Musuvathi 2008] Madanlal Musuvathi, Shaz Qadeer, Thomas Ball, Gérard Basler, Piramanayagam Arumuga Nainar, and Iulian Neamtiu. Finding and reproducing Heisenbugs in concurrent programs. In *Symp. on Operating Systems Design and Implementation*, 2008.

[RedHat] RedHat. RedHat security. http://www.redhat.com/security/updates/classification, 2005.

[Staats 2010] Matt Staats and Corina Păsăreanu. Parallel symbolic execution for structural test generation. In *Intl. Symp. on Software Testing and Analysis*, 2010.

[Stern 1997] Ulrich Stern and David L. Dill. Parallelizing the Murφ verifier. In *Intl. Conf. on Computer Aided Verification*, 1997.

Striking a New Balance Between Program Instrumentation and Debugging Time

Olivier Crameri

EPFL

olivier.crameri@epfl.ch

Ricardo Bianchini

Rutgers University

ricardob@cs.rutgers.edu

Willy Zwaenepoel

EPFL

willy.zwaenepoel@epfl.ch

Abstract

Although they are helpful in many cases, state-of-the-art bug reporting systems may impose excessive overhead on users, leak private information, or provide little help to the developer in locating the problem. In this paper, we explore a new approach to bug reporting that uses partial logging of branches to record the path leading to a bug. We use static and dynamic analysis (both in isolation and in tandem) to identify the branches that need to be logged. When a bug is encountered, the system uses symbolic execution along the partial branch trace to reproduce the problem and find a set of inputs that activate the bug. The partial branch log drastically reduces the number of paths that would otherwise need to be explored by the symbolic execution engine. We study the tradeoff between instrumentation overhead and debugging time using an open-source Web server, the *diff* utility, and four coreutils programs. Our results show that the instrumentation method that combines static and dynamic analysis strikes the best compromise, as it limits both the overhead of branch logging and the bug reproduction time. We conclude that our techniques represent an important step in improving bug reporting and making symbolic execution more practical for bug reproduction.

Categories and Subject Descriptors D.2.5 [*Software*]: Testing and Debugging

General Terms Design, Experimentation, Reliability

Keywords Debugging, Bug Reporting, Symbolic Execution, Static Analysis

1. Introduction

Despite considerable advances in testing and verification, programs routinely ship with a number of undiscovered

EuroSys'11, April 10–13, 2011, Salzburg, Austria.
Copyright © 2011 ACM 978-1-4503-0634-8/11/04... $10.00

bugs. Some of those bugs are later uncovered and reported by users. Debugging is an arduous task in general, and it is even harder when bugs are uncovered by users. Before the developer can start working on a fix, the problem must be reproduced. Reporting systems are meant to help with this task, but they need to strike a balance between privacy concerns, recording overhead at the user site, and time for the developer to reproduce the cause of the bug.

Reporting systems. The current commercial state of the art is represented by the Windows Error Reporting System [Glerum 2009], which automatically generates a bug report when an application crashes. The bug report includes a per-thread stack trace, the values of some registers, the name of the libraries loaded, and portions of the heap. While that information is helpful, the developer must manually find the path to the bug among an exponential number of possible paths.

Furthermore, the information contained in the report may leak private information, which is undesirable and in certain circumstances prevents the use of the tool altogether. Recently, Zamfir and Candea have shown that symbolic execution can be used to partially automate the search for the path to the bug. However, as the manual approach, symbolic execution suffers from a potentially exponential number of paths to be explored [Zamfir 2010].

An alternative is to record the inputs to the program at the user site. Inputs leading to failures can be transmitted to the developer, who uses them to replay the program to the occurrence of the bug. While avoiding the problem of having to search for the path to the bug, divulging user inputs is often considered unacceptable from a privacy viewpoint. Castro *et al.* generate a set of inputs that is different from the original input but still leads to the same bug [Castro 2008]. Their approach does not transfer the original input to the developer, but requires input logging, whole-program replay, and invocation of a constraint solver at the user site.

A more direct approach is to record the path to the bug at the user site, for instance, by instrumenting the program to record the direction of all branches taken. In its naive form, this approach is infeasible because of the CPU, storage and transmission overhead incurred, but in this paper we demon-

```
int main(int argc, char **argv) {        1
  char option = read_option(input);       2
  int result = 0                          3
  if (option == 'a')                      4
    result = fibonacci(20);               5
  else if (option == 'b')                 6
    result = fibonacci(40);               7
                                          8
  printf("Result: %d\n", result).         9
  return 0;                              10
}                                        11
```

Listing 1. A simple program computing the fibonacci sequence.

strate that the approach can be optimized by instrumenting only a limited number of branches.

Our approach. We base our work on the following three observations. First, a large number of branches do not depend on the program input, and therefore their outcome need not be logged, because it is known a priori. We denote branches whose outcome depends on program input as *symbolic*. Other branches are denoted as *concrete*.

Consider the example program in Listing 1. Depending on the input parameter, the program computes the fibonacci number F_n for one of two different values of n. The only input to this program is the parameter that indicates for which value to compute the fibonacci number. While this program may have many branches (especially in the *fibonacci* function, not shown for conciseness), it is sufficient to know the outcome of the branches at line 4 and 6 to fully determine the behavior of the program. Indeed, those two branches are the only symbolic ones, and it suffices to record their outcome.

A second, related observation is that application branches are typically either always symbolic or always concrete. In other words, it is rare that a particular branch at some point in the execution depends on input and at other points does not. An example of this can also be seen in Listing 1. The branches at lines 4 and 6 always depend on input, the others never do. This property is almost always true for branches in the application, and often, albeit not always, true for branches in the libraries.

Restricting our attention to branches that do depend on the input, the third observation is that it is not strictly necessary to record the outcome of all of those. When we record all such branches, the result is a unique program path, and therefore no search is required at the developer's site. When we record a subset of those branches, their outcomes no longer define a unique path but a set of possible paths, among which the developer has to search to find the path to the bug. In other words, there is a spectrum of possibilities between (1) no recording at the user site and a search at the developer site among all possible paths, and (2) complete recording of the outcome of all branches at the user site and no search at the developer site. Various points in this spectrum constitute different tradeoffs between instrumentation overhead at the user and bug reproduction time at the developer site. (We de-

fine bug reproduction as finding a set of inputs that leads the execution to the bug, or, equivalently, finding the direction of all branches taken so that they lead the execution to the bug.) It is this tradeoff that is explored further in this paper.

In particular, we propose three approaches for deciding the set of branches to instrument. The first approach uses dynamic analysis to determine the set of branches that depend on input. This approach is constrained in its effectiveness by the limited coverage of the program that the symbolic execution engine used for dynamic analysis can achieve in a given amount of time. It tends to under-estimate the number of branches that need to be instrumented, therefore leading to reduced instrumentation cost but increased bug reproduction time. The time that the symbolic execution engine is allowed to execute gives the developer an additional tuning knob in the tradeoff: the more time invested in symbolic execution, the better the coverage can be, and therefore the more precise the analysis. The second approach is based on static analysis of the program. Data flow analysis is used to determine the set of branches that depend on the input. This approach is limited by the precision of the static analysis, and in general tends to over-estimate the number of branches to be instrumented. Thus, this approach favors increased instrumentation cost in exchange for reduced bug reproduction time. The third approach combines the above two. It uses symbolic execution for a given amount of time, and then marks the branches that have not yet been visited according to the outcome of the static analysis.

When a bug in the program occurs, the developer runs the program in a modified symbolic execution engine that takes the partial branch trace as input. At each instrumented branch, the symbolic execution engine forces the execution to follow the branch direction specified by the log. In case a symbolic branch has not been logged, the engine explores both alternatives. Symbolic execution along the incorrect alternative eventually causes a further branch to take a direction different from the one recorded in the log. The symbolic execution engine then aborts the exploration of this alternative, and turns its attention to the other, correct branch direction.

Non-deterministic events add another dimension to the tradeoff between logging overhead and bug reproduction time. Either we log all non-deterministic events during execution so that we can reproduce them exactly during replay, but we add overhead at runtime. Or we do not log all of them, but then a non-deterministic event during replay may produce an outcome different from the one during actual execution. The importance of this tradeoff is amplified if some branches are not logged. If all branches are logged, then with high likelihood a different outcome of a non-deterministic event during replay is detected quickly, because a subsequent branch takes a direction different from the one logged during execution. If, however, not all branches are logged and in particular branches that follow the non-deterministic event are not logged, then the replay may require considerable searching to discover the path followed during the actual execution. We explore this tradeoff for system calls, one of the principal sources of non-determinism during sequential execution.

Overview of results. We have implemented the three branch instrumentation methods described above, in addition to the naive approach that logs all branches. We explore the tradeoff between instrumentation overhead and bug reproduction time using an open-source Web server, the *diff* utility, and four coreutils programs. We find that the combined approach strikes a better balance between instrumentation overhead and bug reproduction time than the other two. Moreover, it enables bug reproduction times that are only slightly higher than those for the approach based solely on static analysis. In contrast, this latter approach only marginally reduces logging overhead compared to logging all branches, whereas the approach based solely on dynamic analysis leads to excessively long times to reproduce bugs. More concretely, our results show that the combined approach reduces instrumentation overhead between 10% and 92%, compared to the approach based solely on static analysis. At the same time, it always reproduces the bugs we considered within the allotted time (1 hour), whereas the dynamic approach fails to do so in 6 out of 16 cases. In all circumstances, we find that selectively logging the results of system calls is advantageous: it limits bug reproduction time, and adds only marginally to instrumentation overhead.

Contributions. The contributions of this paper are:

1. The use of symbolic execution prior to shipping the program to discover which branches depend on input and which not.
2. An optimized approach to symbolic execution for bug reproduction that is guided by a symbolic branch log collected at the user's site.
3. The exploration of the tradeoffs between the amount of pre-deployment symbolic execution, the instrumentation overhead resulting from logging the outcome of branches, and the time necessary to reproduce a bug at the developer's site.
4. A combined static-dynamic method for deciding which branches to instrument. The method leads to a better tradeoff than previous systems, making the approach of logging branches more practical and reducing the debugging time.
5. A quantification of the impact of logging the result of selected system calls, demonstrating that it only marginally increases the instrumentation overhead, but considerably shortens the bug reproduction time.

Roadmap. The rest of this paper is organized as follows. Section 2 describes the program analysis and instrumentation methods we study. Section 3 shows how we modify a symbolic execution engine to take as input a partial branch recorded at the user site to reproduce a bug. Section 4 describes some implementation details and our experimental methodology. Section 5 presents the results for an open-source Web server, diff, four coreutils programs, and two microbenchmarks. Section 6 discusses the results and our future work. Section 7 summarizes the related work. Finally, Section 8 concludes the paper.

2. Program Analysis and Instrumentation

Our approach involves analyzing the program to find the symbolic branches and instrument them for logging. We study both dynamic and static analyses. Our instrumentation may use the results of the dynamic analysis only, those of the static analysis only, or combine the results of both analyses. Next, we describe our analyses and instrumentation methods.

2.1 Dynamic Analysis

Our dynamic analysis is based on symbolic execution. We mark input to the program as symbolic and then use symbolic execution to determine whether or not a branch condition depends on input.

Symbolic execution repeatedly and systematically explores all paths (and thus branches) in a program. In the particular form of symbolic execution used in this paper (sometimes called concolic execution [Sen 2005]), the symbolic engine executes a number of runs of the program, each with a different concrete input. Initially, it starts with a randomly chosen set of values for the input and proceeds down the execution path of the program. At each symbolic branch, it computes the branch condition for the direction of the branch followed, and adds this condition to the constraint set for this run. When the run terminates, the overall constraint set describes the set of branch conditions that have to be true for the program to follow the path that occurred in this particular run. One of the conditions is then negated, the constraint set is solved to obtain a new input, and a new run is started.

We initially mark $argv$ as symbolic, as well as the return values of any functions that return input. During symbolic execution, we track which variables depend on input variables, and mark those as symbolic as well. When we execute a branch for the first time, we label it symbolic if any of the variables on which the branch condition depends is symbolic, and concrete otherwise. If during further symbolic execution we revisit a branch labeled symbolic, it stays that way. If, however, we revisit a branch labeled concrete, and now its branch condition depends on at least one symbolic variable, we relabel that branch as symbolic.

Symbolic execution tries to explore all program paths, and is therefore very time-consuming. If it is able to explore all paths, then all branches are visited, with some marked symbolic and some marked concrete. However, this usually can only be done for very small programs. For others, it is necessary to cut off symbolic execution after some amount of time. As a result, at the end of the analysis, in addition to branches labeled symbolic and concrete, some branches remain unlabeled.

All branches marked symbolic are indeed symbolic, but some branches marked concrete may actually be symbolic, because symbolic execution was terminated before the branch was visited with a condition depending on input. The unvisited branches may be either symbolic or concrete.

2.2 Static Analysis

We use interprocedural, path-sensitive static analysis, in which we use a combination of dataflow and points-to analysis. The basic idea of the algorithm is to identify the sources of input (typically I/O functions or arguments to the program), and construct a list of variables whose values depend on input and are thus symbolic. Symbolic branches are then identified as the branches whose condition depends on at least one symbolic variable.

The algorithm works by maintaining a queue of functions to analyze. Initially, the queue only contains the $main$ function. New entries are queued in as function calls are discovered. The set of symbolic variables is initialized to $argv$. In the initialization, the functions that are normally used to read input are marked as returning symbolic values.

Algorithms 1 and 2 show a simplified version of the algorithm that analyzes each function. Each instruction in the program is visited, and the $doInst$ method is called. The dataflow algorithm takes care of loops by using a fixed-point algorithm so that instructions in a loop body are revisited (and $doInst$ is called) only as long as the algorithm output changes.

For an assignment instruction (i.e., an instruction of the form $variable = expression;$), $doInstr$ resolves the variables to which the expression may be pointing, and checks whether any of these is already known to be symbolic. If this is the case, it adds $variable$ to the list of symbolic variables, otherwise it continues. If the instruction is a function call, the algorithm first checks whether the function has already been analyzed or not. If it has, it looks up the results to determine whether with the current set of parameters the function can propagate symbolic memory or not, and updates the list of symbolic variables accordingly. If the function has not yet been visited [1], the algorithm enqueues it for analysis with a reference to the current instruction, so that analysis of the current function can resume when the function has been visited.

The $doStatment$ method in algorithm 2 is called by the dataflow framework on each control flow statement. For if statements, it resolves the list of variables to which the

---

[1] More precisely, if the function has not been visited with the particular combination of symbolic and concrete parameters encountered here.

Algorithm 1: Static analysis algorithm propagating symbolic information (simplified)

```
    /* called on each instruction in the program
       as many times as required by the fixed
       point dataflow algorithm            */
1   method doInstr(instruction i):;
2   begin
3      match i with begin
          /* assignment                     */
4         case target_variable = expression;
5         begin
             /* If any of the variables
                referenced by expression symbolic
                is symbolic mark targe_variable
                symbolic                    */
6            if isSymbolic(e) then
7               makeSymbolic(target_variable);
8            end
9         end
          /* function call                  */
10        case: target_variable = fun_name(parameters);
11        begin
12           symbolic_params =
                getSymbolicParameters(parameters);
             /* If we already visited fun_name
                with this combination of symbolic
                parameters, propagate symbolic
                flag                        */
13           if alreadyVisited(fun_name,
                symbolic_params) then
14              if returnsSymbolicMemory(fun_name)
                then
15                 makeSymbolic(target_variable);
16              end
17           end
18           else
                /* fun_name not visited yet.
                   Queue it and stop analysis of
                   current function. The
                   algorithm will return to this
                   location once fun_name has
                   been visited               */
19              queueFunction(fun_name,
                   symbolic_parameters, i);
20              return abort
21           end
22        end
23     end
24     return continue
25  end
```

condition expression may be pointing. If any of them is symbolic, the branch is labeled symbolic.

While the algorithm in Figure 1 is simplified for the sake of clarity, our actual implementation handles the fact (1) that symbolic variables can be propagated to global variables; (2)

Algorithm 2: Static analysis algorithm identifying symbolic branches (simplified)

```
    /* called on each control flow statement of
       the program.                             */
1   method doStatement(statement s): begin
2   |   match s with;
3   |   begin
4   |   |   Branch(condition_expression)  begin
5   |   |   |   if isSymbolic(condition_expression) then
6   |   |   |   |   logThisBranch();
7   |   |   |   end
8   |   |   end
9   |   end
10  |   return continue;
11  end
```

that the state of global variables changes depending on the path that is being analyzed; and (3) that functions may propagate symbolic variables not only to their return variables, but also to their parameters (passed by reference) or to global variables.

Static analysis is imprecise because the points-to analysis tends to over-estimate the set of aliases to which a variable may point. As a result, all symbolic branches in the program are labeled symbolic by the static analysis, but some concrete branches may also be labeled symbolic. All branches labeled concrete are indeed concrete.

2.3 Program Instrumentation

The developer instruments the branches in the program before the code is shipped. We consider four methods for instrumentation:

- *dynamic* instruments branches according to dynamic analysis.
- *static* instruments branches according to static analysis.
- *dynamic+static* instruments branches according to a combination of dynamic and static analysis.
- *all branches* instruments all branches.

Regardless of which method is used, the list of instrumented branches is retained by the developer, because it is needed to reproduce the bug (Section 3).

Dynamic method. After dynamic analysis, branches are labeled symbolic or concrete, or remain unlabeled. The *dynamic* method only instruments the branches labeled as symbolic. By the nature of the dynamic analysis, we are certain that these branches are symbolic. We do not instrument the branches labeled as concrete, since application branches are typically either always symbolic or always concrete. We also do not instrument the unlabeled branches. The *dynamic* method thus potentially underestimates the number of branches that need to be instrumented. In essence, *dynamic* favors reducing instrumentation overhead at the expense of increased bug reproduction time.

Static method. After static analysis, branches are labeled symbolic or concrete. We instrument the branches labeled as symbolic. By the nature of static analysis, the *static* method guarantees that all symbolic branches are instrumented, but it may instrument a number of concrete branches as well. *Static* therefore favors bug reproduction time at the expense of increased instrumentation overhead.

Dynamic+static method. In the combined method, we run both the static and the dynamic analysis, the latter for a limited time. The dynamic analysis labels branches as symbolic or concrete, or they may remain unlabeled. The static analysis labels branches as symbolic or concrete. The combined method instruments the branches (1) that are labeled symbolic by the dynamic analysis, and (2) that are labeled symbolic by the static analysis, with the exception of those labeled concrete by the dynamic analysis. In other words, when a branch is not visited by dynamic analysis, we instrument it based on the outcome of the static analysis, because this is the only information about this branch. When a branch is visited by dynamic analysis, we instrument it based on the outcome of this analysis. For branches labeled symbolic by dynamic analysis, this is obvious as they are guaranteed to be symbolic and have been labeled symbolic by static analysis as well. For branches labeled concrete by dynamic analysis, this means that we potentially override the outcome of static analysis which may have labeled these branches symbolic. The reasons for this decision are that (1) static analysis may conservatively label concrete branches symbolic, due to an imprecise points-to analysis, and (2) application branches are typically always concrete or always symbolic, as mentioned above.

Dynamic+static may be imprecise in two ways. Symbolic branches may or may not be instrumented. The latter case occurs if they are left concrete by the symbolic execution (for instance, due to the limited coverage of the symbolic execution). Concrete branches may or may not be instrumented. The former case occurs when the static analysis mistakenly labels them as symbolic and they are not visited during symbolic execution.

Although seemingly suffering from a greater degree of imprecision than the other methods, our evaluation shows that this method actually leads to the best tradeoff between instrumentation overhead and time necessary to reproduce the bug.

Logging system calls. In addition to deciding which branches to instrument, we also consider the choice of whether or not to log the results of certain system calls. Doing so adds to the runtime overhead, but can be very beneficial for system calls that can produce a large number of possible outcomes during replay. For example, consider a *select()* system call for reading from any of N file descriptors. Without information about which descriptors became ready and when, symbolic execution during replay would have to explore all combinations of N available descriptors upon each return from *select()*. To avoid having to explore all possible combinations, we instrument the code to log the descriptors that are available when a call to *select()* returns. During replay, we simply re-create these conditions. For the same reasons, it makes sense to instrument calls to *read* to log the number of bytes read.

We log the results of all system calls for which logging considerably simplifies replay, including *select()* and *read()*. The input data itself is never logged. In principle, all instrumentation methods can be combined with logging system calls.

3. Reproducing a Bug

3.1 Replay Algorithm

We use a modified symbolic execution engine [Crameri 2009] to reproduce bugs. The following information is available to the engine prior to bug reproduction: a list of all instrumented branches (saved when the program was instrumented – see Section 2.3), and the bitvector indicating which way the instrumented branches were taken (one bit for each instrumented branch taken during execution at the user site). When the symbolic execution engine encounters a branch, it immediately knows whether or not the branch is symbolic, because it can check whether the branch condition depends on the input.

We refer to a run of the symbolic execution engine as the execution of the engine with a single set of inputs, until it either finds the path to the bug or aborts. A run is aborted when the engine discovers that it is on a path that deviates from the path described by the received bitvector. Each run is started with the bitvector as received from the user site. A constraint set is associated with each run, describing the path followed by the run through the program, and consisting of the conjunction of the conditions for the branch directions taken so far in the run. To later explore alternative paths should the current run abort, the engine also maintains a list of pending constraint sets, describing these alternative unexplored paths.

The engine performs a number of these runs. The initial run is done with random inputs. Subsequent runs use an input resulting from the solution of a constraint set by the constraint solver.

For example, in Listing 1, the set of constraints: {*not (option == 'a'); option == 'b'*} needs to be solved to take the program in the `else if` branch at line 6.

When visiting the `else if` branch at line 6, the engine puts in the pending list the following constraint set: {*not (option == 'a'); not (option == 'b')*}.

During a run, the engine proceeds normally for instructions other than branches. For branches, because of the imprecision of the decision of which branches to instrument, the following four cases have to be distinguished.

1. **The branch is symbolic and not instrumented.** The constraint for the particular direction of the branch taken is added to the constraint set for the run, and symbolic execution proceeds. In addition, a new constraint set is formed by adding the negated constraint to the constraint set for the run. The new constraint set is put on the list of pending constraint sets. The bitvector is left untouched.

2. **The branch is symbolic and instrumented:** The engine takes the next bit out of the bitvector, and compares the direction that was taken during recorded execution to the direction the symbolic execution would take with its input.

   (a) If the two are the same, the constraint is added to the constraint set for the run, and the symbolic execution proceeds.

   (b) If not, the constraint implied by the direction of the branch taken during recorded execution is negated and added to the constraint set for the run. This constraint set is added to the list of pending constraint sets, and the run aborts.

3. **The branch is concrete and instrumented:** The engine takes the next bit out of the bitvector, and compares if the direction that was taken during recorded execution is the same as the direction the symbolic execution would take with the current input.

   (a) If yes, it proceeds further.

   (b) If not, this run of the symbolic execution is aborted.

   The latter case can only occur as a result of the run earlier having taken the wrong direction on a branch that (because of insufficient instrumentation) was not instrumented but should have been.

4. **The branch is concrete and not instrumented:** The engine proceeds. The bitvector is left untouched.

When a run is aborted, the engine looks at the pending list of constraint sets, picks one, solves it, and starts another run with the input resulting from the solver. When this new run passes the branch at which it was produced, the new input, by construction, causes the symbolic execution to take the direction opposite from the one followed in the run during which the constraint set was produced.

3.2 Replay Under Different Instrumentation Methods

The *all branches* instrumentation method instruments all symbolic branches (and all concrete ones as well). Thus, cases 1 and 4 above cannot occur with this method. Furthermore, case 3(b) cannot occur either. The reason is that the engine always proceeds past a symbolic branch in the same direction as recorded during execution. When the run hits a concrete branch, since this branch does not depend on input, the engine is bound to follow the correct direction.

The *static* method also instruments all symbolic branches, since imprecision in the points-to analysis is resolved conservatively (if a pointer might depend on input, it is flagged symbolic). Under this method, cases 1 and 3(b) cannot occur. The latter case cannot occur for the same reason above.

The *dynamic* method may not instrument all symbolic branches, because it may not run long enough to find them. Similarly, the *dynamic+static* method may fail to instrument all symbolic branches, but only when symbolic execution does not run long enough *and* static analysis is inaccurate. In these cases, a run can take the wrong direction at a symbolic branch that was not instrumented, and later hit a concrete branch for which the input fails to satisfy the branch condition. In this case, the engine needs to back up and explore an alternative direction at a symbolic but uninstrumented branch. The constraint sets for these alternative directions reside on the pending list. The search can use any heuristic for deciding which constraint set to pick from the pending list. We currently use a simple depth-first approach.

3.3 Replaying System Calls

As mentioned in Section 2.3, we consider scenarios with and without instrumentation for logging the results of certain system calls.

When these system calls are not logged, we replay them using models of their behavior. The models use symbolic variables to allow the engine to reproduce any behavior that may be produced by the kernel. For instance, for the *read()* system call, we use a symbolic variable for the return value that determines how much input is read. This symbolic variable is constrained to be between -1 and the amount of input requested. During replay, a program executing the *read()* call initially returns the amount of (symbolic) input requested, and execution carries on. Because the return value of the system call is symbolic, if the program checks it in a branch and if the branch has been logged, the log specifies which direction needs to be taken. If that direction fails, the symbolic execution engine aborts the run. Eventually, the number of bytes actually read at the user's site is found.

When these system calls are logged, we replay their execution based on the logs. During replay, the calls for which there are logged results always return exactly the recorded value. Thus, the symbolic execution engine need not search for the actual call results.

4. Implementation and Methodology

Software. For program instrumentation and analysis, we use CIL (C Intermediate Language [Necula 2002]), which is a collection of tools that allow simple analysis and modification of C code.

For static analysis we start by merging all the source code files of the program in one large C file. This allows us to run the analysis on the whole program, making the results more accurate.

We then use two CIL plug-ins for the dataflow and points-to analysis.

For symbolic execution, we use a home-grown concolic execution engine for C programs [Crameri 2009]. The engine instruments the C program and links it with a runtime library for logging constraints.

We also use CIL to instrument the branches in the program. The instrumentation simply uses a bit per branch in a large buffer, and flushes the buffer to disk when it is full. We use a buffer of 4KB in order to avoid writing to disk too often. We do not use any form of online compression, as this would impose additional CPU overhead. Moreover, we could have used a simple branch prediction algorithm to avoid logging all instrumented branches, but this would have required recording the program location for each logged branch. This approach would have required at least another 32 bits of storage per branch, probably ruining any savings obtained by the prediction algorithm.

Benchmarks. We first evaluate the instrumentation overhead in isolation using two microbenchmarks. Next, we reproduce real bugs previously reported in the coreutils programs [Cadar 2008]. Then, we evaluate the tradeoff between instrumentation overhead and bug reproduction time using an open-source Web server, the *uServer* [Pariag 2007]. The *uServer* was designed for conducting performance experiments related to Internet service and operating system design and implementation. We use version 0.6.0, which has approximately 32K lines of code. In our final experiment, we again evaluate the tradeoff between instrumentation overhead and bug reproduction, but this time with the *diff* utility. *Diff* is an input-intensive application that pinpoints the differences between two files provided as input. It contains about 22K lines of code. For all experiments we link the programs with the uClibc library [uClibc].

We study five configurations of each benchmark: four resulting from our four instrumentation methods plus a configuration called *none*, which involves no instrumentation. For the instrumented benchmarks, unless mentioned otherwise, selective system call logging is turned on.

Hardware. Our experimental setup consists of two machines with two quad-core 2.27-GHz Xeon CPUs, 24 GBytes of memory, and a 160-GByte 7200rpm hard disk drive.

5. Evaluation

5.1 Microbenchmarks

We evaluate the cost of the instrumentation by using two simple microbenchmarks. The first microbenchmark comprises a loop that increments a counter 1 billion times. The loop condition (checking the loop bound) consists of a single branch, executing as many times as there are iterations.

We compare the *none* (no instrumentation) and *all branches* versions of this microbenchmark using the Linux *perf* profiler. The results show that the branch logging instrumentation takes 17 instructions on average, including the periodic flushing of the log to disk. In terms of running time, we see an average cost of 3 nanoseconds per instrumented branch (with an average of 2.1 instructions per cycle), for a total overhead of 107%. While considerable, this overhead is still lower than that reported in ODR [Altekar 2009] (about 200% just for the branch recording). The most likely reason is that our instrumentation only logs one bit per branch, while ODR uses a more complicated branch prediction algorithm.

We run a second microbenchmark consisting of the example in Listing 1, which computes the fibonacci sequence for one of two numbers. We instrument this program using the four configurations of our system. Not surprisingly, the configurations other than *all branches* instrument only the symbolic branches corresponding to lines 4 and 6. The results are consistent with our previous microbenchmark: an average overhead of 17 instructions per instrumented branch or about 3 nanoseconds. The *all branches* configuration suffers from a total overhead of 110%, whereas the three others do not incur any noticeable overhead because only two branches are logged.

5.2 Coreutils

We now evaluate our approaches using four real bugs in different programs from the Unix coreutils set: *mkdir*, *mknod*, *mkfifo*, and *paste*. We ran the programs with up to 10 arguments, each 100 bytes long.

Branch behavior. Recall that our approach is based on two assumptions about branches: (1) that there are many (concrete) branches whose outcomes do not depend on input, and (2) that if a branch is executed once with a concrete (symbolic) branch condition, it is most likely always executed with a concrete (symbolic) condition.

To check these assumptions, we modify our symbolic execution engine to run with concrete inputs (instead of generating concrete inputs itself to explore different paths). In addition, at each executed branch, we record whether it is executed with a symbolic or a concrete branch condition.

We show the results of a sample run of *mkdir* in Figure 1; the graphs for the other 3 programs are similar. The figure shows per-branch-location statistics for this experiment. We use the term "branch location" to mean the location of a branch in the source code, and "branch execution" to mean the actual execution of a branch location during run time. Each point on the x axis denotes a branch location that is executed at least once. The y axis shows how many times each branch is executed. The gray bars denote the overall number of branch executions, whereas the black bars denote the number of executions with a symbolic condition. The black bars are therefore a subset of the gray bars.

As the figure illustrates, only a limited number of branch locations is responsible for all the symbolic branch executions. Furthermore, where black bars are present, they completely cover the gray bars. This shows that a particular branch is executed either always with concrete conditions or always with symbolic conditions. These observations support our two assumptions.

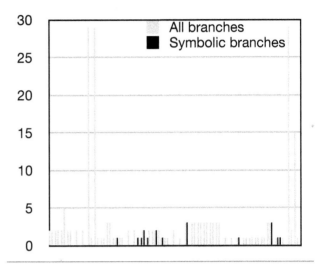

Figure 1. Number of executions of each branch in a sample run of *mkdir*. The overlaid black bars represent the branches executed with symbolic conditions.

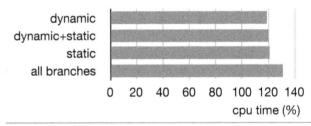

Figure 2. CPU time of *mkdir* instrumented with the four configurations of our system. Results are normalized to the non-instrumented version.

Instrumentation overhead. Figure 2 shows the CPU time associated with the instrumentation for *mkdir* (again, the results for the other programs are similar). Those results are obtained by running the program in the symbolic execution engine for one hour. The figure shows that the time is almost identical for the *dynamic*, *dynamic+static*, and *static* configurations. The static and dynamic analyses produce accurate results in those programs. The *all branches* configuration is the slowest, with an overhead of 31%.

Reproducing bugs. Each program suffers a crash bug that only manifest itself when a very specific combination of arguments is used. For instance, the bug in *paste* occurs with the following command line:

```
paste -d\\ abcdefghijklmnopqrstuvwxyz
```

Table 1 shows the time needed to reproduce the crash bugs in the four programs. The programs being relatively small, symbolic execution is able to cover most of the important branches in a very short amount of time, and static analysis produces accurate results. Thus, we can reproduce the bug in less than two seconds in all of the four instrumented configurations.

Program	Replay time
mkdir	1 sec
mknod	1 sec
mkfifo	1 sec
paste	1.5 sec

Table 1. Time needed to replay a real bug in four coreutils programs. The results are the same with all four configurations of our system.

These bugs have also been used to evaluate ESD [Zamfir 2010]. Interestingly, ESD took significantly more time to reproduce the bugs (between 10 and 15 seconds, albeit with no runtime overhead). This can be attributed to the fact that our system essentially knows the exact path to the bug (during bug reproduction), whereas ESD needs to search many paths.

5.3 uServer

We further evaluate our system using a much larger application (32K lines of code), the *uServer* [Pariag 2007], an open-source Web server sometimes used for performance studies. Unlike the coreutils programs, which are very small, this benchmark elucidates the differences between the approaches for deciding which branches to instrument, and the tradeoff between instrumentation overhead and bug reproduction time.

Branch behavior. We run the *uServer* in our modified symbolic execution environment with 5,000 HTTP requests to demonstrate the nature of the branches (symbolic or concrete). In total, approximately 18 million branches are executed, out of which only 1.8 million or roughly 10% are symbolic. These 1.8 million symbolic branches correspond to multiple executions of the same 53 branch locations in the program.

Figure 3 shows per-branch-location statistics for this experiment. As we can see, most of the black bars entirely cover their corresponding gray bars. This means that these particular branch locations are executed either always with concrete conditions, or always with symbolic conditions. However, the situation is slightly different for the branches in uClibc, where in some cases the black bars almost but not completely cover the gray bars. This situation corresponds to library functions that are sometimes called with concrete values. In this experiment, the number of those cases was very small.

The figure also shows that most branches are executed in the library (81%). However, only 28% of the symbolic branches are executed in the library.

Version	# of instrumented branch locations	
	LC	HC
dynamic	78	246
dynamic + static	1654	1490
static	2104	
all branches	5104	

Table 2. Number of branch locations instrumented in the *uServer* with the different configurations of our system.

Identifying symbolic branch locations. In total, there are 5104 branch locations in the *uServer* code (and 8516 in uClibc).

Table 2 shows the number of instrumented branches in the *uServer* for each configuration of our system. We symbolically execute the *uServer* using 200 bytes of symbolic memory for each accepted connection, and for each file descriptor. We stop the symbolic execution phase after one hour and two hours, obtaining a coverage of 20% (denoted LC for lower coverage) and 33% (HC for higher coverage), respectively. Running longer does not significantly improve branch coverage.

Out of a total of 5104 branches, *static* marks 2104 as symbolic, *dynamic* 246, and *dynamic+static* 1490, in the HC configuration. In addition, with *dynamic*, 1434 branches are marked as concrete, and the remaining branches are not visited.

For the static analysis tool, we need to merge all the source files of both the application and the library. Unfortunately, doing so resulted in a file too large for the points-to analysis to handle.[2] Therefore we perform static analysis only on the *uServer* application code. All branches in the library are treated as symbolic by the static analysis.

With less coverage, there are more branches left unvisited, and therefore more opportunity for the static phase to mark them symbolic. This is why there are fewer instrumented branches in the *dynamic* configuration and more in the *dynamic+static* configuration. Those numbers are used to highlight the effect of branch coverage in our approach.

Instrumentation overhead. We use the httperf [Mosberger 1998] benchmarking tool to compare the performance of the configurations for the *uServer*. We run httperf with a static workload saturating the CPU core on which the *uServer* is running. We consider three performance metrics: number of instrumented branches executed, CPU time, and storage requirement of the instrumentation. All three are roughly proportional to each other.

Figure 4(a) shows CPU time (relative to the non-instrumented version). As we can see, the overhead of *all branches* is significant. The results of *static* are only marginally better, since it instruments all branches in the uClibc library.

---

[2] After six hours, the analysis had made little progress and we aborted it.

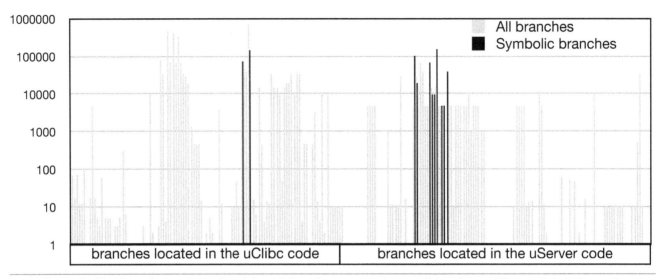

Figure 3. Number of executions of each branch location in a sample run of the *uServer*. The y axis is in log scale.

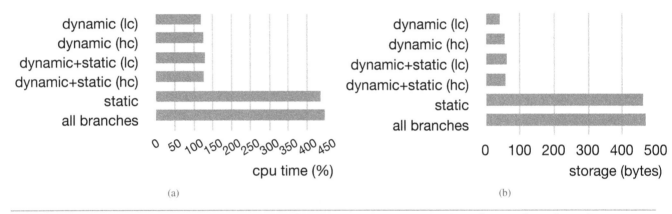

(a)   (b)

Figure 4. CPU time (a) and storage requirement (b) of the *uServer* instrumented with the four configurations of our system.

The two configurations using dynamic analysis perform notably better. The overhead is 17% and 20%, respectively, for the *dynamic* and *dynamic+static* configurations. This is not surprising for *dynamic*, as it instruments only 284 branch locations. The *dynamic+static* configuration instruments many more branch locations, but still far fewer than *all* or *static*.

The coverage obtained with symbolic execution affects the instrumentation overhead of *dynamic* and *dynamic+static*. Increased coverage increases the overhead of *dynamic*, since symbolic execution instruments more branches. In contrast, increased coverage leads to reduced instrumentation in *dynamic+static*, corresponding to branches marked symbolic by the static analysis, left unvisited by the dynamic analysis with low coverage, and marked concrete by the dynamic analysis with high coverage.

Figure 4(b) shows the storage requirements per HTTP request processed by the Web server. The storage overhead is reasonable; around 50 bytes per request in the *dynamic* and *dynamic+static* configurations. This is roughly the same number of bytes in a typical entry in the access log of the Web server.

Both the processing and storage overheads are more significant in the *static* and *all branches* configurations. These configurations represent worst-case scenarios for the areas of the code that are not covered by the symbolic execution.

When a bug occurs, the branch log has to be transferred to the developer. Compression can be used to reduce the transfer time. We observe a compression ratio of 10-20x using gzip. In Section 6, we discuss how to deal with long-running executions, which could generate extremely large branch logs.

Reproducing bugs. To evaluate the amount of effort required to reproduce a bug, we run the *uServer* with five different input scenarios. To demonstrate the impact of code coverage on our approach, we design those scenarios to hit different code areas of the HTTP parser. More specifically, we use HTTP queries of various lengths (between 5 to 400 bytes), with different HTTP methods (e.g., GET, POST) and parameters (e.g., Cookies, Content-Length). We crash the

Version	Exp. 1		Exp. 2		Exp. 3		Exp. 4		Exp. 5	
	LC	HC	LC	HC	LC	HC	LC	HC	LC	HC
dynamic	27s	27s	2877s	79s	∞	170s	∞	287s	∞	168s
dynamic+static	27s	27s	79s	79s	532s	170s	175s	175s	248s	168s
static	27s		79s		170s		175s		168s	
all branches	27s		79s		170s		175s		168s	

Table 3. Time in seconds needed to reproduce each of the five input scenarios to the *uServer* with the four instrumented configurations of our system. The infinity symbol means that the experiment did not terminate in one hour.

	Version	# of symbolic branch locations logged / corresponding # of executions		# of symbolic branch locations NOT logged / corresponding # of executions	
		LC	HC	LC	HC
Exp. 1	dynamic	18 / 112	18 / 112	0	0
	dynamic+static	18 / 112	18 / 112	0	0
	static	18 / 112		0	
	all branches	18 / 112		0	
Exp. 2	dynamic	25 / 129913	39 / 2215	11 / 23062	0
	dynamic+static	36 / 2105	39 / 2215	3 / 110	0
	static	39 / 2215		0	
	all branches	39 / 2215		0	
Exp. 3	dynamic	25 / 554617	42 / 28848	17 / 48485	0
	dynamic+static	45 / 10971	42 / 28848	3 / 1023	0
	static	42 / 28848		0	
	all branches	42 / 28848		0	
Exp. 4	dynamic	24 / 236608	43 / 24012	21 / 185945	6 / 268
	dynamic+static	46 / 11089	48 / 12111	3 / 1023	1 / 1
	static	49 / 12112		0	
	all branches	49 / 12112		0	
Exp. 5	dynamic	25 / 410723	44 / 29136	15 / 45706	0
	dynamic+static	46 / 54539	44 / 28785	3 / 3391	0
	static	44 / 28785		0	
	all branches	44 / 28785		0	

Table 4. Number of symbolic branch locations and symbolic branch executions logged and not logged for each configuration of each experiment of Table 3.

server by sending it a SEGFAULT signal after sending it the input, making sure it crashes at the same location in the code for all four versions. After replay, we verify that each configuration produced input that correctly leads to the same location.

Table 3 shows the bug reproduction times for all four instrumented versions of the five scenarios. Table 4 shows the corresponding number of symbolic branch locations logged and not logged, as well as the number of actual symbolic branch executions. Both tables include the configuration with low coverage (LC) and high coverage (HC).

Unsurprisingly, the *all branches* and *static* versions, which instrument all symbolic branches, perform best. Of course, these versions do so at high runtime and storage overheads (Figure 4).

Dynamic+static in most cases performs only slightly worse than *static*, despite the much lower instrumentation overhead of the former configuration. *Dynamic* comes last, with many LC experiments not finishing in one hour. This is not surprising, as the number of branches identified as symbolic, and therefore the amount of logging, is very low. In fact, Tables 3 and 4 show that the number of symbolic branch locations not logged is well correlated with the replay time. As soon as replay encounters more than a dozen symbolic branch locations that are not instrumented, the replay time exceeds one hour. An approach that does not instrument the code at all, would result in even longer bug reproduction times.

Dynamic+static obtains similar results regardless of coverage. The reason is that symbolic execution may incorrectly

Version	Exp. 1		Exp. 4	
	LC	HC	LC	HC
dynamic	112s	112s	∞	712s
dynamic+static	112s	112s	991s	694s
static	87s		362s	
all branches	56s		343s	

Table 5. Time in seconds needed to reproduce two input scenarios with the *uServer* when not logging system call results. The infinity symbol means that the experiment did not terminate in one hour.

classify some branches as concrete, and thus slow down the search. When running longer, those branches may later be marked symbolic, therefore correcting the error. In our experiment, those differences have a marginal effect on *dynamic+static*, which again suggests that this does not happen frequently.

Impact of logging system calls.

By default, we log the results for some key system calls (Section 2.3). For instance, we log the return value of the *read()* system call and the order of ready file descriptors from *select()* calls.

The measurements in Figure 4 include the overhead of logging these return values. As we only log a limited number of values for a few system calls, logging these values introduces little extra work compared to the logging of the branches. As a result, when not logging system call results, the overhead is reduced by a marginal 0.2%.

Tables 5 and 8 present the bug reproduction times of two of our experiments without logging any system calls (we omit the three other experiments for brevity; their results are similar). All configurations of our system take significantly longer to replay, as the symbolic execution engine needs to determine the exact return values of the selected system calls. The *dynamic* and *dynamic+static* configurations are further penalized, since the back-tracking needed by the unlogged symbolic branches compounds the search for the return values. Interestingly, the *static* configuration performs slightly slower than *all branches*, whereas when logging system calls it performs identically. The reason is that fewer concrete branches are logged, therefore the engine takes slightly more time to realize a wrong turn due to a system call.

5.4 Diff

We now consider the *diff* utility. *Diff* is more challenging than the *uServer* for our dynamic analysis, as its behavior depends more heavily on input. Moreover, *diff* generates very long constraint sets, placing a heavy burden on the constraint solver. For these reasons, dynamic analysis attains a coverage of only 20% of branches, during 1 hour of symbolic execution. In total, there are 8840 branches in the program. *dy-*

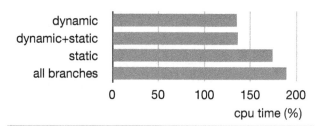

Figure 5. CPU time of *diff* instrumented with the four configurations of our system. Results are normalized to the non-instrumented version.

Version	Exp. 1	Exp. 2
dynamic	∞	∞
dynamic + static	1s	12s
static	1s	12s
all branches	1s	12s

Table 6. Time in seconds needed to reproduce two input scenarios to *diff*. The infinity symbol means that the experiment did not terminate in one hour.

	Version	symbolic branch locations logged / corresponding executions	symbolic branch locations NOT logged / corresponding executions
Exp. 1	dynamic	3 / 2125686	32 / 2369765
	dyn.+static	13 / 904	0
	static	13 / 904	0
	all branches	13 / 904	0
Exp. 2	dynamic	3 / 2478280	24 / 2102506
	dyn.+static	21 / 54623	0
	static	21 / 54623	0
	all branches	21 / 54623	0

Table 7. Number of symbolic branch locations and symbolic branch executions logged and not logged for two input scenarios to *diff*.

namic identifies 440 of them as being symbolic, *static* 4292, and *dynamic+static* 3432.

Instrumentation overheads. We run *diff* on two sample text files. Figure 5 shows the CPU time of the four configurations of our system, normalized against the non-instrumented execution. Consistent with our prior results, *dynamic* and *dynamic+static* perform best with an overhead of approximately 35%.

Reproducing bugs. We replay two executions of *diff* comparing relatively small but different text files. Tables 6 and 7 list the results of these experiments. Because the coverage obtained during dynamic analysis is relatively low, the *dynamic* configuration is unable to finish the experiments in 1

Version	# of symbolic branch locations logged / corresponding # of executions		# of symbolic branch locations NOT logged / corresponding # of executions	
	LC	HC	LC	HC
Exp. 1 — dynamic	40 / 722	43 / 725	8 / 173	5 / 170
Exp. 1 — dynamic+static	40 / 722	43 / 725	8 / 173	5 / 170
Exp. 1 — static	49 / 871		0	
Exp. 1 — all branches	49 / 465		0	
Exp. 4 — dynamic	43 / 245017	65031 (61)	21 / 107650	7 /1950
Exp. 4 — dynamic+static	64 / 88739	64612 (67)	3 / 7397	1 / 1021
Exp. 4 — static	50 / 39581		0	
Exp. 4 — all branches	50 / 37346		0	

Table 8. Number of symbolic branch locations and symbolic branch executions logged and not logged for two input scenarios with the *uServer* when not logging system call results.

hour. The few tens of unlogged symbolic branch locations quickly create a very large number of paths to explore, making it impossible for this approach to finish within the allotted time. In contrast, the three other configurations, and in particular *dynamic+static*, do not suffer from any unlogged symbolic branches and therefore replay quickly.

Collectively, these results again demonstrate that *dynamic+static* strikes the best balance between instrumentation overhead and bug reproduction time.

6. Discussion and Future Work

Branch coverage. Our current results suggest that using a branch trace, even partial, is effective at reducing the amount of searching needed to reproduce a specific buggy execution path. To maintain low instrumentation overhead, it is necessary to obtain sufficient coverage with the initial symbolic execution phase. This problem has received attention in the literature ([Cadar 2008; Godefroid 2008]), and significant progress has been made in recent years. While it is not always possible to achieve 100% coverage, this is a typical goal when testing an application *prior to shipping*. Therefore, the testing effort can be leveraged to identify the symbolic branches at the same time. Moreover, manual test cases can be used in conjunction with symbolic execution to boost code coverage, and many applications already have test suites covering most of their codes.

Constraint solving. Symbolic execution is limited by the ability to solve the resulting constraints. In particular, certain types of programs generate constraints that current state of the art solvers cannot solve. Constraint solving is an active research topic and our approach should directly benefit from any advances in this field.

Non-determinism. Two approaches exist for dealing with non-determinism, resulting, for instance, from system calls or random number generators. One can either log the outcome of the non-deterministic event or one can treat it as (symbolic) input during replay. In this paper, we strike a middle ground between these approaches, logging the out-

come of non-deterministic events that are likely to cause a great deal of search during replay. The results in Section 5 validate this approach, but a more comprehensive treatment of non-deterministic events could be explored.

Multithreading. We can extend our system to support multi-threaded applications by modifying it in two ways. First, the branch trace needs to be split into multiple traces, one per thread. Second, the ordering of thread execution needs to be recorded as well. Implementing the first modification is trivial, and is unlikely to impose any significant additional overhead. The second modification is more difficult. Others [Altekar 2009; Zamfir 2010] have experimented with ideas for recreating a suitable thread scheduling to find race conditions. Our approach of logging a partial trace of branches is complementary to those efforts and could considerably speed up the replay of multi-threaded programs with races.

Long-running applications. The storage overhead and replay time of long-running applications can be problematic. Consider, for example, the case of a Web server running for weeks before crashing. Our current approach reduces storage overhead as much as possible, but with these applications this overhead may still be high. Furthermore, replaying such a long trace may be infeasible, as it will be longer than the original run. Pushing the concept of a partial branch trace further, our approach could be extended to simplify this problem by implementing support for checkpointing. An instrumented application could periodically take a checkpoint of its state and discard the current branch log. Logging branches would then continue from the checkpoint only. The checkpoint would include enough information on the data structures of the program (but not its content). With this information, a symbolic execution engine can treat their content as symbolic, and replay the branch log starting from there. We leave the implementation and the associated research questions of the checkpointing mechanism for future work.

Concolic vs. symbolic. The particular form of symbolic execution we use in this paper is called concolic execution [Sen

2005]. The main difference from pure symbolic execution (as in [Cadar 2008], for instance) is that the engine repeatedly executes the program from beginning to end with concrete inputs, instead of exploring multiple paths in parallel. This implementation difference has no fundamental impact on our system, as in both cases the engine can select the paths in the same order. On one hand, the fact that the application is rerun from the beginning for every path imposes some additional overhead. On the other hand, because concrete inputs are always used, it makes the work of the solver easier. In many instances, branch conditions are already satisfied by the random input chosen. To the best of our knowledge, no comparative studies of the impact of the different implementations have been published yet.

7. Related Work

At one end of the spectrum between instrumentation overhead and bug reproduction effort are record-replay systems that try to capture the interactions between the program and its environment. Different systems capture interactions at different levels. Most systems capture them at the system call or library level. For example, ReVirt [Dunlap 2002] logs interactions at the virtual machine level. All record-replay systems suffer from the overhead of logging the interactions. To reduce this overhead, R2 [Guo 2008] asks the developer to manually specify at what interfaces to capture the program's interactions with its environment.

At the other end of the spectrum are conventional bug reporting systems, which provide a coredump, but no indication of how the program arrived at the buggy state. Obviously, there is no recording overhead, but it takes considerable manual search to find out how the problem came about.

ESD (Execution Synthesis Debugger [Zamfir 2010]) is an attempt to automate some of that search. It tries to do so without recording any information about the program execution. Instead, it uses the stack trace at the time of the program crash, and symbolically executes the program to find the path to the bug location. Although it uses static analysis and other optimizations to reduce the number of paths it needs to explore, it remains fundamentally limited by the exponential path explosion of symbolic execution. Our approach instead performs logging of a set of judiciously chosen branches. This allows us to speed up the automated search with limited runtime overhead. As our results with the coreutils show, our methods reproduce bugs faster, albeit at some modest cost in runtime.

BBR (Better Bug Reporting, [Castro 2008]) investigates the same tradeoff, although more from the perspective of maintaining privacy when a bug is reported to the developer. During execution it logs the program's inputs. After a crash, it replays the entire execution on the user machine based on those inputs. Replay uses an instrumented version of the program that collects the constraints implied by the direction of the branches taken in the program. An input set that satis-

fies these constraints is then returned to the developer. Unlike BBR, we do not log the users' inputs, or require whole-program replay and the execution of a constraint solver on the user machine. Instead, we incur limited logging overhead at the user site and some exploration of alternative paths at the developer site.

Another approach for maintaining privacy is explored by the Panalyst system [Wang 2008]. After a runtime error, the user initially reports only public information. The developer tries to exploit this information using symbolic execution, but can query the user for additional information. The user can choose whether or not to respond to the developer queries. In the limit, Panalyst's effectiveness is constrained by the exponential cost of symbolic execution.

Triage [Tucek 2006] explores yet another way of debugging errors at user sites. It periodically checkpoints programs, and after a failure, it restarts the program from a checkpoint. Heavyweight instrumentation, exploration of alternatives (delta-debugging), and speculative execution may be used during replay. Some applications were successfully debugged using this approach, but the checkpoint may have to be far back in time to allow meaningful exploration.

The same tradeoff between instrumentation overhead and bug reproduction time has also been explored for debugging multithreaded programs. To faithfully replay the execution of a thread, the shared memory interactions with other threads need to be logged.

The cost of doing so is very high, and therefore the PRES [Park 2009] debugger selectively omits logging certain interactions, but requires multiple replay runs before it can recreate a path to a bug. Similarly, in order to avoid logging all shared memory accesses, ODR [Altekar 2009] allows some degree of inconsistency between the actual execution and the replay, provided that the inconsistencies do not affect the output of the program. The techniques used by PRES and ODR could be combined with partial logging of branches as presented in this paper.

Logging of branches has been used to report a program's behavior before a bug in Traceback [Ayers 2005]. The system uses this behavior to reconstruct the control flow leading to the problem. To reduce the instrumentation overhead, Traceback uses static analysis to minimize the number of instructions required to instrument branches, and only logs the most recent branches. Our system goes further by combining dynamic and static analyses to reduce the number of instrumented branches, and reproducing the entire path to the bug.

Symbolic execution for program testing was proposed as far back as 1976 [King 1976], but has recently received renewed attention as increased compute power and new optimizations offer opportunities to tackle the path explosion problem. Klee [Cadar 2008] has been used to symbolically execute the coreutils, and whitebox fuzzing [Godefroid 2008] has been used to allow programs with extensive input parsing to be executed symbolically. We present a new use for symbolic execution, namely to discover which branches in a program are symbolic. In addition, we propose to speed up symbolic execution for bug reproduction by using a symbolic branch log collected at the user site.

8. Conclusion

In this paper, we consider the problem of instrumenting programs to reproduce bugs effectively, while keeping user data private. In particular, we focus on the tradeoff between the instrumentation overhead experienced by the user and the time it takes the developer to reproduce a bug in the program.

We explore this tradeoff by studying approaches for selecting a partial set of branches to log during the program's execution in the field. Specifically, we propose to use static analysis (dataflow and points-to analysis) and/or dynamic analysis (time-constrained symbolic execution) to find the branches that depend on input. Our instrumentation methods log only those branches to limit the instrumentation overhead. When a user encounters a bug, the developer uses the partial branch log to drive a symbolic execution engine in efficiently reproducing the bug.

Our results show that the instrumentation method that combines static and dynamic analyses strikes the best compromise between instrumentation overhead and bug reproduction time. For the programs we consider, this combined method reduces the instrumentation overhead by up to 92% (compared to using static analysis only), while limiting the bug reproduction time (compared to using dynamic analysis only).

We conclude that our characterization of this tradeoff and our combined instrumentation method represent important steps in improving bug reporting and optimizing symbolic execution for bug reproduction.

References

[Altekar 2009] Gautam Altekar and Ion Stoica. ODR: Output-deterministic Replay for Multicore Debugging. In *SOSP '09: Proceedings of the ACM SIGOPS 22nd Symposium on Operating Systems Principles*, pages 193–206, 2009.

[Ayers 2005] Andrew Ayers et al. TraceBack: First Fault Diagnosis by Reconstruction of Distributed Control Flow. In *PLDI '05: Proceedings of the 2005 ACM SIGPLAN Conference on Programming language design and implementation*, pages 201–212, 2005.

[Cadar 2008] Cristian Cadar, Daniel Dunbar, and Dawson R. Engler. KLEE: Unassisted and Automatic Generation of High-Coverage Tests for Complex Systems Programs. In *Proceedings of the 8th USENIX Conference on Operating systems design and implementation*, pages 209–224, 2008.

[Castro 2008] Miguel Castro, Manuel Costa, and Jean-Philippe Martin. Better Bug Reporting with Better Privacy. In *ASPLOS XIII: Proceedings of the 13th International Conference on Architectural support for Programming Languages and Operating Systems*, pages 319–328, 2008.

[Crameri 2009] Olivier Crameri et al. Oasis: Concolic Execution Driven by Test Suites and Code Modifications. Technical report, EPFL, 2009.

[Dunlap 2002] George W. Dunlap et al. ReVirt: Enabling Intrusion Analysis Through Virtual-Machine Logging and Replay. *SIGOPS Oper. Syst. Rev.*, 36(SI):211–224, 2002.

[Glerum 2009] Kirk Glerum et al. Debugging in the (Very) Large: Ten Years of Implementation and Experience. In *SOSP '09: Proceedings of the ACM SIGOPS 22nd Symposium on Operating Systems Principles*, pages 103–116, 2009.

[Godefroid 2008] Patrice Godefroid, Adam Kiezun, and Michael Y. Levin. Grammar-based Whitebox Fuzzing. *SIGPLAN Not.*, 43: 206–215, June 2008.

[Guo 2008] Zhenyu Guo et al. R2: an Application-level Kernel for Record and Replay. In *OSDI'08: Proceedings of the 8th USENIX Conference on Operating Systems Design and Implementation*, pages 193–208, 2008.

[King 1976] James C. King. Symbolic Execution and Program Testing. *Commun. ACM*, 19(7):385–394, 1976.

[Mosberger 1998] D. Mosberger and T. Jin. httperf: A Tool for Measuring Web Server Performance. In *The First Workshop on Internet Server Performance*, pages 59—67, June 1998.

[Necula 2002] George C. Necula et al. CIL: Intermediate Language and Tools for Analysis and Transformation of C Programs. In *Proceedings of Conference on Compiler Construction*, 2002.

[Pariag 2007] David Pariag et al. Comparing the Performance of Web Server Architectures. In *EuroSys '07: Proceedings of the 2nd ACM SIGOPS/EuroSys European Conference on Computer Systems 2007*, pages 231–243, 2007.

[Park 2009] Soyeon Park et al. PRES: Probabilistic Replay with Execution Sketching on Multiprocessors. In *SOSP '09: Proceedings of the ACM SIGOPS 22nd Symposium on Operating systems principles*, pages 177–192, 2009.

[Sen 2005] Koushik Sen, Darko Marinov, and Gul Agha. CUTE: a Concolic Unit Testing Engine for C. In *ESEC/FSE-13: Proceedings of the 10th European Software Engineering Conference*, pages 263–272, 2005.

[Tucek 2006] Joseph Tucek et al. Automatic On-line Failure Diagnosis at the End-user Site. In *HOTDEP'06: Proceedings of the 2nd Conference on Hot Topics in System Dependability*, pages 4–4, 2006.

[uClibc] uClibc. The uClibc Library, a C Library for Linux. http://www.uclibc.org/.

[Wang 2008] Rui Wang, XiaoFeng Wang, and Zhuowei Li. Panalyst: Privacy-aware Remote Error Analysis on Commodity software. In *SS'08: Proceedings of the 17th Conference on Security Symposium*, pages 291–306, 2008.

[Zamfir 2010] Cristian Zamfir and George Candea. Execution Synthesis: a Technique for Automated Software Debugging. In *EuroSys '10: Proceedings of the 5th European Conference on Computer Systems*, pages 321–334, 2010.

Finding Complex Concurrency Bugs
in Large Multi-Threaded Applications

Pedro Fonseca, Cheng Li, and Rodrigo Rodrigues
Max Planck Institute for Software Systems (MPI-SWS)
firstname.lastname@mpi-sws.org

Abstract

Parallel software is increasingly necessary to take advantage of multi-core architectures, but it is also prone to concurrency bugs which are particularly hard to avoid, find, and fix, since their occurrence depends on specific thread interleavings. In this paper we propose a concurrency bug detector that automatically identifies when an execution of a program triggers a concurrency bug. Unlike previous concurrency bug detectors, we are able to find two particularly hard classes of bugs. The first are bugs that manifest themselves by subtle violation of application semantics, such as returning an incorrect result. The second are latent bugs, which silently corrupt internal data structures, and are especially hard to detect because when these bugs are triggered they do not become immediately visible. PIKE detects these concurrency bugs by checking both the output and the internal state of the application for linearizability at the level of user requests. This paper presents this technique for finding concurrency bugs, its application in the context of a testing tool that systematically searches for such problems, and our experience in applying our approach to MySQL, a large-scale complex multi-threaded application. We were able to find several concurrency bugs in a stable version of the application, including subtle violations of application semantics, latent bugs, and incorrect error replies.

Categories and Subject Descriptors D [2]: 5

General Terms Algorithms, Reliability

Keywords Concurrency bugs, Latent bugs, Linearizability, Semantic bugs

1. Introduction

As processors become more and more parallel, applications must become increasingly concurrent to take advantage of this additional processing capacity. However, concurrent systems are notoriously difficult to design, test, and debug, since they are prone to concurrency bugs whose occurrence depends on specific thread interleavings.

Recent advances in the area of testing concurrent systems have provided a series of new techniques to systematically explore different thread interleavings and maximize the chances of exposing concurrency bugs [Burckhardt 2010b, Musuvathi 2008]. However, these techniques assume that concurrency bugs manifest themselves either by causing the program to crash (e.g., due to an illegal memory access) or by triggering assertions written by the developer.

This assumption, however, prevents such tools from capturing some bugs, namely those that fall into the following two important classes. The first class are bugs that manifest themselves as any violation of the application semantics which is not caught by the assertions that the programmer wrote, such as returning an incorrect result to the user. The second class are latent bugs, which silently corrupt internal data structures, and only manifest themselves potentially much later when they are triggered by a subsequent input. These two classes constitute a non-trivial fraction of the concurrency bugs found in important concurrent applications [Fonseca 2010].

In this paper we propose a new technique for finding latent and/or semantic concurrency bugs. Our thesis is that it is possible to implicitly extract a specification, even for large multi-threaded server applications, by testing if the application obeys linearizable semantics [Herlihy 1990]. Intuitively, linearizability means that concurrent requests behave as if they were executed serially, in some order that is consistent with the real-time ordering of the invocations and replies to the requests. While similar ideas have been applied to the design of tools for testing concurrency bugs, they have been limited to testing the atomicity of small sections of the program or library functions with at most hundreds of lines of code [Burckhardt 2010a, Vafeiadis 2010, Xu 2005]. We push this idea to an extreme by postulating that even a complex multi-threaded server with hundreds of thousands of lines of code can come close to obeying linearizable semantics.

By systematically testing if linearizability is upheld, we can find subtle violations of the application semantics without having to write a specification for each concurrent application. Furthermore, by checking if both the output and the internal state of the application obey the inferred semantics, we can identify not only the bugs that manifest themselves immediately as a wrong output, but also those that silently corrupt internal state. However, achieving a meaningful state comparison requires abstracting away many of the low-level details of the state representation. We accomplish this by means of simple annotations that are provided by the tester. This approach also allows the tester to progressively increase the chances of finding latent concurrency bugs by incrementally annotating the state.

We implemented PIKE, a testing tool that brings together these principles and state of the art techniques for the systematic exploration of thread interleavings. We describe the design and implementation of PIKE, and our experience in applying it to MySQL, a multi-threaded database server with hundreds of thousands of lines of code, which represents a share of 40% of the database market [Oracle a].

Our experience demonstrates that, despite the size and complexity of MySQL, in practice the semantics it provides are sufficiently similar to linearizability for our detector to be effective. Although we used only a simple battery of inputs for testing (based on the testing inputs that shipped with the application) we were able to find a considerable number of concurrency bugs in a stable version of the database. Furthermore, the effort to provide the required annotations was small, and after installing simple filters we also found the number of false positives to be modest. All of this was achieved without having to figure out which were the correct outputs (or final states) for any given inputs, since PIKE automatically extracts a specification by comparing the outputs and states of different interleavings.

The remainder of the paper is organized as follows. Section 2 presents the problem and gives an overview of our approach. In Section 3 we introduce PIKE, the tool that we built to find concurrency bugs. Section 4 describes our experience applying PIKE to MySQL. Section 5 presents the results that we obtained from our experience. Finally, Section 6 reviews the existing related work and then we conclude in Section 7.

2. Overview

This section gives an overview of the specific classes of concurrency bugs we are targeting, and the main insights behind our solution.

2.1 Problem statement

The focus of this work is to improve testing techniques for detecting concurrency bugs. By concurrency bugs we mean any deviation from the intended behavior of the application that is not triggered deterministically: triggering concurrency bugs requires that the application is given not only a specific set of inputs that cause the unintended behavior, but also that the operating system executes the application with a specific schedule of thread interleavings. This non-determinism that is introduced by thread scheduling makes the application more error-prone, on the one hand, and these concurrency bugs harder to find, on the other.

In this paper we focus on two categories of concurrency bugs that are particularly hard to detect: semantic and latent bugs.

Semantic concurrency bugs manifest themselves by violations of the application semantics (e.g., by providing a wrong reply to clients). This category of bugs excludes those that crash the application or otherwise generate an exception, for example, by performing illegal memory accesses. Semantic bugs are hard to detect because it is hard to create a specification for reasonably complex applications, and some of the deviations from the intended behavior can be quite subtle. A partial form of specification that can be leveraged to capture some of these problems are assertions, but these only capture some deviations from the intended semantics and are therefore of limited use.

The second category of concurrency bugs that we focus on are *latent* concurrency bugs [Fonseca 2010]. A latent bug is one whose effects are exposed to clients at a significantly different point from the point where it is triggered (i.e., where the problematic thread interleaving leads to the application deviating from the internal behavior that was intended by the developer). Note that the two categories are not mutually exclusive, i.e., a bug can be latent and expose itself to clients by returning a wrong reply.

Because the main focus of our paper are multi-threaded server applications, we can be more specific about what it means for triggering and exposing the bug to occur at different points. In particular, server applications follow a pattern where they receive requests from clients, start the process of handling them, and conclude handling the requests when they issue the replies. Given this pattern, we define a concurrency bug as latent when the set of requests that trigger the bug does not include the subsequent request that causes the bug to be exposed to the clients.

The fact that latent bugs require an extra request to become visible means that testing tools can easily miss these bugs. In particular, this would happen if the set of inputs used by these tools would cause the state corruption to take place but not include the input that subsequently exposes the misbehavior. For example, in many applications, write operations typically only return to clients a status code specifying whether the write was successful or not. In this case, observing the application response might not be sufficient to infer whether the write operation was properly executed or not. Furthermore, as we will see in some of the cases we found

216

in a real application, it is often not simple to determine which subsequent inputs would be required to expose the incorrect internal state.

This supports the idea that, given the nature of latent concurrency bugs, detection tools should inspect the internal state of the application and not limit themselves to analyzing the outputs. However, it is hard to analyze the state of an application to check for its correctness, since it requires application-specific knowledge of what it means for the state to be incorrect.

2.2 Linearizability: The spec from within

To address the absence of a specification capturing these deviations from the intended application semantics we propose extracting such a specification from the behavior of the same application but under different conditions.

In particular, our thesis is that, even for a complex server application with hundreds of thousands of lines of code, the semantics that are intended by the programmer are normally close enough to linearizability [Herlihy 1990] that we can use it as a good first approximation of a specification.

To formally define linearizability, we must first define the notion of history, which is a finite sequence of events that can be either invocation of operations or responses to operations. A history is classified as sequential if its first event is an invocation, each invocation is immediately followed by a matching response, and each response is followed by an invocation. Two histories H and H' are defined as equivalent if, for every process P, the sequence of invocations and responses performed by P is the same, i.e., $H|P = H'|P$. A history H induces an irreflexive partial order $<_H$ on operations such that $o_0 <_H o_1$ if the response event for o_0 precedes the invocation of o_1. Given these definitions, a history of events in a concurrent system H is linearizable if there is an equivalent sequential history S (called a linearization of H) such that $<_H \subseteq <_S$.

Intuitively, this means that, despite its internal concurrency, the server behaves as if requests were processed in sequence, and that this processing took place instantaneously some time between the moment when the client invoked the request and received the respective reply.

Therefore, assuming the application tries to follow linearizable semantics, a testing methodology can be devised by comparing each concurrent execution of the application with all possible linearizations (i.e., all possible sequential executions of the requests) for the same input. If none of the linearizations matches the behavior of the concurrent execution, then a concurrency bug is suspected to have been triggered and an error is flagged.

Testing for linearizability would only require us to inspect the outputs of the concurrent execution against the outputs of the linearizations. This would be sufficient to capture semantic bugs, but not to capture latent bugs. To handle latent concurrency bugs, we can resort to the same principle of testing for linearizability but applying it to the state of

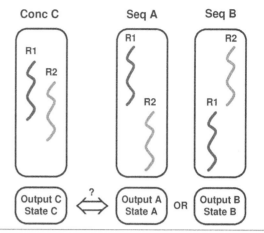

Figure 1. Checking for linearizability of state and outputs of two concurrent requests.

the application. This testing methodology is summarized in Figure 1. It shows two concurrent requests, R_1 and R_2, whose execution overlaps in time (Conc C). To check if the concurrent execution is linearizable we must compare the it to all possible linearizations, namely R_1 followed by R_2 (Seq A) and R_2 followed by R_1 (Seq B). Linearizability is obeyed when both the state and the output of the concurrent execution match both components in at least one of the two linearizations.

Note that by using linearizability as a specification, we are not necessarily extracting a correct specification of the system, not only because the programmer might not have intended the application to obey linearizable semantics, but also because the sequential execution may be buggy, and consequently the deviation to the expected behavior could go undetected. The latter issue is not problematic in the case of concurrency bugs, though, since these arise from the lack of proper synchronization among multiple threads, which does not arise when executing requests without concurrency.

2.3 Capturing application state

As we mentioned, to be able to find latent bugs we need to compare both the output and the state of different executions of the application.

While outputs are fairly straightforward to compare, the same cannot be said about the state of the application. In particular, the naïve approach of simply comparing the state of the various executions bit-by-bit is doomed to fail. The reason is that by changing thread interleavings, the low-level state of the executions will quickly diverge. For instance, if we consider operations such as dynamic memory allocation, slight changes in the thread interleaving could easily change the relative order of allocation requests, and therefore the memory layout of allocated heap space would likely be different as well.

We address this by asking the tester or the developer of the application to provide a *state summary function* which captures an abstract notion of the state in a way that takes into consideration the semantics of the state and allows for a logical comparison, instead of a low-level physical comparison. As an example, a data structure that represents a set of elements should be compared across different executions in such a way that is not only oblivious to the memory layout, but, given that sets can be stored in data structures that imply an ordering such as a list, but the order in which the elements of a set are listed is irrelevant, the state summary function must be oblivious to this order.

While writing this extra code could be a burden for the tester or the developer, we found that in practice these functions are simple to write in part because the internal API of the application is reasonably well defined. Additionally we provide a small library that assists programmers in writing state summary functions for the most common types of data structures. Finally, we note that in our testing framework the state summary functions will always be scheduled until completion, without the possibility of being preempted, and therefore do not have to be synchronized with respect to the existing code nor vice-versa.

2.4 Maintaining the summary functions

Annotating the application undoubtedly requires some effort from testers. During the life-cycle of the application it might not suffice to annotate it once – it might be necessary for testers to revise the annotations when there are new versions of the application. Major updates to the application (which typically involve substantial code rewrites) are likely to require some effort to update the summaries.

But, in practice, we expect that many upgrades to the application will maintain most of the properties of the data structures as well as the interface that is used to access them. In these cases, no changes to the annotations would be required.

3. PIKE: A concurrency bug finding tool

In this section we describe how we combine our linearization approach, which analyzes both the output and the state of different interleavings for linearizability violations, with state of the art testing techniques. The result is a bug finding tool geared towards finding concurrency bugs that are traditionally hard to detect.

3.1 Systematic schedule exploration

The distinguishing property of concurrency bugs, in comparison with non-concurrency bugs, is the fact that only specific interleavings trigger these bugs. This implies that testing concurrent applications requires finding mechanisms to explore multiple thread schedules. However, with the exception of very small applications, it is not feasible to explore all possible thread interleavings because of the state explosion problem.

The traditional approach for exploring interleavings relies on stress testing and noise generation [Ben-Asher 2006]. De-

spite their widespread use, these techniques suffer from three important limitations. First, such techniques are not systematic, i.e., they do not try to avoid redundant or similar interleavings. Second, such techniques do not attempt to prioritize interleavings that are more likely to trigger concurrency bugs. And finally, when a bug is found, these approaches may not allow for reliable replay of the interleaving that triggered the bug.

To overcome these shortcomings, researchers have developed tools to explore thread interleavings in a more controlled manner [Burckhardt 2010b, Eytani 2007, Musuvathi 2008]. These tools try to avoid redundant interleavings, prioritize some interleavings over others, and are able to replay previously run schedules. Such features greatly contribute to improving the ability of developers to explore relevant thread schedules and uncover concurrency bugs. However, to detect when a bug is triggered the developer still has to rely on techniques like programmer-written assertions or the program generating an exception.

PIKE combines our proposed linearizability detector with the *random scheduler* algorithm proposed by Burckhardt et al. which is used by PCT [Burckhardt 2010b]. Here, we briefly describe the general idea of the *random scheduler* algorithm. We refer the reader to the original paper for a more detailed explanation.

At the start of the test run, the *random scheduler* assigns a fixed random priority to each thread and, to control the outcome of data races, it will only allow one thread to run at a time. During execution it makes sure that, at any point in time, the highest priority unblocked thread is the one that is allowed to run. Additionally to explore bugs of different *depth*, at a few random points during the execution it changes the priority of the threads. This simple approach, which offers probabilistic guarantees, has been shown both analytically and empirically to work well at uncovering concurrency bugs [Burckhardt 2010b].

3.2 Handling false positives

One of the challenges we expected to face when deploying PIKE is that linearizability would not necessarily hold for a large, complex application with rich semantics and hundreds of thousands of lines of code. These cases, if not appropriately dealt with, could lead to the tool outputting a large number of false positives.

An example of a data structure that we found to sometimes not obey linearizability is an application-level cache. In particular, this happened in situations where the application logic detected that two requests were being handled concurrently and that would cause a cache entry that one of them would create to be invalidated. In these cases, the application would conservatively not insert that entry into the cache. This behavior might have an impact on performance but does not affect correctness, i.e., an application can always choose not to insert an entry into the cache. However, if the application were to execute the same requests sequen-

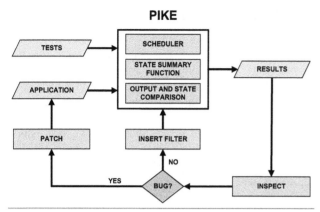

Figure 2. Overall architecture of PIKE. The system receives as inputs a multi-threaded application and a test suite, and contains a feedback loop that can be used by testers to insert filters to avoid false positives when the application deliberately violates linearizability.

tially, because no possible conflict would exist, the last request would be inserted into the cache.

To handle these cases, the state summary functions break the state up into separate components; e.g., an application-level cache would be an individual component. Furthermore, we allow the tester to write a rule that enables the linearizability test to check for inclusion, instead of equality, among the set of entries in some of the state components. In the case of the application-level cache, this rule might allow for checking whether the set of elements in the cache for the concurrent execution are contained in set of elements in the cache for at least one of the sequential executions. We found this approach to work well in practice in reducing the number of false positives to a reasonable level.

Therefore, our final system design contains a feedback loop where testers can add rules that describe such exceptions to linearizability, thus avoiding most false positives and making the problem tractable.

Figure 2 illustrates the overall process. Developers provide PIKE with the application and the testing inputs. PIKE will then run the application multiple times exploring different thread interleavings and checking for linearizability of both state and output. To conclude whether a bug was found, the developer then inspects the results produced by PIKE which include the output, the state and information about the interleaving of the various executions. In case the developer finds various cases of similar false positives he can simply insert a rule to adjust the comparison functions and re-run PIKE.

3.3 Implementation

As Figure 2 also shows, the implementation of PIKE is composed of three components: the scheduler, the *state summary function*, and the component to compare the state and output of the application.

We implemented the scheduler in about 3,000 lines of C code. Our scheduler controls the thread interleaving by intercepting the library calls of the target application and forcing a single thread to run at a time which is randomly chosen according to the *random scheduler* algorithm.

Our scheduler takes control of the application using the *LD_PRELOAD* environment variable and intercepts the pthread library calls made by the application; i.e, the scheduling granularity is at the level of the pthread library calls. Similar levels of granularity have previously been found to produce good results at finding concurrency bugs [Musuvathi 2008].

We require application writers to identify the location where the handling code of each request begins and ends. The scheduler needs to know about these locations to force interleavings that translate into sequential executions. This information also helps in debugging the application when bugs are flagged. Since our scheduler only takes control of the application when it makes pthread calls, it could happen that the running thread (i.e., the runnable thread with highest priority) invokes a system call that does not return. In such a situation, the entire application would block – the highest priority thread would be blocked on a system call and the other threads would have previously been blocked by the scheduler. A situation where this would occur is in the location where the main thread of MySQL spawns new threads to handle new client sessions. To avoid this, we make the scheduler aware of that particular location in the MySQL code and make the scheduler block the main thread as soon as it creates all the expected client-session threads (which is dependent on the input). In comparison with the effort to annotate the application for the purpose of capturing the application state, the effort required to identify these three locations was negligible.

The random scheduler algorithm requires a few parameters to be specified [Burckhardt 2010b]. In our experiments we used the value 50,000 as the maximum number of execution steps per run (after the initialization phase) and we used a single priority inversion point (i.e., we tuned the scheduler to find bugs with depth one). These values were empirically found to produce good results for the application we studied.

The random scheduler algorithm also requires an anti-starvation mechanism. Without this mechanism if the highest priority thread enters a busy wait cycle it would never relinquish the processor and would prevent the entire application from progressing. Examples where such situations could occur are the instances where ad-hoc synchronization methods are used [Xiong 2010]. We implemented the anti-starvation mechanism simply by reducing the priority of the running thread if it runs uninterrupted for more than a certain number of execution steps. We found this mechanism to be particularly useful during initialization periods.

Our implementation also includes a generic library for assisting in capturing the state of the application, however the exact code to capture the state is dependent on the application. In Section 4 we describe our experience with applying PIKE to MySQL.

4. Experience

This section reports on the experience of applying PIKE to find concurrency bugs in MySQL.

4.1 MySQL overview

MySQL represents a challenging case study for our testing tool for several reasons. First, it is a large, complex codebase, with about $360,000$ lines of (mostly C and C++) code and rich application semantics. Second, databases are a critical component of the IT infrastructure of many organizations and therefore it is important to maintain and improve their robustness. In particular, MySQL represents a share of 40% of the database market [Oracle a], and is by far the most popular open-source database server. Finally, MySQL is a mature application with a quality development and maintenance process. The results presented here report on applying our technique to a stable version of MySQL (version 5.0.41).

One of the characteristics of MySQL is that it supports different mechanisms, which are called storage engines, for internally representing and manipulating the state of the database. Users can control which storage engine to use dynamically by parameterizing certain requests during runtime (e.g., *Create Table*) or specifying configuration options set by an administrator. Storage engines represent a significant fraction of the source code of MySQL and implement important parts of the database functionality such as support for indexes and caches, the granularity of locks, and support for compression, replication, or encryption.

To validate our detector we chose to apply it to the MyISAM storage engine. MyISAM [Oracle c] is considered to be one of the most popular storage engines of MySQL [Oracle b] and it has also traditionally been the default storage engine [Oracle c]. In comparison to other engines, MyISAM is optimized for throughput, and is distinctive in that it does not provide the ability to group multiple operations into transactions: instead users have at their disposal explicit locking mechanisms to enforce consistency among groups of operations.

An important point to clarify regarding the semantics of the database server is that we are using PIKE to test for linearizability at the level of individual client requests (i.e., SQL operations), which may differ from the semantics that are provided at higher levels, such as transactions. In particular, it is possible for a database server to offer semantics that are weaker than linearizability, e.g., snapshot isolation at the level of transactions, but still be linearizable at the level of client requests.

4.2 MySQL internal state

As explained in Section 2, PIKE checks whether the application exhibits a linearizable behavior by comparing the internal state of different executions of the application. To achieve this goal, PIKE needs to generate a high-level representation of the internal state which is done in an application-specific way.

By analyzing the source code and based on existing studies [Fonseca 2010] we were able to identify the following data structures, which we believe capture the most important components of the application state.

The *query cache* structure contains pairs of recent instructions that read the state of the database (*SELECT* statements) and their respective results. This structure has been found by its developers to be critical for servers to achieve good performance in many common scenarios. The *query cache*, as one would expect from a cache, should invalidate the relevant entries when they become obsolete due to subsequent and conflicting writes. If the invalidation logic in the application is incorrect it is likely that such mistakes will lead to bugs in which the application returns the wrong results to clients.

The *table cache* stores a set of descriptors, each of which is an in-memory representation of a table schema. When a new thread wants to manipulate a table, it first queries the *table cache* to get a table instance directly if available. Otherwise, in case of a miss, the table schema will be loaded from disk and a new entry will be inserted into the *table cache* structure.

Another type of data structure that we annotated were the *data files*. A *data file* is a critical data structure that stores the actual records for a particular table and is maintained in persistent storage.

To quickly perform searches and find the relevant records in a table, avoiding sequentially scanning the whole table, MySQL also maintains for each table an *index file* which consists of a set of indexes. Each entry in the *index file* consists of a pair of elements. The first element is a *key* (or a group of keys) while the second element is a pointer to the appropriate record in the *data file*.

The *key cache* is a repository for frequently used blocks from the *index files* of all tables. The index block will be loaded into the *key cache* before the first access to a table. From that moment on all subsequent operations will be performed on *key cache* data and will be flushed back to disk at the appropriate time.

Finally, the *binary log* is another important data structure that we annotated. It stores a sequence of all operations that changed the database state, in their order of execution. This structure is critical for replication. Replicas keep their state in sync by shipping the *binary log* between them and re-executing the requests in the order they appear. Missing entries, wrong entries or entries in the wrong order will likely cause replicas to diverge and therefore it can seriously affect the correctness of the service. Additionally the *binary log* is important for recovery purposes.

4.3 State summary functions

To write the summary functions, which capture different parts of the state, we analyzed these different state components and classified them in two categories according to what type of data structures they represent.

Most data structures fall into the *set* category, since they are collections of elements where their order does not matter, except for the *binary log* structure which is an append-only *sequence* where the order in which the elements are added needs to be captured by the summary function.

Starting with the state components that describe sets, their summary function needs to be invoked in all places in the source code where elements are added, removed or modified to or from any of these data structures. Despite the complexity of the state, locating these turned out not to be too complicated since the source code of MySQL is reasonably well structured and there are functions that encapsulate these operations which are called from different points in the code.

At each of these points we invoke a generic summary function for sets, which is designed to provide an efficient update and comparison operation. This function maintains a cumulative hash value for the set (S) which is initialized to zero at the beginning of the execution. Then, upon adding or removing an element e, the summary function captures a hash of the deterministic parts of the element being added or removed (H_e). In this step it is important to remove sources of non-determinism like timestamps that would lead to state divergence. Some of the data structures annotated contain elements contain pointers. In these cases instead of hashing the pointers we hash the elements they point to.

Then, the value of H_e is either added or removed to the cumulative set value S. Both adding and removing is done by XORing the new value with the previous cumulative value, i.e.:

$$S_{new} = S_{old} \oplus H_e. \quad (1)$$

This leads to a compact representation of the state of the set that allows for a trivial comparison operator simply by comparing hashes.

Operations that modify elements are handled by treating them as a sequence of an add and a remove operation.

For the *binary log*, this representation does not work because it does not capture the order in which elements were added to the sequence. Therefore, we change the above equation to capture this order by hashing the concatenation of the previous cumulative value with the new element.

$$S_{new} = SHA1(S_{old}||e). \quad (2)$$

Finally, we also needed to extend this scheme to support containment instead of equality checks for sets. This can be easily achieved by replacing the cumulative XOR of the hash values with a counting Bloom filter [Bonomi 2006]. Alternatively, we can just list all the elements in the set and compare them exhaustively, which is what is done by our implementation.

4.4 Input generation

Like other dynamic bug finding tools, our testing technique requires exploring different inputs in an attempt to find situations in which the application behaves incorrectly. Therefore we must find a diverse set of concurrent database operations that stand a good chance of triggering bugs. Again, the rich semantics and wide interface of MySQL make it particularly challenging given that we can only practically explore a small subset of all possible inputs.

We considered different options for generating test inputs. The obvious option is to generate the inputs manually; however this can be tedious and impractical for applications like MySQL. Another option is to randomly generate inputs, possibly with the aid of grammars that steer the input generation into generating inputs that are considered more useful. This option suffers from the problem that it is not straightforward to instrument the grammar in such a way that it creates multiple concurrent requests that are likely to cause contention for some particular part of the state of the application. A third option is to use tools that analyze the application, try to understand its behavior, and then attempt to automatically generate useful inputs [Cadar 2008a, Godefroid 2005]. However, while these tools work well for small and medium size applications, it is unclear if they can currently scale to the size of a codebase like MySQL.

Therefore we pursued a fourth option. MySQL already contains a large test suite, which has been manually created by the developers and testers of the application. Some of the tests were added specifically to prevent previous bugs from recurring in subsequent versions of the application. However, these tests are sequential tests and therefore would not be useful for finding concurrency bugs. Our solution was to convert these sequential tests into concurrent tests by breaking up the sequence of requests contained in a test and executing them concurrently by separate clients.

When deciding how many concurrent clients to use in our tests, we took into account that studies show that a significant amount of the concurrency bugs found only require a small number of threads to be triggered (typically two) [Lu 2008]. A separate study also showed that only a small number of requests is sufficient to expose bugs [Sahoo 2009]. Taking these factors into consideration, and to make the process more efficient, we generated tests involving two clients and with a limited number of requests per client (typically less than ten requests and starting from an empty database).

The original complete test suit contained approximately $50,000$ requests, as counted by the number of semicolons. Using our approach we manually converted around 5% of those requests into concurrency tests, thus generating 1550 pairs of inputs from concurrent threads.

One could imagine extending MySQL's traditional testing approach to also include concurrency bugs tests instead of just deterministic tests. Similarly to Pike, the extension to the traditional testing approach would also require the use of tools to explore thread interleavings. One of the problems with this extension is that testers would have to manually specify (and update) the set of expected outputs for each test case. Pike instead finds that set automatically for arbitrary inputs. Furthermore Pike analyzes the application internal state to detect latent concurrency bugs.

In the future, we plan to explore other approaches for generating inputs and use them with PIKE.

5. Results

In this section we present the results of our experience of applying PIKE to MySQL.

5.1 Development effort

The first result we report on was the the amount of effort needed to understand the code of MySQL and develop the *state summary functions*. The annotations we inserted added up to 600 lines of code, as counted by the number of semicolons. This represents less than 0.2% of the number of semicolons in the MySQL source code.

While annotating the source code of MySQL, most of the effort was spent understanding the source code. We spent a total of about two man-months in the process of understanding both the structure and semantics of the application and annotating the source code.

5.2 Bugs found

We ran PIKE on MySQL opportunistically in a shared cluster using multiple machines (up to 15 machines). Each machine in the cluster had an AMD Opteron 2.6 GHz processor, 3 GB of RAM and was running a distribution of Linux with kernel version 2.6.32.12.

We tested MySQL by running it on 1550 inputs and for each input we configured PIKE to explore 400 different interleavings using its scheduler. The experiment lasted for about one month. Our implementation of PIKE could be optimized to reduce the computational cost in several ways. In particular, we could avoid going through the initialization phase of MySQL for each run by taking advantage of snapshoting techniques. Another way of speeding up testing could be to run PIKE on the target application previously compiled with optimization flags. The few inputs for which suspicious behavior is observed could then be re-executed with additional debugging support (on the version of the target application not optimized and with application-level debugging options enabled).

During our testing experiments PIKE was able to identify a total of 12 inputs that triggered concurrency bugs. Table 1 presents an overview of the inputs that we found to trigger incorrect behavior and in the following subsections we present our findings in more detail for different types of bugs, categorized according to their effects.

In some cases, we had different inputs that triggered bugs that showed similar effects. Because it was difficult for us to classify whether they correspond to the same bug or not, we decided to present the results in a more objective way by presenting in detail all of the inputs and effects of the bugs we found, instead of trying to count the number of distinct bugs. We then speculate about which of those inputs are likely to be triggering what could be considered the same bug.

Table 2 lists the various inputs that were flagged as positives by PIKE and that we confirmed to be caused by concurrency bugs. The table presents the requests that were concurrently executed in the test cases that triggered concurrency bugs together with the number of distinct thread schedules in which the program exhibited the incorrect behavior. Additionally, we also present information about the state and the output that were observed. Specifically, the table indicates whether the output of the concurrent execution matches the output of the sequential executions (O_A and O_B) and whether the state at the end of the concurrent execution matches the state at the end of either of the sequential executions (S_A and S_B).

Given the linearization algorithm, PIKE flags a concurrent execution as having triggered a concurrency bug if it cannot find a sequential execution (X) that produces both an output and a final state that match its own (i.e., that has O_X="Yes" and S_X="Yes"). We can see that all entries in Table 2 fail to meet this condition.

In addition to discrepancies in the output or the state of the different interleavings, we also found some cases where the execution of the application blocked, which might have been caused by deadlocks, and cases where the application crashed. We have not analyzed these cases, but they are less interesting from our standpoint since these potential bugs would also have been found by other tools like Chess [Musuvathi 2008], or tools that are designed to find deadlock bugs [Naik 2009].

In our experiments, we did not come across non-concurrency bugs, and this is not surprising for two reasons. First, we used inputs that were based on the existing regression tests contained in the MySQL source code, and therefore MySQL should have been previously tested for these or very similar inputs. Second, a non-concurrency bug, if triggered would have likely produced the same wrong results in all interleavings, regardless of the interleaving being sequential or not, and therefore our detector would not have flagged it.

One point we would like to highlight about these results is that we used a testing suite that has been applied repeatedly, albeit in a way that runs inputs sequentially. We postulate that it might be possible to be even more effective if we use a different set of inputs. The downside is that, because

External effect	Non-latent	Latent	Total
Error	2	0	2
Semantic	2	8	10

Table 1. Number of inputs found to trigger concurrency bugs according to latency and external effects.

we focused on what is not the latest version of MySQL, we found that some of the bugs have already been fixed, as we will detail next.

Next, we analyze in more detail the results for the two categories of bugs that our technique is aimed at: violations of the application semantics, and latent bugs. We further divide the first category into semantic bugs and error bugs, depending on whether the violation of the intended semantics corresponds to an incorrect but non-error reply, or a more explicit error.

5.2.1 Semantic bugs

Figure 3 illustrates a representative example of a semantic concurrency bug in MySQL that was found by our detector. In the figure the arrow indicates the interleaving that triggers the bug. This bug is triggered when the server receives a specific *SHOW TABLE* request and a *DROP* request concurrently as shown in Table 3. Figure 3 shows a simplified snippet of the source code that is involved in this concurrency bug. The first thread, while executing the *SHOW TABLE* request obtains a list of names of tables. According to the semantics of the database this returned list should contain the names of all the tables in the database whose name contains the string "t1". But, if before the first thread processes the list of tables names the second thread is able to execute the *remove_table()* function, the open table list becomes obsolete. This in turn means that when the first thread resumes execution it will try to call the *open_tables()* function with an argument that contains obsolete data and will not be able to access the table that was dropped. The result is that the second thread will return to the user a success message for the *DROP* request. However, the first thread will return an entry, for the now non-existent table, indicating that it exists but some of the entries will contain the value NULL.

We note that this particular instance of a semantic bug was eventually reported in the MySQL bug report database, and patched in a version that succeeded the one we tested. However, it is important to note that we did not use that information during the process of generating inputs.

Other semantic bugs provided wrong results in even more subtle ways. For example, there were bugs where the application would simply provide wrong results based on stale data.

5.2.2 Error bugs

A sub-class of the semantic bugs that we found can be labeled as error bugs. We considered bugs to be error bugs if they manifest themselves by returning to the client an ex-

Request 1	SHOW TABLE STATUS LIKE 't1';
Request 2	DROP TABLE t1;

Table 3. Requests responsible for triggering the sample semantic bug

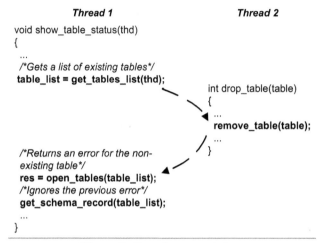

Figure 3. Sample semantic bug

plicit error message, but an error message which is not appropriate given the requests that were executed. During our experiments we found two cases in which error concurrency bugs were triggered.

Table 4 presents the concurrent requests that were found to be responsible for one of the error bugs. This bug occurs when one of the threads attempts to execute a *CREATE LIKE* request, which is supposed to create a new and empty table with a schema that is identical to another existing table, and a specific *INSERT* request that copies data from the existing table into the new table. As illustrated in Figure 4, the first thread, while handling the *CREATE* request, first copies the definition file containing the schema for the existing table. According to the synchronization logic in MySQL, the second thread is allowed to execute the *INSERT* request even before the first thread creates the index file and data file. Because of this, while executing the *INSERT*, the second thread is unable to open the data file and returns an error to the user stating that the *data file* does not exist instead of either succeeding (by writing data) or returning a different error stating that the *table* does not exist.

This example illustrates an important point that error bugs can also be subtle and difficult to distinguish from a correct execution, despite the fact that they return an error. This is because very often an error message is a legitimate outcome of the operation, but the concurrent execution returns the wrong error message. Therefore, and unlike a situation where the application crashes or an assertion fails, we must know application-specific semantics to determine if an error reply is incorrect or not, and PIKE has proven to be effective in determining this.

Requests	EXs	Output			State		
		O_A	O_B	Effect	S_A	S_B	Latent
CREATE TABLE t2 LIKE t1; ‡ INSERT INTO t2 SELECT * FROM t1; ‡	9	No	No	Error	No	Yes	Non-latent
INSERT INTO t3 VALUES (1,'1'),(2,'2'); SELECT DISTINCT t3.b FROM t3,t2,t1 WHERE t3.a=t1.b; †	1	No	Yes	Semantic	No	No	Latent
CREATE TABLE t2 LIKE t1; ‡ INSERT INTO t2 SELECT * FROM t1; ‡	2	No	No	Error	No	Yes	Non-latent
TRUNCATE TABLE t1; SELECT * FROM t2;	35	Yes	No	Semantic	No	No	Latent
INSERT INTO t1 (a) VALUES (10),(11),(12); SELECT a FROM t1;	2	No	Yes	Semantic	No	No	Latent
INSERT INTO t2 VALUES (2,0); SELECT STRAIGHT_JOIN* FROM t1, t2 FORCE (PRIMARY); †	3	Yes	No	Semantic	No	No	Latent
DROP TABLE t1; SHOW TABLE STATUS LIKE 't1';	238	No	No	Semantic	Yes	Yes	Non-latent
INSERT INTO t1 VALUES (1,1,"00:06:15"); † SELECT a,SEC_TO_TIME(SUM(t)) FROM t1 GROUP a,b; †	1	No	Yes	Semantic	No	No	Latent
CREATE TABLE t2 SELECT * FROM t1; DROP TABLE t2;	17	No	No	Semantic	No	No	Non-latent
INSERT INTO t1 (a) VALUES (REPEAT('a', 20)); SELECT LENGTH(a) FROM t1;	3	No	Yes	Semantic	No	No	Latent
INSERT INTO t1 VALUES (80,'pendant'); SELECT COUNT(*) FROM t1 WHERE LIKE '%NDAN%'; †	2	No	Yes	Semantic	No	No	Latent
OPTIMIZE TABLE t1; DROP TABLE t1;	25	Yes	No	Semantic	No	Yes	Latent

Table 2. Properties of the triggered concurrency bugs that PIKE found. The table presents the number of concurrent executions that were flagged as positive for each of the inputs (EXs). Additionally it indicates whether the output of the concurrent executions matched the output of the sequential executions (O_A and O_B) and similarly for the state of the sequential executions (S_A and S_B). (Requests marked with † have been simplified for presentation purposes, the two identical pairs of requests marked with ‡ operate on distinct states)

Request 1	CREATE TABLE t2 LIKE t1;
Request 2	INSERT INTO t2 SELECT * FROM t1;

Table 4. Requests responsible for triggering the sample error bug

5.2.3 Latent bugs

Surprisingly, PIKE was able to find eight different situations that triggered latent concurrency bugs. All of the latent bugs we found had the external effect of providing wrong results in subtle ways and involved the *query cache* structure. As we will describe in Section 5.3, we also found situations where the *binary log* appeared to contain an incorrect state, but we were not confident that these represented bugs (i.e., that the incorrect state would lead to incorrect behavior visible by users) and so we did not flag them as such.

As an example, one of the cases where a latent concurrency bug is triggered occurs when the requests in Table 5 are executed concurrently. The simplified source code relevant to this example is shown in Figure 5. While executing the *SELECT* request, the first thread opens the table, locks it, and in the process makes a copy for itself of the state

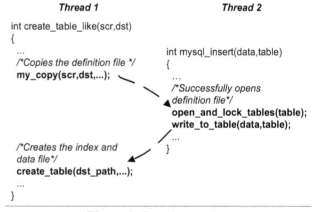

Figure 4. Sample error bug

of the table. The logic of the application allows the second thread to then concurrently insert entries at the logical end of the table. However, when the first thread resumes execution it will rely on its local (and now stale) copy of the state of that table to fetch data. In the process, the first thread will skip the newly inserted entry and provide the old results to the client without immediately violating the semantics of the

| Request 1 | SELECT a FROM t1; |
| Request 2 | INSERT INTO t1 (a) VALUES (10), (11), (12); |

Table 5. Requests responsible for triggering the sample latent bug

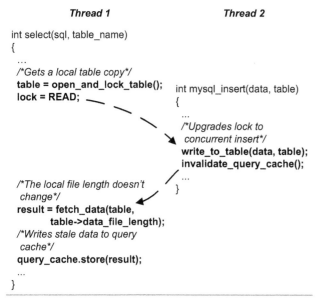

Figure 5. Sample latent bug

application (i.e., the returned value would be consistent with the first thread having executed before the second thread). In this bug, the actual semantic violation arises from the fact that the first thread also stores the stale data which the second thread does not invalidate in the *query cache*. This means that a third thread could, at a later point in time, read the stale data from the *query cache* and expose it to the clients, violating the expected semantics of the application.

We saw the same pattern of latent bugs causing stale entries to be left in the query cache in other test cases, and we again stress that it is likely that some of the situations that triggered latent concurrency bugs could be triggering what could be considered the same bug. However, given the complexity of the application logic to both invalidate the *query cache* and to prevent certain specific concurrent requests from inserting simultaneously entries into the *query cache*, it is hard to state whether we are dealing with the same bugs objectively. Nevertheless it should be noted that the various cases that triggered latent bugs can be caused by very distinct types of requests, as can be seen in Table 2.

5.3 False positives

After the initial tests, approximately one third of the inputs generated potential false positives. Since this high fraction of false positives would make the analysis of the results impractical, we had to insert two filters to reduce the number of false positives which proved to be very effective. These

filters allow testers to avoid false positives when the application deliberately violates linearizability.

The first filter we inserted was related to the table cache. Concurrent requests that try to open the same table concurrently will create distinct but identical entries in the table cache, whereas the same requests executing in sequence can reuse each other's entry. Therefore we inserted a filter that stated that the entries in the table cache for the linearized execution need to be contained in the concurrent one.

The second filter was related to the query cache, and the fact that MySQL sometimes conservatively decides not to cache entries in the query cache when two concurrent requests are executed, one of them is a query, and the other would invalidate the entry for that query in the query cache. In this case our filter says that the query cache entries in the concurrent execution must be contained in the set of entries in the linearization. Note that these may be considered performance bugs (or, at least, missed opportunities for a performance optimization), and this shows that PIKE might also be useful for analyzing and improving performance issues that may affect the application.

After inserting these two filters, the total number of false positives reported was 27. Of these, 22 are related to unexpected interactions between the framework and the application. In particular, some requests took a longer amount of time to complete, which in turn caused an execution timeout in our framework to expire. In other cases false positives were caused by non-determinism in the reply that we had not caught (e.g., calls to the current time or random number generation). A third type of false positives was caused by timeouts in the NFS volume in which our results were written, which affected the output. All of these types of false positives were reasonably easy for us to diagnose.

The remaining five false positives involved a more careful analysis. These were caused by *binary log* entries being reordered (i.e., MySQL would change some internal structures in one order and the *binary log* in another order). This turned out to be acceptable under some circumstances. Typically this happened with pairs of concurrent requests in which one of the requests executed an optimization or maintenance task (e.g., *OPTIMIZE* and *FLUSH* requests). The fact that these operations affect the performance but not the results implies that, when the *binary log* state is required (normally when a replica recovers from a fault), repeating these entries in the wrong order will not affect the output of the operations, but only the moment in the sequence of re-execution of these operations when the performance optimizations are performed.

6. Related Work

The goal of program verification is to guarantee that an implementation complies with its specification. Assuming the specification is correct (i.e., it specifies what the programmer intended), a verified program is guaranteed to be bug free. Model checking [Musuvathi 2004] is a promising technique that follows this approach by exhaustively exploring all possible states of the program. However, currently model checking has difficulty scaling to large programs.

Bug finding tools, on the other hand, although they do not guarantee that all bugs are found, are more scalable. Bug finding tools can be divided into static analysis and dynamic analysis tools depending, respectively, on whether they simply analyze the source code or actually execute the code. Static tools such as RacerX [Engler 2003] and others [Boyapati 2002, Naik 2006] have the advantage of not being limited in their analysis to the execution path determined by the input. On the other hand, dynamic analysis tools, since they actually run the code, have the advantage of having more information about the context of the execution and therefore can potentially achieve a higher accuracy (i.e., fewer false positives). PIKE is an example of a dynamic analysis tool, as are FastTrack [Flanagan 2009], LiteRace [Marino 2009] and Eraser [Savage 1997].

For testing to be successful, developers need to have good test cases. But given that manually generating tests is, in general, a tedious and difficult task, researchers have tried to automate this process by developing tools and methodologies that automatically generate test cases [Cadar 2008a;b, Godefroid 2005]. There have also been attempts to generate test cases specifically for databases [Microsoft, Mishra 2008]. However, automatic tools for test generation typically have difficulty scaling to large and complex applications.

Given the specifics of concurrency bugs, researchers have developed specific tools for handling this special class of bugs. One class of tools attempts to help programmers explore different thread interleavings. A different class of tools which also also specifically target concurrency bugs are data race detectors. We compare to each of these two classes in turn.

Typically, when a multi-threaded application runs natively, the operating system will tend to choose similar thread interleavings for different executions. To make testing more efficient, it is important to test a more diverse set of interleavings. One way of achieving this is by stress testing the application, possibly in combination with noise generators [Ben-Asher 2006]. A more sophisticated approach is to use custom schedulers that try to avoid redundant thread interleavings, prioritize some thread interleavings over others, and allow the programmer to replay a thread interleaving once it finds one that interests him (e.g., a thread interleaving that triggers bugs). Examples of tools that explore different thread interleavings in a smarter way are ConTest [Eytani 2007], CHESS [Musuvathi 2008] and PCT [Burckhardt 2010b]. However, these tools still rely on external mechanisms (e.g., assertion violations) to detect the occurrence of concurrency bugs. PIKE makes use of this approach, in particular the random scheduler algorithm of PCT, to explore different thread interleavings in a controlled way, but is complementary to them in that it enables new ways of finding

bugs that do not rely on capturing exceptions or traditional assertion violations.

Some frameworks allow programmers to specify complex assertions for multi-threaded applications [Burnim 2009]. These typically enable programmers to specify invariants and to specify which parts of the code the invariants apply to. PIKE instead proposes an implicit correctness condition that relies on comparing the application behavior to the behavior during serializable executions.

A second class of tools are the data race detectors which can be roughly divided into two sub-classes depending on which algorithm they use. The first sub-class of data race detectors rely on the lockset algorithm [Savage 1997] to infer whether the programmer protected all accesses to a specific shared variable with a fixed lock. The second sub-class of data race detectors rely on the happens-before algorithm [Flanagan 2009, Marino 2009]. Recently, Erickson et al. have proposed a different data race detector that is not based on either of these algorithms, but is instead based on sampling and the use of breakpoints [Erickson 2010].

Like PIKE, data race detectors are also tools that can be useful for detecting concurrency bugs, however they have distinct features. First, these tools detect data races instead of directly detecting concurrency bugs. Since programs often contain benign data races, simply detecting data races easily leads to false positives. Furthermore the absence of data races is not a guarantee of correct synchronization [Artho 2003, Lu 2006], and hence false negatives can result. Another difference is that race detectors typically operate at the lower-level of individual memory accesses. In contrast, PIKE analyzes the actual output of the application as well as a high-level digest of the state, potentially uncovering bugs that are not triggered by low-level data races and also facilitating the process of inspecting the results.

In order to reduce the number of false positives in data race finding tools and thus reduce the burden on testers, researchers have developed heuristics. By using heuristics some systems attempt to identify scenarios that frequently lead to false positives. DataCollider [Erickson 2010], for example, tries to detect benign data races caused by counters and accesses to different bits of the same variable. One approach is to use heuristics that rely on looking at the instructions at or near the problematic accesses or on manually whitelisting variables. The disadvantage of this approach is that it also increases the risk of missing erroneous data races. Another interesting approach to distinguish erroneous data races from benign data races relies on replaying the execution [Narayanasamy 2007]. It relies on trying to trigger the opposite outcome of the data race and then comparing the low-level results obtained with both data race outcomes. This approach, however, still aims at finding low-level data races.

There have been prior approaches for checking the linearizability of code to improve robustness [Burckhardt

2010a, Vafeiadis 2010, Vechev 2009, Xu 2005]. We differ from these approaches in two ways. First, they typically ignore the internal state of the application, which is important for the detection of latent bugs. Second, they check for the atomicity of smaller sections of code such as code blocks or library calls, which poses fewer challenges than testing the linearizability of large server applications.

AVIO [Lu 2006] detects atomicity violations at the level of individual memory accesses. AVIO achieves this by learning from a large set of runs (which are assumed to be correct) the valid memory access patterns (e.g., when are two consecutive accesses from a thread allowed to be interleaved by an access from another thread). AVIO shares our goal of attempting to find concurrency bugs without relying on finding data races, but in contrast AVIO works at a low-level and relies on training.

Finally, an entirely different approach for dealing with concurrency bugs is by using tools that prevent or make it less likely for programmers to make mistakes. One such approach is to use special programming languages [Vaziri 2006], while another is to use special hardware or frameworks such as transactional memory [Herlihy 1993, Shavit 1995].

7. Conclusion

This paper presented PIKE, a tool for testing concurrent applications. PIKE is able to find two particularly challenging types of bugs: semantic bugs and latent bugs. Semantic bugs generate subtle deviations from the expected behavior of the application, while latent bugs silently corrupt internal data structures, and manifest themselves to clients possibly long after the requests that triggered the bug are executed. PIKE detects these two types of bugs by testing if the application obeys linearizable semantics, both in terms of its outputs and its internal state. Our experience in applying PIKE to find concurrency bugs in MySQL was a positive one. We found that it was simple to write the necessary annotations to capture an abstract view of the service state, and that it was easy to make the number of false positives tractable by writing simple filtering rules for common violations of linearizability at the level of the application state. More importantly, we were able to find several semantic and latent concurrency bugs in a stable version of MySQL. Currently, we are applying PIKE to the current development release, and we are analyzing the potential crash and deadlock bugs we have found with our current test suite.

Acknowledgments

We are grateful for the feedback provided by the anonymous reviewers and for the help provided by our shepherd, Leonid Ryzhyk. Pedro Fonseca was supported by a grant provided by FCT.

References

[Artho 2003] Cyrille Artho, Klaus Havelund, and Armin Biere. High-level data races. *Software Testing, Verification and Reliability*, 13(4):207–227, 2003.

[Ben-Asher 2006] Yosi Ben-Asher, Yaniv Eytani, Eitan Farchi, and Shmuel Ur. Producing scheduling that causes concurrent programs to fail. In *Proc. of Parallel and Distributed Systems: Testing and Debugging (PADTAD)*, pages 37–40, 2006.

[Bonomi 2006] Flavio Bonomi, Michael Mitzenmacher, Rina Panigrahy, Sushil Singh, and George Varghese. An improved construction for counting bloom filters. *Lecture Notes in Computer Science*, 4168:684–695, 2006.

[Boyapati 2002] Chandrasekhar Boyapati, Robert Lee, and Martin Rinard. Ownership types for safe programming: Preventing data races and deadlocks. In *Proc. of Object-Oriented Programming, Systems, Languages, and Applications (OOPSLA)*, pages 211–230, 2002.

[Burckhardt 2010a] Sebastian Burckhardt, Chris Dern, Madanlal Musuvathi, and Roy Tan. Line-up: A complete and automatic linearizability checker. *SIGPLAN Not.*, 45(6):330–340, 2010. ISSN 0362-1340.

[Burckhardt 2010b] Sebastian Burckhardt, Pravesh Kothari, Madanlal Musuvathi, and Santosh Nagarakatte. A randomized scheduler with probabilistic guarantees of finding bugs. *SIGARCH Comput. Archit. News*, 38(1):167–178, 2010. ISSN 0163-5964.

[Burnim 2009] Jacob Burnim and Koushik Sen. Asserting and checking determinism for multithreaded programs. In *Proc. of the European Software Engineering Conference and the Symposium on the Foundations of Software Engineering (ESEC/FSE)*, pages 3–12, 2009.

[Cadar 2008a] Cristian Cadar, Daniel Dunbar, and Dawson Engler. KLEE: Unassisted and automatic generation of high-coverage tests for complex systems programs. In *Proc. of Operating System Design and Implementation (OSDI)*, pages 209–224, 2008.

[Cadar 2008b] Cristian Cadar, Vijay Ganesh, Peter M. Pawlowski, David L. Dill, and Dawson R. Engler. EXE: Automatically generating inputs of death. *ACM Trans. Inf. Syst. Secur.*, 12(2):1–38, 2008. ISSN 1094-9224.

[Engler 2003] Dawson Engler and Ken Ashcraft. RacerX: Effective, static detection of race conditions and deadlocks. *SIGOPS Operating Systems Review*, 37(5):237–252, 2003. ISSN 0163-5980.

[Erickson 2010] John Erickson, Madanlal Musuvathi, Sebastian Burckhardt, and Kirk Olynyk. Effective data-race detection for the kernel. In *Proc. of Operating System Design and Implementation (OSDI)*, pages 1–16, 2010.

[Eytani 2007] Yaniv Eytani, Klaus Havelund, Scott D. Stoller, and Shmuel Ur. Toward a framework and benchmark for testing tools for multi-threaded programs. *Conc. & Comp.: Practice & Experience*, pages 267–279, 2007.

[Flanagan 2009] Cormac Flanagan and Stephen N. Freund. FastTrack: Efficient and precise dynamic race detection. *SIGPLAN Not.*, 44(6):121–133, 2009. ISSN 0362-1340.

[Fonseca 2010] Pedro Fonseca, Cheng Li, Vishal Singhal, and Rodrigo Rodrigues. A study of the internal and external effects of concurrency bugs. In *Proc. of International Conference on Dependable Systems and Networks (DSN)*, pages 221–230, 2010.

[Godefroid 2005] Patrice Godefroid, Nils Klarlund, and Koushik Sen. DART: Directed automated random testing. *SIGPLAN Not.*, 40(6):213–223, 2005. ISSN 0362-1340.

[Herlihy 1993] Maurice Herlihy and J. Eliot B. Moss. Transactional memory: Architectural support for lock-free data structures. *SIGARCH Computer Architecture News*, 21(2):289–300, 1993.

[Herlihy 1990] Maurice P. Herlihy and Jeannette M. Wing. Linearizability: A correctness condition for concurrent objects. *ACM Trans. Program. Lang. Syst.*, 12(3):463–492, 1990. ISSN 0164-0925.

[Lu 2008] Shan Lu, Soyeon Park, Eunsoo Seo, and Yuanyuan Zhou. Learning from mistakes: A comprehensive study on real world concurrency bug characteristics. *SIGARCH Computer Architecture News*, 36(1):329–339, 2008. ISSN 0163-5964.

[Lu 2006] Shan Lu, Joseph Tucek, Feng Qin, and Yuanyuan Zhou. AVIO: detecting atomicity violations via access interleaving invariants. In *Proc. of International Conference on Architectural Support for Programming Languages and Operating Systems (ASPLOS)*, pages 37–48, 2006.

[Marino 2009] Daniel Marino, Madanlal Musuvathi, and Satish Narayanasamy. LiteRace: Effective sampling for lightweight data-race detection. In *Proc. of Programming Languages Design and Implementation (PLDI)*, pages 134–143, 2009.

[Microsoft] Microsoft. Generating test data for databases by using data generators. http://msdn.microsoft.com/en-us/library/dd193262.aspx.

[Mishra 2008] Chaitanya Mishra, Nick Koudas, and Calisto Zuzarte. Generating targeted queries for database testing. In *Proc. of International Conference on Management of Data (SIGMOD)*, pages 499–510, 2008.

[Musuvathi 2004] Madanlal Musuvathi and Dawson R. Engler. Model checking large network protocol implementations. In *Proc. of Networked Systems Design and Implementation (NSDI)*, pages 155–168, 2004.

[Musuvathi 2008] Madanlal Musuvathi, Shaz Qadeer, Thomas Ball, Gérard Basler, Piramanayagam A. Nainar, and Iulian Neamtiu. Finding and reproducing heisenbugs in concurrent programs. In *Proc. of Operating System Design and Implementation (OSDI)*, pages 267–280, 2008.

[Naik 2006] Mayur Naik, Alex Aiken, and John Whaley. Effective static race detection for java. In *Proc. of Programming Languages Design and Implementation (PLDI)*, pages 308–319, 2006.

[Naik 2009] Mayur Naik, Chang-Seo Park, Koushik Sen, and David Gay. Effective static deadlock detection. In *Proc. of International Conference on Software Engineering (ICSE)*, pages 386–396, 2009.

[Narayanasamy 2007] Satish Narayanasamy, Zhenghao Wang, Jordan Tigani, Andrew Edwards, and Brad Calder. Automatically classifying benign and harmful data races using replay analysis. In *Proc. of Programming Languages Design and Implementation (PLDI)*, pages 22–31, 2007.

[Oracle a] Oracle. MySQL :: Market share. http://www.mysql.com/why-mysql/marketshare/.

[Oracle b] Oracle. Storage engine poll. http://dev.mysql.com/doc/refman/5.0/en/storage-engines.html.

[Oracle c] Oracle. The MyISAM storage engine. http://dev.mysql.com/doc/refman/5.0/en/myisam-storage-engine.html.

[Sahoo 2009] Swarup K Sahoo, John Criswell, and Vikram S. Adve. An empirical study of reported bugs in server software with implications for automated bug diagnosis. Tech. Report 2142/13697, University of Illinois, 2009.

[Savage 1997] Stefan Savage, Michael Burrows, Greg Nelson, Patrick Sobalvarro, and Thomas Anderson. Eraser: A dynamic data race detector for multi-threaded programs. *SIGOPS Oper. Syst. Rev.*, 31(5):27–37, 1997. ISSN 0163-5980.

[Shavit 1995] Nir Shavit and Dan Touitou. Software transactional memory. In *Proc. of Symposium on Principles of Distributed Computing (PODC)*, pages 204–213, 1995.

[Vafeiadis 2010] Viktor Vafeiadis. Automatically proving linearizability. In *Proc. of International Conference on Computer Aided Verification (CAV)*, pages 450–464, 2010.

[Vaziri 2006] Mandana Vaziri, Frank Tip, and Julian Dolby. Associating synchronization constraints with data in an object-oriented language. In *Proc. on Principles of Programming Languages (POPL)*, pages 334–345, 2006.

[Vechev 2009] Martin Vechev, Eran Yahav, and Greta Yorsh. Experience with model checking linearizability. In *Proc. on SPIN Workshop on Model Checking Software (SPIN)*, pages 261–278, 2009.

[Xiong 2010] Weiwei Xiong, Soyeon Park, Jiaqi Zhang, Yuanyuan Zhou, and Zhiqiang Ma. Ad hoc synchronization considered harmful. In *Proc. of Operating System Design and Implementation (OSDI)*, pages 1–8, 2010.

[Xu 2005] Min Xu, Rastislav Bodík, and Mark D. Hill. A serializability violation detector for shared-memory server programs. *SIGPLAN Not.*, 40(6):1–14, 2005. ISSN 0362-1340.

Operating System Support for Application-Specific Speculation

Benjamin Wester   Peter M. Chen   Jason Flinn

University of Michigan
Ann Arbor, MI, USA
{bwester,pmchen,jflinn}@umich.edu

Abstract

Speculative execution is a technique that allows serial tasks to execute in parallel. An implementation of speculative execution can be divided into two parts: (1) a *policy* that specifies what operations and values to predict, what actions to allow during speculation, and how to compare results; and (2) the *mechanisms* that support speculative execution, such as checkpointing, rollback, causality tracking, and output buffering.

In this paper, we show how to separate policy from mechanism. We implement a speculation mechanism in the operating system, where it can coordinate speculations across all applications and kernel state. Policy decisions are delegated to applications, which have the most semantic information available to direct speculation.

We demonstrate how custom policies can be used in existing applications to add new features that would otherwise be difficult to implement. Using custom policies in our separated speculation system, we can hide 85% of program load time by predicting the program's launch, decrease SSL connection latency by 15% in Firefox, and increase a BFT client's request rate by 82%. Despite the complexity of the applications, small modifications can implement these features since they only specify policy choices and rely on the system to realize those policies. We provide this increased programmability with a modest performance trade-off, executing only 8% slower than an optimized, application-implemented speculation system.

Categories and Subject Descriptors D.4.7 [*Operating Systems*]: Organization and Design; D.4.8 [*Operating Systems*]: Performance

General Terms Design, Performance

Keywords Policy, Mechanism, Speculative execution

EuroSys'11, April 10–13, 2011, Salzburg, Austria.

1. Introduction

Speculative execution has been widely used as a method for increasing parallelism by allowing serial tasks to execute concurrently. It has been used to improve performance in many hardware and software systems, including processor branch predictors [Smith 1981], distributed file systems [Nightingale 2005], remote displays [Lange 2008], fault-tolerant protocols [Wester 2009], virtual-machine replication [Cully 2008], discrete event simulators [Jefferson 1987], and JavaScript interpreters [Mickens 2010].

To execute speculatively, the system predicts the outcome of a particular operation and continues its execution based on that prediction. When the operation completes, the actual result is compared with the predicted result. If the prediction was correct, the system commits the speculative state. Otherwise, the system corrects the state produced by the misprediction, usually by rolling back to a prior point in time.

Each implementation of speculative execution can be divided in two parts: (1) a *policy* that specifies what operations and values to predict, what actions to allow while speculating, and how to compare results; and (2) the *mechanisms* that support speculative execution, such as checkpointing, rollback, causality tracking, and output buffering.

Existing systems typically implement mechanism and policy together within a single layer, such as the processor, operating system, or application. Unfortunately, no single layer is well-suited to implement both policy and mechanism. Policy decisions are best done by higher layers in the system, such as applications, that understand the semantics of the actions that are being predicted and can more accurately predict a value and compare it with the actual result. In contrast, mechanisms that support speculation policies are best implemented at lower layers in the system (e.g., operating systems). Lower layers exercise more control over the entire system, enabling them to to propagate or coordinate speculations between applications. Implementing the mechanisms for speculative execution in the lower layer also frees application writers from re-implementing speculation for each application.

In this paper, we show how to separate policy from mechanism in a speculation system. We implement a mechanism for speculation in the operating system, where it can easily propagate speculations between multiple applications, control the output from speculative applications, and be shared by multiple applications. We delegate policy decisions to applications, which have the semantic information needed to specify which operations to execute speculatively, what values to predict, what operations to allow during speculation, and what criteria to use when comparing predicted and actual values.

Separating mechanism from policy opens up a new design space regarding what behaviors a policy should specify and how to best describe them. An application-specific policy can address a number of issues at each phase of speculative execution:

- *Starting* the speculation: What actions are predictable? When should each speculation begin?

- *Performing* the speculation: How should output be handled? What data can be marked as speculative? How many resources should be used?

- *Ending* the speculation: Which results should be considered correct? How should the system recover from a misprediction?

Allowing applications to specify their own policies about when and how to speculate enables them to use speculative execution in ways that are difficult to implement using generic policies provided by lower layers. We demonstrate this by building a prototype speculation mechanism at the operating system layer with policies specified in user-space programs. Within this system, we modify three existing applications to demonstrate our approach:

- *Predictive application launching*: The *Bash* shell predicts the next command a user will type and executes it speculatively. An *X11 proxy* permits graphical applications to interact with the X server while being launched speculatively.

- *Firefox* performs certificate revocation checks while continuing to establish an SSL/TLS connection with a server.

- A *Byzantine fault-tolerant (BFT) client* assumes that the first reply to a request is correct without waiting on consensus.

An OS-implemented speculation system lacks the abstractions needed to specify these features, while an application-implemented speculation system limits the scope of each speculation and complicates the development effort. To address these issues, our separated system allows custom policies to be specified in these applications by adding localized changes that reuse a common mechanism. Our changes allow predicted applications to hide 85% of their start time, reduce Firefox's SSL connection latency by 15%, and increase the BFT client's request rate by 82%. We do impose a trade-off on developers: an application using an optimized speculation implementation can improve on our results by 8%, although it must give up system support to do so.

This paper makes the following specific contributions: first, we present a discussion of the rationale for separating speculative policy from the mechanism that implements it. Second, we use this discussion to design and implement a speculation system that places mechanism in the operating system and gives user-space processes control over policy. Finally, we demonstrate that our approach permits existing programs to use speculation for increased performance without requiring extensive modifications.

The rest of this paper is laid out as follows. Section 2 describes how speculative execution works when implemented below the application. Section 3 explores the different behaviors a policy can customize, and Section 4 describes two issues that arise when applications control their own speculation. Section 5 describes what mechanisms we implement in the operating system to support custom speculation policies. Section 6 discusses the process used to locate and implement custom policies. Section 7 describes three case studies for using customized speculation policies and evaluates the performance improvements that custom policies enables. Section 8 describes related work, and Section 9 concludes.

2. Generic Speculation

This section describes how speculative execution works when it is implemented below the application and thus does not understand the program's semantics. We refer to this as a *generic* speculation system.

Speculation can be implemented at many layers below the application, such as in hardware, a virtual-machine monitor, the operating system, or a language runtime. The layer at which speculation is implemented determines the natural unit of execution that the speculation system controls. For example, speculation implemented in a virtual machine monitor would control the execution of virtual machines, while speculation implemented inside an operating system would control the execution of processes. To make the discussion more concrete, our description assumes speculation is implemented in the operating system; the same principles apply to other layers below the application.

Implementing speculation in the operating system provides a good balance of semantic information and scope. The operating system understands the semantics of useful objects like processes, users, and files, yet is low enough in the software stack to control the execution of all applications. The natural unit of computation for OS-level speculation is a process, and the natural unit of state is a process's address space. The OS also sees objects such as files and sockets and manages related state (such as the file table) on behalf of processes. Processes communicate with the OS mainly through the system call interface.

Figure 1 illustrates the generic approach to operating system speculation. A speculation starts when the operating system predicts the results of an *action* (*A*). Actions are units of computation that have definite start and end points. An ac-

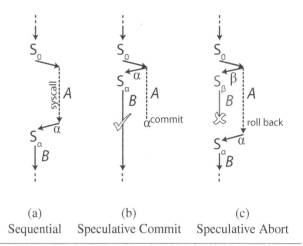

(a)	(b)	(c)
Sequential	Speculative Commit	Speculative Abort

Figure 1. Generic OS speculation. Part (a) shows a process in state S_0 execute syscall A with result α. When the syscall returns, the process continues to execute program action B. In (b) and (c), the system predicts the result of A and returns to user space speculatively while executing A in parallel. If the prediction is correct, the system commits the speculation (b). Otherwise, it aborts (c).

tion causes a process's state to transition from one state to another; actions also may produce output. We refer to the difference in states as the action's *result*. An action is considered *predictable* if its result can be guessed at some point in time before the action completes. For OS-level speculation, actions are typically individual system calls.

Speculative execution allows predictable actions to execute in parallel with the program's future actions. When the operating system can guess the results of a process's action before it executes, the operating system marks the process as speculative, returns the predicted result to the process, and allows it to continue executing speculatively. In parallel, the operating system carries out the action and determines the actual result.

When speculating, a generic speculation system must ensure that no effects resulting from a missed speculation are visible outside the system. To hide misspeculations, the system must roll back all effects of the speculation. To support rollback, the system takes a checkpoint of the process (usually copy-on-write for efficiency) when the speculation begins.

We define the *boundary* of a speculation to be the collection of all objects whose state depends on a speculation. Initially, this boundary will include only the state of the process that initiated the speculative action. As the process interacts with the system, it may try to modify state outside its bounds by generating output. A generic speculation system may handle output in one of three ways, each of which meets the requirement of completely hiding misspeculations.

- *Expand*: the boundary of speculation is expanded to include the receiver of the output, and then the output is sent. When a new object (e.g., a process or file) becomes included in a speculation, a checkpoint of its state must be taken so it can be rolled back if the speculation fails.

- *Defer*: the write is deferred until the speculation commits.

- *Block*: the modifier's execution is halted until the speculation commits.

When the system finishes executing the action, it compares the predicted result with the actual result. If the actual result matches the predicted result, the system commits the speculation and releases any deferred output (Figure 1b). Otherwise, it aborts the speculation and rolls back all state within the speculation's boundary (Figure 1c).

3. Custom Policies

Because an application has more semantic information about its own behavior, its performance can be improved by using a speculation policy that is customized for that application.

A custom, application-specific policy can vary from a generic policy in several ways: creating speculations, managing output, evaluating results, and controlling the commit. Overall, custom policies benefit an application by letting it make more predictions, helping those predictions be more accurate, and allowing it to achieve more work while speculative.

Figure 2 shows an overview of how a sequential execution is parallelized using speculative execution with custom policies. An important distinction between OS generic speculation and custom speculation is which level controls the speculation. In OS generic speculation, the *operating system* executes the action that is being predicted and evaluates the result. In custom speculation, the *application* executes the action that is being predicted and evaluates the result. To allow the application to control the speculation, the system forks the process when the speculation begins. One copy of the process (left side of Figures 2b and 2c) incorporates the predicted result of the action and continues executing speculatively; we call this copy the *speculative process*. The other copy of the process (right side of Figures 2b and 2c) executes the action and compares the actual result with the predicted result; we call this copy the *control process*.

We explore three axes along which an application can provide a customized policy: creating speculations, handling output, and handling commits.

3.1 Creating Speculations

The most basic task in speculative execution is determining where to start and end the speculation, along with what value to predict for that interval. A generic speculation system is not suited to identify the best places to start and end a speculation. First, it sees only a subset of events issued by the process, e.g., system calls. Second, it has little information

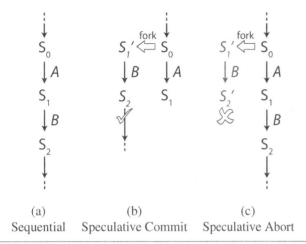

(a)	(b)	(c)
Sequential	Speculative Commit	Speculative Abort

Figure 2. Speculation with custom policies. Part (a) shows a sequential process in initial state S_0 that executes actions A and B, moving its state to S_1 and then to S_2. Parts (b) and (c) show the same process predicting the result of action A and forking a speculative copy of the process that runs B in parallel with A. If S_1 and S_1' are equivalent, the speculation can be committed (b). Otherwise, the speculation is aborted (c), and the process continues from state S_1. We call the left process in Figures (b) and (c) the *speculative process*, and we call the right process in Figures (b) and (c) the *control process*.

by which to determine which of these events are predictable: a certain system call may have a predictable result for one application but not another (e.g., reading a configuration file is more predictable than reading a user document). A generic speculation system may also fail to predict the result of the action, since the same action will often have different results for different applications.

An application sees and understands much more about its own behavior and semantics. For example, a program can start and end speculations at any line of code, rather than only at system calls. We define actions at this level to be an interval of program statements. This definition allows system calls to still count as actions, but it also lets the program speculate over many more regions, including arbitrary function calls. With so many additional actions visible, there is a greater opportunity to find predictable actions.

A custom policy on creating speculations lets the program specify which intervals of code are worthwhile to speculate on and how to predict the intervals' results. The program can pick its actions to be those at an abstraction layer that is easily predictable.

Selecting the right abstraction layer is crucial to locating predictable actions. Interfaces often exist to hide implementation details from the higher layers of a program, and we can take advantage of them to minimize the amount of state that must be predicted. Lower-level actions, themselves unpredictable, may work together to construct a high-level ac-

tion whose effects are well-defined and whose outcome is predictable. Defining a high-level action can filter out the unpredictability of lower-level events and intermediate state changes that are not actually relevant to the overall task.

As an example, consider a program that calls the function `get_user_option()` to display a menu, specify a default choice, and wait for the user to interactively select an option. If we implemented our speculation in a slightly-lower layer of abstraction, the available actions concern the interaction with the menu itself. The program might find itself predicting which menu item the user would select next. At still a lower level, the program might try to speculate on the return value of the `read()` call that gets the user's next keystroke. (Note that this is all the generic system would see.)

By understanding the semantics of the high-level action, a custom policy would let the program speculate over the entire `get_user_option()` function to predict that the user will take the default option. The exact sequence of keystrokes that a user took to make a selection and the internal menu state are irrelevant details that get abstracted away to make the action predictable.

3.2 Output Policy

An application next needs to determine how its output should be handled while it runs speculatively. Recall that a generic speculation system must handle output by expanding the boundary of speculation, deferring the output, or blocking the speculation. Each of these handling strategies has drawbacks in certain situations:

- Expanding the speculative boundary involves more objects in the speculation. This increases complexity and increases the cost of a rollback. For example, if a heavily-shared object such as the X11 server or /etc/passwd became speculative, the speculation would quickly spread among other objects, and the entire system could become speculative (and thus non-responsive).

- Deferring the output prevents the receiver from getting the output and starting useful work. If the speculative sender is waiting on a reply from the recipient, it too will stop making any forward progress.

- Blocking on output is the safest, easiest option, but it performs the worst because it limits how far the application can speculate.

A generic system lacks information about the purpose of the output, the sharing patterns of objects that receive the output, how quickly that output needs to be sent, and how far the application could proceed speculatively after sending the output.

By specifying a custom output policy, an application can choose the best way to handle its output from among these options. With its knowledge of which actions are safe, what it is writing to, and whether it needs a reply, the application is in a better position to make this choice. For instance, an application can avoid deferring writes when it will spin waiting for a reply, and it can expand its speculative bounds only when it would not involve many other objects in the speculation.

In addition, the application may be able to violate the conservative restriction of completely hiding misspeculations, because the application may not care if the output produced by a misspeculation is rolled back. Hence, a custom output policy can specify a fourth strategy in addition to those available to the generic system: *allow* the output without expanding the boundary of speculation and without rolling back the receiver upon misspeculation. An application can safely follow this output strategy in the following scenarios:

- *The output does not modify external state*: Many networking applications use requests that return data without modifying important server state, such as HTTP GET or SQL SELECT requests. Since no state change needs to be undone on a rollback, these requests are safe to allow off the system.

- *The application provides its own safety guarantee*: Even if the system cannot roll back the effects of an output, the application may be able to ensure that on a rollback, the effects of its output will be undone. To guarantee this behavior, a networking application might implement a distributed speculation system by tagging its messages with its outstanding speculations and informing recipients when a rollback occurs.

- *Inconsistent output can be tolerated on a rollback*: A study by Lange shows that users are able to tolerate a limited amount of speculative and inconsistent information being displayed on their screen in exchange for faster performance [Lange 2008].

By customizing its output policy, an application can ensure that its safe output does not cause it to prematurely halt forward progress. A customized output policy also directs the system to handle the unsafe output using the most efficient and appropriate strategy.

3.3 Committing

When the action whose result is being predicted finishes, the system must decide whether to commit or abort the speculation. If the actual result is identical to the predicted result, the speculation can be committed. Without knowing what the application uses the predicted values for, this is the only condition under which a speculation can commit. If a generic system detects any differences between the actual and predicted result, it cannot determine if that difference is significant, so it must be conservative and abort the speculation.

However, some applications can tolerate differences between the predicted and actual states. Custom commit policies let the application specify what differences can be tolerated and how to to deal with those differences. Thus, custom commit policies can broaden the criterion for correct pre-

dictions from being *identical* to being *equivalent*. A custom commit policy can use this flexibility to commit more speculations, thus reducing the number of speculations that roll back and preserving more work. We consider four ways that differences can be equivalent while not being identical.

First, some differences in process state are not semantically important to a valid execution and may be ignored. For example, different patterns of malloc() calls may result in data structures being allocated in different locations. This is safe to ignore if there are no inconsistent pointers to these structures. Likewise, the exact contents of unused stack frames can differ if two executions take different code paths, but these are not significant.

Second, other differences in results may be unused and can also be ignored. For example, a reply to an RPC may convey the complete metadata of a shared object, but the application may only examine its time stamp. Differences in other parts of the reply can be ignored. Another example is when the application uses the time stamp only by comparing with another value. All results where the time stamp is less than the other value are in the same equivalence class.

Third, some state differences may affect execution, but the semantics of the changed state may permit updates to be lost. For instance, a cache may acquire an entry that is not predicted, but the cache's semantics allow it to drop the entry when needed.

Finally, some differences matter but can be imported into the speculative state. To do this, the control process can *forward* the difference to the speculative process to be merged into its current state. Continuing the previous example, although it may be valid for a cache to lose unpredicted entries, the program may wish to preserve them for better performance. If the speculative process has not read or written the differing state before it is forwarded by the control process, the merge can be performed easily without worrying about read/write conflicts. If the speculative process *has* read or written the differing state before it is forwarded by the control process, the updated state can be passed as a message to the speculative process.

4. Issues with Separation

Despite the benefits mentioned in the previous section, splitting the mechanism and policy into different layers causes two new issues that must be addressed. Both issues arise because the application participates in controlling its own speculation.

4.1 Committing State

Our control processes lacks effective isolation between two logically distinct portions of application state: the state used to *control* the speculation is co-mingled with the state *effected* by running the predicted action. The logic carrying out the predicted action and the logic controlling the speculation execute within the same address space (the control process). As a result, there is no easy way for the system to separate the state used by the logic controlling the speculation (which should not be preserved), the predicted results (which must be checked for equivalence), and other unpredicted state (which could be discarded, forwarded, or cause a rollback). That is, if a particular change is detected, it is not obvious how to handle that change.

This issue does not arise in a single-level system. For instance, when an OS speculates on a system call, the process switches to a separate kernel stack, isolating the speculative control logic from the application state. Furthermore, the effects of the system call on the process's state are well-defined. As a result, it is easier to check for equivalent results, and there should never be any completely unpredicted state changes.

It is left to the application's commit policy to decide how to disentangle these pieces of state. In the applications we have modified to use custom speculation, this has not been a significant burden. Still, it is an added complexity that we would avoid if possible.

4.2 Multi-threaded Speculation

The issue of separating state is compounded by multi-threaded processes. Our description of custom speculation so far has assumed that an application has only a single thread, which both executes the action and uses the result. With speculation, we fork this thread into a control thread and a speculative thread. The speculative copy of the thread continues using the predicted result, while the control copy of the thread executes the action and commits or aborts the speculation.

In contrast, with multi-threaded processes, a single address space is shared among many different threads. Some of these threads execute the predicted action and control the speculation; other threads use the results of the predicted action; and some threads may be independent of the predicted action. Forking a process when a speculation begins causes all threads within that process to be copied.

We designed a solution to these issues that lets us speculate using multi-threaded programs. The key issue is deciding which threads to start in each process.

We first consider the case in which the predicted action involves only a single thread (A); other threads may use the predicted result or be independent of the action. Thread A starts a speculation by predicting a result for the action and forking a speculative process. Thread A must run in the control process to execute the predicted action and control the speculation; thread A may also continue in the speculative process after skipping over the portion of its execution that is being predicted.

Threads that use the predicted result must run in the speculative process to achieve the desired parallelism; they cannot run in the control process since they would have to wait for the predicted result.

Threads that are independent of the predicted action may run in both of the control and speculative processes, but this duplicates their work and wastes computing resources. To avoid wasting work, the independent threads are allowed to run in only one of the processes; they are blocked in the other process. Most speculations are more likely to commit than abort, so we run the independent threads in the speculative process (which is more likely to survive the speculation than the control process). While we could merge the changes from the independent threads into the surviving process, this would require us to modify the independent threads to deal with the speculation.

In the general case, multiple threads may be involved in the predicted action. All threads involved in the predicted action must run in the control process, and they must skip the predicted action in the speculative process (if they run in the speculative process). Threads that cooperate on the predicted action must also cooperate on starting and controlling the speculation.

When one thread in a multi-threaded process starts a speculation, our system relies on that thread to determine which of the other currently-running threads should also be started in the control process. Unless otherwise specified, all other threads are assumed to be independent, and remain running only in the speculative process.

5. Mechanism Design & Implementation

Custom policies were introduced in Section 3 by describing what behaviors an application should be allowed to customize. In this section, we discuss how our mechanism layer is built and how applications can express those policies.

5.1 Overview

Our mechanism for speculative execution is implemented at the operating system layer and is based on Speculator, a modified Linux 2.6.26 kernel that provides process-level speculative execution [Nightingale 2005]. Speculator allows processes to continue to interact with the system after becoming speculative. In particular, speculative state can be propagated through several forms of IPC: forking, waiting on children, signals, and file system operations (through files, pipes, and sockets). Each of these kernel objects can be checkpointed and rolled back as needed.

We introduce custom policies to this system by creating a group of new system calls, which are described in Section 5.2. At a high level, policy decisions are executed from within the process, and the system calls are used to direct the mechanism appropriately.

When a speculation starts, the system creates two separate processes (a control process and a speculative process), each with their own address space. The isolation provided by having separate address spaces is crucial: state from the speculative process should not violate causality by influencing the execution of the control process. If a conflict is de-

```
spec_fork(out status, out spec_id)
    Begin a speculation.
commit(in spec_id)
    Commit a speculation.
abort(in spec_id)
    Abort a speculation.
set_policy(in fd, in new_pol, out old_pol)
    Set the file's output policy, returning the old one.
get_specs(out spec_id_list)
    List the current process's uncommitted speculations.
spec_barrier(in spec_id)
    Block until the given speculation has committed.
start_threads(in thread_id_list)
    Start additional threads in the control process.
```

Figure 3. System call API used by an application to construct custom policies.

tected, the system may conservatively abort the speculation. The application should gracefully resume executing as if no speculation were created.

5.2 Policy API

Applications implement custom policies through a new set of system calls. An overview of each call is given in Figure 3.

To create a new speculation, the application invokes spec_fork() and splits into control and speculative processes. After the control process executes its predicted action, it can call commit() or abort() for the speculation.

Custom output policies are specified using the function set_policy(fd, policy). Each write operation can use a per-file policy or, if that is unspecified, a per-thread default policy. Policies specify one of the strategies for handling output described in Section 3.2 or DEFAULT. A single write operation can use its own policy by wrapping it with set_policy().

We permit processes to view the status of ongoing speculations. A process can get a list of its current speculative dependencies by calling get_specs(). The kernel also provides a socket that broadcasts spec_id-s as they are created, committed, and aborted. We also found it useful to allow a speculative process to voluntarily limit its own resource usage. By calling spec_barrier(), a process can halt its execution until some or all of its dependencies have committed.

In a multi-threaded application, only the thread that called spec_fork() is initially started in the control process. All other threads in that process start blocked. If the action requires that other threads be running in the control process, they can be explicitly woken by calling start_threads(). If the speculation aborts, all threads will automatically be woken in the control process. (Note that all threads in the speculative process are active by default.)

6. Design Process

We envision the use of custom speculations as a design process consisting of three steps. First, a developer must locate interesting speculation points in a program. Second, custom speculations must be implemented safely. Finally, the system as a whole should be examined for additional optimization.

6.1 Determining Actions

There are three generic guidelines that should be followed when locating a suitable action to predict. First, executing the action should take longer than the overhead of creating a speculation (i.e., the cost of a fork). Blocking I/O operations (e.g., waiting for user input or network messages) often greatly exceed the overhead cost—our own case studies focus on these operations. Lengthy computations may be appropriate, as long as there are available cores to do the work in parallel. Second, it is important that the speculative process be able to make forward progress. Using a custom output policy may remove some blocking points, thus allowing more progress. Finally, the result of the action must be predictable. By using a custom commit policy, it is sufficient to predict an equivalent result rather than an identical one.

Our system imposes additional constraints on the selection of an appropriate action. We rely on the program to explicitly verify that all effects of the action were predicted. To do this correctly, the developer must be able to understand precisely the effects of the action on the local process's memory. Clean, narrow interfaces for accessing and modifying local state significantly aid the developer in performing this task. An ideal interface cleanly separates pure functions that do not change local state from the mutating functions. In our experience, suitable interfaces are often found at the boundary between program modules. If an action seems too convoluted, it may be more reasonable to look at a different abstraction layer.

6.2 Implementing Custom Policies

Once a suitable action has been located, it is necessary to implement the policy in code as API calls and state modifications. We use the code in Listing 1 as a running example of how to use our API to predict the results of running the function foo().

We found it useful to work with actions defined by a single function. When the code is structured in this way, we can write a wrapper function (spec_foo()) that isolates our policy implementation from the action and surrounding code.

The developer is responsible for deciding how to predict the return value and side effects of executing an action (lines 10–11). Once that prediction is made, spec_fork can be called to split the application into speculative and control processes. The speculative process should update local state as if the action had completed with the predicted result

```
1   int count;  /* Global state */
2
3   int* foo() {
4     ...
5     count++;
6     return ptr;
7   }
8
9   int* spec_foo() {
10    int p_cnt = count + 1;
11    int p_ret = 1;
12    (stat, spec_id) = spec_fork();
13    if (stat == SPEC) {
14      count++;
15      result = new int(p_ret);
16    } else if (stat == CONTROL) {
17      result = foo();
18      if (count == p_cnt && *result == p_ret)
19        commit(spec_id);
20      else
21        abort(spec_id);
22    }
23    return result;
24  }
25
26  void work() {
27    x = spec_foo();  /* Replaces foo() */
28    p = set_policy(fd, ALLOW);
29    send(*x);
30    set_policy(fd, p);
31  }
```

Listing 1. Basic structure for predicting the result of simple function call.

(ln. 14–15). The control process should execute the action (ln. 17) and then, to implement the commit policy, explicitly verify that the changed state matches the prediction (ln. 18).

It is important for correctness that all relevant side effects of the action be predicted (in the speculative process) and verified (in the control process). In the example, if count were ignored in the prediction (i.e., by omitting ln. 14), it would lead to odd program semantics where count appears to increment only when the speculation aborts. Not all differences are relevant: foo might have different dynamic memory allocation patterns from the speculative process's fast update (ln. 15). This difference does not affect program semantics. Hence in the example, only the value of the returned object is checked. It is a challenge to decide which state is relevant. For this reason, it is crucial that the developer be able to understand the behavior of the action.

To selectively allow speculative output on a per-message basis, a program may wrap its I/O functions with calls to set_policy (ln. 28–30). The developer should ensure that the receiver can handle potentially-incorrect data. Also note that after rolling back, messages sent while speculative might be retransmitted.

Although we support executing multiple threads in the control process, a single thread that makes blocking operations on local data is preferred. Acquiring a lock to access shared data may introduce a deadlock if the lock holder is not running in the control thread. If the system can detect the deadlock, the speculation can be aborted, freeing all other threads. We suggest grabbing needed locks or making local copies of data structures before starting the speculation. If multiple threads are required to run in the control process, they should synchronize with each other first before executing a spec_fork. The prediction must include the state changes due to *all* threads' executions.

6.3 Optimization

By examining the behavior of the system in a few key areas, it may be possible to further optimize performance. When a speculation fails, it might be the result of an overly-precise commit policy. An expanded definition of "equivalence" might allow a greater number of speculations to commit. When a speculative process blocks, it could indicate the need for a more permissive output policy. If the process is waiting for output to be released, it could be worthwhile to consider whether it is safe to allow that output. However, the system's performance as a whole may suffer if the boundary of speculation expands too far. If a highly-shared system object becomes speculative, this may suggest that a more-restrictive output policy is needed.

7. Case Studies

To evaluate the effectiveness of our split-layer speculation system, we look at three case studies. We modify each application in the study to add a feature that uses custom policies to achieve greater parallelism. Table 1 shows the applications and which policies they use. For comparison, we discuss the difficulties involved when implementing each feature in single-layer systems at both operating system and application layers. To quantify the changes needed to implement these features, we measure the Lines of Code (LoC) added and modified in each application (excluding blank lines, comments, and braces). Finally, we quantify the improvement in performance due to each feature. Our test system uses dual single-core Xeon 3 GHz processors with 8 GiB of RAM.

Application	Custom Policy		
	Start/End	Output	Commit
Predictive launch:			
Bash	Y		Y
X Proxy		Y	Y
Firefox	Y	Y	Y
BFT	Y	Y	

Table 1. Speculative applications. A "Y" indicates that the application has a custom policy defined for that category.

7.1 Predictive Application Launching

We make use of custom speculation policies to improve perceived application startup time by predicting the launch

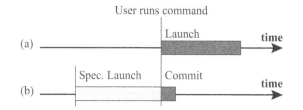

Figure 4. Process Execution Time. In part (a), a process is launched normally at the time the user invokes it. In part (b), the program is launched speculatively ahead of time. We measure the execution time after the user invokes a command (dark bar).

Application	Normal Launch (s)		Speculative Launch (s)
	Warm \$	Cold \$	
LaTeX build[†]	2.66 ± 0.03	4.72 ± 0.07	0.092 ± 0.001
Bash `make`[†]	45.1 ± 0.02	49.0 ± 0.04	0.19 ± 0.001
GIMP*	5.1 ± 0.3	8.4 ± 0.5	0.72 ± 0.03
OpenOffice*	3.33 ± 0.05	11.8 ± 0.08	0.29 ± 0.03

Table 2. Application Run Times[†] and Load Times* for non-interactive and interactive programs, respectively. Normal launches are examined with both warm and cold disk caches; speculative launches were not affected by cache state. Each value is given in seconds and is the mean of 10 runs, with 95% confidence intervals.

of an application and speculatively starting it. This will not decrease the actual time needed to launch the application, but part of that time may be overlapped with the user's think time. As a result, the system will appear more responsive.

We first quantify the potential performance benefit from this technique. When it is possible to successfully predict the next program far in advance, how much work of the program launch can be hidden? Figure 4 illustrates our method for examining the capacity of non-interactive programs to launch speculatively. In a normal launch (Figure 4a), a program starts executing when it is invoked by the user's shell. We measure the run time of the process from its invocation to termination. In a speculative launch (Figure 4b), we begin executing the program before it is requested. Once the speculative program quits making progress, we invoke the application—committing the speculation—and measure the program's run time from that point. We examine two non-interactive applications: building a LaTeX paper, and building the Bash shell via `make`.

Interactive graphical applications do not automatically terminate, so we examine their load time from invocation instead of their run time. We end our load time measurement when the rate of X11 messages sent by the application falls below 200 messages in a 100 ms period. This threshold is arbitrary, but it effectively distinguishes drawing splash screens and main windows from handling smaller incidental actions, like redrawing buttons as the pointer moves across a window. We examine two interactive applications: GIMP 2.2 and OpenOffice 3.1.1.

Table 2 shows the run times and load times of our test applications when launched normally and speculatively. Because of the high impact on load times, we also varied the state of the disk cache. When launching speculatively, we did not find a significant difference in run/load time due to cache state. Although application load times are significantly decreased when using a warm cache, they are not eliminated. When applications are speculatively launched before invocation, almost all execution time spent running/loading the program can be performed before the program is invoked. Compared to a normal launch with a cold cache, at least 91%

of the run/load time is capable of being hidden. Even with warm caches, 85% can be hidden.

Section 7.1.1 describes our modifications to the Bash shell that lets it take advantage of this potential by predicting the user's next command line and executing it speculatively. By itself, the changes to Bash are sufficient to benefit non-interactive commands. Section 7.1.2 describes how we implement an X11 proxy that lets graphical programs benefit from a speculative launch.

7.1.1 Bash

We modified a Bash 3.2.48 shell to predict the next full command line the user will type and begin running it speculatively. Bash predicts one command at a time, starting when the shell prompt is first displayed.

To perform the prediction, we re-implemented the EMA online machine learning algorithm [Madani 2009], which predicts the next line based on the command history. One could also imagine developing an algorithm that alters its guess as the user types. Finding the best predictor is an orthogonal problem; our concern is how to effectively design a system to make use of the predictions.

Following our design process, we identified the interface between Bash and the Readline library as an ideal modification point. We used the basic pattern described in Listing 1 to wrap Bash's call to `readline()` (in Bash's `yy_readline_get()`), which accepts user input and returns it in a new buffer. Other program state is not modified. Our wrapper calls into EMA to generate a predicted buffer. The speculative process returns a copy of this buffer. The control process makes the call to `readline()` and compares the two strings. Note that the two executions return different memory allocations. In this program, only the buffer contents are relevant, so the commit policy makes only that comparison. Other state is assumed, without verification, to not have changed.

Later observation led us to implement two additional changes. First, we found that when a user hits Ctrl-C to interrupt Bash, the signal handler uses `longjmp()` to (incorrectly) bypass our wrapper. We modified the function `throw_to_top_level()` on the interrupt control path to abort outstanding speculations when this happens. Second, we found that tab completion could add spaces to the end of command lines. In response, we added a custom equivalence policy that normalizes commands before comparison.

Overall, only two function in Bash needed modification to permit speculative launching. Basic command prediction used 56 LoC inside Bash to invoke our EMA predictor (433 LoC). The equivalence policy added 36 LoC, mostly text manipulation functions, for a total of 525 LoC. Because Bash relies on the system's default output policy to maintain safety for arbitrary applications, no code was needed to implement the output policy. To put these numbers in perspective, the full source code for Bash is over 100K LoC.

7.1.2 X Proxy

Graphical applications send and receive messages over a socket to communicate with the X server. Following the generic policy used by Bash, a speculative application that attempts to use this socket will either have all of its messages buffered, preventing it from loading, or it will force the X server to become speculative, preventing further user interaction. Neither result is desirable for speculative launching.

The generic policy is unnecessarily restrictive. While loading, an X application issues many requests that read global state or modify application state without resulting in any user-visible output. These messages can be safely exposed to the X server. In particular, applications can create windows and set their properties without exposing those windows to the user (*mapping* the window, in X terminology). The X protocol is designed to operate asynchronously, so those few messages that do result in a visible change can be buffered and released only when the speculation commits.

We design and implement an X proxy that sits between the application and the X server to selectively permit messages through the boundary of speculation. By placing this functionality in a proxy, we can support arbitrary unmodified applications and avoid modifying the core X server.

For ease of development, we modify an existing proxy: xtrace 1.0.2 [Link 2010]. When a new application connects to it, the proxy forks a new server, which becomes speculative immediately after accepting the connection. The proxy takes advantage of system support for buffering output to avoid complicating its own message-handling code. Using custom output policies, requests to map, unmap, or delete widows are deferred. All other requests are allowed. The proxy rewrites sequence numbers in each message to correct for the buffered messages.

When the speculation commits, the system releases the buffered messages, and the application begins to draw its main window. The proxy is notified of the commit and performs a custom commit action: it adjusts its sequence number rewriting algorithm for the newly-released messages. If the speculation aborts, the proxy will exit, breaking its connection with the X server. The X server can recover by releasing application-held resources without rolling back.

Implementing these changes added 280 LoC to xproxy (itself 7K LoC). Most code additions are used for sequence number rewriting.

7.2 Firefox Certificate Checks

Verification can be a slow process whose outcome is often predictable. We use the Firefox 3.5.4 web browser as an example of how to execute verification tasks in parallel with the rest of an application. The task we speculate on is Firefox's verification of a server's public certificate.

Many Internet protocols use the SSL/TLS protocol to establish a secure link between client and server. To establish a session, a Firefox sends a handshake and receives the server's public certificate. It then validates this certificate by contacting the certificate's issuer. Finally, if the certificate is valid, Firefox exchanges random data with the server to derive a session key. Encrypted data can then be sent. We modify Firefox to predict that certificates are valid and speculatively agree on a session key. The data stream should be delayed until the validation is committed.

It would be difficult for a generic speculation system to provide this feature. First, the generic speculation system would need to distinguish the requests used to verify the certificate from other network messages. Second, it would need to predict the entire reply to the client's verification request, which is especially difficult if this certificate has not been previously verified. Furthermore, once the speculation has started, the generic system must treat further output conservatively and prevent it from leaving the system.

Speculation could also be implemented entirely within Firefox. However, this would require the programmer to implement a custom checkpoint mechanism, and such a mechanism would require extensive code modifications throughout the program because Firefox is not written to isolate its state. Furthermore, the programmer would need to manually block most output while speculative.

To express this feature using custom speculations, we create a variant of the `ocsp_GetOCSPStatusFromNetwork()` function in the NSS component, which requests the status of a certificate from a remote server and caches the result. Our speculative process assumes the verification succeeds, so it places a fake success record in the cache before returning. We also use a custom output policy that allows SSL handshake data to be sent: socket output is allowed around some calls to `ssl3_GatherData()`. Certificate prediction and cache modification used 122 LoC, and the output policy was specified in 27 LoC. For comparison, the certificate validation code alone takes 8.5K LoC.

We encountered two difficulties during development. First, by default the validation request is handed off to a dedicated thread that performs simple requests. We did not expect multiple threads to be involved, and the dependency prevented our speculation from succeeding. The easiest fix

Site	Spec. (ms)	Normal (ms)	Speedup
Google Accounts	297.6 ± 31.9	330.3 ± 32.7	9.9%
Windows Live ID	416 ± 46	501 ± 43	17%
Chase home page	310 ± 51	382 ± 46	19%

Table 3. SSL Connection Establishment Time. Time taken to establish the first SSL connection to various sites, for speculative vs. unmodified Firefox. Error values show 95% confidence intervals. Despite the high variance, a T-Test confirms with 94% confidence that there is latency reduction when using speculation.

was to eliminate the dependency by sending the request in the validating thread. Second, sometimes a chain of certificates must be validated. Since the speculative process only inserted a fake cache record for the first certificate, subsequent cache modifications by the control process were being lost. To preserve the data, we implemented a custom commit policy that forwards (via a message buffer in shared memory) the verification response for all certificates from the control process to the speculative process. Forwarding added 90 LoC, for a total of 239 LoC changed in Firefox.

To evaluate the impact of this feature on performance, we used a packet analyzer to measure the amount of time taken to establish an SSL connection with and without speculation. Note that certificate verification is only one step in session establishment. Our results are presented in Table 3. Overall, our improvement decreases the time it takes to establish an initial SSL connection by an average of 15% when certificates have not been revoked.

7.3 BFT Client

We next examine a client in the PBFT-CS protocol [Wester 2009], a program already implemented to use an application-level speculation system. This allows us to compare our generic mechanism against one that has been designed for a single application.

The PBFT-CS protocol was designed to decrease the perceived latency of executing requests on a Byzantine fault-tolerant (BFT) cluster. A complete characterization of this problem is present in the cited work. Here, we summarize the system and discuss how applications used PBFT-CS.

BFT services are accessed through a shared library using an RPC interface: clients submit requests and wait for the service to return a reply. Because each reply may come from a faulty server, it is necessary to wait until a quorum of authenticated matching replies is received before the client can determine the correct reply. Servers must coordinate their execution of requests; consequently each operation typically has high latency. PBFT-CS observes that the first reply is usually correct and allows a client capable of speculation to continue executing before the reply is known to be correct. Further requests encode speculative dependencies so that the service can squash aborted requests. As a result, the client

sees lower latencies for its requests and it can pipeline requests to increase its own local throughput.

To evaluate PBFT-CS, a client was constructed that implements its own lightweight checkpoint system. This is a single-purpose application-implemented speculation system. In this work, we compare that client against our own client designed to use custom policies.

We can see several examples of custom policies in this client description. The BFT code decides when to make a prediction (after receiving one reply), what to predict (that the first reply will be validated), which output to allow (additional BFT requests, with modification), and when to commit (after receiving enough replies).

Our policy-based client implements its speculation logic entirely within the BFT shared library. We modified the inner message-handling routine to expose intermediate results. Then, from a layer between the internal functions and the client, we use spec_fork() to implement our own custom start policy. As results are returned, our layer associates the reply with the current dependency set (from get_specs()) to be encoded on future requests. We set the output policy on BFT sockets to allow all messages to be sent. We implement a default commit policy by requiring the actual reply to be identical to the predicted reply. These internal changes and policies were implemented in 221 LoC, out of 17K LoC for the full library.

By using custom policies, our modified BFT library can be used by any existing BFT application without further modification. Those applications can also specify their own policies for other uses without conflicting with those set by the BFT library.

In contrast, the application-implemented client is tied to the service that is using it. Instead of being written as a sequential process that uses blocking operations (the normal RPC interface), this client uses a main event loop. Making the logic event-based forces state to be isolated and saved outside of the stack, so checkpoints can be safely taken between events using memcpy(). Other applications have far more state (that may extend into the OS, if open files are considered), which will require more complex checkpointing logic.

To interpose on output, the client logic is written not to perform output directly. BFT requests are queued and handled by the mechanism so that checkpoints can be created correctly. Other application output must be queued so it can be released only when its dependencies have committed.

We see the policy-based library as an improvement in programmability. It is also necessary to consider the performance trade-offs involved when selecting between using a policy-based system or an application-implemented system. From PBFT-CS, we evaluate a simple shared counter service with a single operation that increments the counter and returns its value. The client simply executes a fixed number of requests in a tight loop.

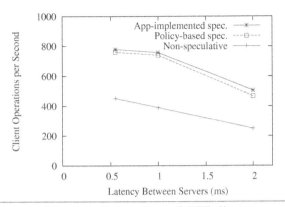

Figure 5. Comparison of BFT clients.

We examine this application from two perspectives. First, we consider the improvement of our client's performance due to speculative execution. Second, we compare two different implementations of speculation: our policy-based system and an application-implemented system tuned for the application. This comparison lets us quantify the performance cost we incur by relying on heavier, generic checkpoints.

The benefit of an application-implemented speculation system is a small performance advantage over our speculation system. Figure 5 compares a non-speculative client against speculative clients implemented in both policy-based and application-implemented systems. We vary the amount of network latency between each server and see how it affects each client's throughput when accessing a lightly-loaded server.

Both speculative clients perform much faster than the non-speculative one. The policy-based client allows the client to issue 82% more requests per second than the non-speculative client with latencies above 0.5 ms. The client using application-implemented speculation employs a checkpoint and restore mechanism that is tuned to the application. Hence, it has less overhead and is able to issue 90% more requests per second than the non-speculative client (an 8% improvement over our generic mechanism). In exchange, the development effort for the client is greatly increased, and it cannot expand its speculative boundary beyond the process itself. A developer must balance these trade-offs when deciding how to implement a feature speculatively.

8. Related Work

Fast Track [Kelsey 2009] is a speculative runtime environment that allows applications to direct speculations over their own execution in a similar style to our custom policies. A programmer invokes `FastTrack()` to fork and and let one branch become speculative, like `spec_fork()`. Each side executes different version of the same action that are predicted to be *identical*: a fast but unsafe version and a slow, correct one. We go beyond the Fast Track model by giving the programmers greater control over when to commit and abort speculations in the presence of state differences that may be irrelevant. Our system also allows applications to specify a custom output policy and to speculate based on the actions of multiple coordinating threads. Fast Track, being implemented in the language compiler and runtime, cannot expand its boundary of speculation beyond its own process.

Prospect [Süßkraut 2010] is a compiler-based platform to generate programs that execute a fast program variant speculatively along with a slow variant that can include additional safety checks. Speculative system calls are allowed, although their effects are only made visible to other processes after a commit. Prospect also commits on equivalent, rather than identical, states. However, this is not verified in current implementations. In the context of our work, applications modified by Prospect could have benefited from the existence of a shared kernel mechanism to handle speculative system calls that would have allowed it to specify a default *defer* output policy. One could also view this project as an implementation of speculative mechanisms and policy at a low layer (the language and runtime) without considering the application semantics.

Crom is another framework that allows applications to control their own speculations [Mickens 2010]. This mechanism is implemented as a JavaScript library that lets web application developers predict upcoming UI events. Developers flag individual events and provide lists of likely values for input controls. Equivalence functions are specified to let the system determine which speculative executions could match the user's actual event. The programming model for JavaScript is simpler than that for arbitrary binaries. Hence, custom policies must deal with a wider range of actions. Crom does not provide an analogue to custom output policies for its two I/O actions: network requests generated by the speculative code are sent and writes to the screen are kept hidden until a commit. Speculations capture the full state of the DOM tree and are isolated from each other, so causality tracking is not needed in this system.

We broadly categorize other work by considering what it is predicting, how much control it gives to applications, and what layer in the software stack implements the mechanism.

Speculative parallelism. Our work is closest to other systems that are designed to execute sequential code segments concurrently. Thread-level speculation (TLS) systems execute blocks of sequential code in parallel on separate threads, predicting that there are are *no memory conflicts* between the blocks [Steffan 1998]. TLS systems provide fine-grained parallelism, and the selection of the blocks is often driven by automated program analysis. The mechanism needed to support speculations at this granularity often has problems rolling back in the presence of system calls or I/O operations, so these are disallowed while speculative. Our system is built to support speculations at a much coarser granularity, and we consider system calls and I/O to be good sources for predictable actions. Because our system predicts *state* instead of *read/write sets*, the programmer can specify what value should be read by future reads.

Transactions. Speculative execution is similar in many ways to atomic transactions, and thus our system is similar to systems that provide operating system support for application transactions, such as QuickSilver [Schmuck 1991] and TxOS [Porter 2009]. Both transactions and speculation execute actions in parallel with other code, and both can commit or abort the action. The difference between transactions and speculation is the relationship between the action and other code. With transactions, other code executes in parallel with the action (with varying degrees of isolation [Gray 1993]). In speculation, the outcome of an action is being predicted, and other threads are continuing based on that prediction.

Transactional memory and optimistic concurrency control are uses of transactions that also leverage a prediction [Herlihy 1993]. As with TLS, these uses of transactions predict that there are no read/write conflicts between concurrently executing threads.

Generic speculation. There are many examples of generic low-level systems that do not take advantage of application semantics. Speculator originally predicted only system-level events such as NFS calls and disk syncing [Nightingale 2005, 2006]. Pulse speculatively resumes threads that are waiting for a resource to see if they will deadlock [Li 2005]. The Time Warp system lets processes in a distributed system run speculatively under the assumption that all their messages arrived in the correct program order [Jefferson 1987]. Țăpuș et al. also performed similar speculations for a distributed shared memory system [Țăpuș 2003]. These systems begin speculations only on system-visible events, and either disallow other output or handle it conservatively.

Systems offering customization. The Atomos programming language offers open transactions, which allow a thread to commit its writes back to memory while inside an uncommitted transaction [Carlstrom 2006]. Our custom output policies also allow for the same behavior, though we also consider blocking and expanding speculative boundary. The Mojave compiler also exposes an interface to start, commit, and abort speculations [Smith 2007]. During a speculation, isolation is preserved, and since this is a runtime-based system, most system calls are not allowed. In Fast Track, the application customizes the actions being predicted and the predicted result, but not other policies.

Custom speculation implementations. The work by Lange et al. on speculative remote displays is an example of a program that uses an application-implemented speculation system [Lange 2008]. They built a remote VNC viewer that predicts screen updates and displays the speculative view to the user. The authors also found that RDP events are also predictable, but they did not attempt to build a viewer, citing RDP's reliance on client state.

9. Conclusions

In this paper, we explored the advantages of separating the mechanism to support speculative execution from the policy that describes what needs to be done. Applications that wish to use speculative execution are freed from the burden of implementing their own mechanisms such as checkpointing, rollback, causality tracking, and output buffering. Instead, they can focus on defining when to begin speculating, what results to predict, how output should be handled when speculative, and when to commit the speculation.

We demonstrate the effectiveness of our mechanism/policy split by examining three different applications that can be easily modified using our shared mechanism. First, our system reduces the startup time of programs by at least 85% when the program's launch can be predicted. Secondly, the latencies of establishing secure connections on Firefox are reduced by 15%, as our new mechanism/policy split allows it to perform certificate verification in parallel, partially removing it from a critical path.

Finally, the BFT client shows the low trade-off between performance and convenience in our system. While using an optimized application-level speculation mechanism gives an 8% performance improvement over our separated speculation system, its use prevents the application from interacting with the rest of the system while speculative.

Acknowledgments

We would like to thank our shepherd Frans Kaashoek and the many anonymous reviewers who provided valuable feedback to improve this paper. This work is supported by Intel and the National Science Foundation under awards CNS-0905149 and CNS-0614985. The views and conclusions contained in this document are those of the authors and should not be interpreted as representing the official policies, either expressed or implied, of NSF, the University of Michigan, or the U.S. government.

References

[Carlstrom 2006] Brian D. Carlstrom, Austen McDonald, Hassan Chafi, JaeWoong Chung, Chi Cao Minh, Christos Kozyrakis, and Kunle Olukotun. The Atomos transactional programming language. In *Proc. 2006 ACM SIGPLAN Conference on Programming Language Design and Implementation*, pages 1–13, Ottawa, Ontario, Canada, June 2006.

[Cully 2008] Brendan Cully, Geoffrey Lefebvre, Dutch Meyer, Mike Feeley, Norm Hutchinson, and Andrew Warfield. Remus: High availability via asynchronous virtual machine replication. In *Proc. 5th USENIX Symposium on Networked Systems Design and Implementation*, pages 161–174, San Francisco, CA, April 2008.

[Gray 1993] Jim Gray and Andreas Reuter. *Transaction Processing: Concepts and Techniques*. Morgan Kaufmann Publishers, Inc., 1993.

[Herlihy 1993] Maurice Herlihy and J. Eliot B. Moss. Transactional memory: Architectural support for lock-free data structures. In *Proc. 20th Annual International Symposium on Computer Architecture*, pages 289–300, San Diego, CA, May 1993.

[Jefferson 1987] D. Jefferson, B. Beckman, F. Wieland, L. Blume, M. DiLoreto, P.Hontalas, P. Laroche, K. Sturdevant, J. Tupman, Van Warren, J. Weidel, H. Younger, and S. Bellenot. Time Warp operating system. In *Proc. 11th ACM Symposium on Operating Systems Principles*, pages 77–93, Austin, TX, November 1987.

[Kelsey 2009] Kirk Kelsey, Tongxin Bai, Chen Ding, and Chengliang Zhang. Fast Track: A software system for speculative program optimization. In *Proc. 7th Annual IEEE/ACM International Symposium on Code Generation and Optimization*, pages 157–168, Seattle, WA, March 2009.

[Lange 2008] John R. Lange, Peter A. Dinda, and Samuel Rossoff. Experiences with client-based speculative remote display. In *Proc. 2008 USENIX Annual Technical Conference*, pages 419–432, Boston, MA, June 2008.

[Li 2005] Tong Li, Carla S. Ellis, Alvin R. Lebeck, and Daniel J. Sorin. Pulse: A dynamic deadlock detection mechanism using speculative execution. In *Proc. 2005 USENIX Annual Technical Conference*, pages 31–44, Anaheim, CA, April 2005.

[Link 2010] Bernhard R. Link. XTrace - trace X protocol connections. http://xtrace.alioth.debian.org/, September 2010.

[Madani 2009] Omid Madani, Hung Bui, and Eric Yeh. Efficient online learning and prediction of users' desktop actions. In *Proc. 21st International Joint Conference on Artificial Intelligence*, pages 1457–1462, Pasadena, CA, July 2009.

[Mickens 2010] James Mickens, Jeremy Elson, Jon Howell, and Jay Lorch. Crom: Faster web browsing using speculative execution. In *Proc. 7th USENIX Symposium on Networked Systems Design and Implementation*, San Jose, CA, April 2010.

[Nightingale 2005] Edmund B. Nightingale, Peter M. Chen, and Jason Flinn. Speculative execution in a distributed file system. In *Proc. 20th ACM Symposium on Operating Systems Principles*, pages 191–205, Brighton, United Kingdom, October 2005.

[Nightingale 2006] Edmund B. Nightingale, Kaushik Veeraraghavan, Peter M. Chen, and Jason Flinn. Rethink the sync. In *Proc. 7th Symposium on Operating Systems Design and Implementation*, pages 1–14, Seattle, WA, October 2006.

[Porter 2009] Donald E. Porter, Owen S. Hofmann, Christopher J. Rossbach, Alexander Benn, and Emmett Witchel. Operating system transactions. In *Proc. 22nd ACM Symposium on Operating Systems Principles*, pages 161–176, October 2009.

[Schmuck 1991] Frank Schmuck and Jim Wylie. Experience with transactions in QuickSilver. In *Proc. 13th ACM Symposium on Operating Systems Principles*, pages 239–253, Pacific Grove, CA, October 1991.

[Smith 1981] James E. Smith. A study of branch prediction strategies. In *Proc. 8th Annual International Symposium on Computer Architecture*, pages 135–148, May 1981.

[Smith 2007] Justin D. Smith, Cristian Ţăpuş, and Jason Hickey. The Mojave compiler: Providing language primitives for whole-process migration and speculation for distributed applications. In *Proc. International Parallel and Distributed Processing Symposium*, pages 1–8, March 2007.

[Steffan 1998] J. Gregory Steffan and Todd C. Mowry. The potential for using thread-level data speculation to facilitate automatic parallelization. In *Proc. 1998 Symposium on High Performance Computer Architecture*, pages 2–13, Las Vegas, NV, February 1998.

[Süßkraut 2010] Martin Süßkraut, Thomas Knauth, Stefan Weigert, Ute Schiffel, Martin Meinhold, and Christof Fetzer. Prospect: A compiler framework for speculative parallelization. In *Proc. 8th Annual IEEE/ACM International Symposium on Code Generation and Optimization*, pages 131–140, Toronto, Ontario, Canada, April 2010.

[Ţăpuş 2003] Cristian Ţăpuş, Justin D. Smith, and Jason Hickey. Kernel level speculative DSM. In *Proc. 3rd IEEE/ACM International Symposium on Cluster Computing and the Grid*, pages 487–494, May 2003.

[Wester 2009] Benjamin Wester, James Cowling, Edmund B. Nightingale, Peter M. Chen, Jason Flinn, and Barbara Liskov. Tolerating lagency in replicated state machines through client speculation. In *Proc. 6th USENIX Symposium on Networked Systems Design and Implementation*, pages 245–260, Boston, MA, April 2009.

SRM-Buffer: An OS Buffer Management Technique to Prevent Last Level Cache from Thrashing in Multicores

Xiaoning Ding *

The Ohio State University

dingxn@cse.ohio-state.edu

Kaibo Wang

The Ohio State University

wangka@cse.ohio-state.edu

Xiaodong Zhang

The Ohio State University

zhang@cse.ohio-state.edu

Abstract

Buffer caches in operating systems keep active file blocks in memory to reduce disk accesses. Related studies have been focused on how to minimize buffer misses and the caused performance degradation. However, the side effects and performance implications of accessing the data in buffer caches (i.e. buffer cache hits) have not been paid attention. In this paper, we show that accessing buffer caches can cause serious performance degradation on multicores, particularly with shared last level caches (LLCs). There are two reasons for this problem. First, data in files normally have weaker localities than data objects in virtual memory spaces. Second, due to the shared structure of LLCs on multicore processors, an application accessing the data in a buffer cache may flush the to-be-reused data of its co-running applications from the shared LLC and significantly slow down these applications.

The paper proposes a buffer cache design called Selected Region Mapping Buffer (SRM-buffer) for multicore systems to effectively address the cache pollution problem caused by OS buffer. SRM-buffer improves existing OS buffer management with an enhanced page allocation policy that carefully selects mapping physical pages upon buffer misses. For a sequence of blocks accessed by an application, SRM-buffer allocates physical pages that are mapped to a selected region consisting of a small portion of sets in LLC. Thus, when these blocks are accessed, cache pollution is effectively limited within the small cache region. We have implemented a prototype of SRM-buffer into Linux kernel, and tested it with extensive workloads. Performance evaluation shows SRM-buffer can improve system performance and decrease the execution times of workloads by up to 36%.

Categories and Subject Descriptors D.4.2 [*Operating Systems*]: Storage Management—Main memory

---

* Currently working at Intel Labs Pittsburgh.

General Terms Design, Performance

1. Introduction

CPU cache and operating system buffer cache are two critical layers in the memory hierarchy to narrow the speed gap between CPU and memory and the speed gap between memory and disks. Since CPU cache is at the hardware level while the buffer cache is a part of operating system, these two layers are designed independently without necessary awareness of each other. However, with the prevalence of multicore architecture and increasingly large capacity of main memory, if these two layers do not work cooperatively, severe performance degradation may be incurred due to mutual impact to each other. On a multicore processor, a thread accessing a large set of data cached in OS buffer may significantly slow down its co-runners because it can easily pollute the shared hardware cache(s) in the processor. This problem needs to be paid serious attention by OS researchers.

An OS buffer cache keeps recently accessed file system data blocks in memory. Thus, future accesses to these blocks can be satisfied from the main memory without long-latency disk accesses. It also buffers recently generated data blocks to delay corresponding disk writes and to absorb rewrites. As the main memory size continues to grow, an increasingly large amount of data blocks can be buffered in memory to serve file accesses of running applications.

Data in OS buffer cache usually have much weaker temporal localities than that in virtual memory spaces [Leung 2008, Roselli 2000]. One of the major reasons is that files are often used by applications as a data storage rather than a working space. For example, an application processing an array of records stored in a file iteratively reads a record from the file into an object in its virtual space and works on the object before it moves forward to the next record. In the application, the object is repeatedly used and has strong temporal locality, and the records stored in the file have weak temporal locality (they are accessed only once during every execution of the application). For simplicity, in this paper, we refer to the data in buffer cache as *buffer data* and the data in application virtual spaces as *VM data*, and we refer to CPU cache as *cache* and OS buffer cache as *buffer*.

File accesses are usually bursty [Gribble 1998]. Thus, when an application accesses buffer data or generates a large amount of buffer data, it evicts to-be-reused VM data from caches and pollutes the caches with buffer data. CPU cache pollution may significantly increase the number of cache misses and lead to serious performance degradation.

Though the cache pollution problem incurred by OS buffer also exists on single-core processors, it is particularly serious on multicore processors, where the shared last level cache architecture is a conventional design. As we will show later in section 2, a thread reading or writing blocks in OS buffer can slow down another thread co-running with it significantly by a factor of two. The main reason for the performance degradation is that cache pollution leads to severe thrashing in the cache shared by the threads. When the thread accessing buffer data evicts to-be-reused VM data of its co-running threads from the shared cache, the co-running threads keep reloading the evicted VM data into the shared cache upon their reuses. The cache thrashing caused by evicting and reloading the to-be-reused VM data significantly degrades application performance.

On a multicore system, through a shared cache, a thread can influence and at the same time be influenced by the execution of multiple co-running threads sharing the cache with it. This intensifies the degree of cache thrashing, as well as the degree of performance degradation. For example, on an Intel Core i7 processor, four cores share the same L3 cache. Thus any thread accessing data in OS buffer may slow down multiple threads. On future processors, a cache may be shared by many cores, and the cache pollution may degrade performance even more seriously.

Cache pollution incurred by OS buffer in existing systems can trace back its root of the page allocation policies, which target to minimize disk accesses caused by paging and OS buffer misses, without taking into account how much cache pollution may be incurred by accessing the blocks in the allocated pages. Upon buffer misses, OS buffer management allocates physical pages to hold the demanded disk blocks. It selects physical pages holding the virtual pages or disk blocks that are not likely to be used in near future, and evicts the virtual pages or blocks to vacate the physical pages for the demanded blocks. The selected physical pages are rather "randomly" mapped to different regions of the shared cache, and make it difficult to control the cache pollution caused by accessing the data in the pages.

To address the cache pollution problem, we have designed a buffer management scheme called *Selected-Region-Mapping-Buffer* (SRM-buffer). SRM-buffer effectively reduces cache pollution with an enhanced page allocation policy. Unlike the buffer management in existing OS's, SRM-buffer carefully selects physical pages as buffer misses happen. For the related blocks that might be repeatedly accessed in sequence, SRM-buffer refers to them as a sequence and allocates physical pages that are mapped to a small region

consisting of a small portion of sets in the shared cache, leveraging the fixed mappings between physical pages and cache sets. Thus, when these blocks are re-accessed later on, cache replacement is limited within this cache region, and cache pollution can be alleviated. Operating systems assign a distinct color to each cache region and the physical pages mapped to the region. Thus, physical pages allocated to the blocks in the same sequence are in the same color in SRM-buffer.

Two technical issues are addressed in SRM-buffer. One issue is how SRM-buffer predicts which blocks are related and accessed together in sequence so that SRM-buffer selects physical pages in the same color for them. To achieve our goal, we have developed a few heuristics based on the information passed down from OS scheduler and file system.

The other issue is how to coordinate different requirements of buffer management and virtual memory management on physical page allocation, while retaining the hit ratio of OS page cache. To reduce cache conflicts, virtual memory management prefers to evenly allocate physical pages in all available colors. To reduce paging activity and OS buffer misses, OS page replacement prefers to reclaim the physical pages holding inactive virtual pages and stale blocks that might not be reused. However, to reduce CPU cache pollution incurred by OS buffer, SRM-buffer prefers to allocate physical pages in the same color. As we will explain later in Section 3, exhaustively allocating pages in one color may hurt page hit ratio because memory for active virtual pages or non-stale blocks may be reclaimed. At the same time, it causes an uneven color distribution among available physical pages, and thus increases the difficulty to evenly allocate physical pages to virtual memory.

To satisfy and balance the above requirements on physical page allocation, SRM-buffer makes a trade-off among OS buffer reducing cache pollution, virtual memory management reducing cache conflicts, and page replacement maintaining high hit ratios. SRM-buffer does not exhaustively allocate physical pages in one color. Instead, SRM-buffer uses different colors for blocks in different sequences, and for a long sequence, SRM-buffer switches colors regularly each time when an appropriate number (e.g. 256 in our current design) of physical pages in a color have been allocated. As we will show in section 4, the trade-off has a minimal effect to the capability of SRM-buffer to reduce cache pollution, but it can effectively avoid the negative effects on virtual page allocation and OS page replacement.

The contributions of the paper are threefold. First, this is the first work that identifies and studies the performance degradation problem caused by bursty OS buffer accesses on multicore systems. Second, we have proposed an effective solution to address the problem by carefully selecting physical pages to buffer disk blocks. Finally, with a prototype implementation based on Linux kernel 2.6.30, we have tested our solution with extensive experiments, which show our design can effectively improve system performance and decrease the execution times of the workloads.

The rest of the paper is organized as follows. Section 2 briefly introduces the CPU cache designs in common multicore processors and explains how bursty OS buffer accesses may cause performance degradation on multicore systems with an illustrative example. Section 3 describes the design of the SRM-buffer. Section 4 provides a comprehensive evaluation of SRM-buffer. This is followed by a related work session where we also discuss other techniques that may address the problem. Section 6 concludes the paper.

2. Background and Motivation

In general, multicore processors have shared last level caches (LLC). Examples include Intel Core 2 and Nehalem, AMD Opteron and Phenom, Sun Niagara, and IBM Power7. Figure 1 illustrates the shared cache structures with Intel Xeon 5355 processors (Core 2 architecture) and Intel Core i7 860 processors (Nehalem architecture), which are used in our experiments. On a Xeon 5355 processor, each core has private L1 caches, and every two cores share an L2 cache. On a Core i7 processor, four cores share the same L3 cache, and each core has an L1 data cache, an L1 instruction cache, and an L2 cache.

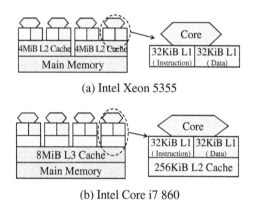

(a) Intel Xeon 5355

(b) Intel Core i7 860

Figure 1: Hardware caches in Intel Xeon 5355 (Core 2) and Core i7 (Nehalem) multicore processors

Compared to private cache architectures on single-core processors, shared cache architectures reduce cache coherence overhead and increase cache space utilization with space sharing and data redundancy reduction. However, pollution in a shared cache may cause cache thrashing, and lead to more significant performance degradation than it does in private cache architectures. In this section, we present an experimental case to demonstrate that the cache pollution problem caused by visiting buffer data in OS buffer can significantly degrade application performance. In the experiment, we run two applications on a Xeon 5355 processor. Application *grep* looks for a randomly-generated character string in a tarball containing Linux kernel source code. Before the experiment, we warm up the OS buffer so that accesses to the tarball are satisfied in OS buffer. The other

application *mergesort* sorts an array of 64-byte records with a non-recursive 2-way mergesort algorithm. In the experiment, we vary the array size from 256KiB to 1.5MiB by increasing the number of records. The working set size of *mergesort* is about two times of the array size, which is smaller than the L2 cache size on the Xeon 5355 processor.

To measure how much *mergesort* is slowed down by *grep*, for each size of the array, we run *mergesort* and *grep* together for two times. We first run the applications on two cores that do not share L2 cache, and collect an execution time T_1 of *mergesort*. Then we run them on two cores sharing the same L2 cache, and collect another execution time T_2 of *mergesort*. We also collect the number of last level cache misses (denoted *LLCM*) incurred by every million instructions of *mergesort*. Finally, we calculate the slowdown of *mergesort* (i.e. $(T_2 - T_1)/T_1$), and show the changes of the slowdown and LLCM in Figure 2a and Figure 2b, respectively.

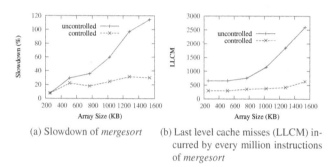

(a) Slowdown of *mergesort*

(b) Last level cache misses (LLCM) incurred by every million instructions of *mergesort*

Figure 2: A comparison of the performance degradation of *mergesort* due to the co-running *grep* polluting the shared L2 cache on the existing Linux system that does not control the pollution (solid lines) and on an improved system that controls the pollution (dotted lines). The array size is varied from 256KiB to 1.5MiB in *mergesort*.

When we increase the number of records in *mergesort*, it becomes more possible for the records to be evicted from the last level cache by *grep*, because the temporal locality of the records gets weaker. As a result, we see the *LLCM* of *mergesort* increases significantly and the slowdown increases accordingly. When the array size is increased to 1.5MiB, *mergesort* is slowed down by as much as 114%.

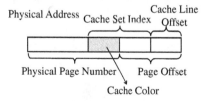

Figure 3: Physical address in the page coloring technique

The performance degradation can be reduced by carefully selecting physical pages to hold the blocks in the tarball visited by *grep*. The selection is based on the page coloring technique, which was proposed in operating systems to reduce cache misses through careful mapping between virtual pages and physical pages [Kessler 1992]. The logic of the page coloring technique is shown in Figure 3. Memory management in operating systems uses the most significant bits of a physical address as the physical page number. When the address is used in a cache lookup operation, some bits in the middle (*cache set index* in the figure) are used to determine the cache set to look up. There are several common bits between the cache set index and the physical page number. These bits are referred to as *cache color*, or *color* for brevity. Cache sets with the same *color* value in their cache set indexes form a *cache region*. Physical pages mapped to the same cache region also have the same color.

Due to the fixed mapping between physical pages and cache sets, there are a fixed number of physical pages in each color. For example, there are 64 different cache colors on Intel Xeon 5355 processor with a page size of 4KiB. These colors evenly divide the last level cache into 64 non-overlapping regions and divide physical pages into 64 disjoint groups. If the physical memory size is 4GiB, there are 16384 pages (i.e. 64MiB physical memory for 4KiB pages) in each color.

Virtual memory management prefers to allocate physical pages in different colors to virtual pages consecutively accessed by an application, so that these pages can use different cache regions to reduce cache conflicts. In the experiment, instead of manipulating the mapping between physical pages and virtual pages, we improve the mapping between physical pages and buffer pages. We use physical pages in the same color to hold the file blocks visited by *grep*. Thus, *grep* only pollutes the corresponding cache region, and other cache regions are not affected.

To confirm the effectiveness of this method, we re-run the applications on two cores sharing the same LLC. We show the slowdowns in Figure 2a with the dotted line. We observe that the slowdowns are significantly reduced, compared to the cases in which the improvement has not been made. At the same time, the execution of *grep* is not affected. Figure 2b explains the reason by showing the significant LLC miss reduction for *mergesort*. When the array size is 1.5MiB, the LLC misses can be reduced by 78%. As a result, the slowdown is reduced from 114% to 30%.

The above example shows the serious performance degradation caused by visiting weak locality buffer data and demonstrates the effectiveness of improving OS buffer page allocation scheme to address this problem. This case study motivates our SRM-Buffer design that will be described in the next section.

3. Selected Region Mapping Buffer Design

In the simple example in last section, cache pollution is restricted within a small cache space by allocating physical pages in the same color to hold buffered blocks. However, to use the similar method in production systems, some practical issues must be addressed. These issues arise because there

are only a fixed number of physical pages in each color. In this section, we first describe these issues and explain how we address them. Then we introduce the design of the SRM-buffer.

3.1 Technical Issues in SRM-buffer Design and Implementation

A straight-forward method for OS buffer to reduce cache pollution is to use only physical pages in a few dedicated colors for OS buffer and physical pages in other colors for other purposes. Though this solution can limit the cache pollution within the corresponding cache regions, it is not practical due to the large sizes of OS buffers and the fixed number of physical pages in each color. OS buffer usually accounts for a large percent of memory usages in modern computer systems. We have examined the memory usage on 24 machines in our department with the *free* command. On average, OS buffers occupy 62% of the physical memory on these machines. This is consistent to the observation in previous work [Bi 2010, Lee 2007]. Given the large sizes of OS buffers, it is not viable to reserve a small number (e.g. 1 or 2) of colors for an OS buffer because there are not enough physical pages in these colors to hold the buffer. Nevertheless, it is not reasonable to assign a sufficient amount of colors (e.g. half of the colors available on the system) dedicatedly to OS buffer either, because it would shrink the cache space available to virtual memory and hurt the performance of computation-intensive applications accessing last level caches with strong localities.

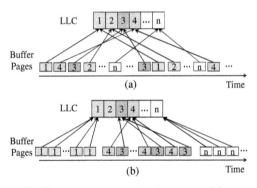

Figure 4: Buffer pages consecutively accessed by an application with a conventional OS buffer (in subfigure a) or an SRM-buffer (in subfigure b). Each subfigure shows the cache regions in the LLC on the top, the buffer pages accessed by the application at the bottom, and the mappings between the cache regions and buffer pages in the middle. The numbers distinguish the colors.

SRM-buffer does not reserve dedicated colors for OS buffer pages. To control cache pollution, for the consecutively accessed buffer pages over a period of time, it tries to minimize the number of colors of the physical pages assigned to them, and thus minimizes the number of cache regions being polluted. This is illustrated by Figure 4 with an example comparing the colors of buffer pages accessed by an application in a conventional OS buffer (Figure 4a) and in an SRM-buffer (Figure 4b). In an SRM-buffer, the application accesses a batch of buffer pages in color #1 in the first time period. Then, in another time period, it accesses pages in colors #3 and #4. After that, it accesses a batch of pages in #n. Thus, only one or two cache regions are polluted in each time period. In comparison, in a conventional OS buffer, the buffer pages visited by the application are rather randomly mapped to the cache regions in LLC. In each time period, any cache region may be polluted.

To achieve the objective described above, SRM-buffer identifies groups of related blocks that are usually accessed in sequence. Then it assigns physical pages in the same color to the blocks in the same group. For example, files are usually accessed sequentially from beginning to end. Thus the blocks in the same file are usually in the same group. In Figure 4b, blocks in color #1 are in one group, and blocks in color #4 are in another group. For simplicity, we call each block group as a *sequence* and the group size as *sequence length* in the rest of the paper. Because accessing the blocks in the same sequence only pollutes a single cache region, SRM-buffer is more effective to control cache pollution with longer sequences. When the length of each sequence is 1, an SRM-buffer becomes a conventional OS buffer.

With appropriate sequence lengths, SRM-buffer reduces cache pollution in three ways: (1) If an application only accesses a few sequences, cache pollution is limited within a few cache regions. For the example in Figure 4b, if the application does not visit other buffer data besides that in four sequences, it pollutes only four cache regions with the SRM-buffer, instead of all the cache regions with a conventional OS buffer. (2) If an application accesses many block sequences (e.g. more than the number of cache regions in the LLC), in each time period, cache pollution is limited within a few cache regions because consecutively accessed blocks are usually correlated and are in same sequences. Thus, most cache regions can still be efficiently used without suffering from cache pollution at any time. (3) The total size of the buffer pages in a sequence is usually larger than the size of a cache region. Mapping the pages in the same sequence to the same cache region makes later-accessed buffer data evict previously-accessed buffer data from the cache region, and thus reduces the chance to replace VM data.

The implementation of SRM-buffer into operating systems need to address two technical issues. One is how to detect block sequences. To avoid the high cost associated with re-mapping a page to a new cache region [Zhang 2009], SRM-buffer must identify blocks belonging to same sequences when they are loaded into OS buffer and provides them with physical pages in appropriate colors. Though an OS buffer can observe which and in which order file blocks are accessed, it is not effective to detect sequences based only on the temporal order of the accesses. For example, when multiple running applications access file blocks con-

currently, their access streams are mixed together. The consecutively accessed blocks in the mixed stream may not be taken as a sequence, due to the lack of logical correlations. They are accessed consecutively only because the applications happen to run together. Next time, if the application mix changes or the applications proceed with different speeds, the temporal order of the blocks will change accordingly. To detect sequences, SRM-buffer uses a few heuristics based on some scheduling and file system information, which are to be introduced in Section 3.2.

The other issue that SRM-buffer must address is how to coordinate different requirements of buffer management and virtual memory management on physical page allocation, while retaining a high hit ratio of OS page cache. In many current systems, OS buffer and virtual memory system are integrated into a *page cache* [Pai 1999, Silvers 2000]. SRM-buffer allocates physical pages in the same color to the blocks in the same sequence. When SRM-buffer allocates physical pages to long sequences, active pages in desired colors may be reclaimed prematurely, despite the availability of free pages or inactive pages in other colors. This reduces the hit ratio of page cache.

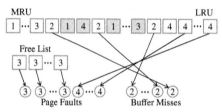

Figure 5: An example illustrating the conflicting requirements on page allocation among different components in OS memory management

Figure 5 illustrates this issue with an example, in which physical pages are allocated upon a series of page faults and buffer misses. In the figure, each rectangle is a physical page. The number in each rectangle is the color of the page (we assume there are four different colors in the system indexed from 1 to 4). An LRU list is on the top of Figure 5, which is used by the page replacement component to organize non-free pages in page cache and to make replacement decisions [1]. On the LRU list, active pages are organized on the MRU end and inactive pages are on the LRU end. Below the LRU list, the figure shows a free list, which organizes free physical pages [2]. Pages on the free list are allocated to hold virtual pages on page faults or to hold file blocks on OS buffer misses. When the number of free pages drops below a threshold, inactive pages on the LRU list are reclaimed to refill the free list. The page faults and buffer cache misses are shown at the bottom of Figure 5 with circles. The arrows show which physical pages are allocated to resolve the page faults and buffer misses. The numbers in the circles show the colors of the physical pages allocated. As shown in Figure 5, when SRM-buffer allocates physical pages in color #2 to a sequence, active pages are reclaimed though there are inactive pages in other colors (shadowed rectangles in the figure) that are less likely to be reused in the future. This may hurt the hit ratio of the page cache.

Virtual memory management prefers to evenly allocate physical pages in different colors to consecutively accessed virtual pages to reduce cache conflicts. However, the allocation of physical pages in same colors in SRM-buffer causes an uneven color distribution among the physical pages available to virtual memory. In SRM-buffer, blocks in the same sequence usually have same access pattern. They are loaded into OS buffer together. Allocating physical pages in the same color to them may exhaust free pages or inactive pages in that color. At the same time, their pages are accessed and may be released together. When the blocks become inactive, their pages move gradually to the LRU end of the LRU list. When the blocks are released, their pages move together onto the free list. Thus, when the blocks in a few long sequences become inactive or are released, the LRU end of the LRU list or the free list may be dominated by pages in the corresponding colors. This makes it difficult for virtual memory management to evenly allocate physical pages in different colors on page faults.

This problem is also illustrated in Figure 5. In the figure, the LRU end of the LRU list and the free list are dominated by pages in color #4 and pages in #3. Thus, only pages in color #3 and #4 are allocated to resolve the page faults. This increases cache conflicts, because the accesses to these pages can only use cache regions in colors #3 and #4, instead of all the cache regions. To prevent this problem, an OS may search the LRU list for pages in desired colors and reclaim them. However, the overhead can be high, and the hit ratio of OS page cache may be reduced if active pages are reclaimed prematurely.

3.2 The Design of SRM-buffer

The allocation of physical pages in same colors to sequences may increase cache conflicts and reduce page cache hit ratio. To minimize the impact, SRM-buffer first sets a threshold (T_l) on sequence length. It breaks down sequences longer than T_l into multiple shorter sequences, and allocates pages in different colors to them. We will show in Section 4 how to select an appropriate T_l to minimize cache pollution and at the same time to avoid impacts on page cache hit ratio. Then SRM-buffer uses a data structure called *colored zone* to coordinate the allocation of physical pages for buffer cache and for virtual memory.

SRM-buffer divides physical memory space into two zones, as shown in Figure 6. The *normal zone* is managed by OS replacement with its data structures (e.g. the active list and inactive list in Linux OS). Active pages and most in-

---

[1] A production system may use other replacement algorithms, e.g. CLOCK. We use LRU in the example is just to simplify the illustration without losing the generality.

[2] Buddy allocation system may be used in production systems. Using free list is just for illustration purpose.

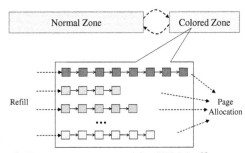

Figure 6: Data structures used in SRM-buffer to organize physical pages.

active pages are managed in the normal zone to maintain the hit ratio of OS page cache. The *colored zone* manages free pages and a small number of inactive pages. To facilitate the physical page allocation for OS buffer and virtual memory, SRM-buffer organizes the pages in the colored zone into multiple lists, and each list links the pages in the same color. When a page in the colored zone is hit, it is moved to the *normal zone*. On page faults or OS buffer misses, SRM-buffer reclaims pages in the colored zone.

```
DATA STRUCTURE:
  #define THRESHOLD 256
  #define NR_COLORS 128 /* Num of colors in system*/
  #define FILE_SIZE_THRESHOLD (THRESHOLD*4096)
  STRUCT process{ ...
    int color, vm_color, nr_block; }
  STRUCT file{ ...
    int color, nr_block; }
FUNCTION:
  update_color(int *c, int *nr_block) {
    (*nr_block)++;
    IF (*nr_block mod THRESHOLD == 0) {
      remove the mark on the list in color c;
      search colored zone for a unmarked list with \
        more pages than THRESHOLD
      if nothing is found, replenish colored zone \
        and repeat the search;
      l = the list found;
      mark l;
      *c = the color of l;
      *nr_block = 0;
    }
  }
ON A BUFFER CACHE MISS:
  f = the file the demanded block is in;
  IF ( sizeof(f) > FILE_SIZE_THRESHOLD) {
    /* Same-file Heuristic */
    update_color(&(f->color), &(f->nr_block));
    alloc page in color f->color;
    return;
  }
  /* Same-application Heuristic */
  p = the process read/write the block;
  update_color(&(p->color), &(p->nr_block));
  alloc page in color p->color;
  return;
ON A VM PAGE FAULT:
  p = current process;
  allocate a physical page in color p->vm_color;
  p->vm_color = (p->vm_color++) mod NR_COLORS;
```

Figure 7: The page allocation algorithm in SRM-buffer

Figure 7 shows the algorithm that SRM-buffer uses to allocate physical pages. On page faults, it reclaims physical pages on different lists (i.e. in different colors) for virtual memory management to reduce cache conflicts. On a buffer miss, it first determines which sequence the demanded block is in. Then it allocates a physical page in the same color as that of other blocks in the sequence. If the demanded block is the first block in a sequence, SRM-buffer selects a list in the colored zone with more pages than T_l, and reclaims a page on the list to hold the demanded block. When such a list cannot be found or a list becomes empty, SRM-buffer refills the colored zone by fetching some inactive pages from the normal zone till the shortest list has more pages than T_l. For example, in Linux OS, SRM-buffer moves inactive pages from the LRU end of the inactive list to the colored zone.

SRM-buffer relies on the following two heuristics to form sequences.

- *Same-file Heuristic*: blocks are considered to be in the same sequence because they are in the same file. Most files, especially large files, are sequentially accessed [Leung 2008, Roselli 2000]. Thus, SRM-buffer allocates physical pages in the same color to every T_l consecutively accessed blocks in the same file to prevent accessing these blocks from polluting CPU cache. SRM-buffer only applies this heuristic to files larger than T_l blocks. This is to prevent SRM-buffer from generating too many short sequences in case applications access a large number of small files, because short sequences negate SRM-buffer's ability to reduce cache pollution.

- *Same-application Heuristic*: blocks that are consecutively accessed by the same process are considered to be in the same sequence. In a system, the same application might be executed multiple times on the same set of data. Thus the blocks consecutively accessed by the same process might be accessed together again in the next run. Based on this heuristic, blocks in multiple small files consecutively accessed by the same application can be viewed as a sequence.

4. Performance Evaluation

We have implemented a prototype of the SRM-buffer into Linux kernel version 2.6.30. With the prototype implementation, we tested the performance of SRM-buffer against a set of micro-benchmarks, real applications, and database workloads. In this section, we first introduce our experiment setup. Then we present the experimental results.

4.1 Experiment Setup

We carried out our experiments on two machines. One machine is a Dell PowerEdge 1900 workstation with two 2.66GHz quad-core Xeon X5355 processors, and the other is a Dell Precision T1500 workstation with an Intel Core i7 860 processor. The architectures of the processors are described in Section 2. The memory sizes of the two machines are 16GiB and 8GiB, respectively. The operating system is 64-bit Red Hat Enterprise Linux AS release 5. The file system is ext3. We used pfmon [HP Corp. 2010] to collect

performance statistics such as last level cache misses. The sequence length threshold T_l is 256.

4.2 Experiments with Database Workloads

In this subsection, we test SRM-buffer with a PostgreSQL database server [PostgreSQL 2008] supporting data warehouse workloads. Most data warehouses use a star or snowflake schema, where join and scan are two most common operations [Ailamaki 2001, Qiao 2008, Stonebraker 2007]. Thus, we created a database in a star schema structure, consisting of a large fact table and several small dimension tables. The size of the fact table is about 4GiB, and the record length is 128 bytes. The numbers of records in the dimension tables range from 100,000 to 600,000, with each record of 256-byte length. Thus the sizes of the dimension tables are from about 24MiB to 146MiB.

We used a client program to issue both hash-join-based queries and sequential-scan-based queries to the database server. A sequential-scan-based query (*sequential-scan* for simplicity) is to summarize the statistic information over the fact table. It searches the records in the fact table, and carries out a hash aggregate over the records that satisfy the conditions specified in the WHERE clause. A hash-join-based query (*hash-join* for simplicity) works on the fact table and a dimension table. It selects the records in the dimension table that satisfy the conditions in the WHERE clause. The selectivity is about 10%. For each record in the fact table, it finds a matching record among those selected from the dimension table based on the join predicate, and carries out an aggregation operation over the matching record pair. To accelerate the join operation, it builds a hash table for the selected records from the dimension table with the key specified in the join predicate so that a matching record can be quickly located.

When a hash-join co-runs with a sequential-scan, the PostgreSQL backend process serving the sequential scan keeps loading the blocks in the fact table from OS buffer to its virtual space [3], and scans the records in the blocks. Thus it pollutes the last level cache in the processor it runs on. If the hash-join and the sequential scan run on the cores sharing the same last level cache, due to the cache pollution, the data in the hash table built for the dimension table may be frequently evicted from the last level cache, and the performance of the hash-join is degraded.

To show the effectiveness of SRM-buffer in reducing cache pollution, we run hash joins and sequential scans on the cores sharing the same last level cache with the vanilla Linux 2.6.30 kernel and the kernel with SRM-buffer enhancement, respectively. We collected the response times and compared them against those collected when each of them runs alone (solo-runs). Due to the cache pollution and

---

[3] PostgreSQL use a buffer ring replacement strategy, which gives sequential scans a very small space in its buffer pool to avoid sequential scans flushing to-be-reused pages in its buffer pool and to minimize double buffering.

the contention for shared resources (e.g. last level cache space, memory controller and bandwidth), the response times collected when the queries co-run are larger than those of their solo-runs. We show the slowdowns of the sequential scans and the hash-joins with different dimension tables in Figure 8.

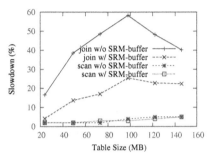

(a) Slowdowns of *hash-join* and *sequential-scan* on PowerEdge 1900

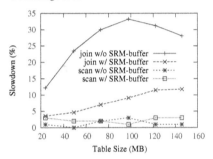

(b) Slowdowns of *hash-join* and *sequential-scan* on Precision T1500

Figure 8: The slowdowns of *hash-join-based queries* and *sequential-scan-based queries* (relative to their solo-runs) when they co-run on the vanilla Linux kernel and the kernel with SRM-buffer enhancement. The sizes of the dimension tables are from 24MiB to 146MiB.

As the figure shows, compared to the sequential-scans, hash-joins are slowed down by much larger percentages. On average, when the Linux kernel without SRM-buffer is used, hash-joins are slowed down by 42% on PowerEdge 1900 and by 26% on Precision T1500, and sequential-scans are slowed down by 3% on PowerEdge 1900 and by 1% on Precision T1500. This is because sequential-scans have weak temporal localities, and they are not affected by cache pollution. The slowdowns of sequential-scans are due to the contention on the hardware resources other than the shared last level cache, e.g. the memory controller and memory bus. The slowdowns of hash-joins are largely caused by the to-be-reused data in hash tables being evicted from the last level cache.

We observed that hash-joins with smaller dimension tables are slowed down by smaller percentages. For example, on PowerEdge 1900, the hash-join with the 24MiB dimension table is slowed down by 17%, but the hash-join with the 98MiB dimension table is slowed down by 58%. This is because hash table sizes are roughly proportional to the di-

mension table sizes. Hash tables built for smaller dimension tables are smaller and less likely to be evicted from the last level cache than those built for larger dimension tables.

On both machines, SRM-buffer reduces the slowdowns of hash-joins by reducing cache pollution caused by sequential-scans. The slowdowns of hash-joins are reduced significantly by up to 33% on PowerEdge 1900 and by up to 24% on Precision T1500. On average, the response times of hash-joins are reduced by 17% on PowerEdge 1900 and by 14% on average on Precision T1500.

While in most data warehouses queries are simple due to the use of star/snowflake schemes and materialized views, there are complex data warehouse queries that need to be paid a special attention due to their long execution times. To study the performance of SRM-buffer with complex database queries, we use TPC-H queries as our workload [TPC 2010]. The data set size of TPC-H benchmark is about 2GiB (scale factor is 2). Two groups of queries are selected. Queries Q6 and Q15 are in the first group. They spend most of their execution time on sequentially scanning the fact table *lineitem*, which incurs a large number of OS buffer accesses. Other queries (Q5, Q7, Q8, Q10, Q11, and Q18) are in the second group. These queries have mixed features. They spend a significant portion of their execution time on operations such as multi-level joins, index scans, and sortings. These operations mainly access PostgreSQL virtual memory space. On the leaf nodes of their execution plan trees, there are operations that sequentially scan *lineitem* or major dimension tables, which incur OS buffer accesses.

Before the experiment, we vacuum the database so that query executions can be more efficient and OS buffer becomes warm. In the experiment, we run combinations of two queries on PowerEdge 1900. Each of the first three combinations consists of a query from the first group and a query from the second group. Each of the other combinations consists of two queries from the second group. We run each query combination on both vanilla Linux kernel and the kernel with SRM-buffer, and collect query execution times. Then we calculate the slowdowns of each query relative to its solo-run.

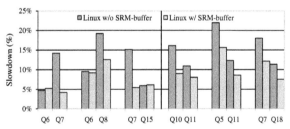

Figure 9: The slowdowns of TPC-H queries on PowerEdge 1900.

We show the slowdowns in Figure 9. Generally, SRM-buffer cannot achieve as good performance with complex queries as it does with simple queries. This is because the query plan trees of complex queries are more complex than those of simple queries, and complex queries spend more time on the operations on the non-leaf nodes of their query plan trees, which mostly access virtual memory. Thus, they cause less intensive cache pollution and incur smaller slowdowns than simple queries. This limits the potential that SRM-buffer improves performance.

In the first three combinations, only the queries from the second group (Q7 and Q8) suffer from cache pollution. Thus, SRM-buffer reduces their slowdowns by 7%~10%. The queries from the first group (Q6 and Q15) have similar performance on both kernels. In the other three combinations, both queries cause cache pollution and suffer from cache pollution at the same time. SRM-buffer can reduce the slowdowns for both of them.

4.3 Experiments with Other Workloads

In this subsection, we selected the following benchmarks, which are briefly described as follows, and measured their execution times in varying scenarios.

- *grep* is a tool to search a collection of files for the lines matching a given regular expression. We run it to look for a randomly generated word from the directory we have used to compile and install PostgreSQL 8.3. The total size of the files in the directory is about 100MiB.
- *tar* is a tool that puts multiple files into the single archive. We run it to put the files under the above mentioned *PostgreSQL* directory into a tarball.
- *PostMark* is an industry-standard benchmark from Network Appliance Inc, which is designed to emulate Internet applications such as e-mail server and news groups [Katcher 1997]. It conducts file accesses and operations, such as *reading or appending files*, and *creating or deleting files* over a pool of files. At the beginning of our experiment, PostMark creates 1,000 files whose sizes range from 128KiB to 384KiB. Then a number of accesses and operations are performed on these files. About 20% of file accesses are reads and the rest are appending. Half of file operations create new files, and the other half delete files.
- *mergesort* is a synthetic benchmark as described in Section 2. The array size is 1.5MiB.
- *FFT, MM,* and *LU* are from NIST SciMark2 benchmark [Pozo 2000]. *FFT* performs one-dimensional fast Fourier transformation. *MM* multiplies two sparse matrices in compressed-row format. *LU* computes the LU factorization of a dense matrix.

Among the benchmarks, *mergesort, FFT, MM,* and *LU* focus on the data in their virtual memory spaces. For convenience, we call these benchmarks *VM-intensive applications*. In comparison, benchmarks *grep, tar,* and *PostMark* more focus on the data saved in files. These benchmarks read data into their VM spaces from files, carry out some processing, and save newly-generated data into files if there is any. We call these benchmarks *file-intensive applications*.

- **Experiments on PowerEdge 1900**

We first carry out experiments on PowerEdge 1900. We select two cores sharing the same last level cache to run the benchmarks. In each experiment, we co-run a VM-intensive application with a file-intensive application on the cores. Before the experiment, we run each of the applications individually multiple times without the interference from the other application to warm-up the OS buffer and to collect its solo-run execution time.

Due to the resource contention and cache pollution, co-running the applications increases the execution times. In Figure 10a, we show the performance slowdowns of the VM-intensive applications on the vanilla Linux 2.6.30 kernel and on the kernel with an SRM-buffer enhancement. Without SRM-buffer, the VM-intensive applications are slowed down by 64% on average due to the co-running with file-intensive applications. With SRM-buffer, the slowdowns can be reduced to 28% on average. The reduction of the slowdowns corresponds to the reduction of execution times from 16% to 29% (by 22% on average), which is shown in Figure 10b. For the file-intensive applications, we have not observed noticeable performance differences between the executions on the kernels with and without SRM-buffer.

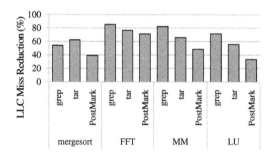

Figure 11: The reduction of last level cache misses for *VM-intensive applications* on PowerEdge 1900.

The performance improvement comes from the significant reduction of extra last level cache misses incurred by cache pollution. We have collected the number of last level cache misses with pfmon and found that last level cache misses can be reduced by 34% to 85% for the VM-intensive applications after adopting SRM-buffer in the kernel. This is shown in Figure 11.

- **Experiments on Precision T1500**

On Precision T1500, there are four cores sharing the same last level cache. Thus, we co-run four applications on them. Before each experiment, we run the applications individually to warm-up OS buffer and to get their solo-run execution times as we have done in the experiments on PowerEdge 1900.

We select *FFT, MM, LU, grep,* and *tar* in the experiments. We first run the three selected VM-intensive applications with each of *grep* and *tar* on both the vanilla Linux

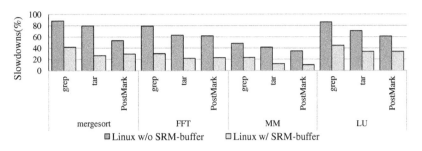

(a) Slowdowns of the VM-intensive applications compared to their solo runs

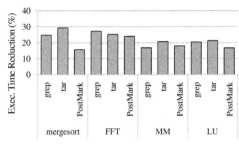

(b) Performance improvement achieved by SRM-buffer

Figure 10: The slowdowns of *VM-intensive applications* compared to their solo runs, and the performance improvement of these applications achieved by using SRM-buffer to reduce cache pollution on PowerEdge 1900.

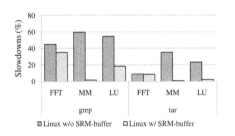

(a) Slowdowns of the VM-intensive applications compared to their solo runs

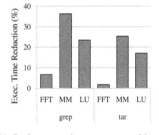

(b) Performance improvement achieved by SRM-buffer

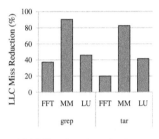

(c) LLC cache miss reduction

Figure 12: The slowdowns of *VM-intensive applications* compared to their solo runs, and the execution time reduction and last level cache miss reduction of these applications achieved by Linux with SRM-buffer on Precision T1500.

kernel and the kernel with SRM-buffer. The slowdowns of the VM-intensive applications on both kernels (compared to their solo-runs) are shown in Figure 12a. When a file-intensive application shares the same last level cache with multiple VM-intensive applications, it may slow down all the VM-intensive applications, as shown in the figure. For both the co-runnings with *grep* and *tar*, *FFT* has the smallest slowdowns, and *MM* has the largest slowdowns. This is because *FFT* accesses its data with the strongest temporal locality among the three VM-intensive applications, and *MM* has the weakest temporal locality. When the three VM-intensive applications co-run with *grep* or *tar*, the data sets in the working set of *MM* are most likely to be evicted by the buffer data loaded by *grep* or *tar*, and the data sets in the working set of *FFT* are least likely to be disrupted.

Figure 12a also shows that SRM-buffer helps reducing the slowdowns of all the VM-intensive applications, but at different degrees. SRM-buffer reduces the slowdown of *MM* by the largest degree because *MM* suffers most from the cache pollution caused by OS buffer. As illustrated in Figure 12c, for the co-running with *grep*, SRM-buffer reduces the LLC cache misses of *MM* by 90%. This reduces the slowdown of *MM* from 60% to 2%, which corresponds to a 36% execution time reduction, as illustrated in Figure 12b.

To demonstrate how the performance of a VM-intensive application is degraded by multiple file-intensive applications. We co-run a VM-intensive application with a varied

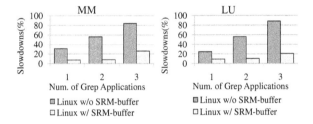

Figure 13: The slowdowns of *MM* and *LU* (compared to their solo-runs) when each of them co-runs with a varied number of *grep* (from 1 to 3).

number of file-intensive applications. In Figure 13, we show the slowdowns of *MM* and *LU* when each of them co-runs with *grep* on both the vanilla Linux kernel and the kernel with SRM-buffer enhancement. When we increase the number of *grep* processes from 1 to 3, the slowdowns of *MM* and *LU* increases accordingly. We see the slowdown of *MM* is increased from 32% to 84%, and the slowdown of *LU* is increased from 24% to 88%. With SRM-buffer, the slowdowns can be significantly reduced. Even co-running with three *grep* applications, the slowdowns of *MM* and *LU* can still be limited below 26% with SRM-buffer.

4.4 Experiments on Access Pattern Changes

The performance of SRM-buffer relies on allocating physical pages in the same color to hold the blocks in the same sequence. The sequence detection and subsequent buffer page allocation in a prior run of an application are expected to effectively reduce cache pollution for its later runs, because the blocks in the same sequence are likely accessed in the same order from the OS buffer as that in the sequence. However, a reasonable speculation is that the sequences detected for an application and the corresponding page allocation do not fit the access patterns of other applications, and cannot reduce cache pollution they incur.

To investigate the interference effect caused by the different access patterns in different applications, we designed experiments in which two applications access the same set of data with different access patterns on PowerEdge 1900. We use *grep* and *diff* as file-intensive applications, and use *FFT*, *MM*, and *LU* as VM-intensive applications. *Diff* is a tool that compares two files or two directories in a byte-by-byte manner. *Grep* and *diff* scans files in different access patterns. *Grep* scans files basically in the order of their layout in the file system, but *diff* visits files in the alphabetic order of directory names and file names. In this subsection, we use *diff* to compare two identical PostgreSQL directories, which are also used by *grep* to search a randomly generated key from. Most files in the PostgreSQL directories are source files and object files with sizes less than 100KiB.

In the experiments, we run one of *grep* and *diff* to load the file blocks in the two PostgreSQL directories into the OS buffer, and then use the other to co-run with each VM-intensive application on the cores sharing the same last level cache. Specifically, in one experiment, for each of the vanilla Linux kernel and the kernel with SRM-buffer, we execute *diff* to load the file blocks into OS buffer. The sequence detection and buffer page allocation are based on the access patterns in *diff*. Then we co-run *grep* with each VM-intensive application, and collect the execution times of the applications. For each VM-intensive application, we calculate the execution time reduction achieved by the kernel adopting SRM-buffer. In the other experiment, we switch *grep* and *diff*. On each kernel, we execute *grep* to load the file blocks into the OS buffer. Then we co-run *diff* with each VM-intensive application, and collect the execution times. We also calculate the execution time reductions for VM-intensive applications.

Figure 14 compares the execution time reductions obtained in the above two experiments against those obtained in the cases where the same file-intensive application loads the file blocks into OS buffer and then co-runs with a VM-intensive application. The comparison confirms that access pattern changes can reduce the effectiveness of SRM-buffer. However, SRM-buffer can still achieve decent performance improvements.

We have analyzed the reason. In the experiments, the blocks in each file are accessed in the same order by *grep* and *diff*, though files are accessed in different orders. Thus, to some degree, the sequences detected during *grep* warming up OS buffer can help reducing the cache pollution incurred

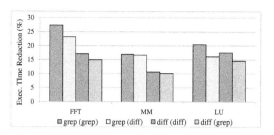

Figure 14: The execution time reductions of *FFT*, *MM*, and *LU* achieved by SRM-buffer. The names in brackets indicate the file-intensive applications loading the data set into OS buffer, and the names outside of brackets indicate the file-intensive applications co-running with VM-intensive applications.

by *diff* accesses these blocks, and vice versa. In the experiments, most files in the PostgreSQL directories are small. With larger files, the access pattern changes would have less impact on SRM-buffer's performance. Please note that in most cases files are accessed sequentially from beginning to end [Roselli 2000]. This makes SRM-buffer have good resistance to access pattern changes under normal workloads.

4.5 Experiments with Contrived Adverse Workloads

To demonstrate the extent to which SRM-buffer could be ill-behaved, we have designed a few arguably worst-case scenarios. In the first scenario, a synthetic workload reads the blocks in a 512MiB file with a random access pattern. Each time it reads 4096 bytes from a file, and there is no think time between consecutive reads. In this case, no matter how SRM-buffer assigns physical pages, it cannot reduce the cache pollution incurred by the workload. In the experiment, we first scan the file sequentially to warm up the OS buffer. Then we co-run the workload with VM-intensive applications, including *MM*, *LU*, and *FFT*. Our experiments show that the performance of the VM-intensive applications does not change no matter whether SRM-buffer is enabled or not. Though SRM-buffer cannot improve performance in this scenario, it does not cause performance degradation because of its low overhead.

In the second scenario, we use an application to repeatedly scan a file. The file size is 3.4MiB on PowerEdge 1900 and 7.4MiB on Precision 1500. Without SRM-buffer, the data set in the file can fit into the last level cache. Thus, accessing it does not incur last level cache misses after the data set is loaded into the cache during the first scan. However, with SRM-buffer, the buffer pages are mapped to only a few cache regions (4 on PowerEdge 1900 and 8 on Precision 1500), which cannot hold the data set. Thus, each scan may incur a large number of misses in the last level cache, and the execution is slowed down.

We compare the execution times of the application on the vanilla Linux kernel and the Linux kernel with SRM-buffer. The application is slowed down by 12% on PowerEdge 1900 and by 17% on Precision 1500 with SRM-buffer enabled, which represent a substantial performance loss. However, the performance loss could often be avoided at application level by copying the data set to virtual memory space first and using the copy in virtual memory space.

When the system is close to running out of free pages, SRM-buffer moves inactive pages to fill the colored zone. These inactive page may be reclaimed prematurely. When physical memory size is large, the colored zone accounts for a small percentage of physical memory space. For example, on PowerEdge 1900, when T_l is equal to 256, the colored zone usually occupies less than 0.8% of the physical memory. Thus only a relatively small amount of coldest pages may be moved to the colored zone because SRM-buffer moves coldest pages first (e.g. pages on the LRU end of the inactive list in Linux OS). Changing the reclamation order of these pages will not hurt OS page cache hit ratio. However, if the physical memory size on a system is small, the colored zone may account for a nontrivial portion of physical memory. For example, when physical memory size is 1024MiB, the colored zone can occupy about 12% of the physical space, assuming the number of colors is 64 and T_l is 256. Thus active pages may be used to fill the colored zone and be replaced prematurely.

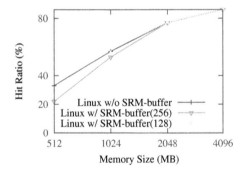

Figure 15: Page cache hit ratios of *PostMark* on Linux systems with and without SRM-buffer when memory size is varied from 512MiB to 4GiB. The numbers in brackets are sequence length thresholds.

To study how SRM-buffer affects OS page cache hit ratio, in the last scenario, we reduce physical memory size gradually from 4GiB to 512MiB and compare the page cache hit ratios of *PostMark* benchmark on the Linux kernels with and without SRM-buffer. In order to change the physical memory size, we carry out the experiments in a virtual machine. To stress the page cache, we increased the number of files in *PostMark* to 32,000.

Figure 15 shows the hit ratios. When memory size is large, SRM-buffer has little impact on hit ratio. However, when memory size is smaller than 1GiB, the hit ratio may be reduced by a non-trivial percentage. For example, the hit ratio is reduced by 11% when memory size is 512MiB and sequence length threshold T_l is 256. To reduce the penalty

on page cache hit ratio on the systems with small amounts of physical memory, one can reduce the sequence length threshold T_l. As shown in Figure 15, when memory size is 512MiB, with the T_l lowered to 128, the hit ratio difference between Linux systems with and without SRM-buffer can be significantly reduced to 3.8%. We will show in the next subsection that SRM-buffer can achieve decent performance improvement even when T_l is reduced to 16.

4.6 Parameter Sensitivity

In SRM-buffer, the sequence length threshold T_l is an important parameter. With larger thresholds, SRM-buffer can form longer sequences, and thus can reduce cache pollution more effectively. However, forming long sequences may reduce the hit ratio of OS page cache. With the experiments in this subsection, we show that there is a range of T_l that can effectively reduce cache pollution with minimum impacts on page cache hit ratio.

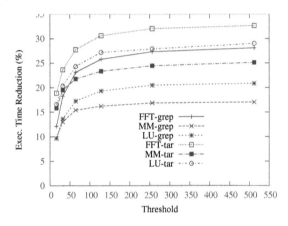

Figure 16: The execution time reductions of *FFT*, *MM*, and *LU* achieved by SRM-buffer in their co-runnings with *grep* or *tar*. The sequence length threshold T_l is varied from 16 to 512.

In the experiments, we vary the value of T_l from 16 to 512, and re-run the experiments of co-running each of *FFT*, *MM*, and *LU* with *grep* or *tar* on PowerEdge 1900, because their performance is more sensitive to the value of T_l than others. In Figure 16, we show the execution time reductions of the VM-intensive applications achieved by SRM-buffer reducing cache pollution. When T_l is less than 128, increasing T_l can significantly improves the performance of SRM-buffer. As shown in the figure, the average execution time reduction increases quickly from 14% to 24% when T_l is increased from 16 to 128. However, increasing T_l to more than 128 only yields incremental performance improvements for SRM-buffer. When T_l is increased from 128 to 512, the average execution time reduction is increased by 1.7%. Based on these experiments, a system with a reasonable physical memory size can choose a desired sequence length threshold between 128 and 512 to effectively reduce cache pollution without reducing OS page cache hit ratio.

4.7 Experiments with More Cores

When more cores share the same cache, more applications may access OS buffer and pollute the cache. Thus, application performance is degraded more seriously. To show this trend, we select a micro-benchmark, which sorts the records saved in files. For each file, it reads the records from the file, sorts them with a blocked merge sort algorithm, and saves the sorted records into the file. File sizes are from 32MiB to 64MiB, and record size is 32B. We select a Core i7 processor with a 8MiB L3 cache shared by 4 cores and a Xeon 7560 processor with a 24MiB L3 cache shared by 8 cores. The cores in the processors have same private cache resources. On each processor, we run multiple instances of the benchmark simultaneously, one on each core. When one instance is reading or writing records, it pollutes the shared cache and slows down other instances. To calculate the slowdowns, we also run one instance of the benchmark alone on each processor. When the block size is 0.5MiB, the average slowdown is 11% on the Core i7 processor and is 13% on the Xeon 7560 processor. When the block size is increased to 0.75MiB, the average slowdown increases to 15% on the Core i7 processor and to 25% on the Xeon 7560 processor. The slowdowns are higher on the Xeon 7560 processor because cache pollution happens more frequently with more instances running on the Xeon 7560 processor.

5. Related Work

With the prevalence of multicores, the performance issues with the shared resources on multicores, especially the shared last level caches, have attracted much attention. To address these issues, a few researchers develop sophisticated scheduling policies for multicores to co-schedule threads that can efficiently use the shared resources [Fedorova 2005, Knauerhase 2008, Zhuravlev 2010]. Cache pollution incurred by OS buffer can be reduced by improving OS scheduling policy to avoid co-running VM-intensive applications and file-intensive applications or to run them on the cores that do not share the last level cache. However, such policy makes VM-intensive applications share the last level cache and the caused space contention degrades their performance. Meanwhile, enforcing such scheduling policies depends on the availability of cores not sharing caches and the classification of applications, which is not always available.

Besides improving thread scheduling, some efforts have been focused on cache partitioning to provide each of the running threads with a chunk of dedicated cache space to avoid interference from other co-running threads [Lin 2008, Tam 2007, Zhang 2009]. They only target private data sets in each thread's virtual space. The data sets shared by multiple threads, including the data sets in OS buffer, are largely ignored in their designs.

Based on the page coloring technique, there are methods proposed to reduce cache pollution introduced by visiting weak locality data in application virtual memory [Lu 2009, Soares 2008]. These methods keep weak locality data within physical pages in a few dedicated colors. These methods are proposed for virtual memory management, and cannot be adapted to alleviate the cache pollution problem caused by an OS buffer, because OS buffer is in kernel space and does not belong to any applications. At the same time, these methods require the total size of weak locality data not exceeding a small percentage of physical memory size. This requirement is not practical for an OS buffer, which usually accounts for a large percentage of physical memory in a computer system.

There are other techniques to avoid memory accesses polluting CPU caches with the facilities provided by processors, such as noncacheable memory supports and streaming load/store supports [Intel 2010]. Cache pollution can also be reduced by invalidating the cache lines holding weak locality data promptly after their data usage. However, our experiments discover that using the techniques to reduce cache pollution incurred by a buffer cache can significantly impact the throughput of the buffer cache (Limited by space, results are not included in the paper). There are various reasons. For example, setting buffer pages noncacheable reduces throughput because memory controller prefetching is refrained and spatial locality cannot be exploited.

Improving I/O performance of OS buffer has been one of the most actively researched area. Intelligent replacement algorithms have been proposed to keep active blocks in memory to minimize I/O operations (e.g. 2Q [Johnson 1994], MQ [Zhou 2004], ARC [Megiddo 2003], and LIRS [Jiang 2002]). A few works reduce the power consumption of OS buffers [Bi 2010, Lee 2007]. However, for our best knowledge, we have not see any work focusing on addressing the CPU cache pollution problem caused by OS buffer.

6. Conclusion

On a multicore system, a thread accessing a large data set in buffer cache can slow down its co-running threads significantly by a factor of two because it flushes the to-be-reused data of its co-running threads from the shared last level cache on the processor. SRM-buffer addresses the problem by enhancing the page allocation policies in OS buffer. It leverages the fixed mapping between physical pages and cache regions. On OS buffer misses, it carefully selects physical pages to resolve the misses. The selection is to ensure that the buffer pages in an OS buffer are mapped to appropriate cache regions, such that, when a thread accesses a large data set in buffer cache, every group of buffer pages consecutively accessed by the thread are mapped to a few cache regions. Thus, during the time period in which the buffer pages in the same group are accessed, the cache pollution is limited within the corresponding cache regions. To achieve the objective, SRM-buffer detects block sequences and allocates physical pages mapped to the same cache region to the blocks in each sequence. Our evaluation with a prototype implementation in Linux kernel shows SRM-buffer can improve application performance and decrease the execution times of workloads by up to 36%.

7. Acknowledgments

We are grateful to Dr. Herbert Bos for his comments and suggestions on the final version of the paper. We thank the anonymous reviewers for their constructive comments. This research was supported by National Science Foundation under grants CNS0834393, CCF0913150, and a CIFellowship subaward CIF-B-173.

References

[Ailamaki 2001] Anastassia Ailamaki, David J. DeWitt, Mark D. Hill, and Marios Skounakis. Weaving relations for cache performance. In *VLDB'01*, pages 169–180, 2001.

[Bi 2010] Mingsong Bi, Ran Duan, and Chris Gniady. Delay-hiding energy management mechanisms for DRAM. In *HPCA'10*, pages 1–10, 2010.

[Fedorova 2005] Alexandra Fedorova, Margo Seltzer, Christoper Small, and Daniel Nussbaum. Performance of multithreaded chip multiprocessors and implications for operating system design. In *USENIX'05*, pages 26–26, 2005.

[Gribble 1998] Steven D. Gribble, Gurmeet Singh Manku, Drew Roselli, Eric A. Brewer, Timothy J. Gibson, and Ethan L. Miller. Self-similarity in file systems. pages 141–150, 1998.

[HP Corp. 2010] HP Corp. Perfmon project, 2010. URL http://www.hpl.hp.com/research/linux/perfmon.

[Intel 2010] Intel. intel 64 and ia-32 architectures optimization reference manual, 2010.

[Jiang 2002] Song Jiang and Xiaodong Zhang. LIRS: an efficient low inter-reference recency set replacement policy to improve buffer cache performance. In *SIGMETRICS'02*, pages 31–42, 2002.

[Johnson 1994] Theodore Johnson and Dennis Shasha. 2Q: A low overhead high performance buffer management replacement algorithm. In *VLDB'94*, pages 439–450, 1994.

[Katcher 1997] Jeffrey Katcher. PostMark: a new file system benchmark. Technical report, Network Appliance Inc., 1997. TR 3022.

[Kessler 1992] R. E. Kessler and Mark D. Hill. Page placement algorithms for large real-indexed caches. *ACM Trans. Comput. Syst.*, 10(4):338–359, 1992.

[Knauerhase 2008] Rob Knauerhase, Paul Brett, Barbara Hohlt, Tong Li, and Scott Hahn. Using os observations to improve performance in multicore systems. *IEEE Micro*, 28(3):54–66, 2008.

[Lee 2007] Min Lee, Euiseong Seo, Joonwon Lee, and Jin-soo Kim. Pabc: Power-aware buffer cache management for low power consumption. *IEEE Trans. Comput.*, 56(4):488–501, 2007.

[Leung 2008] Andrew W. Leung, Shankar Pasupathy, Garth Goodson, and Ethan L. Miller. Measurement and analysis of large-scale network file system workloads. In *USENIX'08*, pages 213–226, 2008.

[Lin 2008] Jiang Lin, Qingda Lu, Xiaoning Ding, Zhao Zhang, Xiaodong Zhang, and P. Sadayappan. Gaining insights into multicore cache partitioning: Bridging the gap between simulation and real systems. In *HPCA'08*, pages 367–378, Salt Lake City, UT, 2008.

[Lu 2009] Qingda Lu, Jiang Lin, Xiaoning Ding, Zhao Zhang, Xiaodong Zhang, and P. Sadayappan. Soft-OLP: Improving hardware cache performance through software-controlled object-level partitioning. In *PACT'09*, pages 246–257, 2009.

[Megiddo 2003] Nimrod Megiddo and Dharmendra S. Modha. ARC: A self-tuning, low overhead replacement cache. In *FAST'03*, pages 115–130, 2003.

[Pai 1999] Vivek S. Pai, Peter Druschel, and Willy Zwaenepoel. IO-lite: a unified I/O buffering and caching system. In *OSDI'99*, pages 15–28, 1999.

[PostgreSQL 2008] PostgreSQL. PostgreSQL: The world's most advanced open source database, 2008. URL: http://www.postgresql.org/.

[Pozo 2000] Roldan Pozo and Bruce Miller. SciMark 2.0, 2000. URL: http://math.nist.gov/scimark2/.

[Qiao 2008] Lin Qiao, Vijayshankar Raman, Frederick Reiss, Peter J. Haas, and Guy M. Lohman. Main-memory scan sharing for multi-core CPUs. *PVLDB*, 1(1):610–621, 2008.

[Roselli 2000] Drew Roselli, Jacob R. Lorch, and Thomas E. Anderson. A comparison of file system workloads. In *USENIX'00*, pages 41–54, 2000.

[Silvers 2000] Chuck Silvers. UBC: an efficient unified I/O and memory caching subsystem for netbsd. In *USENIX'00*, pages 54–54, 2000.

[Soares 2008] Livio Soares, David Tam, and Michael Stumm. Reducing the harmful effects of last-level cache polluters with an OS-level, software-only pollute buffer. In *MICRO'08*, pages 258–269, 2008.

[Stonebraker 2007] Michael Stonebraker, Chuck Bear, Ugur etintemel, Mitch Cherniack, Tingjian Ge, Nabil Hachem, Stavros Harizopoulos, John Lifter, Jennie Rogers, and Stan Zdonik. One size fits all? part 2: benchmarking results. In *CIDR'07*, pages 173–184, 2007.

[Tam 2007] David Tam, Reza Azimi, Livio Soares, and Michael Stumm. Managing shared L2 caches on multicore systems in software. In *WIOSCA*, pages 26–33, 2007.

[TPC 2010] TPC. TPC Benchmark H. URL: http://www.tpc.org/tpch/spec/tpch2.13.0.pdf, 2010.

[Zhang 2009] Xiao Zhang, Sandhya Dwarkadas, and Kai Shen. Towards practical page coloring-based multicore cache management. In *EuroSys'09*, pages 89–102, 2009.

[Zhou 2004] Yuanyuan Zhou, Zhifeng Chen, and Kai Li. Second-level buffer cache management. *IEEE Trans. Parallel Distrib. Syst.*, 15(6):505–519, 2004.

[Zhuravlev 2010] Sergey Zhuravlev, Sergey Blagodurov, and Alexandra Fedorova. Addressing shared resource contention in multicore processors via scheduling. In *ASPLOS'10*, pages 129–142, 2010.

Is Co-scheduling Too Expensive for SMP VMs?

Orathai Sukwong

Electrical and Computer Engineering
Carnegie Mellon University, Pittsburgh, PA, USA
osukwong@ece.cmu.edu

Hyong S. Kim

Electrical and Computer Engineering
Carnegie Mellon University, Pittsburgh, PA, USA
kim@ece.cmu.edu

Abstract

Symmetric multiprocessing (SMP) virtual machines (VMs) allow users to take advantage of a multiprocessor infrastructure. Despite the advantage, SMP VMs can cause synchronization latency to increase significantly, depending on task scheduling. In this paper, we show that even if a SMP VM runs non-concurrent applications, the synchronization latency problem can still occur due to synchronization in the VM kernel.

Our experiments show that both of the widely used open source hypervisors, Xen and KVM, with the default schedulers are susceptible to the synchronization latency problem. To remediate this problem, previous works propose a co-scheduling solution where virtual CPUs (vCPUs) of a SMP VM are scheduled simultaneously. However, the co-scheduling approach can cause CPU fragmentation that reduces CPU utilization, priority inversion that degrades I/O performance, and execution delay, leading to deployment impediment. We propose a *balance scheduling* algorithm which simply balances vCPU siblings on different physical CPUs without forcing the vCPUs to be scheduled simultaneously. Balance scheduling can achieve similar or (up to 8%) better application performance than co-scheduling without the co-scheduling drawbacks, thereby benefiting various SMP VMs. The evaluation is thoroughly conducted against both concurrent and non-concurrent applications with CPU-bound, I/O-bound, and network-bound workloads in KVM. For empirical comparison, we also implement the co-scheduling algorithm on top of KVM's Completely Fair Scheduler (CFS). Compared to the synchronization-unaware CFS, balance scheduling can significantly improve application performance in a SMP VM (e.g. reduce the average TPC-W response time by up to 85%).

Categories and Subject Descriptors D.4.1 [**Process Management**]: Scheduling.

General Terms Algorithms, Experimentation, Performance.

Keywords Virtualization, Synchronization.

1. Introduction

Virtualization provides a flexible computing platform for cloud computing (e.g. Amazon EC2) and server consolidation. It helps to maximize physical resource utilization and simplify system and infrastructure management. Virtualization mainly consists of a software layer between an operating system (OS) and hardware called a *hypervisor*, and a software version of a machine called a *virtual machine* (VM) or *guest*. Examples of hypervisors are VMware ESXi [VMware2010c], Xen [Barham2003] and KVM [KVM2008]. Like a real machine, a VM can run any application, OS or kernel without modifications. A VM can be configured with different hardware settings, such as the number of virtual CPUs (vCPUs) and the size of hard disk and memory. A VM with multiple vCPUs, which behave identically, is called a symmetric multiprocessing (SMP) VM. As a rule of thumb, a SMP VM should not have more vCPUs than available physical CPUs [VMware2010a], a practice followed in this paper.

Virtualization can cause problems which do not exist in a non-virtualized environment. For instance, a spinlock (used for kernel synchronization) in a non-virtualized environment is assumed to be held for a short period of time and does not get preempted. But a spinlock held by a VM can be preempted due to vCPU preemption [Uhlig2004], vastly increasing synchronization latency and potentially blocking the progress of other vCPUs waiting to acquire the same lock. Combined with preemptible synchronization in a concurrent application inside a SMP VM, the synchronization latency problem in the VM can be severe, resulting in significant performance degradation.

Deducing from the results of our experiments (Section 5.3.1), the default schedulers of several current hypervisors (Xen and KVM) still seem to be unaware of the synchronization latency problem. To reduce synchronization latency, previous works propose a co-scheduling solution where vCPUs of a SMP VM are scheduled simultaneously. Recently proposed systems [Weng2009, Bai2010] selectively apply co-scheduling only to SMP VMs running concurrent applications because a non-concurrent application has no application synchronization and thus may not significantly benefit from co-scheduling. In this paper, we show that a SMP VM running a non-concurrent application, such as a single-threaded, synchronization-free and I/O-bound task, can also benefit from co-scheduling due to synchronization in the guest OS.

Nonetheless, co-scheduling can still cause CPU fragmentation, priority inversion [Lee1997] and execution delay. These drawbacks can hinder deployment of various SMP VMs. For example, VMware's co-scheduling solution [VMware2010b] tries to maintain synchronous progress of vCPU siblings by deferring the advanced vCPUs until the slower ones catch up. This can be too rigorous for a SMP VM running a minimal synchronization application, as shown in Section 5.3.1.

We propose the *balance scheduling* algorithm which provides application performance similarly to or better than that of the traditional co-scheduling approach without the co-scheduling drawbacks. The concept of balance scheduling is simple – balancing vCPU siblings on different CPUs without precisely scheduling the vCPUs at the same time. This is easily accomplished by dynamically setting CPU affinity of vCPUs so that no two vCPU siblings are in the same CPU runqueue. We implement the balance scheduling algorithm on top of KVM's scheduler (Completely Fair Scheduler – CFS [Molnar2007]) in the Linux kernel.

For empirical comparisons, we also implement a co-scheduling algorithm, called dynamic time-slice (DT) co-scheduling, based on CFS to avoid the impact of different resource optimizations found in different hypervisors. DT co-scheduling should perform similarly to classic co-scheduling, despite the differences in implementation. Our co-scheduling implementation is based on CFS with dynamic time slice, while the previous implementation relies on a scheduler with static time slice. Compared to the co-scheduling algorithm [Ousterhout1982], DT co-scheduling has less computational complexity and does not incur CPU fragmentation. The expected synchronization latency of the DT co-scheduling algorithm is theoretically the lower-bound of the co-scheduling algorithm (details in Section 5.2.2).

Because VMs can run many types of programs, we extensively evaluate the scheduling algorithms against both non-concurrent and concurrent applications with various degrees of synchronization. We also test them with different workloads (CPU-bound, I/O-bound and network-bound). The empirical results show that balance scheduling can significantly improve application performance (e.g. reducing the average TPC-W response time by up to 85% compared to CFS). Balance scheduling also yield similar or better application performance (e.g. up to 8% higher X264 throughput) than co-scheduling without the drawbacks of co-scheduling, thus benefiting many SMP VMs.

We also evaluate balance scheduling against affinity-based scheduling [Vaddagiri2009]. Both balance scheduling and affinity-based scheduling similarly manipulate each vCPUs' CPU affinity. Unlike affinity-based scheduling (static configuration), balance scheduling can potentially adapt to load changes. Balance scheduling dynamically sets CPU affinity before a scheduler assigns a runqueue to a vCPU, allowing the vCPU to run on the least-loaded CPU where there is no vCPU siblings. Load is measured as the number of runnable tasks in a per-CPU runqueue.

This paper makes the following contributions:

- We show that a SMP VM running a non-concurrent application can also suffer from the synchronization latency problem due to synchronization in the guest OS.
- We propose the *balance scheduling* algorithm and present its performance analysis. We also compare the computational complexity of the balance scheduling and co-scheduling algorithms.
- We implement the balance scheduling and co-scheduling algorithms on top of CFS for empirical comparison. We theoretically and empirically show that our co-scheduling implementation is a refined variation of classic co-scheduling.
- We perform a thorough evaluation on the balance scheduling, co-scheduling and affinity-based scheduling algorithms, in addition to CFS.

The rest of this paper is organized as follows. Section 2 elaborates on the synchronization latency problem. Section 3 describes the co-scheduling approach. Section 4 presents our proposed balance scheduling algorithm. Section 5 discusses the evaluation. Section 6 describes related work. Section 7 is the conclusion.

2. Synchronization in SMP VMs

2.1 Lock Primitive

In a concurrent program, a lock primitive is used to provide synchronization among concurrent threads. Different OSes may support different types of locking. Typically there are two major types of lock primitives [Fischer2005].

Semaphore/Mutex (non-busy-wait). The thread that is waiting for this lock can be blocked and go to sleep, allowing the scheduler to context switch to another runnable thread. This lock primitive is normally used in applications where synchronization may take long to complete (e.g. waiting to receive a network packet).

Spinlock (busy-wait). A spinlock is used when synchronization is expected to take only a short amount of time. Thus, it is inefficient to perform context switching. The lock-waiter thread keeps spinning CPU cycles until it successfully acquires the lock. Spinlocks are simple and usually used in kernel. An OS kernel typically does not preempt a kernel thread which is holding a spinlock. With virtualization, a spinlock in a VM may be preempted due to vCPU preemption.

2.2 Synchronization latency

Synchronization latency is the amount of time it takes a thread to successfully acquire a lock. Synchronization latency in a SMP VM is simply the lock latency experienced by vCPUs of a VM. There are two causes of synchronization latency: task scheduling and preemption or blocking. The hypervisor scheduler can preempt vCPUs at any time, regardless of what they are executing.

Synchronization latency depends on task scheduling when two or more vCPUs simultaneously want the same lock and this lock is blocked or preempted, as shown in Figure 1B. Otherwise, the latency is equal to or less than the amount of time it takes the lock-holder thread to finish synchronization and release the lock (T_H) as shown in Figure 1A.

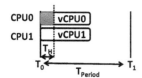

Figure 1A. Synchronization latency without preemption.

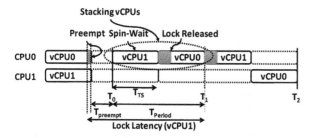

Figure 1B. Synchronization latency with preemption.

Normally, a hypervisor scheduler, such as CFS or Xen's Credit Scheduler [Yaron2007], allows vCPUs to be scheduled to run on any CPU. It is possible that the lock-waiter thread can be scheduled before the lock-holder thread when a lock-holder thread is preempted, as shown in Figure 1B. We call this task scheduling situation *vCPU stacking*. In the worst case scenario, vCPU1 has to wait $T_{preempt} + T_{period}$, as opposed to $T_{preempt} + T_H$. $T_{preempt}$ is measured from the time that vCPU0 is preempted until one of these vCPUs is re-scheduled, and T_{TS} is a time slice of a vCPU, assuming all time slices are the same. Normally, T_H is in the order of microseconds and T_{period} is in the order of milliseconds. The worst case latency may increase to several milliseconds. When waiting for a spinlock, many CPU cycles will also be wasted.

3. Co-scheduling

Ousterhout proposed a co-scheduling algorithm [Ousterhout1982] that schedules a set of concurrent threads simultaneously to reduce synchronization latency. Several previous works [VMware2008, Weng2009, Bai2010] apply co-scheduling to SMP VMs. As shown in Figure 2, co-

scheduling can significantly reduce synchronization latency (from $T_{preempt} + T_{period}$ to $T_{preempt} + T_{H'}$). Note that co-scheduling cannot prevent preemption and eliminate $T_{preempt}$, as shown in Figure 2.

A simple way to co-schedule a set of tasks is finding a time slice that has a sufficient number of available physical CPUs to run all tasks, assuming every time slice has the same size. These tasks are delayed until such a time slice is found. This approach causes CPU fragmentation and priority inversion [Lee1997, VMware2008].

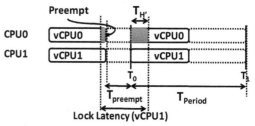

Figure 2. Synchronization latency with co-scheduling.

3.1 CPU fragmentation

As shown in Figure 3, with the co-scheduling approach, vCPU0 and vCPU1 cannot be scheduled until T_1, although both become runnable at T_0 because there is only one CPU idle at T_0. This is called *CPU fragmentation*, which can reduce CPU utilization and also delay the vCPU execution.

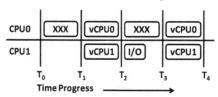

Figure 3. CPU fragmentation in co-scheduling.

3.2 Priority Inversion

Priority inversion is where a higher priority task is scheduled after a lower priority task. For example, an I/O-bound job is given a priority to run whenever it is ready. However, it cannot run because all CPUs are allocated to the co-scheduled tasks. This problem can adversely affect interactive or I/O-bound jobs, and under-utilize other resources (e.g. disks). As seen in Figure 3, when an I/O-bound job is ready between T_0 and T_1, the I/O job has to wait until T_2 because the scheduler already assigns the slot T_1 on both CPUs to vCPU0 and vCPU1, given that both vCPUs are runnable since T_0. The longer the time slice of vCPU1 (T_2-T_1), the longer the disk sits idle and the higher the I/O latency, for example.

4. Balance Scheduling

4.1 Description

To alleviate the synchronization latency problem, we propose the *balance scheduling* algorithm which balances vCPU siblings on different physical CPUs without precisely scheduling the vCPUs simultaneously. It is simply achieved by dynamically setting CPU affinity of vCPUs so that no two vCPU siblings are in the same CPU's runqueue. Unlike co-scheduling, it does not incur CPU fragmentation, priority inversion or execution delay.

4.2 Severity of the vCPU-Stacking Problem

Balance scheduling can be considered a probabilistic type of co-scheduling. It increases the chance of vCPU siblings being scheduled simultaneously by reducing the likelihood of the vCPU-stacking situation (described in Section 2.2). To estimate the probability of the vCPU-stacking occurrence, we empirically measure how often KVM's CFS scheduler places vCPU siblings in the same CPU's runqueue when running one or more CPU-intensive SMP VMs. We run three experiments: one, two and three four-vCPU VMs with our CPU-bound workload (described in Section 5.1) in a four-CPU host. Each CPU runqueue is examined every 700 microseconds to see what tasks are in the queues by inspecting */proc/sched_debug*. We then count the number of samples where the runqueue has two, three and four vCPU siblings in the same runqueue.

As shown in Table 1, the risk of vCPU siblings being stacked grows as the number of VMs increases (the runqueue size also increases). When only one VM is running in the host, the chance that more than one vCPU sibling will be running sequentially is not significant (~6%). When the number of VMs increases to two, the chance substantially increases to 43.13%. Stacking vCPUs can undermine an illusion of synchronous progress of vCPUs, expected from the guest OS [VMware2010]. Without this illusion, the guest OS may malfunction or panic.

# VMs	# vCPUs in the same runqueue			> 1 vCPU in the same runqueue
	2	3	4	
1	5.518%	0.045%	0.001%	5.564%
2	31.903%	10.717%	0.507%	43.127%
3	29.730%	12.091%	4.111%	45.932%

Table 1. The probability of vCPU-stacking.

4.3 Computational Complexity Analysis

We compare the computational complexity of balance scheduling and co-scheduling. Assuming each time slice is the same, the pseudo code of the co-scheduling algorithm for scheduling k vCPUs of a SMP VM is described in Algorithm 1. According to the pseudo code, the computational complexity of the co-scheduling algorithm is $O(NR)$ where N is the number of physical CPUs and R is the runqueue size.

The pseudo code of the balance scheduling algorithm is shown in Algorithm 2. The computational complexity of balance scheduling is $O(N)$ because the number of vCPUs

is always less than or equal to the number of CPUs. By fixing N, the complexity of balance scheduling and co-scheduling becomes $O(1)$ and $O(R)$ respectively. Therefore, balance scheduling has less computational complexity than co-scheduling.

Algorithm 1: Co-scheduling

---

for each time slot i
   available_cpus ← 0
   for each CPU j
      if time slot i on CPU j is idle
         increment *available_cpus* by 1
      end if
      if *available_cpus* ≥ k
         assign vCPUs to available CPUs
         return
      end if
   end for each
end for each

Algorithm 2: Balance Scheduling

---

all_cpus ← set of all physical CPUs
if (task T has not been assigned a runqueue)
   and (task T is a vCPU)
   VMID ← Parent PID of task T
   used_cpus ← {}
   for each vCPU v of *VMID*
      add CPU that v is on in *used_cpus*
   end for each
   CPUS of task T ← *all_cpus* – *used_cpus*
end if

4.4 Performance Analysis

We theoretically show the synchronization latency improvement in balance scheduling compared to CFS, with the different numbers of available physical CPUs. We also estimate the impact on application performance.

As mentioned earlier, task scheduling can affect the lock latency when the lock is needed by two or more vCPUs and also preempted. We calculate the expected lock latency of balance scheduling and CFS using the equations in Appendix A. The following assumptions are made: each task in a runqueue has the same weight, each runqueue has the same size, the average lock holding time is one microsecond and two vCPUs need to acquire the same lock simultaneously. As shown in Figure 4A, the expected lock latency increases as the runqueue size grows. Intuitively, when the runqueue size is one (only one vCPU in the runqueue), balance scheduling and CFS are practically the same. The expected lock latency also lowers as the number of CPUs increases due to decrease in the vCPU-stacking probability. Balance scheduling can reduce the expected latency more than CFS as balance scheduling avoids vCPU stacking. As shown in Figure 4B, balance scheduling can significantly improve the expected lock latency compared to CFS (more than 14.4% for four CPUs), when the runqueue size is less than six. The experiments in Section 5.3.6 show that the average runqueue size is practically about 4-6, even if a host has many threads.

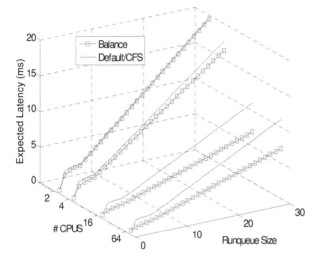

Figure 4A. The expected lock latency in balance scheduling and CFS.

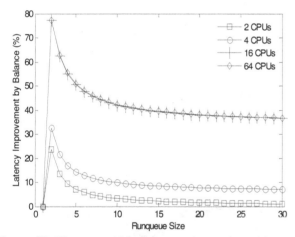

Figure 4B. The expected lock latency improvement in balance scheduling compared to CFS.

Quantifying variation in application performance due to the change in synchronization latency is difficult. Conceptually, the impact of lock latency on application performance should be similar to a step function. As long as lock latency does not exceed a threshold leading to an operation timeout, a change in application performance should appear insignificant. Otherwise, the change can be substantial. For example, an application has to send five TCP packets. We assume the application takes 100 locks per second and the TCP average response time is 10 milliseconds or greater. TCP transmission timeout is 200 milliseconds by default in Linux. If the lock latency

increases from one to two microseconds without any TCP retransmission, the response time of each packet will be increased by at most 20 microseconds, which is 0.2% or less increase in the average response times. But if the lock latency exceeds the threshold causing TCP timeout and a retransmission, then the average response time becomes 50.004 milliseconds ((50 + 200 + 0.02)/5) or greater, which is a 400% increase in the average response time. Balance scheduling is designed to reduce the likelihood that the lock latency becomes exceedingly high. High lock latency usually occurs when a scheduler stacks vCPUs. As shown in Section 5.3.2, balance scheduling causes no TCP retransmission in TPC-W, but CFS does.

5. Evaluation

We extensively evaluate how balance scheduling, co-scheduling, affinity-based scheduling and CFS (KVM's default scheduler) improve application performance. The experiments are conducted with applications ranging from single-threaded and synchronization-free applications to concurrent applications with different degrees of synchronization. The applications also carry different types of workloads (CPU-bound, I/O-bound and network-bound) in various scenarios (combinations of SMP and non-SMP VMs run concurrently in the host).

5.1 Experimental Setup

All experiments run on a physical machine with Intel Core2 Quad CPU Q8400 2.66GHz and 4 GB of RAM with 1Gbit Network card. The physical host runs Fedora Linux kernel 2.6.33 with QEMU 0.11.0. The guest OSes are either Fedora 12 or 13. The selected applications are Pi, HackBench, X.264, Compile, TPC-W, Dell DVDstore, BZip2, Tar, TTCP, Ping, Bonnie++, our synthesized disk and CPU workloads, and our multiple-independent-process workload. Where relevant we use the fourth extended file system (ext4) [Mathur2007] in the experiments.

Pi [Yee2010] is a multi-threaded and CPU-bound program entirely fitting in the memory. It calculates 100,000,000 digits of pi using the Chudnovsky Formula. We use the computing time as a performance metric.

HackBench [HackBench2008] is a multi-threaded program measuring Unix-socket (or pipe) performance. We run HackBench using four threads with 10,000 loops. The completion time (seconds) is used as a metric.

X.264 [Phoronix2010] is a multi-threaded and CPU-bound application, which performs H.264/AVC video encoding. It reports the average throughput in frames per second.

Compile is a compilation test on *libvirt* library using *rpmbuild* tool (a multi-process program). We measure the amount of time it takes to compile (in seconds).

TPC-W [TPC2000] is a transactional web benchmark using multiple web interactions to simulate a retail store's activities. We use Apache HTTP server version 2.2.14 for the proxy server, Tomcat5 version 5.5.27 for the web server and MySQL version 5.1.44 for the database server. These servers are multi-threaded applications.

Dell DVD Store (DVDstore) [Dell2007] is an open source simulation of an online ecommerce site. We use MySQL server 5.1.45 and Apache HTTP Server 2.2.15 for the database and web servers running in the same VM. 100 clients with five-second thinking time concurrently connect from another physical machine located on the same network for three minutes. The average response time is used as a performance metric.

BZip2 and Tar are single-thread data compressor programs. We use *BZip2* to compress a 460 MB file and use Tar to decompress a 1.1GB file (*Untar*). We measure the time it takes to complete the task.

TTCP [TTCP1996] is a single-thread socket-based application that measures TCP and UDP throughput (kB/s) between two systems.

Ping is a single-thread and network-bound program that sends ping packets to another machine located on the same network.

Bonnie++ [Coker2001] is an I/O benchmark measuring hard drive and file system performance. By default, it creates one thread for each test, except the seek test that uses three threads.

Our disk-bound workload is a single-thread and disk-I/O-bound program of our own creation that sequentially creates, writes and deletes small files on a local disk. We measure the time it takes to finish the job.

Our CPU-bound workload is a single-thread and CPU-bound program primarily consuming only CPU resources with minimal memory footprint and I/O usage. It runs infinite loops with simple additions.

Our multiple-independent-process workload consists of multiple processes that independently run a finite number of loops with simple arithmetic calculations.

The host CPU utilization is collected using *dstat* [Wieërs2010]. The I/O statistics are gathered from */sys/block/vda/stat*. We record the runqueue size of each physical CPU by sampling */proc/sched_debug* every second. The sample average is the average runqueue size. We quantify application performance improvement by calculating a performance speed-up. The speed-up metric of a scheduling algorithm (*SpeedUp$_{SCHED}$*) is computed using the following equation, where *Perf$_{SCHED}$* is the application performance result achieved by the scheduling algorithm and *Perf$_{CFS}$* is the application performance result achieved by CFS.

$$SpeedUp_{SCHED} = \frac{Perf_{SCHED} - Perf_{CFS}}{Perf_{CFS}}$$

We create seven experiments for the evaluation. Experiment 1 shows the degree of the synchronization problem in several hypervisors (Xen, VMware and KVM). To eliminate a different resource optimization factor in different hypervisors, we use only KVM hypervisor for the rest of the experiments. Experiment 2 and 3 quantify synchronization latency improvement and efficiency of CPU resources by each scheduling algorithm respectively. Experiment 4 measures performance improvement in both concurrent and non-concurrent applications. Experiment 5 assesses the scalability of the scheduling algorithms. Experiment 6 determines the scheduling performance and CPU runqueue sizes, when the machine hosts many VMs. Experiment 7 shows the performance of SMP and non-SMP VMs running in the same host.

5.2 Implementation

KVM is seamlessly integrated into the Linux kernel. It has a loadable kernel module providing the core of virtualization, and relies on existing Linux kernel modules for the rest of the functionalities (e.g. a scheduler). In KVM, a VM is a regular Linux process with vCPU processes, which require a modified QEMU for device emulation.

We implement the balance scheduling and co-scheduling algorithms based on CFS. Unlike its predecessors, CFS dynamically calculates a time slice for each runnable task. The time slice is calculated as follows, where N_T is the number of tasks in a runqueue, *MinPeriod* is the minimum period and *MinSlice* is the minimum time slice.

$$MinTasks = \frac{MinPeriod}{MinSlice}$$

$$Period = \begin{cases} MinPeriod & \text{if } N_T \leq MinTasks \\ MinSlice \times N_T & \text{if } N_T > MinTasks \end{cases}$$

$$Time\ Slice = \frac{Period}{N_T}$$

In version 2.6.33 of the Linux kernel, by default the minimum time slice is one millisecond, the minimum period is five milliseconds and all tasks in a runqueue have the same weight. The time slice calculation and scheduling decision are made independently on each runqueue (one per CPU). CFS implements a runqueue as a red-black tree [Cormen2001], sorted by each task's *vruntime* (virtual runtime in nanoseconds). The scheduler always selects the task with the smallest *vruntime* to run next.

5.2.1 Balance scheduling

The balance scheduling algorithm can be easily implemented. We modify CFS to dynamically set the *cpus_allowed* field in each vCPU's *task_struct* so that no two vCPU siblings are in the same runqueue. The

cpus_allowed field indicates a set of CPUs that this task can run on. This *cpus_allowed* setting is done before a runqueue is chosen for a vCPU.

5.2.2 Co-scheduling

The classic co-scheduling algorithm (details in Section 4.3) is designed with a static-time-slice assumption. This design cannot be applied to CFS due to its dynamic time slice calculation. In CFS, the second tasks in the different runqueues may not be scheduled at the same time, for example.

We create our version of co-scheduling, called *dynamic time slice (DT) co-scheduling*. To schedule vCPUs simultaneously, we first modify CFS so that it never inserts any two vCPU siblings in the same runqueue, like in balance scheduling. We then force the scheduler to schedule all runnable vCPU siblings simultaneously. However, this step only occurs when the scheduler normally selects the first vCPU sibling from a runqueue. As a result, we still preserve fairness among VMs without keeping track of vCPUs' runtime. To force the scheduler to context switch to the chosen vCPU, we call the *resched_cpu* function with the CPU ID. This function sets *TIF_NEED_RESCHED* flag on the current task and then sends an *smp_send_reschedule* inter-processor interrupt (IPI) to the targeted CPU. We modify the *pick_next_entity* function in *sched_fair.c* so that it can choose the targeted vCPU, instead of the lowest *vruntime* task. Unlike the previous co-scheduling approach, our DT co-scheduling algorithm does not incur CPU fragmentation and execution delay. However, DT co-scheduling may shorten the time slice of the current task due to premature preemption and incur additional context switching.

Our DT co-scheduling algorithm is a refined version of the previous co-scheduling algorithm. It has less computational complexity than the previous co-scheduling algorithm ($O(N)$ versus $O(NR)$). Its expected synchronization latency is the lower-bound of the previous co-scheduling algorithm. The expected synchronization latency of DT co-scheduling is $T_H + T_{INT+CTX}$ where T_H is the lock-holding time and $T_{INT+CTX}$ is the amount of time it takes to send an IPI and perform context switching. Due to CPU fragmentation, the expected latency of the previous co-scheduling algorithm is $T_H + \sum_{i=1}^{\infty} P_i T_{TS}(i-1)$. T_{TS} is a size of time slice. P_i is the probability of having sufficient CPUs to run all vCPU siblings at time slice i and $\sum_{i=1}^{\infty} P_i = 1$. $T_{INT+CTX}$ is normally in the order of microseconds and T_{TS} is in the order of milliseconds. Therefore, $\sum_{i=1}^{\infty} P_i T_{TS}(i-1) \geq T_{TS} > T_{INT+CTX}$. Moreover, the empirical results show that the DT co-scheduling algorithm can improve application performance by up to 6% compared to the previous co-scheduling algorithm. Please see Appendix B for more details. Hence, our DT co-scheduling algorithm should be adequate for the comparative evaluation.

5.2.3 Affinity-based scheduling

We use the *virsh vcpupin* command to modify the CPU affinity of vCPUs. At the beginning of each experiment, we bind each vCPU to a CPU in such a way that the number of vCPUs per physical CPU is relatively the same and vCPU siblings cannot be assigned to the same physical CPU.

5.3 Experimental Results

5.3.1 Experiment 1

Experiment 1 shows the degree of the synchronization problem in several current hypervisors. We run two CPU-intensive workloads: HackBench (intensive synchronization) and the multiple-independent-process workload (no application synchronization) in a four-vCPU VM along with three one-vCPU VMs running our CPU-bound workload.

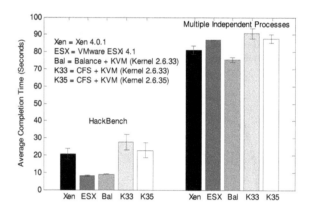

Figure 5. The average completion time with 95% confidence interval assuming the normal distribution.

As shown in Figure 5, Xen (using the Credit scheduler) and KVM (CFS) have higher completion times than balance scheduling on HackBench due to their synchronization-unaware schedulers. They treat all vCPU siblings as independent entities. Although VMware ESXi's scheduler uses a co-scheduling algorithm to mitigate the synchronization problem, their algorithm can be too restrictive for certain applications that barely incur synchronization. As shown in Figure 5, VMware's scheduler has the lowest completion time on HackBench (9.65% less than balance scheduling approach) because VMware's scheduler maintains synchronous progress of vCPU siblings. However, this also causes VMware's scheduler to complete the multiple-independent-process workload (14.88%) slower than the balance scheduling approach. VMware's scheduler stops the advanced vCPUs until the slow vCPUs catch up [VMware2010b], resulting in vCPU-execution delay. Note that we confine the comparison to the CPU-only tests since different hypervisors may have different optimizations on other resources (e.g. network and disk I/O).

We also create a Windows version of HackBench to test on a VM with a different guest OS (i.e. Windows Server 2008). The results are consistent with our findings from the Fedora guest OS. The Windows VM spends 20.23 and 10.46 seconds using CFS/KVM and balance scheduling/KVM respectively, while the Fedora VM spends 20.58 and 9.02 seconds. These results suggest that the balance scheduling approach can benefit other guest OSes than Linux.

5.3.2 Experiment 2

The goal of this experiment is to show the improvement on the synchronization latency by different scheduling algorithms. We run TPC-W benchmark using three four-vCPU VMs for the proxy, web and database servers. The maximum of 250 clients concurrently connect from another physical machine to the proxy server. The average and 90th percentile response times experienced by the clients are reported. We also use *ftrace* [Edge2009] to monitor the amount of time that the vCPUs of the proxy server take to execute the *spin_lock* function, and use *SystemTap* [RedHat2010] to monitor TCP retransmissions in the VMs.

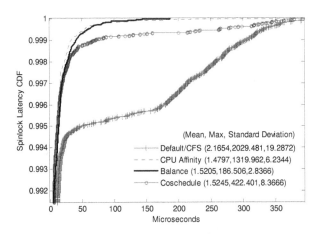

Figure 6A. The spinlock latency CDF of the proxy server.

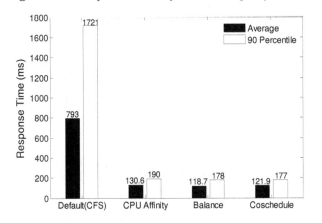

Figure 6B. The response time statistics of TPC-W.

As shown in Figure 6A, balance scheduling, affinity-based scheduling and co-scheduling can similarly improve the average spinlock latency compared to CFS (decreased by 29.78%, 31.67% and 29.60% respectively). The significant increase in the spinlock latency caused by CFS can trigger a TCP timeout leading to TCP retransmissions. The retransmissions in the proxy server can cause disruptions in the subsequent servers (the web and database servers) and eventually affect the overall response time. From the experiment, we find 1,363 retransmissions between the proxy and web servers, 399 retransmissions from the web to database servers, and 38 TCP retransmissions between the clients and the proxy server with CFS, while there is no retransmission with the other scheduling algorithms. These retransmissions severely degrade the TPC-W performance. In balance scheduling, affinity-based scheduling and co-scheduling, the average response time is reduced by 85.04%, 83.53% and 84.62% respectively, as shown in Figure 6B. These results show that balance scheduling can significantly improve the synchronization latency and application performance, compared to CFS. Balance scheduling also performs similarly to co-scheduling (achieving about the same average and 90th percentile response times).

5.3.3 Experiment 3

The synchronization latency problem not only degrades application performance, but also wastes CPU resources due to unnecessary CPU spinning. This experiment shows the improvement in processing efficiency by the different scheduling algorithms. We run the Bonnie++ benchmark in a four-vCPU VM along with a two-vCPU VM running the CPU-bound workload. The two-vCPU VM is used to simulate a background workload. For each I/O test, Bonnie++ reports I/O throughput and CPU utilization in the VM. We use these metrics to calculate throughput per CPU utilization, which is then used to compute the speed-up. This speed-up metric indicates the I/O processing efficiency.

Even though Bonnie++ spawns only a single thread for each test (except the seek test), it can encounter the synchronization problem due to intensive disk I/O processing in the guest OS. Balance scheduling, affinity-based scheduling and co-scheduling can help reduce excessive CPU cycles caused by synchronization-unaware scheduling, thereby having more CPU cycles for useful work. As shown in Figure 7, balance scheduling, affinity-based scheduling and co-scheduling significantly increases the I/O processing efficiency by up to 40%, 45% and 53% for the read operation (SeqInput); 70%, 73% and 63% for the write operation (SeqOutput) and 374%, 439% and 382% for the seek operation respectively. The I/O latency is also improved. As shown in Figure 8, balance scheduling, affinity-based scheduling and co-scheduling reduce the I/O read latency by 48%, 23% and 35%, compared to CFS. The I/O write latency is not improved as

much as the read latency due to the disk being a bottleneck. The gain by each scheduling algorithm, excluding CFS, can be varied at each trial depending on cache performance.

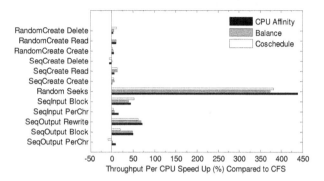

Figure 7. The speed up of I/O throughput per CPU utilization of Bonnie++ benchmark.

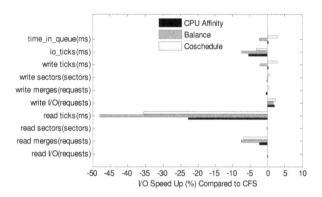

Figure 8. The I/O statistics in the Bonnie VM.

5.3.4 Experiment 4

This experiment shows how much each scheduling algorithm can improve the performance of applications ranging from single-threaded programs without any locking, to multi-threaded programs with different degrees of application synchronization. We also test the algorithms with different types of workloads (CPU-bound, I/O-bound and network-bound). We run two SMP VMs in the host: one four-vCPU VM running an application, except for TTCP, and one two-vCPU VM running our CPU workload. This two-vCPU VM is for simulating a background workload. For TTCP, we run two four-vCPU VMs in the host: one for a TTCP transmitter and the other for a TTCP receiver.

For the multi-threaded applications (Pi, HackBench, X.264, Compile, and DVDstore), affinity-based scheduling, balance scheduling and co-scheduling similarly improve the application performance by up to 85% compared to CFS, as shown in Figure 9. The improvement varies due to the degree of synchronization in the SMP VM. HackBench incurs intensive synchronization due to socket sharing in the guest VM kernel, as opposed to Pi, which incurs a relatively small degree of application synchronization.

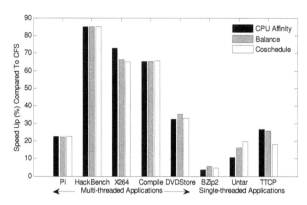

Figure 9. The performance improvement of different applications using affinity-based, balance and co-scheduling.

Modern kernels are capable of servicing multiple applications simultaneously. To understand the impact of kernel synchronization on application performance, we run two independent (synchronization-free) processes of the disk workload in the four-vCPU VM. As shown in Figure 10, balance scheduling, affinity-based scheduling, and co-scheduling reduce the completion time by 35%, 32%, and 31% compared to CFS, respectively, due to file system synchronization. The improvement in file system performance increases I/O aggregation as indicated by the 20% reduction in the average I/O write requests. These results suggest that synchronization can incur in a VM despite running synchronization-free applications. Balance scheduling reduces the completion time by 5% compared to co-scheduling due to additional context switching.

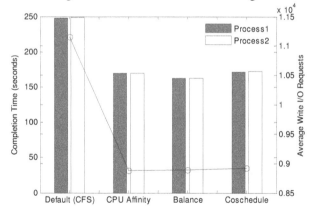

Figure 10. Performance of multiple disk-I/O processes in a SMP VM.

We also run a single-threaded application (no application synchronization) in the VM to understand the effect of synchronization in the guest VM kernel. As shown in Figure 9, balance scheduling, affinity-based scheduling and co-scheduling improve TTCP performance by 26%, 27% and 18%, Untar performance by 16%, 11% and 20%, and BZip2 performance by 6%, 4% and 5% compared to CFS respectively. The improvement depends on the degree of

synchronization in the VM kernel. TTCP mainly relies on the guest kernel for network processing, while Untar processes in both user and kernel spaces (for I/O processing) and BZip2 mainly runs in the user space with minimal kernel assistance. For TTCP, balance scheduling has 6% higher TTCP throughput than co-scheduling due to additional context-switching. For Untar and BZip2, the completion times vary over 20 trials because the improvement due to their small degree of kernel synchronization can be outweighed by cache performance.

Overall, the results show that balance scheduling, affinity-based scheduling and co-scheduling can benefit any application that incurs synchronization in either application or kernel inside a SMP VM. The performance improvement depends on the degree of synchronization in the VM. Balance scheduling can improve application performance up to 6% more than co-scheduling due to additional context-switching.

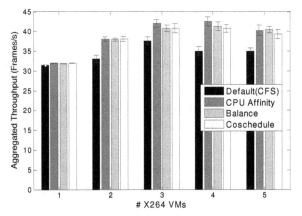

Figure 11A. The aggregated throughput of all X.264 VMs with 95% confidence interval.

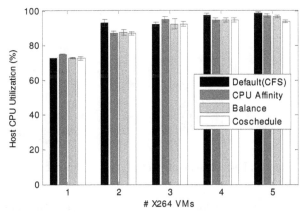

Figure 11B. The CPU utilization on the host with 95% confidence interval.

5.3.5 Experiment 5

This experiment assesses the scalability of each scheduling algorithm as the number of VMs increases. We keep adding more four-vCPU VMs running X.264 until reaching the host's maximum CPU capacity. We use the aggregated throughput of all VMs as a performance metric.

As shown in Figure 11A and B, balance scheduling, affinity-based scheduling and co-scheduling scale better than CFS due to the synchronization latency problem. Their X.264 throughputs increase as the number of VMs and the CPU utilization increases, when the host runs between one to three VMs. The host reaches its maximum capacity, when running 3-4 VMs. As shown in Figure 11A, affinity-based scheduling achieves 3% higher X264 throughput than balance scheduling due to better cache performance. When the host has five VMs, the thrashing effect starts to take place. Performance of all scheduling algorithms decreases, while the CPU utilization does not. As seen in Figure 11A and B, balance scheduling yields up to 4% higher in the X264 throughput than co-scheduling with about the same amount of CPU resources due to additional context switching.

5.3.6 Experiment 6

As discussed in Section 4.4, the performance of balance scheduling theoretically declines as the runqueue size grows. In this section, we show that in practice the average runqueue (per CPU) does not exceed six even if the four-CPU host has more than 24 threads. We run 14 four-vCPU VMs in the host: one X.264 VM and the rest (13 VMs) running the CPU workload with a *CPULimit* program [Marletta2010]. *CPULimit* is used to control CPU usage in the VMs. The maximum CPU usage of the 13 VMs is 8%, bounded by the maximum CPU capacity in the host. 14 VMs is the maximum number of VMs we can run concurrently due to the memory capacity. We measure the X.264 throughput and the runqueue size of each host CPUs.

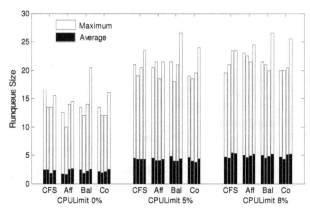

Figure 12. The average and maximum runqueue size of four physical CPUs by each scheduling approach.

In this experiment, there are 56 vCPU threads, in addition to other threads (e.g. QEMU and system threads), alive in the host. One may expect to have at least 14 tasks per runqueue. In fact, a runqueue contains only runnable threads, not threads that are blocked or sleeping. As shown in Figure 12, as the CPU usage in the 13 VMs increases,

the average runqueue size of each CPU increases but still remains less than six, although the maximum runqueue at a certain moment can go up to 26. Due to the limited number of runnable threads in the runqueue, balance scheduling still performs very well even if the host has many VMs running. As shown in Figure 13, balance scheduling improves the X.264 throughput by up to 82.69%, 3.8%, and 4.2%, compared to CFS, affinity-based scheduling and co-scheduling respectively.

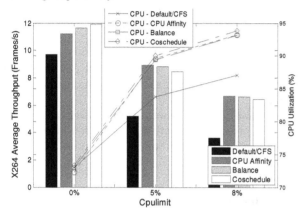

Figure 13. The X.264 performance and the host CPU utilization.

5.3.7 Experiment 7

This experiment shows how the scheduling algorithms affect performance of both SMP and non-SMP VMs running in the same host. We run X.264 in four-vCPU VMs and Ping in a one-vCPU VM. Ping sends an ICMP packet to another machine every one millisecond for 300,000 times. It goes to sleep after a packet is sent. We record the X.264 throughput and the standard deviation of Ping response times (jitter). High jitter can cause an undesirable effect, for example unusable video rendering.

As shown in Figure 14, balance scheduling has better Ping jitter (up to 41%) and more aggregated X.264 throughput (up to 8%) than affinity-based scheduling due to global load balancing. In balance scheduling, the Ping vCPU should always run in the least-loaded CPU, but it is not always the case in affinity-based scheduling. By default, the load balancer is triggered every 60 milliseconds. It is possible that a CPU has more load than the others for a certain period of time. By design, balance scheduling allows a vCPU to move to the least-loaded CPU every time it wakes up, given that the CPU does not have its siblings. In the affinity-based scheduling, the vCPU has to run on the same CPU. Hence, balance scheduling can better adapt to load changes than affinity-based scheduling. The benefit of load adaptation decreases as the number of the available CPUs for vCPU siblings decreases.

Balance scheduling has better the X.264 throughput (up to 8%) and Ping jitter (up to 2.5%) than co-scheduling due to priority inversion and additional context switching.

Balance scheduling yields (up to 12%) higher aggregated X.264 throughput than CFS due to the synchronization latency problem. It also has similar or (up to 27%) higher Ping jitter than CFS. These results suggest that balance scheduling can effectively schedule both SMP and non-SMP VMs without suffering from priority inversion and global load balancing.

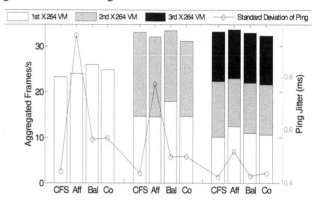

Figure 14. The X.264 performance in SMP VMs and Ping jitter performance in non-SMP VMs.

5.3.8 Discussion

Application performance degradation in a SMP VM depends on the degree of synchronization in both applications and OS inside the VM. As shown in Experiment 3 and 4, a SMP VM running synchronization-free applications (no application locks) can also suffer from the synchronization latency problem because the guest OS is capable of concurrent processing.

For example, the file system in guest OS can process multiple read/write requests simultaneously to reduce the latency perceived by users. Synchronization is required to provide concurrent modifications on the file system structure. In Experiment 4, we simultaneously run two independent disk-I/O processes which continuously create, read and write a number of files in the same directory. In the file system, a file or directory is represented by an *inode* which can be identified by a unique number within a file system. An inode contains file information, such as physical locations of file data, permission, and file size. An inode for a directory also has a list of inodes, identifying files in the directory. When two processes concurrently create new files in the same directory, they need to be synchronized in order to access and update the directory inode.

Similar to the file system, the network processing in guest OS also requires synchronization. For instance, when the networking layer and a device driver access a buffer simultaneously, a lock must be held prior to the access. A buffer (a block of memory) is used to store network packets. In Experiment 4, we run a single-thread network application, TTCP, in the SMP VM. TTCP continuously sends a number of TCP packets to another VM, which will send TCP ACK packets back upon receiving TCP packets. For the packet transmission, the networking layer creates packets and places them in the buffer. The device driver removes packets from the same buffer and sends to the network. The networking layer and the device driver require synchronization to access the shared buffers. Hence, even if a SMP VM runs a synchronization-free and network-bound application, the synchronization is still required in the guest OS.

Due to task scheduling, synchronization latency in a SMP VM can significantly increase, adversely affecting application performance. By design, the co-scheduling algorithm should work exceptionally well with synchronization-intensive applications because it synchronizes the execution of vCPU siblings, as shown in Figure 2. It would be futile to schedule the vCPU siblings in different time slots, if they often contend on the same lock. However, if the synchronization is barely required in a SMP VM, forcing vCPU siblings to be scheduled simultaneously can result in vCPU-execution delay, leading to application performance degradation. As shown in Figure 3, if the vCPUs mostly execute independent jobs, each vCPU should be able to run as soon as a CPU becomes available without unnecessary delay.

Unlike co-scheduling, balance scheduling does not force vCPU siblings to be scheduled simultaneously. It just balances vCPU siblings on different physical CPUs to increase a chance of the vCPUs being scheduled simultaneously. Balance scheduling never delays vCPU execution. Hence, minimal synchronization applications should benefit from balance scheduling more than co-scheduling. As shown in Experiment 1, balance scheduling has the shortest completion time (13% better than co-scheduling) on the multiple-independent-process workload (no application synchronization). But VMware's co-scheduling solution has the smallest completion time (10% better than balance scheduling) on HackBench (synchronization-intensive application). Theoretically, balance scheduling should be preferable to co-scheduling as the degree of synchronization in a SMP VM decreases. We also evaluate co-scheduling and balance-scheduling against other concurrent applications with different degree of synchronization (TPC-W, DVDstore, Compile and X264). As shown in Experiment 2 and 4, balance scheduling can improve application performance similarly to DT co-scheduling with a possible few percentage gain (e.g. reduce the average response times of TPC-W and DVDstore by up to 3%). Overall, balance scheduling exhibits a promising capability in alleviating the synchronization latency problem without the co-scheduling drawbacks.

Balance scheduling can also significantly improve application performance, compared to synchronization-unaware schedulers, such as Xen and CFS. As shown in Experiment 1, balance scheduling can complete HackBench and the multiple-independent-process workload

6% and 56% quicker than Xen/Credit scheduler does, respectively. Balance scheduling can reduce the average TPC-W response time by 85% compared to CFS. It also improves the I/O processing efficiency by up to 40% and 70% for the disk-I/O read and write operations, compared to CFS. Additionally, balance scheduling can effectively schedule more SMP VMs than CFS. As shown in Experiment 5, balance scheduling increases the aggregated X264 throughput as the number of VMs increases (up to four VMs). With CFS, the X264 throughput increases, when the number of VMs increases up to three VMs. Then, the X264 throughput starts to drop. The X264 throughputs by CFS are also consistently less than the throughputs by balance scheduling (up to 15%). The reason is that CFS wastes more CPU cycles due to the synchronization latency problem.

Moreover, balance scheduling can potentially adapt to load changes, unlike affinity-based scheduling (static configuration). As shown in Experiment 7, balance scheduling can improve Ping jitter up to 41%, compared to affinity-based scheduling.

6. Related Work

In the past, without virtualization, Ousterhout [Ousterhout1982] proposed a co-scheduling algorithm which schedules concurrent threads simultaneously to reduce application synchronization latency. Lee et al. [Lee1997] show that the co-scheduling algorithm can cause CPU fragmentation, which reduces CPU utilization, and priority inversion, which reduces I/O performance and other resource utilization. Later works [Feitelson1992, Wiseman2003] try to improve on the co-scheduling algorithm.

With virtualization, the synchronization latency problem becomes severe; spinlocks in a guest OS can get preempted. This never happens in a non-virtualized environment. Uhlig et al. [Uhlig2004] identify this problem as lock-holder preemption (LHP) in SMP VMs. They propose several techniques to prevent LHP. The techniques require augmenting guest OS or installing a special-crafted device driver, and thus may not be feasible in commodity OSes (e.g. Windows). Balance scheduling does not prevent LHP, but alleviates effect of LHP. Even if spinlocks in a SMP VM are no longer preempted, application locks can still benefit from balance scheduling.

To mitigate the synchronization latency problem in SMP VMs, previous works [VMware2008, Weng2009, Bai2010] propose a co-scheduling solution where vCPU siblings are scheduled simultaneously. Unlike co-scheduling, balance scheduling only balances vCPU siblings on different physical CPUs without forcing the vCPUs to be scheduled at the same time. Balance scheduling can be easily implemented and significantly improve application performance without the complexity and drawbacks found in co-scheduling (CPU fragmentation, priority inversion and execution delay).

VMware developed several versions of co-scheduling for VMware ESXi. The first version, called *strict co-scheduling*, is included in VMware ESX 2.x [VMware2008]. Due to CPU fragmentation, VMware created *relaxed co-scheduling* (ESX 3.x) where all vCPU siblings are stopped and only the lagging vCPUs are started simultaneously when they are out of synchronization. The relaxed co-scheduling is further refined in ESX 4.x [VMware2010b] – stopping only advanced vCPUs, instead of all vCPUs. Balance scheduling is similar to the relaxed co-scheduling in a sense that the scheduling operation is per vCPU. But balance scheduling never delays execution of a vCPU to wait for another vCPU in order to maintain synchronous progress of vCPU siblings. Balance scheduling is also simpler. No discrepancy accruing in progress of vCPU siblings is required. To avoid the co-scheduling drawbacks, Weng et al. [Weng2009] limit co-scheduling to a SMP VM with a concurrent application, unlike balance scheduling which does not share any co-scheduling drawbacks, thereby benefiting both concurrent and non-concurrent SMP VMs.

Jiang et al. [Jiang2009] propose several techniques to improve KVM performance, such as temporarily increasing the priority of vCPUs and approximately co-scheduling vCPU siblings by changing their scheduling class from SCHED_OTHER (default scheduling class in CFS) to SCHED_RR (real-time scheduling class). Changing the priority of vCPUs can affect the fairness and performance of other VMs; unlike balance scheduling which never changes scheduling class or priority of vCPUs.

AMD [Langsdorf2010] and Intel [Intel2010] also provide architectural support for heuristically detecting contended spinlocks so that the hypervisor can de-schedule them to reduce excessive CPU cycle use. They add additional fields in the VM data structure (Pause-Filter-Count in AMD and PLE_Gap and PLE_Window in Intel). For example, in Intel, PLE_Gap is an upper bound on the amount of time between two successive executions of PAUSE in a loop. PLE_Window is an upper bound on a guest allowed for a PAUSE loop. According to KVM's codes, PLE_Gap is set to 41 and PLE_Window is 4096. It means that this approach can detect a spinning loop that lasts around 55 microseconds on a 3GHz CPU. As mentioned earlier, the synchronization problem incurs not only by synchronization in applications inside a VM, but also synchronization in the guest kernel. As shown in Figure 6A, most spinlocks in VMs last less than 50 microseconds. Hence, this support should help cease application locks rather than spinlocks in kernel. However, the values of PLE_Gap and PLE_Window should not be too small due to the cost of VM_EXIT, (4-5K cycles [Zhang2008], depending on CPU architectures). VM_EXIT can also cause performance loss due to transition cost (VM exit, VM reads, VM writes, VM entry, and TLB flushing cost).

7. Conclusion

Despite the benefit of parallel processing, SMP VMs can also increase synchronization latency significantly, depending on task scheduling. In this paper, we show that a SMP VM running non-concurrent applications can also need synchronization for concurrent processing in the guest OS.

To mitigate the synchronization problem, previous works have proposed a co-scheduling solution, which rigorously maintains synchronous scheduling of vCPU siblings. This approach can be too expensive for SMP VMs with minimal synchronization due to delay in vCPU execution. We propose the balance scheduling algorithm, which simply balances vCPU siblings on different physical CPUs without strictly scheduling the vCPUs simultaneously. Balance scheduling can improve performance of concurrent SMP VMs similarly to co-scheduling without the co-scheduling drawbacks (CPU fragmentation, priority inversion and execution delay). Unlike co-scheduling, balance scheduling can also effectively schedule SMP VMs with minimal synchronization; thereby benefiting many SMP VMs. In practice, most applications, including concurrent applications, should not demand intensive synchronization. Minimal synchronization usage is encouraged in concurrent applications to promote parallelism. Synchronization serves as the bottleneck in parallel execution. Yet, it is still necessary in many concurrent applications. Additionally, a number of existing and legacy applications are still non-concurrent.

Acknowledgments

We would like to thank Akkarit Sangpetch for insightful discussions on the implementation issues and John Lanyon for comments. We would like to also thank our shepherd, Jacob Gorm Hansen, and the anonymous reviewers for helpful comments on the paper.

Appendix

A. *Expected lock latency calculation*

We use Eq. 1 and 2 to calculate the expected lock latency of CFS and balance scheduling respectively. T_H is the average lock holding time. $|RQ|$ is a runqueue size. $|VW|$ is the number of vCPUs that want to acquire the same lock, and $|CPU|$ is the number of available physical CPUs.

$Eq. 1$ $Expected\ Latency_{default} =$

$$P_{stack}\left[\frac{\binom{|RQ|}{|VW|}(|VW|-1)!|CPU|T_P+|CPU|\alpha}{|CPU|\binom{|RQ|}{|VW|}|VW|!}\right]+$$

$$(1-P_{stack})\left[\frac{T_P\left(\left(\sum_{i=2}^{|RQ|}|CPU|\binom{(i-1)|CPU|}{|VW|-1}(|VW|-1)!\right)-\binom{|RQ|}{|VW|}|CPU|(|VW|-1)!\right)}{\beta}+\right.$$

$$\left.\frac{\sum_{i=2}^{|VW|}\binom{|CPU|}{i}|RQ|\left(\prod_{k=i+1}^{|VW|}((|RQ|-1)|CPU|-(k-i-1))\right)T_H}{\beta}+\frac{|CPU|\alpha}{\beta}\right],\quad where\ \alpha=$$

$$\left(\sum_{i=1}^{|RQ|-(|VW|-1)}\binom{|RQ|-i}{|VW|-1}(|VW|-1)iT_{TS}\right), \beta =$$
$$\left(\frac{|CPU||RQ|}{|VW|}\right)|VW|! - |CPU|\binom{|RQ|}{|VW|}|VM|! = \left(\frac{(|CPU||RQ|)!}{(|CPU||RQ|-|VW|)!}\right.$$
$$\left.\frac{|CPU||RQ|!}{(|RQ|-|VW|)!}\right), and |CPU| \geq |vCPU| \geq |VW|$$

Eq. 2 *Expected Latency*$_{balance}$

$$= \frac{T_P\left(\left(\sum_{i=2}^{|RQ|}|CPU|\binom{(i-1)|CPU|}{|VW|-1}\right) - \binom{|RQ|}{|VW|}|CPU|(|VW|-1)!\right)}{\beta}$$

$$+ \frac{\sum_{i=2}^{|VW|}\binom{|CPU|}{i}|RQ|(\prod_{k=i+1}^{|VW|}((|RQ|-1)|CPU|-(k-i-1)))T_H}{\beta}$$

$$+ \frac{|CPU|\alpha}{\beta}$$

B. Analysis of our DT scheduling

The computation complexity of DT co-scheduling is O(N) where N is the number of CPUs, according to the pseudo code in Algorithm 3.

Algorithm 3: DT Co-scheduling

---

all_cpus ← set of all physical CPUs
if task T is a vCPU
    if task T is not assigned a runqueue
        VMID ← Parent PID of task T
        used_cpus ← { }
        for each vCPU v of *VMID*
            add CPU that v is on in *used_cpus*
        end for each
        CPUs of task T ← *all_cpus* – *used_cpus*
    else if task T is the first vCPU of the VM to be
        scheduled
        for each vCPU sibling v of task T
            if v is not currently scheduled
                send reschedule interrupt
                context switch to v
            end if
        end for each
    end if
end if

We also run three multi-threaded applications (Pi, HackBench and DVDstore) to compare the performance of our DT co-scheduling and the co-scheduling in [Ousterhout1982]. We mimic the co-scheduling on KVM by changing vCPUs' scheduling class from SCHED_OTHER (CFS) to SCHED_RR (RT scheduling) with the priority of 20. RT tasks have higher priority than CFS tasks. By default, the RT period is 1 second and the RT runtime is 950 milliseconds. This reserved RT runtime is given to RT tasks first and the rest is allocated to CFS tasks. We experiment with four combinations of RT runtime and period: 15ms/30ms, 28ms/30ms, 500ms/1000ms and 950ms/1000ms (default). As shown in Figure 15, our DT co-scheduling improves DVDStore (I/O and network-intensive) performance at least 6% better than

the co-scheduling. DT co-scheduling improves HackBench and Pi performance at least 0.7% and 0.3% better than the co-scheduling respectively. These results show that our DT co-scheduling can perform similarly or better than the co-scheduling without tuning the time slice and period parameters.

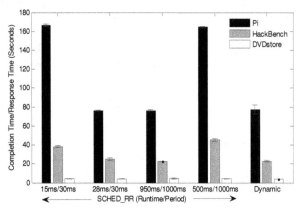

Figure 15. The comparison of application performance between the SCHED_RR-based co-scheduling and our DT co-scheduling.

References

[Bai2010] Y. Bai, C. Xu, and Z. Li. "Task-aware based co-scheduling for virtual machine system", In Proceedings of the 2010 ACM Symposium on Applied Computing. SAC '10. ACM, New York, NY, 181-188.

[Barham2003] P. Barham, B. Dragovic, K. Fraser, S. Hand, T. Harris, A. Ho, R. Neugebauer, I. Pratt, and A. Warfield. 2003. Xen and the art of virtualization. In Proceedings of the nineteenth ACM symposium on Operating systems principles (SOSP '03). ACM, New York, NY, USA, 164-177.

[Coker2001] R. Coker. Bonnie++ version 1.03. http://www.coker.com.au/Bonnie++/, 2001.

[Cormen2001] T. H. Cormen,, C. E. Leiserson, and R. L. Rivest. Introduction to Algorithms, Second Edition. MIT Press and McGraw-Hill, 2001. Chapter 13: Red-Black Trees, pp. 273–301.

[Dell2007] Dell, Inc. The DVD Store Version 2. http://www. dell techcenter.com/page/DVD+Store, December, 2007.

[Edge2009] J. Edge. A look at ftrace. http://lwn.net/Articles/ 322666/, March, 2009. (accessed August 2010).

[Feitelson1992] D. Feitelson, L. Rudolph. Gang scheduling performance benefits for fine-grain synchronization. Journal of Parallel and Distributed Computing, 1992.

[Fischer2005] G. Fischer, C. Rodriguez, C. Salzberg, S. Smolski. Linux Scheduling and Kernel Synchronization. Nov 11, 2005. Prentice Hall Professional.

[HackBench2008] HackBench, http://people.redhat.com/mingo/cfs-scheduler/tools/hackbench.c, September 2008.

[Intel2010] Intel. Intel 64 and IA-32 Architectures Software Developer's Manual. Volume 3B: System Programming Guide, Part 2, June 2010.

[Jiang2009] W. Jiang, Y. Zhou,, Y. Cui, W. Feng, Y. Chen, Y. Shi, and Q. Wu. CFS Optimizations to KVM Threads on Multi-Core Environment. In Proceedings of the 2009 15th international Conference on Parallel and Distributed Systems. ICPADS2009.

[KVM2008] Qumranet. KVM. Kernel Based Virtual Machine. http://www.linux-kvm.org/, September, 2008.

[Langsdorf2010] M. Langsdorf. Patchwork: Support Pause Filter in AMD processors. https://patchwork.kernel.org/ patch/48624/ (accessed May 2010).

[Lee1997] W. Lee, M. Frank, V. Lee, K. Mackenzie and L. Rudolph, Implications of I/O for Gang Scheduled Workloads, Job Scheduling Strategies for Parallel Processing, pp. 215-237, 1997.

[Marletta2010] A. Marletta. CPU Usage Limiter for Linux. http://cpulimit.sourceforge.net/ (accessed August 2010).

[Mathur2007] A. Mathur, M. Cao, S. Bhattacharya, A. Dilger, A. Tomas, L. Vivier. The new ext4 filesystem: current status and future plans. *Proceedings of the Linux Symposium*. Ottawa, ON, CA: Red Hat. 2007.

[Molnar2007] I. Molnar. CFS design. http://people.redhat.co m/mingo/cfs-scheduler/sched-design-CFS.txt, May 2007.

[Ousterhout1982] J. Ousterhout, "Scheduling Techniques for Concurrent Systems,"Proc. 3rd International Conference on Distributed Computing Systems, October 1982.

[Phoronix2010] Phoronix Test Suite. X.264 Benchmark. http: //www.phoronix-test-suite.com/index.php?k=downloads (accessed September 2010)

[RedHat2010] Red Hat, IBM, Hitachi, and Oracle. SystemTap. http://sourceware.org/systemtap/

[TPC2000] TPC. Transaction Processing Performance Council. TPC-W: A transactional web e-Commerce benchmark. http://www.tpc.org/tpcw/, January 2000.

[TTCP1996] TTCP Utility. Test TCP (TTCP) Benchmarking Tool and Simple Network Traffic Generator. http://www .pcausa.com/Utilities/pcattcp.htm, 1996.

[Uhlig2004] V. Uhlig, J. LeVasseur, E. Skoglund, and U. Dannowski. "Towards scalable multiprocessor virtual machines", In Proceedings of the 3rd Conference on Virtual Machine Research and Technology Symposium - Volume 3, 2004. USENIX Association, Berkeley, CA.

[Vaddagiri2009] S. Vaddagiri, B.B. Rao, V. Srinivasan, A.P. Janakiraman, B. Singh, and V.K. Sukthankar. Scaling software on multi-core through co-scheduling of related tasks. In Linux Symp., pages 287–295, 2009.

[VMware2008] Drummonds. VMware, Inc. Co-scheduling SMP VMs in VMware ESX server. May 2, 2008. http://communities.vmware.com/docs/DOC-4960.

[VMware2010] VMware, Inc. VMware vSphere 4: The CPU Scheduler in VMware ESX 4 White Paper. http://www.vmware.com/files/pdf/perf-vsphere-cpu_scheduler.pdf (accessed September 2010).

[VMware2010a] VMware, Inc. Performance best practices for VMware vSphere 4.0. VMware ESX 4.0 and ESXi 4.0. http://www.vmware.com/pdf/Perf_Best_Practices_vSphere4.0.pdf (accessed September 2010)

[VMware2010b] VMware, Inc. VMware vSphere 4: The CPU scheduler in VMware ESX 4.1, September 2010. http://www.vmware.com/files/pdf/techpaper/VMW_vSphere41_cpu_schedule_ESX.pdf (accessed September 2010).

[VMware2010c] VMware, Inc. VMware vSphere Hypervisor (ESXi). http://www.vmware.com/products/vsphere-hypervisor/index.html. (accessed September 2010).

[Weng2009] C. Weng, Z. Wang, M. Li, and X. Lu. "The hybrid scheduling framework for virtual machine systems", In Proceedings of the 2009 ACM SIGPLAN /SIGOPS international Conference on Virtual Execution Environments. VEE '09. ACM, New York, NY, 111-120.

[Wieërs2010] D. Wieërs. Dstat: Versatile resource statistics tool. http://dag.wieers.com/home-made/dstat/.

[Wiseman2003] Y. Wiseman , D. Feitelson, Paired Gang Scheduling, IEEE Transactions on Parallel and Distributed Systems, v.14 n.6, p.581-592, June 2003.

[Yaron2007] Yaron. Xen Wiki. Credit Scheduler. http://wiki.xensource.com/xenwiki/CreditScheduler November, 2007. (accessed August 2010).

[Yee2010] Yee, J. A y-cruncher-A Multi-Threaded Pi-Program. http://www.numberworld.org/y-cruncher/, August 2010.

[Zhang2008] X. Zhang, Y. Dong. Optimization Xen VMM Based on Intel Virtualization Technology. International Conference on Internet Computing in Science and Engineering, 2008 (ICICSE'08).

Kaleidoscope: Cloud Micro-Elasticity via VM State Coloring

Roy Bryant[1]   Alexey Tumanov[1]   Olga Irzak[1]   Adin Scannell[1]
Kaustubh Joshi[2]   Matti Hiltunen[2]   H. Andrés Lagar-Cavilla[2]   Eyal de Lara[1]

[1]University of Toronto, [2]AT&T Labs Research

Abstract

We introduce cloud micro-elasticity, a new model for cloud Virtual Machine (VM) allocation and management. Current cloud users over-provision long-lived VMs with large memory footprints to better absorb load spikes, and to conserve performance-sensitive caches. Instead, we achieve elasticity by swiftly cloning VMs into many transient, short-lived, fractional workers to multiplex physical resources at a much finer granularity. The memory of a micro-elastic clone is a logical replica of the parent VM state, including caches, yet its footprint is proportional to the workload, and often a fraction of the nominal maximum. We enable micro-elasticity through a novel technique dubbed VM state coloring, which classifies VM memory into sets of semantically-related regions, and optimizes the propagation, allocation and deduplication of these regions. Using coloring, we build Kaleidoscope and empirically demonstrate its ability to create micro-elastic cloned servers. We model the impact of micro-elasticity on a demand dataset from AT&T's cloud, and show that fine-grained multiplexing yields infrastructure reductions of 30% relative to state-of-the art techniques for managing elastic clouds.

Categories and Subject Descriptors   D.4.7 [*Operating Systems*]: Organization and Design – Distributed Systems; D.4.1 [*Operating Systems*]: Process Management – Multiprocessing Multiprogramming Multitasking

General Terms   Design, Experimentation, Measurement, Performance

Keywords   Virtualization, Cloud Computing

1. Introduction

Cloud computing caters to bursty Internet workloads with a utility model that emphasizes pay-per-use and elasticity of provisioning. As with all utilities, there is a granularity associated with service delivery and billing. For Infrastructure as a Service (IaaS) clouds, the granule is the virtual machine (VM). Adopting virtualization as a building block yields distinct advantages in security, isolation and ease of management, but this coarse granularity imposes inefficient patterns that harm both users and providers.

In an ideal cloud, 'elastic' servers grow and shrink in tight concert with user demand. Currently, a load balancer adjusts the size of a pool of full-sized worker VMs [Amazon a;b] that are booted from scratch from a template. Unfortunately, this heavy-weight mechanism is a poor match for the operation model of an efficient utility. Creation is slow – new servers take a while to boot because it's a laborious I/O bound task. Moreover, this latency is hard to predict – instantiation latencies in Amazon's EC2 cloud have been observed to fluctuate sharply around a two-minute mean [Hyperic]. Furthermore, once booted, the server's performance-critical application and OS caches are essentially empty, which degrades performance when it is most needed to service demand spikes. Finally, VMs claim a full memory footprint even if they are required for only short periods of time and much of their memory is not actually used.

Therefore, server owners have incentives to keep VMs active for long periods, both to provide slack resources during long instantiation latencies, and because servers with large, warm buffers become crucial to overall performance and are too valuable to sacrifice. This practice may explain why the proportion of EC2 'Extra Large' instances (15 GiBs of RAM) has grown from 12% to 56% in a year, even while the total number of servers has tripled. Further, the proportion of servers running longer than a month has nearly doubled [RightScale]. These behaviors detract from the cloud vision of matching resource usage to actual demand, inflate user costs while still failing to achieve a good QoS if the load exceeds expectations, and curtail providers' ability to consolidate and optimize infrastructure use.

This paper proposes a vision of *cloud micro-elasticity*, in which cloud server elasticity is achieved through short-lived, transient clone VMs, which are copies of a running VM instance and allocate resources (memory, disk) only on demand. To enable this, we introduce color-based fractional

Eurosys '11   April 10–13, Salzburg, Austria.
Copyright © 2011 ACM 978-1-4503-0634-8/11/04... $10.00

VM cloning, a new technique that allows the fine-grained management of VM state, and enables the swift instantiation of stateful VMs that allocate resources in proportion to use. By cloning a warm, running VM instead of booting a new one, our workers inherit their parent VM's state and do not require warming. They come online faster, reach peak performance sooner, and because short-lived worker VMs typically access only a fraction of their state, they can service transient spikes in load from within a smaller footprint.

Color-based fractional VM cloning uses a novel VM state replication technique. Instead of blindly treating the VM as a uniform collection of pages, it bridges the semantic gap between the Virtual Machine Monitor (VMM) and the guest OS by examining architectural information (e.g., page table entries) and other clues to glean a more detailed understanding of the guest's state. This higher-quality knowledge allows the VMM to optimize the propagation of state to clones by identifying semantically related regions. Specifically, we use VM state coloring to tailor the prefetching of kernel vs user space regions, code vs data regions, and to optimize the propagation of the file system page cache. Finally, coloring provides hints that guide memory consolidation by identifying regions with a high likelihood of content similarity.

To evaluate the performance of color-based fractional VM cloning, we implemented *Kaleidoscope*, an elastic server that reacts to transient load spikes by spawning fractional VMs. Experiments using elastic Web and Online Analytical Processing (OLAP) workloads show that Kaleidoscope significantly improves on the current state of the art. First, Kaleidoscope instantiates new stateful clones in seconds, and nearly matches the runtime performance of an idealized cloning strategy that uses zero-latency eager full state replication. Second, by bridging the semantic gap, Kaleidoscope is effective in finding state that is likely to be needed by the new clone. For example, for the OLAP workload, it achieves 2.9 times the query throughput with 43% less waste than color-blind cloning. Third, Kaleidoscope's fractional VM workers grow only as needed to satisfy new allocations or hold newly transferred state. In our experiments, the memory footprint of workers reached only 40% to 90% of their parent's memory allocation.

To further evaluate the advantages of color-based fractional VM cloning, we simulated its effects on traces collected from AT&T's hosting operation. The simulation shows that the finer-grained handling of VM state drastically reduces infrastructure use. With VM cloning to rapidly create new workers, a server can scale faster, and therefore requires much less slack capacity to deal with load increases. Also, the reduced memory footprint translates into a denser packing of VMs on physical infrastructure. The net result is a 30% reduction in infrastructure, which creates energy and money savings that can be shared with end-users via more attractive fine-grained pricing schemes.

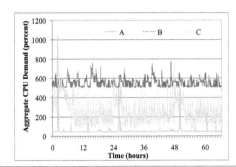

Figure 1. Aggregate CPU demand for sample customers

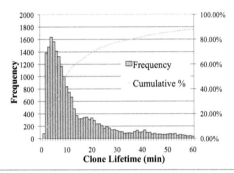

Figure 2. Elastic workers are typically short-lived

We also discuss the ways in which Kaleidoscope ensures correct and consistent behavior. Although VM cloning is not universally applicable without tuning, many legacy server applications, including the OLAP database and Ecommerce Apache Web server used in our evaluation, function correctly without modification.

This paper makes the following contributions. First, we introduce the notion of VM state coloring as a general mechanism to bridge the semantic gap and glean high-quality information on the runtime state of a VM. Second, we show how state coloring can be implemented efficiently by exploiting x86 architectural properties and guest kernel introspection extensions. Third, we present Kaleidoscope, a micro-elastic server that uses state coloring to optimize the replication and sharing of VM state, and delivers a QoS that approaches that of fully over-provisioned servers, while consuming resources proportional to the immediate demand. And fourth, we quantify the benefits of deploying Kaleidoscope servers using a data set of multi-customer demand extracted from AT&T's hosting operation, showcasing substantial savings for users and providers.

2. Real Data Motivates Micro-Elasticity

To use resources efficiently, elastic servers should grow and shrink in tight concert with user demand. We examine the potential for such elasticity using a month of demand data from AT&T's hosting operation (see Section 9 for more details about the data traces). AT&T hosting is a traditional hosting business in which customers buy rack-space to statically provision web-facing multitier applications, such as portals, shopping sites, and enterprise services.

Need for Elasticity. Analysis shows that over the whole month, an average customer tier has a mean demand of only 15.3% of its peak, thus indicating ample long-term fluctuations. Furthermore, demand elasticity also percolates to smaller timescales. Figure 1 shows the total CPU demand for three sample customer tiers. While the characteristics can be very different across customers, they all exhibit significant short-term variations, and thus could benefit from fine-grained elasticity.

Elastic Workers are Short Lived. Figure 2 shows the rate of creation and the lifetimes of elastic server workers, if we were to closely follow demand by maintaining a CPU utilization of between 70% and 90% across all workers. The results show a frequent creation of very short-lived workers: 23,214 workers would be created for a set of only 248 elastic servers. The mean worker lifetime would be only a little over 10 minutes, with over 85% of the workers needed for less than an hour. These workers, therefore, are essentially single-purpose entities that are frequently created to service a narrow workload during short periods of demand pressure, and have limited time to grow their active memory footprint. A mechanism that could allow them to be created cheaply, quickly, and with an allocation proportional to their use would be of great benefit.

3. Designing Efficient Micro-Elasticity

We achieve micro-elasticity by building upon live VM cloning, and augmenting its capabilities through two separate techniques: coloring of VM state to improve its propagation and sharing, and fractional VM allocations to minimize state footprint. Through the combination of these techniques we enable the swift instantiation of fractional, stateful VMs that are virtual copies of an existing server instance, but which are allocated resources (memory, disk, network) proportional to their actual use. In this section we provide background on cloning, illustrate its limitations and motivate the introduction of state coloring and fractional VM allocations.

3.1 Live VM Cloning with SnowFlock

SnowFlock [Lagar-Cavilla 2009] introduced the concept of live VM cloning across a cluster or cloud of physical machines. Cloning requires replicating the state of a VM, and SnowFlock achieves this with on-demand paging. In this approach, a clone is quickly created from a small architectural VM descriptor containing metadata, virtual device (NIC, disk) specifications, and architectural data structures such as page tables, segment descriptors, and virtual CPU (VCPU) registers. The clone VM then triggers page faults as it encounters missing pages of memory or disk, and any referenced state is lazily transferred by a copy-on-demand mechanism.

SnowFlock complements copy-on-demand with multicasting of VM state. Multicast enhances network scalability and results in implicit prefetching, as clones will receive replies to requests issued by sibling clones created at the same time, and presumably accessing similar code or data. SnowFlock's multicast need not guarantee delivery of state to all clients, only to the client explicitly requesting it.

3.2 The Challenges of State Propagation

SnowFlock adopted on-demand paging to minimize instantiation time and optimize resource usage. With on-demand paging, there is space to apply 'late-binding' optimizations that may overlap or hide the overhead of state propagation with useful work performed by the clone. Unfortunately, for many servers, on-demand paging as implemented by SnowFlock results in an extended warmup period in which performance of the new instance is significantly degraded due to blocking waiting for the working set to be fetched (discussed in Section 8.1). Our experimental results show that on-demand fetching is so inefficient that it negates the benefits derived from warm caches.

An alternative to on-demand paging is eager full replication; this approach is similar to traditional VM migration, with the difference that at the end, there are two VMs running. Unfortunately, eager full replication places heavy demands on the network and results in long instantiation times. In addition, it requires that memory be allocated for all the parent's state, much of which may not be used. On the upside, because all state is fetched eagerly, once started, the new worker can quickly achieve peak performance.

3.3 Color-Based Fractional VM Cloning

To achieve the benefits of both eager and on-demand propagation (fast VM instantiation, short warmup period, resource allocation proportional to use) without their respective shortcomings, we use two novel mechanisms that optimize VM cloning performance: *VM state coloring*, which discriminates otherwise uniform VM state into semantically-related regions allowing state to be efficiently prefetched and shared; and *fractional allocation*, which dynamically allocates memory to accommodate only the state that is actually accessed by the new worker.

Color-based fractional VM cloning makes it possible for users to achieve high resource utilization while still accommodating transient load increases at a low rate of service violations. Users keep just enough server instances to deal with the average short-term demand placed on the service (e.g., keep average worker utilization at 80%), and instantiate new transient workers to deal with any sudden load increases. As shown in Section 2, short-term demand spikes are prevalent and typically result in the creation of workers that are needed for ten or less minutes. When demand for the service subsides, the transient workers are shut down and their resources returned to the cloud. Our approach significantly reduces memory footprints relative to static overprovisioning, but it should be noted that this benefit is not entirely free. It incurs a modest network cost for the fractional state transfer.

By cloning a warm, running VM, new workers come online within a few seconds, and by inheriting their parent VM's state, they do not require warming. Using VM state coloring, we efficiently prefetch state to mitigate the page fault blocking associated with on-demand VM cloning, which significantly boosts performance. By allocating memory on demand and sharing identical pages, fractional VM workers save space, and short-lived workers service transient spikes in load from within a smaller footprint.

4. VM State Coloring

Historically, VM state has been treated as uniform binary state, enabling virtualization to simplify many tasks. For example, by saving the entire RAM of a VM as one flat binary file, computation migration can be implemented robustly [Satyanarayanan 2005]. However, the limited information that the VMM has about the state of the guest, referred to as the semantic gap [Chen 2001], can constrain the effectiveness of system services such as I/O scheduling or malware detection [Jones 2006a;b, Litty 2008].

We have devised a set of *VM state coloring* mechanisms that allow us to classify the memory of a VM into a set of semantically meaningful regions. We color VM state by inspecting *architectural information*, such as that contained in page table entries, and performing *introspection* on the guest kernel's data structures. Without adding significant overhead, the discrimination of otherwise opaque VM state into semantically-related regions allows us to optimize VM cloning performance.

4.1 Architecture-based Coloring

Page table entries in x86 contain a wealth of information regarding memory pages. First, pages are tagged as executable or not by the NX bit. Second, pages can be tagged as belonging to the kernel or user space with a 'user' bit. Even if the OS (or the VMM) chooses not to use this bit, a walk of the page tables allows for the reconstruction of a corresponding virtual address: both 32 and 64 bit OSes typically allocate the lower portion of virtual addresses for user space and the higher portion for kernel space. Third, commodity x86 OSes present a bijective mapping between user-space processes and root page table pages, allowing for the discrimination of state unique to a given process.

Using this architectural information we can color memory in roughly four groups: kernel code, kernel data, user code and user data. Further granularity can be achieved by splitting the user colors on a per-process basis, which is left for future exploration. We also note that using x86 architectural information in this fashion makes our coloring robust, as its usefulness is completely independent of the software stack.

A complication arises in 64bit OSes such as Linux, which install a contiguous mapping of the entire physical memory into the kernel data space for expedited access to physical memory addresses. This mapping is called the 'direct map'. As a result, all pages including free ones, have at least one page table mapping. Many kernel data structures are only reached via their address in the direct map, without any other mappings needed. To tell apart kernel data from free (or uncolored) pages, we need to go beyond page table analysis.

4.2 Introspective Coloring

At the root of the semantic gap problem is the fact that the OS has a more complete knowledge of system resources than the VMM. Both the Xen VMM and the Linux OS maintain a 'frame table,' an array of compact records describing the properties of a page. Among other things, a Xen frame table entry indicates the owner VM and the number of page table mappings of a page across all VMs in the system.

A Linux frame table entry has more useful information. First, page records corresponding to frames of memory that belong to a file lead to a radix tree containing all fellow pages mapping the same file. Memory frames can thus be colored as belonging to the file system page cache, and for further granularity, we can group the pages used for each individual file. We underscore that the latter requires no knowledge of the actual file attributes nor file system internals.

Second, the page record indicates whether the page is being used by the guest, or is free. It reflects 'real' usage of the page and is intended to capture all usage, through a combination of a reference count and a set of flags. We thus rely on a page structure record with all of its fields and count reset to identify pages as free. This enables differentiation of free and kernel data pages. Given the circumstances, we find this to be the most conservative method and also the most robust, as the notion of a page structure record is broadly applicable.

We note that the original implementation of VM cloning [Lagar-Cavilla 2009] performs a form of paravirtual coloring. By instrumenting the kernel page allocator it could identify victim pages used for new memory allocations, and prevent the clone VM from issuing a request for a victim page which will be immediately overwritten. There is naturally a very high correlation between victim and unused pages.

4.3 VM State Coloring for Efficient Propagation

We apply VM state coloring at the point of cloning. During the generation of the architectural descriptor, all page table entries have to be processed to turn MFNs (machine frame numbers, i.e., the memory frames of a host) into PFNs (physical frame numbers, i.e., the memory frames of a guest VM) – this translation is later undone when the clone VM is created in a different host. At this point, architectural coloring is applied to partition memory into four disjoint regions; the kernel and user space regions are each subdivided by data vs executable.

Figure 3. Color map Rendering of a memory snapshot of a VM running the SPECweb Support workload. X axis is page number and wraps around for presentation. Legend: Page Cache - yellow; User and Kernel Data - light and dark blue; User and Kernel Code - light and dark red; Free - black.

After the architectural descriptor is generated, a *memory server* is left running on the parent VM's host machine. This memory server keeps a map of the parent VM's memory frozen at the point of cloning (a 'checkpoint'), and uses copy-on-write to allow the parent VM to proceed with execution while serving the frozen image to clones. The memory server is aware of the architectural coloring information, and is able to examine at will the VM's frozen memory, and in particular the guest kernel's frame table. In this way, the memory server performs introspection to identify free pages, which are re-colored to form a fifth region. Further, the memory server identifies pages belonging to the file system page cache, and also re-colors these pages on a per-file basis. For pages that have multiple colors, for example an executable page that is also in the file system page cache, the specific color that will be used is application-dependent and configurable.

To better understand the advantage of coloring state, consider the color map (Figure 3) of a Web server at the point of cloning. For the sake of presentation, we have not refined the coloring to be per-file. A key observation is the interspersing of different colors in the physical memory space of the VM due to virtual-to-physical translations and memory fragmentation.

Each clone is aided by a *memtap* process in charge of obtaining the memory the clone needs. Memtap's objective is to keep the clone's VCPUs blocked as little as possible. A blocked VCPU not only affects the QoS of the request it is serving, but also effectively disables the guest kernel's ability to multiprocess and service any other requests with that VCPU. When a clone faults on a missing page and requests it from the parent, it also asks for suggestions of related pages that are likely to be needed soon. The memory server uses the principle of spatial locality within the color in which the explicit request falls, so the pages prefetched may be scattered across physical memory.

Table 1. Kaleidoscope's prefetching is tuned by color.

Color	Window	Color	Window
Kernel Code	4	Kernel Data	12
User Code	4	User Data	16
		Page Cache Data	8

A naïve alternative to per-color prefetching is 'color-blind' prefetching, which lumps together multiple unrelated colors (including free pages, which cannot be distinguished without the use of introspective coloring) in the same prefetch block. Because prefetched pages are allocated even if they are not used, color-blind's less targeted approach wastes memory. We show in Section 8 how the color map increases the accuracy (fewer 'wasted' fetches of unneeded pages) and efficiency (more faults avoided) of prefetching, and how it outperforms color-blind.

In the Kaleidoscope prototype, we tuned the prefetch strategy by semantic region to further improve efficiency (see Table 1). We obtained our per-color policies by post-processing the state propagation activity on a set of experiments with coloring turned on, but no prefetching enabled. We simulated lookahead and pivot policies with different window sizes, and chose the most effective ones for the evaluation in Section 7. Kaleidoscope's primary distinction is to use a reduced prefetch window for executable pages, regardless of whether they reside in the file cache. Secondarily, we found that by refining the window size by data page properties provided a modest improvement. By increasing the prefetch window for user data and reducing the window for data pages in the file system page cache, efficiency and accuracy improved by 5.5% and 0.7% respectively, compared to a uniform window size of 12 for all data pages. We leave for future work the online prediction of prefetching policies using techniques similar to our post-processing profiling.

5. Fractional Allocation

When a typical VM is created, all of its backing memory is pre-allocated. This is also the case in the SnowFlock VM cloning implementation, where the cloned VM's memory is allocated eagerly, and subsequently populated with the state that is fetched from the parent VM on-demand. In contrast, the fine-grain, per-page usage knowledge we extract through coloring opens up the opportunity to optimize the memory footprint of the cloned VM.

First, on-demand fine-grained propagation of the memory of a clone VM calls for the on-demand allocation of its memory frames, a technique we call *fractional allocation*. Second, the hints extracted through memory coloring can direct content-based sharing of memory pages across VMs with great efficiency and modest effort.

5.1 Implementing Fractional Footprints

Fractional allocation is achieved by allocating on-demand the underlying pages of memory of a cloned VM. This is realized through the concept of a 'ghost MFN'. A ghost MFN has the property of serving as a placeholder that encodes the clone's PFN that it backs, and a flag indicating absence of actual allocation. The ghost MFN is placed in lieu of an allocated MFN in the page tables, and the PFN-to-MFN translation table that each Xen paravirtual guest maintains. The first guest access to the PFN triggers a shadow page fault in the hypervisor, which is trapped and handled by allocating the real MFN to replace the ghost. Note that the very same page fault is already handled to draw missing state from the parent VM. Separately, as state is prefetched by the memtap memory daemon, the daemon itself can request the allocation of the MFNs needed to store prefetched content. Finally, we avoid fragmenting the host's free page heap by increasing the granularity of requested memory chunks to, for example, 2 GiBs or 512 pages at a time, while still replacing ghost MFNs one at a time.

5.2 VM State Coloring for Memory Deduplication

The color map's semantic hints allow clone VMs to significantly reduce their memory footprint for very little cost. Certain colors, specifically page cache pages and executable pages (kernel or user-space), yield a relatively high probability of inter-VM sharing within the same host. Other colors are typically populated with data (stacks, unaligned buffers, heap pointers) that all but nullify the chances of sharing. A similar principle is exploited in related para-virtual sharing work [Milosz 2009].

During construction of the descriptors, we calculate the 128-bit hash values [Hsieh 2004] of pages in candidate colors and include them for use by the clone. Each host maintains a content-addressable store (CAS) of shareable pages it has previously fetched for different clones, with in-use pages stored once in physical memory and referenced by each live clone that needs it. Using coloring to guide content sharing is more efficient than previous work [Gupta 2008, Waldspurger 2002] for three reasons. Because hash values are calculated once on the parent and passed to every clone, we efficiently avoid repetitions of brute force traversals of memory and hash calculations – Section 8 shows that our color-directed sharing captures most of the sharing opportunities among VMs with an order of magnitude less overhead. Also, cloned worker VMs have 'fate determinism': they are single-purpose, transient, and seldom start new processes or significantly change their behavior. Thus the expense of periodically re-scanning the memory for sharing opportunities is not warranted. Finally, sharing is applied only to seldom-updated executable and file system page cache pages, which minimizes the overhead cost incurred by breaking sharing of pages that are updated.

Sharing pages with identical contents complements fractional allocation to reduce the footprint of clone VMs. The net effect is a clone footprint that grows as a function of the workload. Because the footprint reflects state fetched as-needed, minus color-directed sharing, it allows the clone to perform the work of a fully stateful VM with an effective footprint which is much less than what is typically achievable via ballooning or brute-force memory deduplication.

It should be noted that both footprint reduction mechanisms have a welcome performance side-effect. Sharing hits prevent round-trips to the server to fetch the necessary page. Coloring prevents the needless propagation of free pages, as their actual contents are irrelevant – with fractional allocation we can simply take a free page in the host and scrub it. We note that in an environment without fractional allocation, coloring could enable the mapping of all free pages to the same underlying physical frame. This would work as automatic ballooning, without the need for guest collaboration, and with instantaneous self-regulation as shares are broken. It would also yield a higher sharing ratio than content-based sharing by disregarding the actual (unused) contents. We leave exploration of this opportunity for future work.

6. The Kaleidoscope Prototype

Kaleidoscope is a prototype elastic server that uses live VM cloning, state coloring, and fractional footprints to create VM workers in response to bursts in load. New workers are created in seconds and inherit the warm state of their parent VM. We describe the architecture of a Kaleidoscope server, how the worker pool is managed, and how new clones interact with secondary storage. We close the section with a discussion on the mechanisms we provide to guarantee correctness of live-cloned servers.

6.1 Kaleidoscope Server Architecture

A Kaleidoscope elastic server is a dynamically-resizing cluster of VMs. There are three roles in the cluster. First, there is a parent VM which is a traditional (i.e., booted from scratch) VM containing the necessary software stack. Second, fractional worker VMs are cloned from the parent as transient workers to handle load fluctuations. Third, a gateway VM interfaces the cluster with the outside world and manages the load using the Linux IP Virtual Server (IPVS). It routes client requests to workers, monitors the number of incoming client connections, and spawns new clones when a high water-mark threshold is exceeded. In this manner, no server in the pool, parent included, is ever overloaded – provided there are physical resources available to create more clones. Conversely, when the load drops, extraneous workers are starved of new connections and discarded once their work is complete.

Kaleidoscope currently creates a fresh checkpoint for each generation of clones to ensure the warmth of the inherited file system cache, although this could be tuned to reuse 'master' checkpoints to conserve resources if slightly cooler buffers provide sufficient performance. Similarly, its scaling speed can be tuned by configuring how many new workers are created simultaneously. Because Kaleidoscope can multicast VM state, in a highly bursty environment with flash crowds we can aggressively clone multiple workers at each step, and subsequently scale back quickly if the large step proves unwarranted.

The local disk of the parent VM is cloned to all child VMs. Typically, this is the root disk with application binaries and libraries, while high-volume application data is served by a storage backend. The semantics of disk cloning are identical to memory: clones see the same disk, although modifications to it remain private, and are discarded upon clone termination. We have not seen the necessity to improve upon the original disk cloning implementation [Lagar-Cavilla 2009]. The local disk is provided purely on-demand, but is rarely used by transient clones who find most of their requests satisfied by in-memory kernel caches, whose propagation we do optimize. We also note that the virtual disk is implemented as a sparse flat file for clones. This implicitly guarantees fractional allocation of disk blocks as a function of use, mirroring the behavior of a clone's memory footprint.

Multi-tier stacks rely on a variety of engines for data backends: RDBMSes like MySQL, caching layers like memcached, infrastructure key-value stores like SimpleDB [Amazon c], and user-deployed key-value stores such as Cassandra [Lakshman 2009]. With the complexity associated with deploying a data backend for a given application, we have decided to stay clear of any specific storage backend architecture in this work. We assume that, in most cases, a backend can be found that is fast enough to render the processing servers the bottlenecks. In this paper, workers get their static data from an NFS server.

6.2 Correctness and Consistency

Spontaneously cloning random, unsuspecting servers may yield unsatisfactory results. For instance, if an email server were cloned in the middle of sending a queue of messages, the clones would unwantedly send duplicates of the messages still in the queue of the parent at the time of cloning. As another simple example, if an application server keeps a count for the number of sessions it has handled, and periodically commits it to an underlying database, the count will drift off in each clone, and conflicting numbers will be committed by each clone to the database.

So when is it safe to use Kaleidoscope? Kaleidoscope's cloning is not intended to be a provider-driven primitive that is applied to an unmodified server. A power user or system administrator who is familiar with the server's behavior should decide whether Kaleidoscope should be deployed, when it is safe to clone, and whether code must be modified. We assume providers still exert ultimate control by limiting the resources available to users, in order to thwart DoS attacks, 'fork bomb'-analogous attacks, and other threats.

Despite the nominal necessity for expert intervention, there is a large set of applications for which Kaleidoscope cloning is trivial and harmless to apply. In our experience, many common legacy server applications work correctly with minor or no modifications. Our Apache Web server

with static files worked fine, as did OLAP database analytics, and the SPECweb suite of benchmarks, which store their client account information in a back end. Because Kaleidoscope does not hand off live connections from the parent to clone, latency sensitive applications such as video servers should also work well.

6.2.1 In-Flight Requests

Kaleidoscope can be applied only to servers with built-in support for maintaining state consistency amongst a dynamically changing pool of load balanced nodes. Beyond this basic requirement, Kaleidoscope servers need to handle the case where a server instance is processing a request at the time of cloning (a likely scenario in a busy server), because each of the created clones will also be doing so when it comes alive.

Kaleidoscope servers need to ensure that operations that are under way at the time of cloning are treated as follows: (1) read-only operations to clone state need no extra handling; (2) operations that modify state cached on the clone that needs to remain consistent must either finish or abort on all nodes. An alternative is to sidestep the issue by queueing new write operations at the load balancer and cloning only after all ongoing writes have committed at the master (a 'write barrier'); (3) requests that may modify cached data that need not be kept consistent across nodes need to be routed consistently. Many servers distribute load by assigning an account or session persistently to a node, and the load balancer routes requests observing the node-session binding. Sessions handled by the parent VM at the point of cloning need to be discarded by the clones, and the load balancer must still route requests through the parent VM; (4) external side effects (a write to a database server, file system) need to happen only once. Conveniently, this is the default behavior of many applications, such as database servers, which roll back in-flight transactions submitted over dropped connections. Alternatively, an intermediary arbitration layer could handle clone-born duplicates and ensure that that visible external side-effects are emitted only once.

6.2.2 Consistency Mechanisms

Kaleidoscope provides four mechanisms that help applications maintain correctness and consistency across a VM clone operation:

Programatic Integration One of the major appeals of live VM cloning is the ability to integrate cloud fan-out decisions into program logic, with an interface similar to other privileged system calls. Much like UNIX fork(), the process within the server invoking the clone call will receive, upon success, a unique clone ID. This allows programs, modified or written from scratch, to immediately react to cloning as part of their flow control logic. We note that throughout this paper cloning is triggered by a load-balancer; the clones themselves are not modified to perform an explicit call, but can retrieve their ID through a `proc` interface.

IP Address Adjustment Upon cloning, a clone's IP address is automatically reconfigured before any inbound or outbound packets are allowed. Clones share an internal private network with their parent and other select entities such as the load balancer or the backend server. Clones are assigned a new IP address within the private network as a function of their ID. The reconfiguration of IP addresses requires no developer intervention.

On the parent's side, network connections that are open at the point of cloning remain open and working. If the parent VM accepts client connections, Kaleidoscope does not attempt to hand these off to the clone. On the clone VM, the connection is inherited but the assignment of a new IP address during cloning forces all inbound and outbound connections to drop – in many cases, this will result in automatically discarding session state that the parent handles entirely. However, the new clone has to graciously deal with broken connections to system resources, such as the load balancer and backend storage. Once those connections are re-established, the new, cloned workers require no further network-plane intervention to ensure correctness.

Reconfiguration Hook Much like the Linux hotplug infrastructure calls user-space scripts through the creation of `kobjects` and `uevents`, the guest kernel will automatically invoke a reconfiguration script after cloning, if one has been registered. As previously discussed, our only modification to the servers used in this paper was to have clones deal with automatic IP reconfiguration by remounting their NFS connections and subscribing to the load balancer through a simple script invoked through the reconfiguration hook.

SIGCLONE For more involved scenarios, a special asynchronous signal can be sent to processes once a clone comes alive. Handling of this signal will require code level changes, and is only intended for complex situations – none of our experiments in this paper used SIGCLONE. Processes explicitly subscribe to receive SIGCLONE, which will be sent as a one of the available POSIX real-time signal and thus terminate the process if unhandled.

7. Experimental Setup

As we anticipated, it is much faster to clone workers than to boot them (five seconds vs. two minutes). Thus, our experiments look past this major advantage, and instead examine the value of micro-elasticity and warmed caches by comparing cloning to an idealized cloud where new workers boot instantly at no cost. We measure the initial behavior of six worker prototypes under five workloads, and examine the benefits of fractional footprints and page sharing.

7.1 Worker Prototypes

Each of the six prototypes strikes a different balance between performance and efficiency. All prototypes run on identical hardware, and all applications, including Apache and MySQL, run without modification. Where indicated, the prototypes are previously warmed by a related workload (different random seeding). All cloned worker VMs run on a different physical host than their parent. The prototypes are:
Warm Static Previously warmed, this statically overprovisioned server sets the upper bound for achievable performance. This worker is equivalent to an idealized zero-latency cloning strategy that uses eager full replication.
Cold Standby Recently booted and held on cold standby, it represents an idealized version of current elastic clouds, where copying and booting workers is instant and free. Cold standby results, if extended by an average of two minutes, are comparable to results from today's commercial clouds.
Minimal Clone Newly cloned using basic VM cloning, this worker faults on every page it needs with no wasted fetches.
Kaleidoscope Newly cloned worker that uses the color map to prefetch close PFNs within the same semantic region. The prefetch window is tuned by semantic region (see Section 4.3 and Table 1).
Aggressive Clone This color-blind clone is tuned for higher performance and prefetches the next 16 PFNs.
Conservative Clone Color-blind clone tuned for efficiency by constraining its prefetch window to four pages.

7.2 Web and Database Workloads

We tested each prototype under five workloads.

OLAP Server In the absence of an accepted standard OLAP benchmark, we drive the MySQL database server with ad hoc decision-support queries drawn from the TPC-H [TPC-H] benchmark. We use only 17 of the 22 TPC-H queries because five are long running and are poorly suited to evaluate a server's initial, transient performance. Our elastic OLAP server is read-only and does not support database writes, which is consistent with standard industry practice, where complex decision-support queries are typically run against a read-only copy of a business's main OLTP database to avoid performance interference.

Web Server with Static Content Apache Web server is driven by Httperf [Mosberger 1998], requesting 3,000 files selected randomly from 7,500 static HTML files (random sizes, 1 to 128 KiB). Where warmed, all files are loaded in random order into the file system cache.

Web Server with Dynamic Content Apache is driven by requests for dynamic content by the industry-standard SPECweb 2005 [SPECweb 2005] benchmark. This benchmark consists of three standard workloads: *Banking*, where users check balances and manipulate accounts through secure connections; *Ecommerce*, where users search, browse and purchase products; and *Support* where users browse and search product listings and download files through insecure

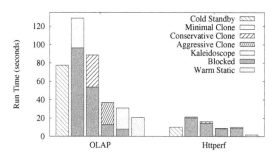

Figure 4. Prefetching reduces page faults and VCPU blocking, so benchmarks finish sooner. Legend top-to-bottom matches columns left-to-right.

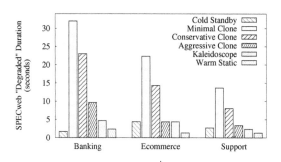

Figure 5. Prefetching reduces the length of time that prototypes fail to meet SPECweb's minimum acceptable QoS. Legend top-to-bottom matches columns left-to-right.

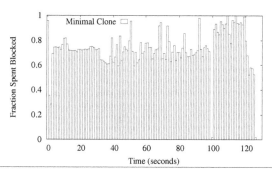

Figure 6. VM cloning copies state on demand, leaving the VCPU frequently blocked while page faults are serviced.

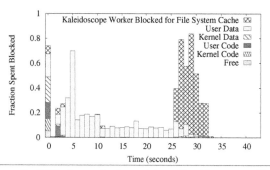

Figure 7. With fewer faults and less blocking, new Kaleidoscope workers run database queries faster. No faults were registered after the first 33 seconds.

connections. Each workload is a closed-loop simulation of user visits based on Markov chains derived from the web logs of typical sites. The workloads draw from many representative files and model a typical distribution of file sizes and overlaps in file access patterns.

All workers are implemented as Xen 3.4.0 VMs running 64-bit Linux (Debian Core 5) on eight identical Sun Fire X2250s (each with eight Xeon cores, 8 GiB RAM, and dual Gigabit Ethernet). Workloads are generated on five Dell servers (four Xeon cores, 4 GiB RAM, Gigabit Ethernet). OLAP workers run MySQL 5.1.47 in a 2 GiB VM. Web workers run Apache 2.2.9 in a 768 MiB VM. All workload data files are accessed through NFS.

8. Results

In this section we discuss the performance, efficiency and resource use of the six prototypes under the five workloads.

8.1 Performance

In our experiments, we measure performance in two ways. For the OLAP and Httperf benchmarks, which impose a fixed volume of work, we measure the run time to assess performance (Figure 4). For SPECweb, however, the run time is fixed because the benchmarks sustain a specified load over time, and instead we measure whether the server's QoS is

'acceptable', which is defined as meeting a minimum latency for a specified percentage of requests. For these benchmarks we measure the 'degraded duration', the number of seconds for which SPECweb finds the instantaneous QoS 'unacceptable' (Figure 5). Another difference between the two groups of benchmarks is the value of their accrued state. For OLAP and Httperf, the warmed memory state speeds future work, as evidenced by the margin at which Warm Static outperforms Cold Standby (factors of 3.7 and 5.6, respectively). For the SPECweb benchmarks, the dynamic content varies for each request, which renders cached state much less valuable and causes Cold and Warm to perform nearly equally.

The Minimal Clone's lazy state propagation exacts a heavy price, resulting in the worst performance on every benchmark. Minimal Clone is unable even to match the Cold Standby performance, showing that the cost of copying state purely on demand negates the entire benefit of warmed memory state. Figure 6 provides insight into Minimal Clone's poor performance. The plot shows that fetching state purely on demand leaves the VCPU frequently blocked while page faults are serviced.

In contrast, Kaleidoscope nearly matches the performance of Warm Static for all benchmarks. For OLAP, it leverages the inherited warm state to outperform Cold Standby by a factor of 2.5, and ran only 10 seconds slower than Warm Static, achieving 67% of its best-possible throughput. By comparison, Cold Standby achieved only 27% throughput during its 77 second run time. (Note that these performance gains would be even more dramatic if the latency and overhead of booting the cold VM were included.) Figure 7 shows that Kaleidoscope's color-directed prefetching largely eliminates the state transfer cost that so hampered Minimal Clone. *In summary, Kaleidoscope's approach to state replication achieves the fast instantiation time associated with on-demand cloning while coming very close to matching the runtime performance of eager full replication.*

Both color-blind prototypes performed better than Minimal Clone, but failed to materially beat Kaleidoscope on any benchmark. The key is that prefetching a page before it is needed eliminates faults, reduces VCPU blocking and boosts performance, whereas prefetching an unneeded page wastes network bandwidth, and in our fractional footprint environment, memory allocation. Figure 9(a) shows the relative fault reduction and waste of blind prefetching for various window sizes, with conservative strategies toward the top left, and aggressive toward the lower right.

By bridging the semantic gap, Kaleidoscope is better at estimating whether pages are likely to be needed (Figure 9(b)), and eliminates more faults with less waste. For the OLAP workload, it outperforms the Conservative Clone and achieves 2.9 times the throughput with 43% less waste. The Aggressive Clone is tuned for higher performance, but still falls 16% short of Kaleidoscope's OLAP throughput and wastes seven times as much prefetch bandwidth and clone memory allocation.

8.2 Scalability

To maintain an acceptable QoS for large spikes in load, it is important that Kaleidoscope's performance scale well with the number of simultaneous clones. Figure 8 shows that, for up to eight simultaneous new workers under the OLAP workload, Kaleidoscope scales well and outperforms Cold Standby. Kaleidoscope's inheritance of the parent's warmed caches is efficient, and multicast is effective in distributing the universally needed pages to the sibling clones. In fact, each clone's fetch act as prefetch for others, which boosts overall performance. In contrast, Cold Static workers tend to slow each other as they contend for the central database files.

Larger-scale Kaleidoscope elasticity can be achieved by cloning multiple parents in parallel. Although not implemented in the current prototype, clones older than several minutes could be converted into cloneable parents by transferring the full VM image. This configuration should support an exponentially increasing load that doubles every minute.

8.3 Lightening the Backend Load

Clones, with their inherited warm caches, sometimes exert less load on the backend storage than cold workers that warm their caches from scratch. This secondary benefit is most pronounced for the Httperf benchmark, where static HTML files stored in the page cache save 200MB of reads,

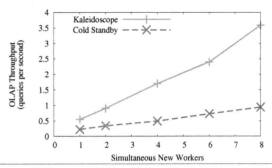

Figure 8. The database's OLAP throughput scales well with the number of simultaneous clones.

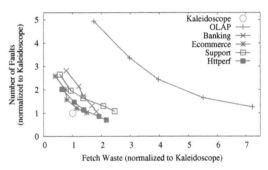

(a) Locality-based fetching fails to match Kaleidoscope's fault reduction to waste ratio. Windows from 4 (top left, less waste more faults) to 16 (bottom right, less faults more waste) pages.

Workload	Pages Needed	Faults Avoided	Pages Fetched Unnecessarily
OLAP	193,788	163,987	8,195
Banking	97,451	76,675	21,341
Ecommerce	75,044	58,734	17,352
Support	65,786	51,387	13,589
Httperf	58,850	39,980	8,625

(b) Kaleidoscope's color-directed prefetching efficiently avoids most faults with relatively few wasted fetches.

Figure 9. Prefetch Efficiency Kaleidoscope's use of semantic hints yields better fault reduction with less waste.

and for OLAP, where MySQL's user space data structures are sufficient to satisfy future queries, eliminating 890MB of backend reads. The least benefit is derived for the SPECweb workloads. Because they randomly selects files from a very large set, cached results are statistically unlikely to help.

8.4 Fractional Footprints

Kaleidoscope's fractional VM workers grow only as needed to satisfy new allocations or hold newly transferred state, as illustrated for the five workloads (with page sharing disabled) in Figure 10(a). For the SPECweb workloads with their large numbers of files and simulated users, the workers grow to claim approximately 90% of their allocation within the first minute. The Httperf and OLAP benchmarks are quite different, and under their intense but narrower loads,

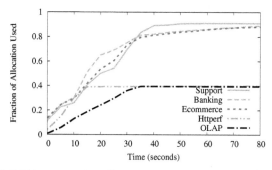

(a) Kaleidoscope workers allocate memory as necessary, an advantage when spikes in load are short lived. Traces include transferred state plus new allocations. Legend matches plots top-to-bottom.

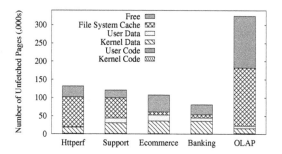

(b) Kaleidoscope workers avoid unneeded pages, mostly from the file system cache and free list. Legend matches bars top-to-bottom.

Figure 10. Fractional Footprints Kaleidoscope's workers efficiently avoid fetching or allocating unnecessary pages.

Kaleidoscope workers reached only 40% and 39% of their nominal memory size, respectively. Figure 10(b) shows the distribution of the unfetched parent state by semantic region.

The 3,000 static HTML files of the Httperf benchmark are easily cached with capacity to spare, and since future requests hit the same files, the allocation remains stable. As may be common given the coarse sizing of VMs available in today's commercial clouds, the worker is oversized for the load requirements, and Kaleidoscope offers the attractive possibility of efficiently retaining the unused capacity for other purposes. The OLAP worker is similar. It, too, is oversized, and clones of the parent VM are even more so – fractional footprint results in over one GiB of savings. Whereas the parent temporarily needed memory to cache the database files while it populated its user space data structures, the cache is later not needed to service requests and therefore remains unfetched and unallocated by the clone.

8.5 Sharing Identical Pages

A 'fair' evaluation of the content-addressable store (CAS) is important because results are dependent on the experimental design. For example, clones of the same parent and from the same, or even different, generations are extremely similar, and for read-only workloads could be nearly identical. This would skew the analysis of sharing opportunities. We

Table 2. Color-Directed vs Blind CAS

	Kaleidoscope	Blind CAS
Size of CAS Cache	22,507	248,717
Pages Hashed	26,166	258,072
Saved by Sharing	5,522	9,355
Lost to Divergence	1,848	2,413
Net Savings	3,674	6,942

therefore assess Kaleidoscope's page sharing with clones of the MySQL and Apache parents as follows. We assume that a Kaleidoscope deployment would avoid placing siblings on the same host, but that hosts might be used for workers of several distinct parents. To determine the 'initial sharing opportunities' we consider two clones and their initial state as inherited from each parent, and count the duplicate pages present in the union of their states. As the clones service their workloads (OLAP and Banking respectively), they update some of their pages, which may require that shares be broken. After both workloads are complete, we consider the 'net savings', the count of duplicate pages that are still shared.

Kaleidoscope provides very high CAS efficiency by tracking hash values only for those pages that are the most likely to be sharable, and are the least likely to subsequently diverge. Table 2 shows the initial and longer-term sharing opportunities for the Banking and OLAP benchmarks running on the same host. By tracking only free, kernel (except the file system cache), and executable pages (including those in the file system cache), *Kaleidoscope captures 53% of the sharing opportunities while computing hashes for only 10% of pages, and performs an average of 7.1 hash calculations per page saved. Blind CAS netted additional pages at 10 times that cost.*. Kaleidoscope is not intended to compete with general CAS systems, but is rather taking advantage of savings it encounters at nearly zero cost. It should be noted that Kaleidoscope has relatively few sharing opportunities because it has already eliminated the wholesale waste from free pages by avoiding their allocation entirely.

As a further benefit, the local CAS cache can be used to service clone page faults without incurring the round trip cost of fetching the page from the parent VM. This could happen frequently where hosts are used for subsequent generations of workers servicing later spikes in load.

9. Implications for Cloud Data Centers

Finally, we examine the implications of deploying Kaleidoscope for the QoS, resource use, and infrastructure requirements of elastic clouds. We conduct a simulation-driven study using one month of CPU and memory demand data collected from AT&T hosting in January 2010. The data is for a subset of 248 customers' tiers hosted on a total of 1,740 statically allocated physical processors (PPs) and collected at 5 minute intervals. The selection of customer tiers was based on those customers who had instrumentation of processor and memory consumption turned on – from that set we retained only those PPs devoted to web and application server tiers. We consider a scenario in which this demand is served by a hypothetical cloud data center that contains identical physical machines (PMs) with 16 CPU threads and 24 GiB of RAM. VMs are packed into PMs using a first-fit bin-packing algorithm - a newly created VM is allocated to the first PM which has sufficient memory and CPU capacity available.

CPU demand (% PP used) is aggregated over all PPs belonging to a customer tier at every time interval and then divided equally amongst the number of VMs. When the aggregate utilization (i.e., aggregate CPU demand/CPU capacity across all VMs belonging to the tier) for a customer exceeds the high threshold T_H, sufficient additional VMs are created to reduce utilization back below T_H. When the aggregate utilization falls below low threshold T_L, enough VMs are removed to bring utilization back between T_L and T_H. The thresholds determine how efficiently the cloud's resources are utilized. High values will delay new VM creation and hasten the destruction of underutilized VMs, and fewer VM's will be needed to satisfy a given demand.

The additional capacity of a newly instantiated VM is brought online only after an *instantiation interval* whose duration depends on the VM creation mechanism. We consider an 'overload' an event in which the CPU demand at that point in time exceeds the current VM allocation. Overloads happen because the VM creation mechanism used is too slow and VMs are not created with sufficient anticipation. We measure QoS degradation due to overloads by the accumulated value of the 'unmet demand' in CPU-seconds - i.e., one CPU-second implies a shortfall of an entire CPU core's worth of demand for the period of one second.

We estimate the seconds of unmet demand, the CPU-hours used by all the customers, and number of physical machines required in the cloud for the following scenarios:
Warm Static: Each physical processor (PP) is mapped to a single VCPU VM (100% demand) with the PP's peak memory allocation. VMs are statically allocated at the beginning of the simulation. This represents traditional over-provisioning with the best performance, and meets all demands at the expense of a large infrastructure commitment.
Elastic Cloud: Simulates current elastic clouds with dynamic addition and removal of VMs. Each VM is allocated memory equal to the peak demand of any PP in the tier over the entire month. An instantiation interval of 2 minutes is used based on current clouds. Page sharing is optimistically simulated by reducing the parent VM's memory by a fixed tunable percentage throughout the whole simulation. Extra workers do not share pages due to their transient nature.
Kaleidoscope: Similar to the Elastic Cloud with an instantiation interval of 5 seconds, based on performance reported in Section 7. A new clone's memory starts at 20% of its final size and grows dynamically for 40 seconds according to one of two memory growth profile's from Figure 10(a) - a

large footprint clone corresponding to the Support, Ecommerce, and Banking workloads that grows to 90% of the full memory allocation, and a small footprint corresponding to the Httperf and OLAP workloads that grows to 40% of the full memory allocation. Finally, the final memory size on all VMs (parents and clones) are reduced by a fixed tunable percentage when simulating memory sharing. To account for missed sharing opportunities in the transient clones, we reduce sharing percentage by 47% for clones (according to Table 2). We do not take into account the improvements in sharing overhead caused by Kaleidoscope's state coloring.

Figure 11(a) shows the substantial impact of Kaleidoscope on QoS by plotting cumulative unmet demand across all customers over the entire one-month period. We used $T_L = T_H - 20\%$ and interpolated the demand data to 5 second intervals. The plot shows that QoS degrades exponentially with the slower VM creation time of the Elastic Cloud and higher CPU thresholds. In practice, VM cold booting used in the Elastic Cloud will result in even higher QoS violations for many applications due to the performance degradation, caused by cold caches, that is ignored in these results. In comparison, Kaleidoscope with $T_H = 90\%$ has over three orders of magnitude less unmet demand than the Elastic Cloud. Thus, slow VM creation imposes the adoption of more conservative VM creation and deletion thresholds, and is, in effect, the modus operandi in today's IaaS operations.

Higher CPU thresholds lead to better utilization of resources and increased IaaS efficiency. Figure 11(b) shows the number of CPU-hours actually allocated to all customers over the one-month period, as a function of the CPU threshold. CPU hours are counted (with a five-second granularity) from the moment an allocation decision is made, and thus are independent from the instantiation latency. The corresponding Warm Static allocation, whose CPU-hour count is independent of creation thresholds, is over an order of magnitude higher (4.2 million CPU-hours) and is not shown in the figure. By combining the results from Figure 11(b) with Figure 11(a) we see that Kaleidoscope with a $T_H = 90\%$ outperforms current Elastic Clouds with $T_H = 50\%$ by resulting in a 98% reduction in the number of seconds of unmet demand, while still requiring 26% fewer resources to be purchased by cloud users.

Finally, Figure 11(c) shows the number of physical machines required as a function of memory page sharing. We compare the Elastic Cloud and Kaleidoscope scenarios. For Elastic Cloud, we consider two creation thresholds, an aggressive one that minimizes resource usage (90%), and a conservative one that minimizes unmet demand (60%). For Kaleidoscope we only consider the aggressive 90% threshold, and compare three different memory growth profiles for clones: Support (growth of up to 90% of the footprint), OLAP (growth of up to 40%), and a 50-50 Mix with customers randomly assigned either profile. For both Elastic Cloud and Kaleidoscope we also factor in an overall

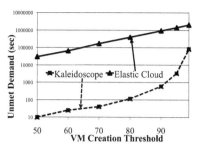
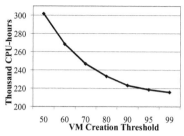
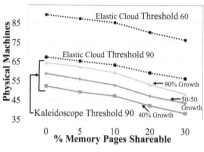

(a) Unmet demand – a proxy for QoS – as a function of VM creation and destruction thresholds. Log scale, higher is worse

(b) VM CPU hours billed as a function of VM creation thresholds. Note that this is independent of instantiation latency.

(c) Infrastructure use, as a function of memory consolidation ratio. For Kaleidoscope we consider three fractional allocation profiles.

Figure 11. Simulations using AT&T Hosting data show that Kaleidoscope improves cloud QoS, resource use, and efficiency.

page sharing success percentage. Even with identical CPU thresholds, Kaleidoscope significantly reduces the number of physical machines needed by the IaaS provider by 5% to 30%. If the VM boot techniques used in the Elastic Cloud setup require a conservative threshold of 60%, then even higher infrastructure reductions of up to 50% are possible.

10. Related Work

We propose VM state coloring as another way of bridging the semantic gap between VM management and OS knowledge, and is very different from 'memory coloring', which has been used in the past for techniques improving the performance of processor memory caches. The problem of the semantic gap in virtualized environments was first formulated by Chen and Noble [Chen 2001]. Patagonix [Litty 2008] and Antfarm [Jones 2006a] perform a form of architecture-based semantic gap bridging, using x86 page table knowledge to identify user-space processes inside a VM. Self-migration [Hansen 2004], is an approach to migration that demands tight collaboration between the host VMM and guest VM, pervasively bridging the semantic gap.

Geiger [Jones 2006b] targets semantic gap issues related to the OS page cache, and presents an OS-independent alternative to our page cache identification mechanism. Geiger, however, requires the use of approximation heuristics to detect when a page in the page cache has been given a different use. It also needs to reverse-engineer the underlying filesystem to detect block device transactions pertaining to the journal or other non-file structures, and to group guest pages according to their backing file.

Potemkin [Vrable 2005] implements a different form of VM cloning: Potemkin clones are short-lived lightweight VMs residing in the same host as the parent (or template) and sharing memory via copy-on-write. The canonical work in memory management of VMs is Waldspurger's [Waldspurger 2002]. Conceptually, Difference Engine [Gupta 2008] extends these ideas by adding sub-page sharing. Both systems perform repeated cycles of brute-force fingerprinting of the entire host memory to achieve memory dedupli-

cation. Satori [Milosz 2009] is similar to our work in that it uses an efficient source of VM introspection, virtual disk DMA operations, to guide the sharing mechanism.

To the best of our knowledge, we are the first to evaluate the use of VM cloning to dynamically scale servers and preserve QoS during spikes in load. Prior related work has focused on optimizing application QoS within a static VM allocation. Approaches include the utilization of VM migration to relieve datacenter hot spots [Wood 2006], workload management and admission control to optimize QoS and resource use [Elnikety 2004, Urgaonkar 2008a], and allowing applications to barter resources [Norris 2004].

We close by highlighting related work in the broad area of dynamic resource provisioning. Typically, dynamic VM provisioning is achieved with copy-and-boot techniques [Murphy 2009, Urgaonkar 2008b]. Another technique is to keep a pool of pre-configured machines on standby, and to bring these generic hot spares to bear as required [Fox 1997]. These are all examples of approaches that tie up computing resources in reserve, or present prolonged instantiation latencies leading in many cases to subpar performance.

11. Conclusions

In this paper we have introduced the notion of cloud micro-elasticity, in which servers react to load by swiftly spawning transient, short-lived cloned VM's with warm application caches and a tightly adjusted fractional memory footprint. Micro-elasticity is achieved by tailoring propagation and sharing policies to the different types of memory in a VM, a technique we call VM state coloring. VM state coloring can spawn stateful clones with warm application caches at a fraction of the cost (and footprint) of state-of-the-art techniques. By simulating cloud micro-elasticity on traces collected from AT&T's multi-tenant hosting environment, we obtain reductions in infrastructure use of roughly 30%, which benefit both providers and users.

There are two important paths for future work. First, the consistency assumptions governing some multi-tiered applications might be subverted by the impromptu addition of cloned VMs. This has not been the case for our Web and OLAP workloads; nonetheless, we need to further explore the suitability of our sanitization mechanisms for other workloads. Second, the principles of VM state coloring can be very useful in WAN VM migration; improvements on the efficiency of content-addressing, prefetching, and even fetch avoidance are all valuable optimizations to migration.

Acknowledgments

We thank Michael Mior for his input throughout this project. We thank Philip Patchin, John Wilkes, Angela Demke-Brown, Bianca Schroeder, and Mary Fernandez for their feedback prior to submission. Eleftherios Koutsofios enabled access to the data used in the simulation study. We thank the anonymous Eurosys reviewers, and our shepherd Andreas Haeberlen for helping improve the quality of the final version of this paper. This research was partially supported by the National Science and Engineering Research Council of Canada (NSERC) under grant number 261545-3, and a hardware donation from Oracle Inc.

References

[Amazon a] Amazon. Auto Scaling. http://aws.amazon.com/autoscaling/.

[Amazon b] Amazon. Elastic Load Balancing. http://aws.amazon.com/elasticloadbalancing/.

[Amazon c] Amazon. SimpleDB. http://aws.amazon.com/simpledb/.

[Chen 2001] P. Chen and B. Noble. When Virtual is Better Than Real. In *Proc. 8th Workshop on Hot Topics in Operating Systems (HotOS)*, Elmau/Oberbayern, Germany, May 2001.

[Elnikety 2004] S. Elnikety, E. Nahum, J. Tracey, and W. Zwaenepoel. A Method for Transparent Admission Control and Request Scheduling in E-Commerce Web Sites. In *Proc. 13th WWW*, pages 276–286, New York City, NY, May 2004.

[Fox 1997] A. Fox, S. D. Gribble, Y. Chawathe, E. A. Brewer, and P. Gauthier. Cluster-based Scalable Network Services. In *Proc. 16th SOSP*, pages 78–91, Saint Malo, France, October 1997.

[Gupta 2008] D. Gupta, S. Lee, M. Vrable, S. Savage, A. C. Snoeren, G. Varghese, G. M. Voelker, and A. Vahdat. Difference Engine: Harnessing Memory Redundancy in Virtual Machines. In *Proc. 8th OSDI*, San Diego, CA, December 2008.

[Hansen 2004] J. A. Hansen and E. Jul. Self-migration of Operating Systems. In *11th ACM SIGOPS European Workshop*, Leuven, Belgium, September 2004.

[Hyperic] Hyperic. CloudStatus. cloudstatus.com.

[Jones 2006a] S. T. Jones, A. C. Arpaci-Dusseau, and R. H. Arpaci-Dusseau. Antfarm: Tracking Processes in a Virtual Machine Environment. In *Proc. Usenix ATC*, Boston, MA, June 2006.

[Jones 2006b] S. T. Jones, A. C. Arpaci-Dusseau, and R. H. Arpaci-Dusseau. Geiger: Monitoring the Buffer Cache in a Virtual Machine Environment. In *Proc. 12th ASPLOS*, San Jose, CA, October 2006.

[Lagar-Cavilla 2009] H. A. Lagar-Cavilla, J. Whitney, A. Scannell, P. Patchin, S. M. Rumble, E. de Lara, M. Brudno, and M. Satyanarayanan. SnowFlock: Rapid Virtual Machine Cloning for Cloud Computing. In *Proc. 4th EuroSys*, pages 1–12, Nuremberg, Germany, April 2009.

[Lakshman 2009] A. Lakshman and P. Malik. Cassandra - A Decentralized Structured Storage System. In *Proc. 3rd LADIS*, Big Sky, MT, October 2009.

[Litty 2008] L. Litty, H. A. Lagar-Cavilla, and D. Lie. Hypervisor Support for Identifying Covertly Executing Binaries. In *Proc. 17th Usenix Security*, San Jose, CA, July 2008.

[Milosz 2009] G. Milosz, D. Murray, S. Hand, and M. Fetterman. Satori: Enlightened Page Sharing. In *Proc. Usenix ATC*, San Diego, CA, July 2009.

[Mosberger 1998] D. Mosberger and T. Jin. httperf—a Tool for Measuring Web Server Performance. *SIGMETRICS Performance Evaluation Review*, 26(3):31–37, 1998.

[Murphy 2009] M. A. Murphy, B. Kagey, M. Fenn, and S. Goasguen. Dynamic Provisioning of Virtual Organization Clusters. In *Proc. 9th CCGRID*, pages 364–371, Shanghai, China, 2009.

[Norris 2004] J. Norris, K. Coleman, A. Fox, and G. Candea. On-Call: Defeating Spikes with a Free-Market Application Cluster. In *Proc. 1st ICAC*, pages 198–205, Washington, DC, USA, 2004.

[Hsieh 2004] P. Hsieh. Super Fast Hash function, 2004. http://www.azillionmonkeys.com/qed/hash.html.

[RightScale] RightScale. More Servers, Bigger Servers, Longer Servers, and 10x of That. http://blog.rightscale.com/2010/08/04/more-bigger-longer-servers-10x/.

[Satyanarayanan 2005] M. Satyanarayanan, M. Kozuch, C. Helfrich, and D. O'Hallaron. Towards Seamless Mobility on Pervasive Hardware. *Pervasive and Mobile Computing*, 1(2), 2005.

[SPECweb 2005] SPECweb. Standard Performance Evaluation Corp., 2005. http://www.spec.org/web2005/.

[TPC-H] TPC-H. Transaction Processing Performance Council. http://www.tpc.org/tpch/.

[Urgaonkar 2008a] B. Urgaonkar and P. Shenoy. Cataclysm: Scalable Overload Policing for Internet Applications. *Network and Computing Applications*, 31(4):891–920, 2008.

[Urgaonkar 2008b] B. Urgaonkar, P. Shenoy, A. Chandra, P. Goyal, and T. Wood. Agile Dynamic Provisioning of Multi-tier Internet Applications. *ACM Transactions in Autonomic Adaptive Systems*, 3(1):1–39, 2008.

[Vrable 2005] M. Vrable, J. Ma, J. Chen, D. Moore, E. Vandekieft, A. Snoeren, G. Voelker, and S. Savage. Scalability, Fidelity and Containment in the Potemkin Virtual Honeyfarm. In *Proc. 20th SOSP*, Brighton, UK, October 2005.

[Waldspurger 2002] C. A. Waldspurger. Memory Resource Management in VMWare ESX Server. In *Proc. 5th OSDI*, Boston, MA, 2002.

[Wood 2006] T. Wood, P. Shenoy, A. Venkataramani, and M. Yousif. Sandpiper: Black-box and Gray-box Resource Management for Virtual Machines. In *Proc. 2nd NSDI*, Boston, MA, 2006.

Scarlett: Coping with Skewed Content Popularity in MapReduce Clusters

Ganesh Ananthanarayanan∓, Sameer Agarwal∓, Srikanth Kandula◇,
Albert Greenberg◇, Ion Stoica∓, Duke Harlan†, Ed Harris†

∓ UC Berkeley   ◇ Microsoft Research   † Microsoft Bing

{ganesha, sameerag, istoica}@cs.berkeley.edu, {srikanth, albert, dukehar, edharris}@microsoft.com

Abstract

To improve data availability and resilience MapReduce frameworks use file systems that replicate data *uniformly*. However, analysis of job logs from a large production cluster shows wide disparity in data popularity. Machines and racks storing popular content become bottlenecks; thereby increasing the completion times of jobs accessing this data even when there are machines with spare cycles in the cluster. To address this problem, we present Scarlett, a system that replicates blocks based on their popularity. By accurately predicting file popularity and working within hard bounds on additional storage, Scarlett causes minimal interference to running jobs. Trace driven simulations and experiments in two popular MapReduce frameworks (Hadoop and Dryad) show that Scarlett effectively alleviates hotspots and can speed up jobs by 20.2%.

Categories and Subject Descriptors D.4.3 [*Operating Systems*]: File Systems Management–Distributed file systems

General Terms Algorithms, Measurement, Performance

Keywords Datacenter Storage, Locality, Fairness, Replication

1. Introduction

The MapReduce framework has become the de facto standard for large scale data-intensive applications. MapReduce based systems, such as Hadoop [Hadoop], Google's MapReduce [Dean 2004], and Dryad [Isard 2007] have been deployed on very large clusters consisting of up to tens of thousands of machines. These systems are used to process large datasets (e.g., to build search indices or refine ad placement) and also in contexts that need quick turn-around times (e.g., to render map tiles [Dean 2009]). These deployments represent a major investment. Improving the performance of

MapReduce improves cluster efficiency and provides a competitive advantage to organizations by allowing them to optimize and develop their products faster.

MapReduce jobs consist of a sequence of dependent *phases*, where each phase is composed of multiple *tasks* that run in parallel. Common phases include map, reduce and join. Map tasks read data from the disk and send it to subsequent phases. Since the data is distributed across machines and the network capacity is limited, it is desirable to *co-locate* computation with data. In particular, co-locating map tasks with their input data is critical for job performance since map tasks read the largest volume of data. All MapReduce based systems go to great lengths to achieve data locality [Dean 2004, Isard 2009, Zaharia 2010].

Unfortunately, it is not always possible to co-locate a task with its input data. The uniform data replication employed by MapReduce file systems (e.g., Google File System [Ghemawat 2003], Hadoop Distributed File System (HDFS) [HDFS]) is often sub-optimal. Replicating each block on three different machines is not enough to avoid *contention* for slots on machines storing the more popular blocks. Simply increasing the replication factor of the file system is not a good solution, as data access patterns vary widely in terms of the total number of accesses, the number of concurrent accesses, and the access rate over time. Our analysis of logs from a large production Dryad cluster supporting Microsoft's Bing, shows that the top 12% of the most popular data is accessed over ten times more than the bottom third of the data. Some data exhibits high access concurrency, with 18% of the data being accessed by at least three unique jobs at a time.

Contention for slots on machines storing popular data may hurt job performance. If the number of jobs concurrently accessing a popular file exceeds the number of replicas, some of these jobs may have to access data remotely and/or compete for the same replica. Using production traces, we estimate that as a direct consequence of contentions to popular files, the median duration of Dryad jobs increases by 16%.

To avoid contentions and improve data locality, we design a system, Scarlett, that *replicates files based on their access patterns and spreads them out to avoid hotspots, while minimally interfering with running jobs.* To implement this approach it is critical to accurately predict data popularity. If we don't,

Dates	Phases $(\times 10^3)$	Jobs $(\times 10^3)$	Data (PB)	Network (PB)
May 25,29	44.3	23.4	35.5	1.55
Aug 20,24	77.1	48.8	47.7	2.10
Sep 15,19	74.4	40.1	54.0	1.82
Oct 15,19	49.0	33.0	69.5	2.17
Nov 16,20	96.4	45.3	54.4	1.70
Dec 10,14	46.4	42.4	51.9	1.40

Table 1: **Details of Dryad job logs collected from Microsoft Bing's cluster.**

we may either create too few replicas thus failing to alleviate contention, or create too many replicas thus wasting both storage and network bandwidth. To guide replication, Scarlett uses a combination of historical usage statistics, online predictors based on recent past, and information about the jobs that have been submitted for execution. To minimize interference with jobs running in the cluster, Scarlett operates within a storage budget, replicates data lazily, and uses compression to trade processing time for network bandwidth. Finally, Scarlett benefits from spreading out the extra replicas, and hence cluster load, on machines and racks that are lightly loaded.

We evaluate the benefits of Scarlett's replication scheme in the context of two popular MapReduce frameworks, Hadoop [Hadoop] and Dryad [Isard 2007]. These frameworks use different approaches to deal with data contention. While Hadoop may run a late-arriving task remotely, Dryad aims to enforce locality by evicting the low priority tasks upon the arrival of higher priority tasks [Isard 2009]. We note that these approaches, as well as new proposals to improve locality in the presence of contention by delaying the tasks [Zaharia 2010], are orthogonal and complementary to Scarlett. Indeed, while these schemes aim to minimize the effect of contentions, Scarlett seeks to avoid contentions altogether. By providing more replicas, Scarlett makes it easier for these schemes to co-locate data with computation and to alleviate contention to popular data.

We have deployed Scarlett on a 100-node Hadoop cluster, and have replayed the workload traces from Microsoft Bing's datacenter. Scarlett improves data locality by 45%, which results in a 20.2% reduction of the job completion times of Hadoop jobs. In addition, by using extensive simulations, we show that Scarlett reduces the number of evictions in the Dryad cluster by 83% and speeds up the jobs by 12.8%. This represents 84% of the ideal speedup assuming no contention. Finally, we show that Scarlett incurs low overhead, as it is able to achieve near-ideal performance by altering replication factors only once in 12 hours, using less than 10% extra storage space, and generating only 0.9% additional network traffic.

The rest of the paper is outlined as follows. In §2, we quantify the skew in popularity and its impact. §3 presents a solution to cope with the popularity skew using adaptive and efficient replication of content. We look at how Dryad and Hadoop are affected by popularity skew in §4. §5 evaluates

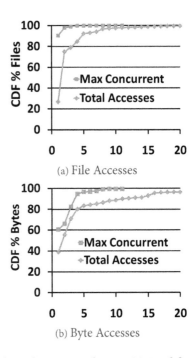

(a) File Accesses

(b) Byte Accesses

Figure 1: **File and Byte Popularity: CDFs of the total numbers of jobs that access each file (or byte) as well as the number of concurrent accesses. The x-axis has been truncated to 20 for clarity, but goes up to 70.**

the performance benefits of Scarlett. We discuss related work in §6 and conclude in §7.

2. Patterns of Content Access in Production

We analyzed logs from a large production cluster that supports Microsoft Bing over a six month period in 2009 (see Table 1). The cluster software, known as Cosmos, is a modified form of Dryad [Isard 2007] that runs jobs written in Scope [Chaiken 2008]. The dataset contains more than 200K jobs that processed over 300 petabytes of data of which over 10PB were moved on the network across racks. The logs contain details of individual tasks as well as dependencies across phases (map, reduce, etc.). For each task, we record its start and end times, the amount of data it reads and writes and the location in the cluster where the task ran and where its inputs were drawn from. We also obtain similar data at the granularity of phases and jobs. The cluster primarily ran production jobs. Some of these are mining scripts that repeatedly ran as new data arrives.

We examine the variation in popularity across files and how popularity changes over time. We also quantify the effects of popularity skew – hotspots in the cluster.

2.1 Variation in Popularity

Since accesses to content are made by jobs, we examine popularity at the smallest granularity of content that can be addressed by them. We colloquially refer to this unit as a *file*. In practice, this smallest unit is a collection of many blocks and

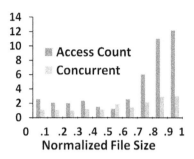

Figure 2: **Popularity of files as a function of their sizes, normalized to the largest file; the largest file has size 1. The columns denote the average value (# accesses, concurrence) of files in each of the ten bins.**

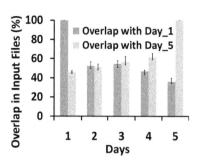

Figure 3: **Overlap in files accessed across five days, for each month listed in Table 1. With the first and fifth day as references, we plot the fraction of bytes accessed on those days that were also accessed in the subsequent and preceding days, respectively.**

often has semantic meaning associated with it such as records within a certain time range from a data stream.

There is a large variation among files in their number of accesses as well as in their number of concurrent accesses. Figure 1a plots CDFs over files of the total number of tasks that access each file and the maximum number of tasks that concurrently access each file. The figure shows that 2.5% of the files are accessed more than 10 times and 1.5% of the files are accessed more than three times concurrently. On the other hand, a substantial fraction of the files are accessed by no more than one task at a time (90%) and no more than once over the entire duration of the dataset (26%).

Files vary in size, so to examine the byte popularity, Figure 1b weights each file by its size. Compared to Figure 1a, we see that both CDFs move to the right, indicating that more fraction of the bytes are in files that have more accesses. We see that 38% of all data is accessed just once in the five-day interval. On the other hand, 12% of the data is accessed more than 10 times, i.e., 12% of the data is 10x more popular than roughly a third of the data. Recall that each block in the file system is replicated three times. The figure shows that 18% of the data have at least three concurrent accesses, i.e., are operating *at brim*, while 6% of them have more concurrent accesses than the number of replicas.

We believe that popularity skew in MapReduce clusters arises due to a few reasons. Due to abundantly available storage, a lot of data is logged for potential future analysis but only a small fraction is ever used. Some other datasets, however, correspond to production pipelines (e.g., process newly crawled web content) and are always used. Their popularity spikes when the data is most fresh and decays with time. The sophistication of analysis and the number of distinct jobs varies across these production datasets. In contrast to the popularity skew observed in other contexts (e.g., of web and peer-to-peer content), we see that the hottest content is not as hot and that there is more moderately hot content. Likely this is because the content consumer in this context (business groups for research and production purposes) have wider at-

tention spans, are more predictable, and less peaky as compared to consumers of web and p2p content.

When multiple tasks contend for a few replicas, the machines hosting the replicas become hotspots. Even if these tasks ran elsewhere in the cluster, they compete for disk bandwidth at the machines hosting the replicas. When the cluster is highly utilized, a machine can have more than one popular block. Due to collisions between tasks reading different popular blocks, the effective number of replicas per block can be fewer as some of the machines hosting its replicas are busy serving other blocks.

We find high correlation between the total number of accesses and number of concurrent accesses with a Pearson's correlation factor of 0.78, implying that either of these metrics is sufficient to capture file popularity.

We also find that large files are accessed more often. Figure 2 bins files by their size, with the largest file having a normalized size of 1, and plots the average number of accesses (total and concurrent) to files in each bin. Owing to their disproportionately high access counts, focusing on just the larger files is likely to yield most of the benefits.

2.2 Change in Popularity

Files change in popularity over time. Figure 3 plots the overlap in the set of files accessed across five consecutive days for each month listed in Table 1, with *day-1* and *day-5* as references. We observe a strong day effect – only 50% of the files accessed on any given day are accessed in the next or the previous days. Beyond this initial drop, files exhibit a gradual ascent and decline in popularity. Roughly 40% of the files accessed on a given day are also accessed four days before or after. The relatively stable popularity across days indicates potential for prediction techniques that learn access patterns over time to be effective.

On an hourly basis, however, access patterns exhibit not only the gradual ascent and decline in popularity that we see over days but also periodic bursts in popularity. Figure 4 plots hourly overlap in the set of files accessed, with two

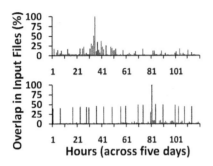

Figure 4: Hourly overlap in the set of files accessed with two sample reference hours (*hour-35* **and** *hour-82*). **The graph on the top shows a gradual change while the bottom graph shows periodically accessed files.**

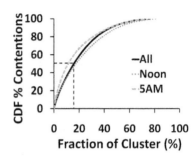

Figure 5: Hotspots: One-sixth of the machines account for half the contentions in the cluster.

illustrative reference hours. The figure on top shows gradual variation while the bottom figure shows that some sets of files are accessed in bursts. We conjecture that the difference is due to the types of files involved – the hour on the top likely consists of a time-sensitive set of files used by many different users or groups, so their popularity decays faster and more smoothly, while the bottom hour likely consists of a set of files used by fewer but more frequent users explaining the periodic bursts.

2.3 Effect of Popularity Skew: Hotspots

When more tasks want to run simultaneously on a machine than that machine can support, we will say a *contention event* has happened. MapReduce frameworks vary in how they deal with contention events. Some queue up tasks, others give up on locality and execute tasks elsewhere in the cluster and some others evict the less preferred tasks. Others adopt a combination approach – make tasks wait a bit before falling back and running them elsewhere. Regardless of the coping mechanism, contention events slow down the job and waste cluster resources.

Figure 5 plots a CDF of how contentions are distributed across the machines in the cluster. The figure shows that 50% of contentions are concentrated on a small fraction of machines (less than $(\frac{1}{6})^{th}$) in the cluster. Across periods of low and high cluster utilization (5AM and 12PM respectively), the pattern of hotspots is similar.

We attribute these hotspots to skew in popularity of files. Hotspots occur on machines containing replicas of files that have many concurrent accesses. Further, current placement schemes are agnostic to correlations in popularity, they do not avoid co-locating popular files, and hence increase the chance of contentions.

2.4 Summary

From analysis of production logs, we take away these lessons for the design of Scarlett:

1. The number of concurrent accesses is a sufficient metric to capture popularity of files.

2. Large files contribute to most accesses in the cluster, so reducing contention for such files improves overall performance.

3. Recent logs are a good indicator of future access patterns.

4. Hotspots in the cluster can be smoothened via appropriate placement of files.

3. Scarlett: System Design

We make the following two design choices. First, Scarlett considers replicating content at the smallest granularity at which jobs can address content. Recall that we call this a *file*. Scarlett does so because a job will access all blocks or none in a file. Even if some blocks in a file have more replicas, the block(s) with the fewest replicas become the bottleneck, i.e., tasks in the job that access these hot blocks will straggle [Ananthanarayanan 2010] and hold back the job. Second, Scarlett adopts a *proactive* replication scheme, i.e., replicates files based on predicted popularity. While we considered the reactive alternative of simply caching data when tasks executed non-locally (thereby increasing the replication factor of hot data), we discarded it because it does not work for frameworks that cope with contention by other means (e.g., by queuing or evicting low priority tasks). In addition, proactively replicating at the granularity of files is simpler to implement; it keeps improvements to the storage layer independent of the task execution logic.

Scarlett captures the popularity of files and uses that to increase the replication factor of oft-accessed files, while avoiding hotspots in the cluster and causing minimal interference to the cross-rack network traffic of jobs. To do so, Scarlett computes a replication factor r_f for each file that is proportional to its popularity (§3.1) while remaining within a budget on extra storage due to additional replicas. Scarlett smooths out placement of replicas across machines in the cluster so that the expected load on each machine (and rack) is uniform (§3.2). Finally, Scarlett uses compression and memoization to reduce the cost of creating replicas (§3.3).

```
Used Budget, $B_{used} \leftarrow 0$
$F \leftarrow$ Set of files sorted in descending order of size
Set $r_f \leftarrow 3 \quad \forall f \in F$                    ▷ Base Replication
for file $f \in F$ do
    $r_f \leftarrow \max(c_f + \delta, 3)$          ▷ Increase $r_f$ to $c_f + \delta$
    $B_{used} \leftarrow B_{used} + f_{size} \cdot (r_f - 3)$
    break if $B_{used} \geq B$
end for
```

Pseudocode 1: Scarlett **computes the file replication factor** r_f **based on their popularity and budget** B. c_f **is the observed number of concurrent accesses. Here files with larger size have a strictly higher priority of getting their desired number of replicas.**

```
Used Budget, $B_{used} \leftarrow 0$
$F \leftarrow$ Set of files sorted in descending order of size
Set $r_f \leftarrow 3 \quad \forall f \in F$                    ▷ Base Replication
while $B_{used} < B$ do
    for file $f \in F$ do
        if $r_f < c_f + \delta$ then
            $r_f \leftarrow r_f + 1$                ▷ Increase $r_f$ by 1
            $B_{used} \leftarrow B_{used} + f_{size}$
            break if $B_{used} \geq B$
        end if
    end for
end while
```

Pseudocode 2: **Round-robin distribution of the replication budget** B **among the set of files** F.

Recall that current file systems [Ghemawat 2003, HDFS] divide files into blocks and uniformly replicate each block three times for reliability. Two replicas are placed on machines connected to the same rack switch, and the third is on a different rack. Placing more replicas within a rack allows tasks to stay within their desired rack. Datacenter topologies are such that there is more bandwidth within a rack than across racks [Kandula 2009]. The third replica ensures data availability despite rack-wide failures. Our analysis in §2 shows sizable room for improvement over this policy of uniform replication.

3.1 Computing File Replication Factor

Scarlett replicates data at the granularity of files. For every file, Scarlett maintains a count of the maximum number of concurrent accesses (c_f) in a *learning window* of length T_L. Once every *rearrangement period*, T_R, Scarlett computes appropriate replication factors for all the files. By default, $T_L = 24$ hours and $T_R = 12$ hours. The choice of these values is guided by observations in the production logs that show relative stability in popularity during a day. It also indicates Scarlett's preference to conservatively replicate files that have a consistent skew in popularity over long periods.

Scarlett chooses to replicate files proportional to their expected usage c_f. The intuition here is that the expected load at a machine due to each replica that it stores be constant – the load for content that is more popular is distributed across a proportional number of replicas. To provide a cushion against under-estimates, Scarlett creates δ more replicas. By default $\delta = 1$. Scarlett lower bounds the replication by three, so that data locality is at least as good as with the current file systems. Hence the desired replication factor is $\max(c_f + \delta, 3)$.

Scarlett operates within a fixed budget B on the storage used by extra replicas. We note that storage while available is not a free resource, production clusters routinely compress data before storing to lower their usage. How should this budget be apportioned among the various files?

Scarlett employs two approaches. In the *priority* approach, Scarlett traverses the files in descending order of their size and

increases each file's replication factor up to the desired value of $c_f + \delta$ until it runs out of budget. The intuition here is that since files with larger size are accessed more often (see Figure 2) and also have more tasks working on them, it is better to spend the limited budget for replication on those files. Pseudocode 1 summarizes this approach. We would like to emphasize that while looking at files sorted by descending order of size is suited for our environment, the design of Scarlett allows any ordering to be plugged in.

The second *round-robin* approach alleviates the concern that most of the budget can be spent on just a few files. Hence, in this approach, Scarlett increases the replication factor of each file by at most 1 in each iteration and iterates over the files until it runs out of budget. Pseudocode 2 depicts this approach. The round-robin approach provides improvements to many more files while the priority approach focuses on just a few files but can improve their accesses by a larger amount. We evaluate both distribution approaches for different values of the budget in §5.

The following desirable properties follow from Scarlett's strategy to choose different replication factors for files:

- Files that are accessed more frequently have more replicas to smooth their load over.

- Together, δ, T_R and T_L track changes in file popularity while being robust to short-lived effects.

- Choosing appropriate values for the budget on extra storage B and the period at which replication factors change T_R can limit the impact of Scarlett on the cluster.

3.2 Smooth Placement of Replicas

We just saw which files are worthwhile to replicate but where to place these replicas? A machine that contains blocks from many popular files will become a hotspot, even though as shown above, there may be enough replicas for each block such that the per-block load is roughly uniform. Here, we show how Scarlett smooths the load across machines.

In current and future hardware SKUs, reading from the local disk is comparable to reading within the rack,

```
for file f in F do
    if r_f > r_f^{desired} then
        Delete Replicas                              ▷ De-replicate
        Update l_m accordingly
    end if
end for
for file f in F do
    while r_f < r_f^{desired} do
        for blocks b ∈ f do
            m* ← arg min(l_m)∀ machines m
            Replicate(b) at m*
            l_{m*} ← l_{m*} + c_f/r_f                ▷ Update load
        end for
        r_f ← r_f + 1
    end while
end for
```

Pseudocode 3: **Replicating the set of files F with current replication factors r_f to the desired replication factors $r_f^{desired}$. l_m is the current expected load at each machine due to the replicas it stores.**

since top-of-rack switches have enough backplane bandwidth to support all intra-rack transfers. Reading across racks however continues to remain costly due to network oversubscription [Kandula 2009]. Hence, Scarlett spreads replicas of a block over as many racks as possible to provide many reasonable locations for placing the task.

Scarlett's placement of replicas rests on this principle: place the desired number of replicas of a block on as many distinct machines and racks as possible while ensuring that the expected load is uniform across all machines and racks.

A strawman approach to achieve these goals would begin with random circular permutations of racks and machines within each rack. It would place the first replica at the first machine on the first rack. Advancing the rack permutation would ensure that the next replica is placed on a different rack. Advancing to the next machine in this rack ensures that when this rack next gets a replica, i.e., after all racks have taken a turn, that replica will be placed on a different machine in the rack. It is easy to see that this approach smooths out the replicas across machines and racks. The trouble with this approach, however, is that even one change in the replication factor changes the entire placement leading to needless shuffling of replicas across machines. Such shuffling wastes time and cross-rack network bandwidth.

Scarlett minimizes the number of replicas shuffled when replication factors change while satisfying the objective of smooth placement in the following manner. It maintains a *load* factor for each machine, l_m. The load factor for each rack, l_r, is the sum of load factors of machines in the rack. Each replica is placed on the the rack with the least load and the machine with the least load in that rack. Placing a replica increases both these factors by the expected load due to that replica ($= c_f/r_f$). The intuition is to keep track of

the current load via the load factor, and make the desired changes in replicas (increase or decrease) such that the lightly loaded machines and racks shoulder more load. In practice, Scarlett uses a slight fuzz factor to ensure that replicas are spread over many distinct racks and machines. This approach is motivated by the Deficit Round Robin scheme [Shreedhar 1996] that employs a similar technique to spread load across multiple queues in arbitrary proportions.

Pseudocode 3 shows how, once every T_R, after obtaining a new set of replication factors $r_f^{desired}$, Scarlett places those replicas. Files whose replication factors have to be reduced are processed first to get an updated view of the load factors of racks and machines. We defer how replicas are actually created and deleted to the next subsection. Traversing the list of files and its blocks, Scarlett places each replica on the next lightly loaded machine and rack. Replicas of the same block are spread over as many machines and racks as possible.

3.3 Creating Replicas Efficiently

Replication of files causes data movement over already oversubscribed cross-rack links [Kandula 2009]. This interferes with the performance of tasks, especially those of network-intensive phases like reduce and join. A skew in the bandwidth utilization of racks leads to tasks that read data over them lagging behind the other tasks in their phase, eventually inflating job completion times [Ananthanarayanan 2010]. While our policy of placing one replica per rack makes cross-rack data movement inevitable during replication, we aim to minimize it. The approximation algorithm in Pseudocode 3 takes a first stab by retaining the location of existing blocks. As a next step, we now reduce the interference caused due to replication traffic. In addition to replication traffic running at lower priority compared to network flows of tasks, we employ two techniques that complement each other – (a) equally spread replication traffic across all uplinks of racks, and (b) reduce the volume of replication traffic by trading network usage for computation using compression of data.

3.3.1 While Replicating, Read From Many Sources

We adopt the following simple approach to spread replication traffic equally across all the racks. Suppose the number of replicas increases from r^{old} to r^{new}. The old replicas equally distribute the load of creating new replicas among themselves. Each old replica is a source for $\lceil \frac{r_{new} - r_{old}}{r_{old}} \rceil$ new replicas. In the case of $\frac{r_{new}}{r_{old}} \leq 2$, each rack with old replicas will have only one copy of the block flowing over their uplinks at a time.

When the increase in number of replicas is greater than 2, Scarlett starts from r_{old} and increases the replication factor in steps of two, thereby doubling the number of sources in every step. This strategy ensures that no more than a logarithmic number of steps are required to complete the replication while also keeping the per-link cost for each block being replicated a constant independent of the number of new replicas being created.

Task *t* wants to run on
machine *M* for data locality

```
Is slot on          YES
M free?          ────────────►    Run t on M
                  (Scarlett)
    │
    │ NO
    ▼
Wait for slot on
M to vacate for   ────────────►    Run t on any
some time τ        (Hadoop)           machine

                  ────────────►    Run t on M
                   (Dryad)         after possible
                                      eviction
```

Figure 6: **Scheduling of tasks, and the different approaches to deal with conflict for slots due to data locality.** Scarlett tries to shift the focus to the "YES" part of the decision process by preferentially replicating popular content.

3.3.2 Compress Data Before Replicating

Recent trends in datacenter hardware and data patterns point to favorable conditions for data compression techniques. These techniques tradeoff computational overhead for network bandwidth [Skibinski 2007]. However, the trend of multiple cores on servers presents cores that can be devoted for compression/decompression purposes. Also, the primary driver of MapReduce jobs are large text data blobs (e.g., web crawls) [Hadoop Apps] which can be highly compressed.

Libraries for data compression are already used by MapReduce frameworks. Hadoop, the open-source version of MapReduce, includes two compression options [Chen 2010]. The *gzip* codec implements the DEFLATE algorithm, a combination of Huffman encoding and Lempel-Ziv 1977 (LZ77). Other popular variants of Lempel-Ziv include LZMA and LZO. Dryad supports similar schemes. Since replication of files is not in the critical path of job executions, our latency constraints are not rigid. With the goal of minimizing network traffic, we employ compression schemes with highest reduction factors albeit at the expense of computational overhead for compression and decompression. We present benchmarks of a few compression schemes (e.g., the PPMVC compression scheme [Skibinski 2004]) as well as the advantages of compressing replication data in §5.

3.3.3 Lazy Deletion

Scarlett reclaims space from deleted replicas lazily, i.e., by overwriting it when another block or replica needs to be written to disk. By doing so, the cost to delete is negligible. Deleted replicas of a block are removed from the list of available replicas.

4. Case Studies of Frameworks

In this section, we describe the problems caused due to contention events in two popular MapReduce frameworks,

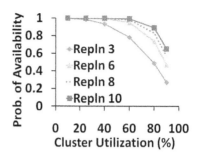

Figure 7: **The probability of finding a replica on a free machine for different values of file replication factor and cluster utilization.**

Dryad [Isard 2007] and Hadoop [Hadoop]. The impact of hotspots on a framework depends on how it copes with contention. As a prelude to evaluating Scarlett, we analyze how hotspots impact these frameworks, quantify the magnitude of problems and the potential gains from avoiding hotspots.

There is a growing trend towards sharing of clusters for economic benefits (e.g., as mentioned in [Cloud Benefits]). This raises questions of resource allocation and MapReduce job managers support weighted distribution of resources between the different jobs [Isard 2009], reflecting the relative importance of jobs. Each job is entitled to use a *legitimate* quota of slots. However, all frameworks allow jobs to use more than their legitimate share subject to availability, called *bonus* slots. This reduces idling of resources and improves the overall throughput of the cluster.

A natural question that arises is, *how to deal with a task that cannot run at the machine(s) that it prefers to run at?* Job managers confront this question more frequently for tasks with data locality constraints. Despite the presence of free slots, locality constraints lead to higher contention for certain machines (§2.3). Many solutions to deal with contention are in use. First, less preferred tasks (e.g., those running in bonus slots) can be evicted to make way [Isard 2007; 2009]. Second, the newly arriving task can be forced to run at a suboptimal location in the cluster [Hadoop, Isard 2007, Zaharia 2010]. Third, one of the contending tasks can be paused until contention passes (e.g., wait in a queue) [Isard 2007]. In practice, frameworks use a combination of these individual approaches, such as waiting for a bounded amount of time before evicting or running elsewhere in the cluster. See Figure 6 for a summary. Each of these solutions are suited to specific environments depending on the service-level agreement constraints, duration of tasks, congestion of network links and popularity skew of input data (described in detail in §6).

Note that Scarlett provides orthogonal benefits. Scarlett minimizes the occurrence of such contentions in the first place by replicating the popular files thereby ensuring that enough machines with replicas are available.

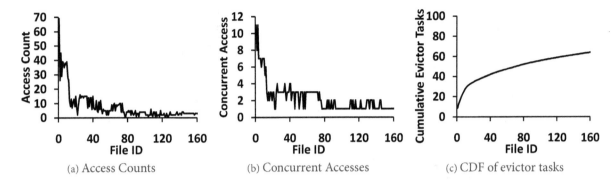

(a) Access Counts   (b) Concurrent Accesses   (c) CDF of evictor tasks

Figure 8: **Correlation between file characteristics and eviction of tasks. We plot only the top 1% of the eviction-causing files for clarity. Popular files directly correlate with more evictions ((a) and (b)). The cumulative number of evictor tasks is plotted in (c). Large files also correlate with evictions – the 1% of the files in this figure account for 35% of the overall storage, and 65% of overall evictions.**

4.1 Benefits from selective replication

We present a simple analysis that demonstrates the intuition behind how increased replication reduces contention. With m machines in the cluster, k of which are available for a task to run, the probability of finding one of r replicas of a file on the available machines is $1 - (1 - \frac{k}{m})^r$. This probability increases with the replication factor r, and decreases with cluster utilization $(1 - \frac{k}{m})$.

Figure 7 plots the results of a numerical analysis to understand how this probability changes with replication factors and cluster utilizations. At a cluster utilization of 80%, with the current replication factor (r=3), we see that the probability of finding a replica among the available machines is less than half. Doubling the replication factor raises the probability to over 75%. Even at higher utilizations of 90%, a file with 10 replicas has a 60% chance of finding a replica on a free machine. By replicating files proportionally to their number of concurrent accesses, Scarlett improves the chances of finding a replica on a free machine.

As described earlier, Scarlett's replication reduces contention and provides more opportunities for map tasks to attain machine or rack locality. Storing more replicas of popular files provides more machine-local slots (§3.1) while spreading out replicas across racks and preventing concentration of popularity (§3.2) facilitates rack locality when machine locality is not achievable.

4.2 Evictions in Dryad

Dryad's [Isard 2007] scheduler pre-emptively evicts a bonus task when a legitimate task makes a request for its slot. We examine an early version of the Cosmos cluster that does so. Here we quantify the inefficiencies due to such evictions. In this version, bonus tasks were given a 30s notice period before being evicted. We refer to the legitimate and bonus tasks as *evictor* and *evicted* tasks respectively.

Likelihood of Evictions: Of all tasks that began running on the cluster, 21.1% of them end up being evicted. An overwhelming majority of the evicted tasks (98.2%) and the evictor tasks (93%) are from map phases. These tasks could have executed elsewhere were more replicas available. Reclaiming resources used by killed tasks will reduce the load on the cluster. As a second-order effect, we see from Figure 7, that the probability of finding a replica on the available machines improves with lower utilization. In addition, the spare resources can be used for other performance enhancers, such as speculative executions to combat outliers.

Correlation of evictor tasks and file popularity: Figure 8 explores the correlation between evictor tasks and the characteristics of the input files they work on – access count, concurrency and size. For clarity, we plot only the top 160 files contributing to evictions, out of a total of 16000 files (or ~1%) – these account for 65% of the evictor tasks. As marked in the figure, we see that the files that contribute the most to evictor tasks are directly correlated with popularity – they have high total and concurrent numbers of accesses (Figures 8a and 8b). High concurrency directly leads to contention and eviction. A larger number of accesses implies that more tasks run overall on the machines containing these files and hence there is a greater probability of evictions. In addition, the worst sources of evictor tasks also are the bigger files. The 1% of the files plotted in Figure 8 contribute to a disproportionate 35% of the overall storage size and account for 65% of all evictions (Figure 8c). This validates our design choice in §3.1 where we order the files in descending order of size before distributing the replication budget.

Inflation of Job Durations: Figure 9 plots the improvement in completion times for the jobs in the cluster in an ideal case wherein all the evicted tasks execute to completion (i.e., do not have to be re-executed) and the evictor tasks achieve locality. The potential median and third-quartile improvements are 16.7% and 34.1% respectively. In large production clus-

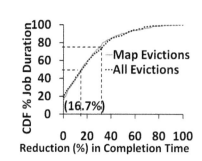

Figure 9: **Ideal improvement in job completion times if eviction of tasks did not happen.**

ters, this translates to millions of dollars in savings. Since map tasks dominate among the evictees, we see that providing this hypothetical improvement (locality without evictions) to just the map tasks is nearly the same (median of 15.2%) as when all evictions are avoided.

That evictions happen even in the presence of idle computational resources points to evictions being primarily due to contention for popular data. Our results in §5 show that Scarlett manages to reduce evictions by 83% in Dryad jobs.

4.3 Loss of Locality in Hadoop

Hadoop's policy of dealing with contention for slots is to force the new task to forfeit locality. Delay Scheduling [Zaharia 2010] improves on this default policy by making tasks wait briefly before deciding to cede locality. The data from Facebook's Hadoop logs in [Zaharia 2010] shows that small jobs (which constitute 58% of all Hadoop jobs) achieve only 5% node locality and 59% rack locality.

Hadoop does not use evictions despite a few reasons in their favor. First, eviction can be more efficient than running the new task elsewhere, for example, if the ongoing task has just started and some other machines on the cluster that the task can run at are available. Second, Quincy [Isard 2009] shows that if bonus tasks were not pre-empted, the cluster's resource allocation can be significantly far away from the desired value causing jobs to be starved and experience unpredictable lags. Our evaluation in §5 shows a 45% increase in locality for map tasks in Hadoop jobs, resulting in three quarters of the jobs speeding up by more than 44%. Half of the jobs improve by at least 20%. Note that these observed gains due to Scarlett are larger for Hadoop than for Dryad.

5. Evaluation

We first present the evaluation set-up and then proceed to the performance benefits due to Scarlett.

5.1 Methodology

We evaluate the benefits of Scarlett using an implementation and deployment of Hadoop jobs (§5.2) as well as extensive simulation of Dryad jobs (§5.3) described in Table 1. In addition, we also check the sensitivity of Scarlett's performance

to the various algorithmic parameters (§5.4), budget size and distribution (§5.5), and compression techniques (§5.6).

Implementation: We implement Scarlett by extending the Hadoop Distributed File System (HDFS) [HDFS]. Our modules in HDFS log the access counters for each file and appropriately modify the replication factors of files using the adaptive learning algorithm described in §3.1. Note that this change is transparent to the job scheduler.

Hadoop Workload: Our workload of Hadoop jobs is constructed out of the traces mentioned in Table 1. We use the same inter-arrival times and input files for jobs, thereby preserving the load experienced by the cluster as well as the access patterns of files. However the file sizes are appropriately scaled down to reflect the reduced cluster size. We replace the Dryad job scripts by Hadoop programs, randomly chosen between wordcount, group by, sort and grep. We believe this approximation is reasonable as the thrust of our work is largely on the advantage of reading data locally as opposed to the specific computation. We replay a 10 hour trace from Table 1. We test our implementation on a 100-node cluster in the DETER testbed [Benzel 2006] each with 4GB memory and 2.1GHz Xeon processors.

Trace-driven Dryad Simulator: We replay the production Dryad traces shown in Table 1 with detailed simulators that mimic job operation. The simulator is extensive in that it mimics various aspects of tasks including distribution of duration and amount of input data read/written, locality of input data based on the placement of replicas, probability of failure, stragglers and recomputations [Ananthanarayanan 2010], and cluster characteristics of when computation slots open up. It also takes evictions into account by verifying if a replica can be found in the unutilized machines.

Metrics: Our primary figure of merit is the reduction in completion time of jobs where,

$$Reduction = \frac{Current - Modified}{Current}$$

We weight the jobs by their duration and use CDFs to present our metric. Weighting jobs by their duration helps differentiate the impact of Scarlett on larger jobs versus smaller jobs. Larger jobs contain more tasks and utilize more cluster cycles, therefore an improvement in a larger job would be as a result of more tasks benefiting. We also consider improvements in locality, defined as the number of tasks that are able to run on the same machines that have their input data.

We first present a summary of our results:

- Scarlett's replication speeds up median Hadoop jobs by 20.2% in our cluster, and median Dryad jobs by 12.8% (84% of ideal) in our trace-driven simulations.

- Revisiting replication factors and placement of files once in 12 hours is sufficient, thereby limiting replication overhead.

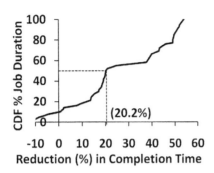

Figure 10: **Improvement in data locality for tasks leads to median and third-quartile improvements of 20.2% and 44.6% in Hadoop job completion times.**

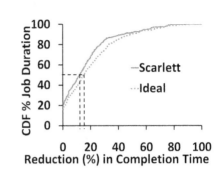

Figure 11: **Increased replication reduces eviction of tasks and achieves a median improvement of 12.8% in job completion times, or 84% of ideal.**

- Performance under a storage budget of 10% is comparable to an unconstrained replication.
- Replication increases network overhead by only 1% due to effective data compression.

5.2 Does data locality improve in Hadoop?

Hadoop's reaction to a request for a slot that is currently in use is to forfeit locality after a brief wait (See §4.3). We measure the improvement in completion times due to higher data locality for Hadoop jobs in our cluster using Scarlett over the baseline of HDFS that replicates each file thrice. We set $\delta = 1$, let T_L range from 6 to 24 hours, set storage budget $B = 10\%$ and rearrange once at the beginning of the ten hour run ($T_R \geq 10$ hours).

Figure 10 marks the reduction in completion times of 500 jobs. We see that completion times improve by 20.2% and 44.6% at the median and 75th percentile respectively. This is explained by the increase in fraction of map tasks that achieve locality. The fraction of map tasks that achieve locality improves from 57% with vanilla HDFS to 83% with Scarlett, in other words a 45% improvement.

5.3 Is eviction of tasks prevented in Dryad?

As described in §4.2, replicating popular files reduces the necessity for eviction and wastage of work, in turn leading to jobs completing faster. The ideal improvement when all evictions by map tasks are avoided is 15.2% at median, and Scarlett produces a 12.8% median improvement. Here, we set $\delta = 1$, $T_R = 12$ hours, let T_L range from 6 to 24 hours, and set storage budget $B = 10\%$. Figure 11 compares the ideal case with our replication scheme where we obtain 84% of ideal performance at median. The ideal case contains no evictions and at the same time assumes that all evictor tasks achieve locality.

A closer look reveals that by replicating popular content, Scarlett avoids 83% of all evictions, i.e., the evictor tasks could be run on another machine containing a replica of their input. Note that this number goes up to 93%, when we consider evictions by tasks operating on the top hundred popu-

lar files, confirming the design choice in Scarlett to focus on the more popular files. Increasing the storage budget provides marginal (but smaller improvements) – with an increased storage budget of 20% Scarlett prevents 96% of all evictions.

5.4 Sensitivity Analysis

We now analyze the sensitivity of Scarlett to the parameters of our learning algorithm– rearrangement window, T_R, and the cushion for replication, δ. T_R decides how often data is moved around, potentially impacting network performance of currently running jobs. δ results in greater storage occupancy and more replication traffic.

Figure 12a compares the improvement in Dryad job completion times for different rearrangement windows, i.e., $T_R = \{1, 12, 24\}$ hours. Interestingly, we see that T_R has little effect on the performance of jobs. Re-evaluating replication decisions once a day is only marginally worse than doing it once every hour. This points to Scarlett's minimal interference on running jobs. It also points to the fact that most of the gains in the observed Dryad workload accrue from replicating files that are consistently popular over long periods. Results from our Hadoop deployment are similar. For T_R values of 1, 5 and 10 hours, the median improvements are 21.1%, 20.4% and 20.2% respectively. By default, we set T_R to 12 hours, or rearrange files twice a day.

The replication allowance δ impacts performance. Changing δ from 0 to 1 improves performance substantially, but larger values of δ have lower marginal increases. Figure 12b shows that for δ values of 0, 1 and 2, the median reductions in Dryad job durations are 8.5%, 12.8% and 13.8%. Note the improvement of 42% as δ changes from 0 to 1. Likewise, our Hadoop jobs see a 56% increase from 12.9% to 20.2% in median improvement in completion time as we shift δ from 0 to 1. We believe this is because operating at the brim with a replication factor equal to the observed number of concurrent accesses is inferior to having a cushion, even if that were only one extra replica.

Figure 13 shows the cost of increasing δ. Values of storage overhead change upon replication, i.e., once every T_R=12

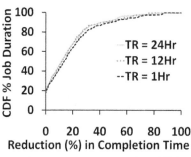

(a) Rearrangement Window (T_R)

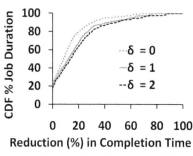

(b) Replication Allowance (δ)

Figure 12: **Sensitivity Analysis of T_R and δ. Rearranging files once or twice a day is only marginally worse than doing it at the end of every hour. We set T_R as 12 hours in our system. On the other hand, δ plays a vital role in the effectiveness of** Scarlett's **replication scheme.**

hours. $\delta = 2$ results in a 24% increase in storage, almost double of the overhead for $\delta = 1$. Combined with the fact that we see the most improvement when moving from δ from 0 to 1, we fix δ as 1.

5.5 Storage Budget for Replication

Figure 14a plots reduction in Dryad job completion times for various budget values, measured with respect to the baseline storage of three replicas per file. Here, we use the priority distribution, i.e., larger files are preferentially replicated over smaller files within the budget. A budget of 10% improves performance substantially (by 88%) over a 5% limit. As expected, the lower budget reduces storage footprint at the expense of fewer files being replicated or a smaller replication factor for some files. The marginal improvement is smaller as the budget increases to 15%. This indicates that most of the value from replication accrues quickly, i.e., at small replication factors for files. Conversations with datacenter operators confirm that 10% is a reasonable increase in storage use.

Note however that the improvement going from a budget of 2% to a budget of 5% is smaller than when going from 5% to 10%. This is likely because the distribution policy used by Scarlett is simple and greedy but not optimal. Likely, there are some files, replicating which yields significantly more benefit per unit extra storage, that Scarlett fails to replicate when

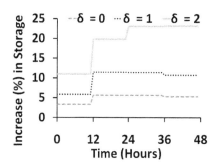

Figure 13: **Increasing the value of the replication allowance (δ) leads to** Scarlett **using more storage space. We fix δ as 1.**

budgets are small. However, these inefficiences go away with a slightly larger budget value of 10%, and we choose to persist with the simpler algorithm.

Hadoop jobs, from Figure 14b, exhibit a similar trend. The increase in median completion time when moving from a budget of 5% to 10% is much higher (120%). This indicates that how Hadoop deals with contentions (by moving tasks elsewhere) is likely more sensitive to the loss of locality when popular files are not replicated.

Priority vs. Round-robin Distribution: Recall from §3 that the replication budget can be spread among the files either in a priority fashion – iterate through the files in decreasing order of size, or distributed iteratively in a round-robin manner. Figure 15a plots the performance of Dryad jobs with respect to both these allocations. For a replication budget of 10%, we observe that the priority allocation gives a median improvement of 12.8% as opposed to 8.4% with round-robin allocation, or a 52% difference. This is explained by our causal analysis in Figure 2 and Figure 8 that shows that large files account for a disproportionate fraction of the evicting tasks while also experiencing high levels of concurrent accesses. Hence, giving them a greater share of the replication budget helps avoid more evictions. Hadoop jobs exhibit a greater difference of 63% between the two distributions showing greater sensitivity to loss of locality (Figure 15b).

However, the difference in advantage between the two distributions are negligible at small replication budgets. As we see in Figure 15a, the limited opportunity to replicate results in there being very little to choose between the two distribution strategies.

5.6 Increase in Network Traffic

For $\delta = 1$, the maximum increase in uncompressed network traffic during rearrangement of replicas is 24%. Using the PPMVC compression scheme [Skibinski 2004], this reduces to an acceptable overhead of 0.9%.

We also present micro-benchmarks of various compression techniques. Table 2 lists the compression and de-compression speeds as well as the compression ratios achieved by a few compression algorithms [Skibinski 2004;

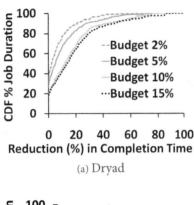

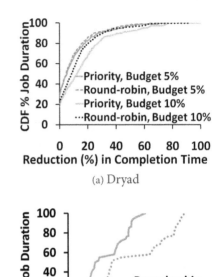

(a) Dryad

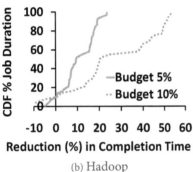

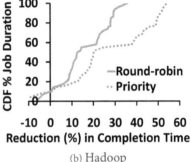

(b) Hadoop

Figure 14: **Low budgets lead to little fruitful replication. On the other hand, as the graph below shows, budgets cease to matter beyond a limit.**

Scheme	Throughput (Mbps)		Compression Factor
	Compress	De-compress	
gzip	144	413	12–13X
bzip2	9.7	88.2	19–20X
LZMA	3.6	375	22–23X
PPMVC	30.2	31.4	26–27X

Table 2: **Comparison of the computational overhead and compression factors of compression schemes.**

2007]. There is a clear trend of more computational overhead providing heavier compression. Given the flexible latency constraints for replication and low bandwidth across racks, Scarlett leans toward the choice that results in the least load on the network.

6. Related Work

The principle of "replication and placement" of popular data has been employed in different contexts in prior work. Our contributions are to (i) identify (and quantify) the content popularity skew in the MapReduce scenario using production traces, (ii) show how the skew causes contention in two kinds of MapReduce systems (ceding locality for Hadoop vs. eviction for Dryad), and (iii) design solutions that operate under a storage budget for large data volumes common in MapReduce systems. While we have evaluated Scarlett pri-

Figure 15: **Priority distribution of the replication budget among the files improves the median completion time more than round-robin distribution.**

marily for map tasks, we believe the principle of proactively replicating content based on its expected concurrent access can be extended to intermediate data too (e.g., as in Nectar [Gunda 2010] that stores intermediate data across jobs).

Much recent work focuses on the tussle between data locality and fairness in MapReduce frameworks. Complementary to Scarlett, Quincy [Isard 2009] arbitrates between multiple jobs. Delay Scheduling [Zaharia 2010] on the other hand supports temporary relaxation of fairness while tasks wait to attain locality. This can alleviate contention by steering tasks away from a hotspot. It however makes some assumptions that do not hold universally: (a) task durations are short and bimodal, and (b) one task queue per cluster (as in Hadoop). In the Cosmos clusters at Microsoft, tasks are longer (median of 145s as opposed to the 19s in [Zaharia 2010]) to amortize overheads in maintaining task-level state at the job manager, copying task binaries etc. Task lengths are also more variable in Dryad owing to diversity in types of phases. Finally, Dryad uses one task-queue per machine, further reducing the load at the job scheduler to improve scalability. Scarlett does not rely on these assumptions and addresses the root cause of contention by identifying and replicating popular content. Furthermore, Scarlett can be beneficially combined with both Delay Scheduling and Quincy.

The idea of replicating content in accordance to popularity for alleviating hotspots has been used in the past. Caching popular data and placing it closer to the application is used in various content distribution networks (CDNs) [Akamai, Coral CDN] in the Internet. Beehive [Ramasubramanian 2004] proactively replicates popular data in a DHT to provide constant time look-ups in peer-to-peer overlays. Finally, dynamic placement of popular data has also been recently explored in the context of energy efficiency [Verma 2010]. To the best of our knowledge, ours is the first work to understand popularity skew and explore the benefits of dynamic data replication in MapReduce clusters. The context of our work is different as file access patterns and sizes in MapReduce significantly differ from web access patterns. It differs from Beehive due to the different application semantics. While Beehive is optimized for lookups, Scarlett aims at parallel computation frameworks like MapReduce. Further, our main goal is to increase performance rather than be energy efficient, so we aim for spreading data across nodes as opposed to compaction.

Bursts in data center workloads often result in peak I/O request rates that are over an order of magnitude higher than average load [Narayanan 2008]. A common approach to deal with such bursts is to identify overloaded nodes and offload some of their work to less utilized nodes [Appavoo 2010, Narayanan 2008]. In contrast, our approach is geared towards a read-heavy workload (unlike [Narayanan 2008]), common to MapReduce clusters. While Dynamo [Appavoo 2010] reactively migrates (not replicate) popular data, we replicate and do so proactively, techniques more suited to our setting. Recent work [Belaramani 2009, Stribling 2009] on providing semantic context to the file system can be leveraged to implement our replication policies.

A wide variety of work has also been done in the area of predictive pre-fetching of popular files based on historical access patterns [Curewitz 1993] as well as elaborate program and user based file prediction models [Yeh 2002]. However, these are in the context of individual systems and deal with small amounts of data unlike our setting with petabytes of distributed storage, the replication and transfer of which require strict storage/network constraints.

Some prior work on dynamic database replication policies [Soundararajan 2006] is very similar in flavor to ours. However, these policies are reactive in reference to application latency requirements. Our work, on the other hand, focuses on designing proactive replication policies.

Finally, much recent work has gone into designs for full bisection bandwidth networks. By suitably increasing the numbers of switches and links in the core, these designs ensure that the network will not be the bottleneck for well-behaved traffic [Al-Fares 2008, Greenberg 2009]. Well-behaved refers to the hose model constraint, which requires the traffic incoming to each machine to be no larger than the capacity on its incoming network link. We note that Scarlett's benefits remain even if networks have full bisection bandwidth, since concurrent access of blocks results in a bottleneck at the source machine that stores them. By providing more replicas (as many as the predicted concurrent access), Scarlett alleviates this bottleneck.

7. Conclusion

Analyzing production logs from Microsoft Bing's datacenters revealed a skew in popularity of files, making the current policy of uniform data replication sub-optimal. Machines containing popular data became bottlenecks, hampering the efficiency of MapReduce jobs. We proposed Scarlett, a system that replicates files according to their access patterns, ageing them with time. Using both a real deployment and extensive simulations, we demonstrated that Scarlett's replication improved data locality in two popular MapReduce frameworks (Dryad and Hadoop) and sped up jobs by 20.2%. Scarlett's guided replication used limited extra storage (less than 10%) and network resources (1%).

Acknowledgments

We would like to thank the Microsoft Bing group for access to Cosmos traces and informative discussions that guided this work. Bikas Saha and Ramesh Shankar contributed domain knowledge and critiques that improved this work. We are also grateful to our shepherd, Robbert van Renesse and the anonymous reviewers whose input greatly improved the paper. For feedback on the draft, we acknowledge Ali Ghodsi, Lucian Popa, Matei Zaharia and David Zats of the RAD Lab. We are also thankful to the DETER team for supporting our experiments.

References

[Akamai] Akamai. Akamai content distribution network. http://www.akamai.com/.

[Al-Fares 2008] M. Al-Fares, A. Loukissas, and A. Vahdat. A Scalable, Commodity Data Center Network Architecture. In *SIGCOMM'08: Proceedings of the ACM SIGCOMM 2008 conference on Data communication*, 2008.

[Ananthanarayanan 2010] G. Ananthanarayanan, S. Kandula, A. Greenberg, I. Stoica, E. Harris, and B. Saha. Reining in the Outliers in Map-Reduce Clusters using Mantri. In *OSDI'10: Proceedings of the 9th USENIX Symposium on Operating Systems Design and Implementation*, 2010.

[Appavoo 2010] J. Appavoo, A. Waterland, D. Da Silva, V. Uhlig, B. Rosenburg, E. Van Hensbergen, J. Stoess, R. Wisniewski, and U. Steinberg. Providing a Cloud Network Infrastructure on a Supercomputer. In *HPDC'10: Proceedings of the 19th ACM International Symposium on High Performance Distributed Computing*, 2010.

[Belaramani 2009] N. M. Belaramani, J. Zheng, A. Nayate, R. Soulé, M. Dahlin, and R. Grimm. PADS: A Policy Architecture for Distributed Storage Systems. In *NSDI'09: Proceedings of the 6th USENIX Symposium on Networked Systems Design and Implementation*, 2009.

[Benzel 2006] T. Benzel, R. Braden, D. Kim, C. Neuman, A. Joseph, K. Sklower, R. Ostrenga, and S. Schwab. Experience with DETER: A Testbed for Security Research. In *International Conference on Testbeds and Research Infrastructures for the Development of Networks and Communities*, 2006.

[Chaiken 2008] Ronnie Chaiken, Bob Jenkins, Perke Larson, Bill Ramsey, Darren Shakib, Simon Weaver, and Jingren Zhou. SCOPE: Easy and Efficient Parallel Processing of Massive Datasets. In *VLDB'08: Proceedings of the 34th Conference on Very Large Data Bases*, 2008.

[Chen 2010] Y. Chen, A. Ganapathi, and R. Katz. To Compress or Not To Compress - Compute vs. IO tradeoffs for MapReduce Energy Efficiency. In *Proceedings of the First ACM SIGCOMM Workshop on Green Networking*, 2010.

[Cloud Benefits] Cloud Benefits. Cloud compute can save govt agencies 25-50% in costs, 2010. http://googlepublicpolicy. blogspot.com/2010/04/brookings-cloud-computin% g-can-save-govt.html.

[Coral CDN] Coral CDN. The coral content distribution network. http://www.coralcdn.org/.

[Curewitz 1993] K. Curewitz, P. Krishnan, and J. S. Vitter. Practical Prefetching via Data Compression. In *SIGMOD'93: Proceedings of the 1993 ACM SIGMOD International Conference on Management of Data*, 1993.

[Dean 2009] J. Dean. Designs, lessons and advice from building large distributed systems, 2009. http://www.cs.cornell.edu/ projects/ladis2009/talks/dean-keynote-ladis20%09.pdf.

[Dean 2004] J. Dean and S. Ghemawat. MapReduce: Simplified Data Processing on Large Clusters. In *OSDI'04: Proceedings of the 6th Symposium on Operating Systems Design and Implementation*, 2004.

[Ghemawat 2003] S. Ghemawat, H. Gobioff, and S. Leung. The Google File System. In *SOSP'09: Proceedings of the 19th ACM Symposium on Operating Systems Principles*, 2003.

[Greenberg 2009] A. Greenberg, N. Jain, S. Kandula, C. Kim, P. Lahiri, D. A. Maltz, P. Patel, and S. Sengupta. VL2: A Scalable and Flexible Data Center Network. In *SIGCOMM'09: Proceedings of the ACM SIGCOMM 2009 conference on Data communication*, 2009.

[Gunda 2010] P. K. Gunda, L. Ravindranath, C. A. Thekkath, Y. Yu, and L. Zhuang. Nectar: Automatic Management of Data and Computation in Data Centers. In *OSDI'10: Proceedings of the 9th USENIX Symposium on Operating Systems Design and Implementation*, 2010.

[Hadoop] Hadoop. http://hadoop.apache.org.

[Hadoop Apps] Hadoop Apps. Applications and organizations using hadoop, 2010. http://wiki.apache.org/hadoop/PoweredBy.

[HDFS] HDFS. Hadoop distributed file system. http://hadoop. apache.org/hdfs.

[Isard 2007] M. Isard, M. Budiu, Y. Yu, A. Birrell, and D. Fetterly. Dryad: Distributed Data-parallel Programs from Sequential Building Blocks. In *EuroSys'07: Proceedings of the European Conference on Computer Systems*, 2007.

[Isard 2009] M. Isard, V. Prabhakaran, J. Currey, U. Wieder, K. Talwar, and A. Goldberg. Quincy: Fair Scheduling for Distributed Computing Clusters. In *SOSP'09: Proceedings of the 22nd ACM Symposium on Operating Systems Principles*, 2009.

[Kandula 2009] S. Kandula, S. Sengupta, A. Greenberg, P. Patel, and R. Chaiken. Nature of Datacenter Traffic: Measurements and Analysis. In *IMC'09: Proceedings of the Ninth Internet Measurement Conference*, 2009.

[Narayanan 2008] D. Narayanan, A. Donnelly, E. Thereska, S. Elnikety, and A. Rowstron. Everest: Scaling Down Peak Loads Through I/O Off-Loading. In *OSDI'08: Proceedings of the 8th USENIX Symposium on Operating Systems Design and Implementation*, 2008.

[Ramasubramanian 2004] V. Ramasubramanian and E. G. Sirer. Beehive: O(1) Lookup Performance for Power-law Query Distributions in peer-to-peer Overlays. In *NSDI'04: Proceedings of the First Symposium on Networked Systems Design and Implementation*, 2004.

[Shreedhar 1996] M. Shreedhar and G. Varghese. Efficient Fair Queuing Using Deficit Round-Robin. In *IEEE/ACM Transactions on Networking*, 1996.

[Skibinski 2004] P. Skibinski and S. Grabowski. Variable-length contexts for PPM. In *DCC'04: Proceedings of IEEE Data Compression Conference*, 2004.

[Skibinski 2007] P. Skibinski and J. Swacha. Fast and Efficient Log File Compression. In *CEUR ADBIS'07: Advances in Databases and Information Systems*, 2007.

[Soundararajan 2006] G. Soundararajan, C. Amza, and A. Goel. Database Replication Policies for Dynamic Content Applications. In *EuroSys'06: Proceedings of the European Conference on Computer Systems*, 2006.

[Stribling 2009] J. Stribling, Y. Sovran, I. Zhang, X. Pretzer, J. Li, F. Kaashoek, and R. Morris. Flexible, Wide-Area Storage for Distributed Systems with WheelFS. In *NSDI'09: 6th USENIX Symposium on Networked Systems Design and Implementation*, 2009.

[Verma 2010] A. Verma, R. Koller, L. Useche, and R. Rangaswami. SRCMap: Energy Proportional Storage Using Dynamic Consolidation. In *FAST'10: Proceedings of the 8th USENIX Conference on File and Storage Technologies*, 2010.

[Yeh 2002] T. Yeh, D. E. Long, and S. A. Brandt. Increasing Predictive Accuracy by Prefetching Multiple Program and User Specific Files. *HPCS'02: Proceedings of the 16th Annual International Symposium on High Performance Computing Systems and Applications*, 2002.

[Zaharia 2010] M. Zaharia, D. Borthakur, J. Sen Sharma, K. Elmeleegy, S. Shenker, and I. Stoica. Delay Scheduling: A Simple Technique for Achieving Locality and Fairness in Cluster Scheduling. In *EuroSys'10: Proceedings of the European Conference on Computer Systems*, 2010.

CloneCloud: Elastic Execution between Mobile Device and Cloud

Byung-Gon Chun

Intel Labs Berkeley

byung-gon.chun@intel.com

Sunghwan Ihm

Princeton University

sihm@cs.princeton.edu

Petros Maniatis

Intel Labs Berkeley

petros.maniatis@intel.com

Mayur Naik

Intel Labs Berkeley

mayur.naik@intel.com

Ashwin Patti

Intel Labs Berkeley

ashwin.patti@intel.com

Abstract

Mobile applications are becoming increasingly ubiquitous and provide ever richer functionality on mobile devices. At the same time, such devices often enjoy strong connectivity with more powerful machines ranging from laptops and desktops to commercial clouds. This paper presents the design and implementation of CloneCloud, a system that automatically transforms mobile applications to benefit from the cloud. The system is a flexible application partitioner and execution runtime that enables unmodified mobile applications running in an application-level virtual machine to seamlessly off-load part of their execution from mobile devices onto device clones operating in a computational cloud. CloneCloud uses a combination of static analysis and dynamic profiling to partition applications automatically at a fine granularity while optimizing execution time and energy use for a target computation and communication environment. At runtime, the application partitioning is effected by migrating a thread from the mobile device at a chosen point to the clone in the cloud, executing there for the remainder of the partition, and re-integrating the migrated thread back to the mobile device. Our evaluation shows that CloneCloud can adapt application partitioning to different environments, and can help some applications achieve as much as a 20x execution speed-up and a 20-fold decrease of energy spent on the mobile device.

Categories and Subject Descriptors C.2.4 [*Distributed Systems*]: Client/server

General Terms Algorithms, Design, Experimentation, Performance

EuroSys'11, April 10–13, 2011, Salzburg, Austria.
Copyright © 2011 ACM 978-1-4503-0634-8/11/04... $10.00

Keywords Mobile cloud computing, partitioning, offloading, migration, smartphones

1. Introduction

Mobile cloud computing is the next big thing. Recent market research predicts that by the end of 2014 mobile cloud applications will deliver annual revenues of 20 billion dollars [Beccue 2009]. Although it is hard to validate precise predictions, this is hardly implausible: mobile devices as simple as phones and as complex as mobile Internet devices with various network connections, strong connectivity especially in developed areas, camera(s), GPS, and other sensors are the current computing wave, competing heavily with desktops and laptops for market and popularity. Connectivity offers immediate access to available computing, storage, and communications on commercial clouds, at nearby wireless hot-spots equipped with computational resources [Satyanarayanan 2009], or at the user's PC and plugged-in laptop.

This abundance of cloud resources and the mobile opportunity to use them is met by the blinding variety of flash-popular applications in application stores by Apple, Google, Microsoft, and others. Now mobile users look up songs by audio samples; play games; capture, edit, and upload video; analyze, index, and aggregate their mobile photo collections; analyze their finances; and manage their personal health and wellness. Also, new rich media, mobile augmented reality, and data analytics applications change how mobile users remember, experience, and understand the world around them. Such applications recruit increasing amounts of computation, storage, and communications from a constrained supply on mobile devices—certainly compared to tethered, wall-socket-powered devices like desktops and laptops—and place demands on an extremely limited supply of energy.

Yet bringing a demanding mobile application to needed cloud resources tends to be inflexible: an application is either written as a monolithic process, cramming all it needs to do

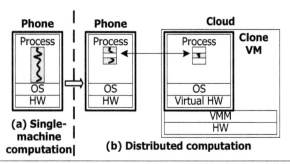

(a) Single-machine computation

(b) Distributed computation

Figure 1. CloneCloud system model. CloneCloud transforms a single-machine execution (mobile device computation) into a distributed execution (mobile device and cloud computation) automatically.

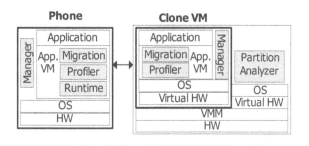

Figure 2. The CloneCloud prototype architecture.

on to the mobile device; or it is split in the traditional client-server paradigm, pushing most computation to the remote server; or it is perhaps tailored to match an expected combination of client (e.g., given browser on particular phone platform), environment (a carrier's expected network conditions), and service. But what might be the right split for a low-end mobile device with good connectivity may be the wrong split for a high-end mobile device with intermittent connectivity. Often the choice is unknown to application developers ahead of time, or the possible configurations are too numerous to customize for all of them.

To address this problem, in this paper we realize our CloneCloud vision [Chun 2009] of a flexible architecture for the seamless use of ambient computation to augment mobile device applications, making them fast and energy-efficient. CloneCloud boosts *unmodified* mobile applications by off-loading the *right* portion of their execution onto *device clones* operating in a computational cloud[1]. Conceptually, our system automatically transforms a single-machine execution (e.g., computation on a smartphone) into a distributed execution optimized for the network connection to the cloud, the processing capabilities of the device and cloud, and the application's computing patterns (Figure 1).

The underlying motivation for CloneCloud lies in the following intuition: as long as execution on the cloud is significantly faster than execution on the mobile device (or more reliable, more secure, etc.), paying the cost for sending the relevant data and code from the device to the cloud and back may be worth it. Unlike partitioning a service statically by design between client and server portions, CloneCloud late-binds this design decision. In practice, the partitioning decision may be more fine-grained than a yes/no answer (i.e., it may result in carving off different *amounts* of the original application for cloud execution), depending on the expected workload and execution conditions (CPU speeds, network performance). A fundamental design goal for CloneCloud is to allow such fine-grained flexibility on what to run where.

Another design goal for CloneCloud is to take the programmer out of the business of application partitioning. The kinds of applications on mobile platforms that are featured on application stores and gain flash popularity tend to be low-margin products, whose developers have little incentive to optimize manually for different combinations of architectures, network conditions, battery lives, and hosting infrastructures. Consequently, CloneCloud aims to make application partitioning automatic and seamless.

Our work in this paper applies primarily to application-layer virtual machines (VMs), such as the Java VM, DalvikVM from the Android Platform, and Microsoft's .NET. We choose application-layer VMs since they are widely used in mobile platforms. In addition, the application-layer VM model has the relative ease of manipulating application executables and migrating pieces thereof to computing devices of diverging architectures, even different instruction set architectures (e.g., ARM-based smartphones and x86-based servers).

The CloneCloud prototype meets all design goals mentioned above, by rewriting an unmodified application executable. While the modified executable is running, at automatically chosen points individual threads migrate from the mobile device to a device clone in a cloud; remaining functionality on the mobile device keeps executing, but blocks if it attempts to access migrated state, thereby exhibiting opportunistic but very conservative concurrency. The migrated thread executes on the clone, possibly accessing native features of the hosting platform such as the fast CPU, network, hardware accelerators, storage, etc. Eventually, the thread returns back to the mobile device, along with remotely created state, which it merges back into the original process. The choice of where to migrate is made by a partitioning component, which uses static analysis to discover constraints on possible migration points, and dynamic profiling to build a cost model for execution and migration. A mathematical optimizer chooses migration points that optimize objective (such as total execution time or mobile-device energy consumption) given the application and the cost model. Finally, the run-time system chooses what partition to use. Figure 2 shows the high-level architecture of our prototype.

The paradigm of opportunistic use of ambient resources is not new [Balan 2002]; much research has attacked appli-

---

[1] Throughout this paper, we use the term "cloud" in a broad sense to include diverse ambient computational resources discussed above.

cation partitioning and migration in the past (see Section 7). CloneCloud is built upon existing technologies, but it combines and augments them in a novel way. We summarize our contributions here as follows.

- Unlike traditional suspend-migrate-resume mechanisms [Satyanarayanan 2005] for application migration, the CloneCloud migrator operates at thread granularity, an essential consideration for mobile applications, which tend to have features that must remain at the mobile device, such as those accessing the camera or managing the user interface.

- Unlike past application-layer VM migrators [Aridor 1999, Zhu 2002], the CloneCloud migrator allows native system operations to execute both at the mobile device and at its clones in the cloud, harnessing not only raw CPU cloud power, but also system facilities or specialized hardware when the underlying library and OS are implemented to exploit them.

- Similar to MAUI [Cuervo 2010], the CloneCloud partitioner automatically identifies costs through static and dynamic code analysis and runs an optimizer to solve partitioning problems, but we go a step further by not asking for the programmer's help (e.g., source annotations).

- We present the design, implementation, and evaluation of an operational system that combines the features in widely-used Android platforms. CloneCloud can achieve up to 20x speedup and 20x less energy consumption of smartphone applications we tested.

In what follows, we first give some brief background on application-layer VMs (Section 2). We then present the design of CloneCloud's partitioning components (Section 3) and its distributed execution mechanism (Section 4). We describe our implementation (Section 5) and experimental evaluation of the prototype (Section 6). We survey related work in Section 7, discuss limitations and future work in Section 8, and conclude in Section 9.

2. Background: Application VMs

An application-level VM is an abstract computing machine that provides hardware and operating system independence. Its instruction sets are platform-independent bytecodes; an executable is a blob of bytecodes. The VM runtime executes bytecodes of methods with threads. There is typically a separation between the *virtual* portion of an execution and the *native* portion; the former is only expressed in terms of objects directly visible to the bytecode, while the latter includes management machinery for the virtual machine, data and computation invoked on behalf of a virtual computation, as well as the process-level data of the OS process containing the VM. Interfacing between the virtual and the native portion happens via native interface frameworks.

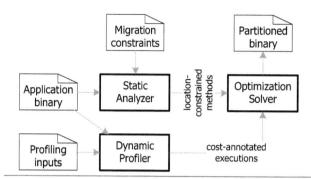

Figure 3. Partitioning analysis framework.

Runtime memory is split between VM-wide and per-thread areas. The *Method Area*, which contains the types of the executing program and libraries as well as static variable contents, and the *Heap*, which holds all dynamically allocated data, are VM-wide. Each thread has its own *Virtual Stack* (stack frames of the virtual hardware), the *Virtual Registers* (e.g., the program counter), and the *Native Stack* (containing any native execution frames of a thread, if it has invoked native functions). Most computation, data structure manipulation, and memory management are done within the abstract machine. However, external processing such as file I/O, networking, using local hardware such as sensors, are done via APIs that punch through the abstract machine into the process's system call interface.

3. Partitioning

The partitioning mechanism in CloneCloud is off-line, and aims to pick which parts of an application's execution to retain on the mobile device and which to migrate to the cloud. Any application targeting the application VM platform may be partitioned; unlike prior approaches, including the recent MAUI project [Cuervo 2010], the programmer need not write the application in a special idiom or annotate it in a non-standard way, and the source code is not needed. The output of the partitioning mechanism is a *partition*, a choice of execution points where the application migrates part of its execution and state between the device and a clone. Given a set of execution conditions (we currently consider network characteristics, CPU speeds, and energy consumption), the partitioning mechanism yields a partition that optimizes for total execution time or energy expended at the mobile device. The partitioning mechanism may be run multiple times for different execution conditions and objective functions, resulting in a database of partitions. At runtime, the distributed execution mechanism (Section 4) picks a partition from the database and implements it via a small and fast set of modifications of the executable before invocation.

Partitioning of an application operates according to the conceptual workflow of Figure 3. Our partitioning framework combines static program analysis with dynamic program profiling to produce a partition.

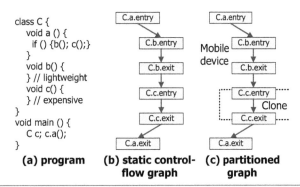

```
class C {
    void a () {
        if () {b(); c();}
    }
    void b() {
    } // lightweight
    void c() {
    } // expensive
}
void main () {
    C c; c.a();
}
```

(a) program   **(b) static control-**   **(c) partitioned**
                     flow graph   **graph**

Figure 4. An example of a program, its corresponding static control-flow graph, and a partition.

The *Static Analyzer* identifies legal partitions of the application executable, according to a set of constraints (Section 3.1). Constraints codify the needs of the distributed execution engine, as well as the usage model. The *Dynamic Profiler* (Section 3.2) profiles the input executable on different platforms (the mobile device and on the cloud clone) with a set of inputs, and returns a set of profiled executions. Profiled executions are used to compose a cost model for the application under different partitions. Finally, the *Optimization Solver* finds a legal partition among those enabled by the static analyzer that minimizes an objective function, using the cost model derived by the profiler (Section 3.3). The resulting partition is stored as a configuration file. At runtime, the chosen partition drives the execution of the application.

3.1 Static Analyzer

The partitioner uses static analysis to identify legal choices for placing migration and re-integration points in the code. In principle, these points could be placed anywhere in the code, but we reduce the available choices to make the optimization problem tractable. In particular, we restrict migration and re-integration points to method entry and exit points, respectively. We make two additional restrictions for simplicity. First, we only allow migration at the boundaries of application methods, not core-system library methods, which simplifies the implementation of a partition at runtime. Second, we only allow migration at VM-layer method boundaries, not native method boundaries, since the techniques required to migrate partial execution state differ vastly for the two types of methods. Note, however, that although we disallow migration while already in native execution, we do allow migrated methods to invoke native ones.

Figure 4 shows an example of a program, relevant parts of its static control-flow graph, and a particular legal partition of the program. Class C has three methods. Method a() calls method b(), which performs lightweight processing, followed by method c(), which performs expensive processing. The static control-flow graph approximates control flow in the program (inferring exact control flow is undecidable as program reachability is undecidable). The approxi-

mation is conservative in that if an execution of the program follows a certain path then that path exists in the graph (but the converse typically does not hold). In the depicted static control-flow graph, only entry and exit nodes of methods are shown, labeled as <class name>.<method name>.<entry | exit>. A possible partition as shown in Figure 4(c) runs the body of method c() on the clone, and the rest of the program on the mobile device. As described above, method c() may not be a system library or a native method, but may itself invoke system libraries or native methods.

3.1.1 Constraints

We next describe three properties of any legal partition, as required by the migration component, and explain how we use static analysis to obtain constraints that express these properties.

PROPERTY 1. *Methods that access specific features of a machine must be pinned to the machine.*

If a method uses a local resource such as the location service (e.g., GPS) or sensor inputs (e.g., microphones) in a mobile device, the method must be executed on the mobile device. This primarily concerns native methods, but also the main method of a program. The analysis marks the declaration of such methods with a special annotation M—for Mobile device. We manually identify such methods in the VM's API (e.g., VM API methods explicitly referring to the camera); this is done once for a given platform and is not repeated for each application. We also always mark the main method of a program. We refer to methods marked with M as the V_M method set.

PROPERTY 2. *Methods that share native state must be collocated at the same machine.*

An application may have native methods that create and access state below the VM. Native methods may share native state. Such methods must be collocated at the same machine as our migration component does not migrate native state (Section 4.1). For example, when an image processing class has `initialize`, `detect`, `fetchresult` methods that access native state, they need to be collocated at the same machine. To avoid a manual-annotation burden, native state annotations are inferred automatically by the following simple approximation, which works well in practice: we assign a unique annotation Nat_C to all native methods declared in the same class C; the set V_{Nat_C} contains all methods with that annotation.

PROPERTY 3. *Prevent nested migration.*

With one phone and one clone, this implies that there should be no nested suspends and no nested resumes. Once a program is suspended for migration at the entry point of a method, the program should not be suspended again without a resume, i.e., migration and re-integration points must be executed alternately. To enforce this property, the static analysis builds the static control-flow graph of an application, capturing the caller-callee method relation; it exports this as two relations, $DC(m_1, m_2)$, read as "method m_1 *Directly Calls* method m_2," and $TC(m_1, m_2)$ read as "method m_1 *Transitively Calls* method m_2," which is the transitive closure of DC. For the example in Figure 4, this ensures that if partitioning points are placed in a(), they are not placed in b() or c(). The remaining legal partitions place migration points at at b(), at c(), or at both b() and c().

3.2 Dynamic Profiler

The profiler collects the data that will be used to construct a cost model for the application under different execution settings. The cost metric can vary, but our prototype uses execution time and energy consumed at the mobile device.

The profiler is invoked on multiple executions of the application, each using a randomly chosen set of input data (e.g., command-line arguments and user-interface events), and each executed once on the mobile device and once on the clone in the cloud. The profiler outputs a set S of executions, and for each execution a *profile tree* T and T', from the mobile device and the clone, respectively. We note that random inputs may not explore all relevant execution paths of the application. In our future work, we hope to explore symbolic-execution-based techniques for high-coverage input generation [Cadar 2008].

A profile tree is a compact representation of an execution on a single platform. It is a tree with one node for each method invocation in the execution; it is rooted at the starting (user-defined) method invocation of the application (e.g., main). Specific method calls in the execution are represented as edges from the node of the caller method invocation (parent) to the nodes of the callees (children); edge order is not important. Each node is annotated with the cost of its particular invocation in the cost metric (execution time in our case). In addition to its called-method children, every non-leaf node also has a leaf child called its *residual node*. The residual node i' for node i represents the residual cost of invocation i that is not due to the calls invoked within i; in other words, node i' represents the cost of running the body of code excluding the costs of the methods called by it. Finally, each edge is annotated with the state size at the time of invocation of the child node, plus the state size at the end of that invocation; this would be the amount of data that the migrator (Section 4.1) would need to capture and transmit in both directions, if the edge were to be a migration point. Edges between a node and its residual child have no cost.

Figure 5 is an example of an execution trace and its corresponding profile tree. a is called twice in main, one a call invoking b and c, and one a call invoking no other method. A tree node on the right holds the execution time of the corresponding method in the trace (the length of the square bracket on the left). main' and a' are residual nodes, and they hold the difference between the value of

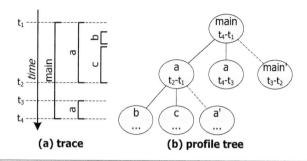

Figure 5. An example of an execution trace (a) and its corresponding profile tree (b). Edge costs are not shown.

their parent node and the sum of their sibling nodes. For example, node main' holds the value $T[\text{main}'] \equiv t_3 - t_2 = (t_4 - t_1) - ((t_4 - t_3) + (t_2 - t_1))$.

To fill in profile trees, we temporarily instrument application-method entry and exit points during each profile run on each platform (recall that system-library and native methods are not valid partitioning points, so we do not instrument them). For our execution-time cost metric, we collect timings at method entry and exit points, which we process to fill in tree node annotations. We compute migration costs (edge weights) by simulating migration at each profiled method: we perform the suspend-and-capture operation of the migrator (Section 4.1) and measure the captured state size, both when invoking the child node and when returning from it; we set the annotation of the parent edge $TE[i]$ of invocation i with that value. Recall that for every execution E, we capture two profile trees T and T', one per platform with different annotations.

For each invocation i in profiling execution E, we define a computation cost $C_c(i, l)$ and a migration cost $C_s(i)$, where l is the location of the invocation. We start with execution time cost. We fill in $C_c(i, l)$ from the corresponding profile tree collected at location l (if $l = 0$, the location is the mobile device and is filled from tree T, and if $l = 1$, the location is the clone and is filled from tree T'). If i is a leaf profile tree node, we set $C_c(i, l)$ to be the annotation of that node (e.g., $C_c(i, 0) \equiv T[i]$); otherwise, we set it to the annotation of the residual node i'. We fill $C_s(i)$ as the cost of making invocation i a migrant invocation. This cost is the sum of a suspend/resume cost and a transfer cost. The former is the time required to suspend a thread and resume a thread. The latter is a volume-dependent cost, the time it takes to capture, serialize, transmit, deserialize, and reinstantiate state of a particular size (assuming for simplicity all objects have the same such cost per byte). We precompute this per-byte cost[2], and use the edge annotations from the mobile-device profile tree to calculate the cost.

---

[2] One could also estimate this per-byte cost from memory, processor, and storage speeds, as well as network latency and bandwidth, but we took the simpler approach of just measuring it.

For energy consumption, we use a simple model that appears consistent with the kinds of energy measurements we can perform with off-board equipment (the Monsoon power monitor [Mon], in our setup). Specifically, we use a model that maps three system variables to a power level. We consider CPU activity (processing/idle), display state (on/off), and network state (transmitting or receiving/idle), and translate them to a power draw via a function P from $\langle CPU, Scr, Net \rangle$ triples to a power value. We estimate function P experimentally, and use it to construct two cost models, one where screen is on, and one where screen is off, as follows. For the cost model with the screen on, we set all $C_c(i, 0) \equiv P(CPUOn, ScrOn, NetIdle) \times T[i]$, i.e., the execution time at the device at power levels consistent with high CPU utilization, display on, but not network activity. We set all $C_c(i, 1) \equiv P(CPUIdle, ScrOn, NetIdle)$, i.e., the execution time at the clone, but at power levels consistent with idle CPU at the device; recall that we do not care about energy at the clone, but about energy expended at the mobile device while the clone is processing. Finally, we set all $C_s(i)$ from the execution-time model $C_s(i)$'s above, which hold the time it takes to migrate for an invocation, multiplied by power $P(CPUOn, ScrOn, NetOn)$. We note that our energy consumption model is a coarse starting point, with some noise, especially for very close decisions to migrate or not (see Section 6).

3.3 Optimization Solver

The purpose of our optimizer is to pick which application methods to migrate to the clone from the mobile device, so as to minimize the expected cost of the partitioned application. Given a particular execution E and its two profile trees T on the mobile device and T' on the clone, one might intuitively picture this task as optimally replacing annotations in T with those in T', so as to minimize the total node and weight cost of the hybrid profile tree. Our static analysis dictates the legal ways to fetch annotations from T' into T, and our dynamic profiling dictates the actual trees T and T'. We do not differentiate among different executions E in the execution set S; we consider them all equiprobable, although one might assign non-uniform frequencies in practice to match a particular expected workload.

More specifically, the output of our optimizer is a value assignment to binary decision variables $R(m)$, where m is every method in the application. If the optimizer chooses $R(m) = 1$ then the partitioner will place a migration point at the entry into the method, and a re-integration point at the exit from the method. If the optimizer chooses $R(m) = 0$, method m is unmodified in the application binary. For simplicity and to constrain the optimization problem, our migration strategy chooses to migrate or not migrate *all invocations* of a method. Despite its simplicity, this conservative strategy provides us with undeniable benefits (Section 6); we leave further refining differentiations depending on calling stack, method arguments, etc., to future work.

Not all partitioning choices for $R(.)$ are legal (Section 3.1.1). To express these constraints in the optimization problem, we define an auxiliary decision variable $L(m)$ indicating the location of every method m, and three relations I, as well as DC and TC computed during static analysis. $I(i, m)$ is read as "i is an invocation of method m," and is trivially defined from the profile runs. Whereas DC and TC are computed once for each application, I is updated with new invocations only when the set S of profiling executions changes.

Using the decision variables $R(.)$, the auxiliary decision variables $L(.)$, the method sets V_M and V_{Nat_C} for all classes C defined during static analysis, and the relations I, DC and TC from above, we formulate the optimization constraints as follows:

$$L(m_1) \neq L(m_2), \qquad \forall m_1, m_2 : DC(m_1, m_2) = 1$$
$$\wedge R(m_2) = 1 \quad (1)$$
$$L(m) = 0, \qquad \forall m \in V_M \quad (2)$$
$$L(m_1) = L(m_2), \quad \forall m_1, m_2, C : m_1, m_2 \in V_{Nat_C} \quad (3)$$
$$R(m_2) = 0, \qquad \forall m_1, m_2 : TC(m_1, m_2) = 1$$
$$\wedge R(m_1) = 1 \quad (4)$$

Constraint 1 is a soundness constraint, and requires that if a method causes migration to happen, it cannot be collocated with its callers. The remaining three correspond to the three properties defined in the static analysis. Constraint 2 requires that all methods pinned at the mobile device run on the mobile device (Property 1). Constraint 3 requires that methods dependent on the native state of the same class C are collocated, at either location (Property 2). And constraint 4 requires that all methods transitively called by a migrated method cannot be themselves migrated (Property 3).

The cost of a (legal) partition $R(.)$ of execution E is defined as follows, in terms of the auxiliary variables $L(.)$, the relation I and the cost variables C_c and C_s from the dynamic profiler:

$$C(E) = Comp(E) + Migr(E)$$
$$Comp(E) = \sum_{i \in E, m} [(1 - L(m))I(i, m)C_c(i, 0)$$
$$+ L(m)I(i, m)C_c(i, 1)]$$
$$Migr(E) = \sum_{i \in E, m} R(m)I(i, m)C_s(i)$$

$Comp(E)$ is the computation cost of the partitioned execution E and $Migr(E)$ is its migration cost. For every invocation $i \in E$, the computation cost takes its value from the mobile-device cost variables $C_c(i, 0)$, if the method m being invoked is to run on the mobile device, or from the clone variables $C_c(i, 1)$ otherwise. The migration cost sums the individual migration costs $C_s(i)$ of only those invocations i whose methods are migration points. Finally, the optimization objective is to choose $R()$ so as to minimize $\sum_{E \in S} C(E)$. We use a standard integer linear programming (ILP) solver to solve this optimization problem with the above constraints. One can extend our optimization formulation to include a constraint limiting total energy consumption while optimizing total execution time or one limiting execution time while optimizing energy consumption.

4. Distributed Execution

The purpose of the distributed execution mechanism in CloneCloud is to implement a specific partition of an application process running inside an application-layer virtual machine, as determined during partitioning (Section 3).

The life-cycle of a partitioned application is as follows. When the user attempts to launch a partitioned application, current execution conditions (availability of cloud resources and network link characteristics between the mobile device and the cloud) are looked up in a database of pre-computed partitions. The lookup result is a partition configuration file. The application binary loads the partition and instruments the chosen methods with *migration* and *re-integration* points—special VM instructions in our prototype. When execution of the process on the mobile device reaches a migration point, the executing thread is suspended and its state (including virtual state, program counter, registers, and stack) is packaged and shipped to a synchronized clone. There, the thread state is instantiated into a new thread with the same stack and reachable heap objects, and then resumed. When the migrated thread reaches a re-integration point, it is similarly suspended and packaged as before, and then shipped back to the mobile device. Finally, the returned packaged thread is merged into the state of the original process.

CloneCloud migration operates at the granularity of a thread. This allows a multi-threaded process (e.g., a process with a UI thread and a worker thread) running on the phone to off-load functionality, one thread-at-a-time. For example, a process with a UI thread and a worker thread can migrate the functionality of the worker thread. The UI thread continues processing, unless it attempts to access migrated state, in which case it blocks until the offloaded thread comes back. Note that the current CloneCloud system does not support a distributed shared memory (DSM) model; when there are two worker threads that share the same state, they cannot be offloaded at the same time. CloneCloud enables threads, local and migrated, to use—but not migrate—native, non-virtualized features of the platform on which they operate: this includes the network and natively implemented API functionality (such as expensive-to-virtualize image processing routines), etc. Furthermore, when the underlying library and OS are designed to exploit unvirtualized hardware accelerators such as GPUs and cryptographic accelerators, the system seamlessly gain benefits from these special features. In contrast, most prior work providing application-layer virtual-machine migration keeps native features and functionality exclusively on the original platform, only permitting the off-loading of pure, virtualized computation.

These two unique features of CloneCloud, thread-granularity migration and native-everywhere operation, enable interesting execution models. For example, a mobile application can retain its user interface threads running and interacting with the user, while off-loading worker threads to the cloud if this is beneficial. This would have been impossible with monolithic process or VM suspend-resume migration, since the user would have to migrate to the cloud along with the code. Similarly, a mobile application can migrate a thread that performs heavy 3D rendering operations to a clone with GPUs, without having to modify the original application source; this would have been impossible to do seamlessly if only migration of virtualized computation were allowed.

CloneCloud migration is effected via three distinct components: (a) a per-process *migrator thread* that assists a process with suspending, packaging, resuming, and merging thread state, (b) a per-node *node manager* that handles node-to-node communication of packaged threads, clone image synchronization and provisioning; and (c) a simple partition database that determines what partition to use.

The migrator functionality manipulates internal state of the application-layer virtual machine; consequently we chose to place it within the same address space as the VM, simplifying the procedure significantly. A manager, in contrast, makes more sense as a per-node component shared by multiple applications, for several reasons. First, it enables application-unspecific node maintenance, including file-system synchronization between the device and the cloud. Second, it amortizes the cost of communicating with the cloud over a single, possibly authenticated and encrypted, transport channel. Finally, it paves the way for future optimizations such as chunk-based or similarity-enhanced data transfer [Muthitacharoen 2001, Tolia 2006]. Our current prototype has a simple configuration interface that allows the user to manually pick out a partition from the database, and to choose new configurations to partition for. We next delve more deeply into the design of the distributed execution facilities in CloneCloud.

4.1 Suspend and Capture

Upon reaching a migration point, the job of the thread migrator is to suspend a migrant thread, collect all of its state, and pass that state to the node manager for data transfer. The thread migrator is a native thread, operating within the same address space as the migrant thread, but outside the virtual machine. As such, the migrator has the ability to view and manipulate both native process state and virtualized state.

To capture thread state, the migrator must collect several distinct data sets: execution stack frames and relevant data objects in the process heap, and register contents at the migration point. Virtualized stack frames—each containing register contents and local object types and contents—are

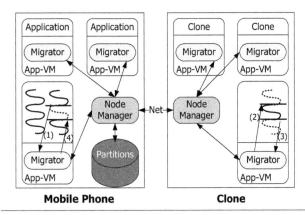

Figure 6. Migration overview.

readily accessible, since they are maintained by the VM management software. Starting with local data objects in the collected stack frames, the migrator recursively follows references to identify all relevant heap objects, in a manner similar to any mark-and-sweep garbage collector. For each relevant heap object, the migrator stores its field values, as well as relevant static class fields.

Captured state must be conditioned for transfer to be portable to adapt to different instruction set architectures (e.g., ARM and x86 architectures). First, object field values are stored in network byte order to allow for incompatibilities between different processor architectures. Second, whereas typically a stack frame contains a local native pointer to the particular class method it executes (which is not portable across address spaces or processor architectures), we store instead the class name and method name, which are portable.

In our prototype, state that is captured and migrated to the clone is marked in the phone VM. Remaining threads continue processing, unless they attempt to access such marked (migrated state), in which case they block until the off-loaded thread returns. This reduces the concurrency of some applications, making our current prototype better suited to applications with loosely-coupled threads. A distributed memory synchronization mechanism may alleviate this shortcoming, but we have yet to study the implications of the additional complexity involved.

4.2 Resume and Merge

As soon as the captured thread state is transferred to the target clone device, the node manager passes it to the migrator of a freshly allocated process. To resume that migrant thread, the migrator overlays the thread context over the fresh address space, essentially reversing the capture process described in Section 4.1. The executable text is loaded (it can be found under the same filename in the synchronized file system of the clone). Then all captured classes and object instances are allocated in the virtual machine's heap, updating static and instance field contents with those from the captured context. As soon as the address space contains

all the data relevant to the migrant thread, the thread itself is created, given the stack frames from the capture, the register contents are filled to match the state of the original thread at the migration point in the mobile device, and the thread is marked as runnable to resume execution.

As described above, the cloned thread will eventually reach a reintegration point in its executable, signaling that it should migrate back to the mobile device. Reintegration is almost identical conceptually to the original migration: the clone's migrator captures and packages the thread state, the node manager transfers the capture back to the mobile device, and the migrator in the original process is given the capture for resumption. There is, however, a subtle difference in this reverse migration direction. Whereas in the forward direction—from mobile device to clone—a captured thread context is used to create a new thread from scratch, in the reverse direction—from clone to mobile device—the context must *update* the original thread state to match the changes effected at the clone. We call this a *state merge*.

A successful design for merging states in such a fashion depends on our ability to map objects at the original address space to the objects they "became" at the cloned address space; object references themselves are not sufficient in that respect, since in most application-layer VMs, references are implemented as native memory addresses, which look different in different processes, across different devices and possibly architectures, and tend to be reused over time for different objects.

Our solution is an *object mapping table*, which is only used during state capture and reinstantiation in either direction, and only stored while a thread is executing at a clone. We instrument the VM to assign a per-VM unique object ID to each data object created within the VM, using a local monotonically increasing counter. For clarity, we call the ID at the mobile device MID and at the clone CID. Once migration is initiated at the mobile device, a mapping table is first created for captured objects, filling for each the MID but leaving the CID null; this indicates that the object has no clone counterpart yet. After instantiation at the clone, the clone recreates all the objects with null CIDs, assigning valid fresh CIDs to them, and remembers the local object address corresponding to each mapping entry. At this point, all migrated objects have valid mappings.

During migration in the reverse direction, objects that came from the original thread are captured and keep their valid mapping. Newly created objects at the clone have the locally assigned ID placed in their CID, but get a null MID. Objects from the original thread that may have been deleted at the clone are ignored and no mapping is sent back for them. During the merge back at the mobile device, we know which objects should be freshly created (those with null MIDs) and which objects should be overwritten with the contents fetched back from the clone (those with non-null MIDs). "Orphaned" objects that were migrated out but died at the

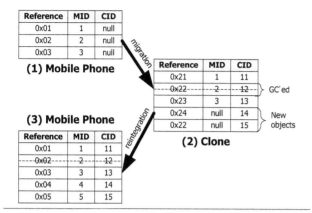

Reference	MID	CID
0x01	1	null
0x02	2	null
0x03	3	null

(1) Mobile Phone

Reference	MID	CID	
0x21	1	11	
~~0x22~~	~~2~~	~~12~~	GC'ed
0x23	3	13	
0x24	null	14	New
0x22	null	15	objects

(2) Clone

(3) Mobile Phone

Reference	MID	CID
0x01	1	11
~~0x02~~	~~2~~	~~12~~
0x03	3	13
0x04	4	14
0x05	5	15

Figure 7. Object mapping example.

clone become disconnected from the thread object roots and are garbage-collected subsequently. Note that the mapping table is constructed and used only during capture and reintegration, not during normal memory operations either at the mobile device or at the clone.

Figure 7 shows an example scenario demonstrating the use of object mapping. During initial migration, objects at addresses 0x01, 0x02, and 0x03 are captured. The migrator creates the mapping table with three entries, one for each object, with the local ID of each object—1, 2, and 3, respectively—in MID, and null CIDs. At the clone, the mapping table is stored, updating each entry with the local address of each object (0x21, 0x22, and 0x23, respectively). When the thread is about to return back to the mobile device, new entries are created in the table for captured objects whose IDs are not already in the CID column (objects with IDs 14 and 15). Entries in the table whose CID does not appear in captured objects are deleted (the second entry in the figure). Remaining entries belong to objects that came from the original thread and are also going back (those with CID 11 and 13). Note that memory address 0x22 was reused at the clone after the original object was destroyed, but the object has a different ID from the original object, allowing the migrator to differentiate between the two. Back at the mobile device, new objects are created for entries with null MIDs (bottom two entries), objects with non-null MIDs are updated with the returned state (first and third entries), and one object (with local address 0x02) is left to be garbage-collected.

We use object mappings for a subtly different purpose, as well. Because new processes are forked as copies of a "template" process —the *Zygote*, in the Android nomenclature— and because that template exists in all booted instances of the Android platform, we can avoid transmitting unchanged system heap objects. However, whereas application objects can be named at the time of creation, Zygote objects are created concurrently defying consistent naming. To address the challenge, we name each system object according to its class name and invocation sequence among all objects of that class—this assumes that objects from each class are con-

structed in the same order at Zygote processes on different platforms, an assumption that holds true in all Zygote instances we have seen so far.

5. Implementation

We implemented our prototype of CloneCloud partitioning and migration on the cupcake branch of the Android OS [AOS]. We tested our system on the Android Dev Phone 1 [ADP] (an unlocked HTC G1 device) equipped with both WiFi and 3G connections, and on clones running within the Android x86 virtual machine. Clones execute on a server with a 3.0GHz Xeon CPU, running VMware ESX 4.1, connected to the Internet through a layer-3 firewall. We modified the Dalvik VM (Android's application-level, register-based VM, principally targeted by a Java compiler front-end) [Dal] for dynamic profiling and migration. These modifications comprised approximately 8,000 lines of C code.

We implemented our static analysis on Java bytecode using JChord [JCh]. We modified JChord to support root methods of analysis that are different from `main`. We modified Dalvik VM tracing to efficiently trace execution and migration cost and to trace only application methods in which we are interested. To profile energy consumption of the phone, we connected the Monsoon power monitor [Mon], to the phone, measured the current drawn from the phone at a 5KHz frequency, and computed power consumption from the current and the voltage we set. We use lp_solve [lps] to solve the optimization problem for each execution environment.

The migrator uses Dalvik VM's thread suspension, a mechanism common in other application VMs. Threads are only suspended at bytecode instruction boundaries. We capture and represent execution state with a modified version of hprof [HPR]. We extend the format to also store thread stacks and class file paths, as well as store CIDs and MIDs to each object; we modified object creation and destruction in DalvikVM to assign those IDs. The object mapping table is a separate hash table inside the Dalvik VM, created only when migration begins, and destroyed after reintegration.

Migration is initiated and terminated in the modified application via two new system operations: `ccStart()` and `ccStop()`, respectively. The application thread calling these operations notifies the migrator thread inside Dalvik, and suspends itself. Once the migrator thread gains control, it checks with the loaded partition if it should migrate, and if so handles the rest.

6. Evaluation

To evaluate our prototype, we implemented three applications. We ran those applications either on a phone—a status quo, monolithic execution—or by optimally partitioning for two settings: one with WiFi connectivity and one with 3G.

We implemented a virus scanner, image search, and privacy-preserving targeted advertising; we briefly describe

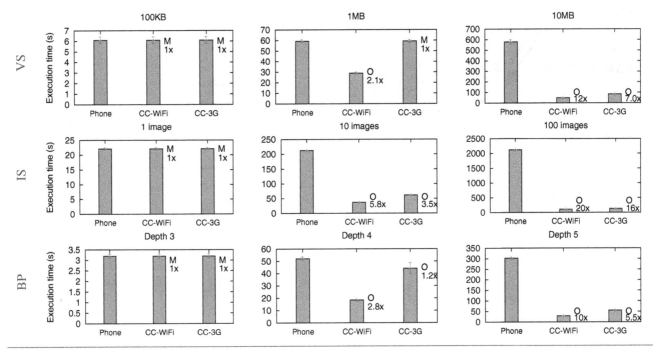

Figure 8. Mean execution times of virus scanning (VS), image search (IS), and behavior profiling (BP) applications with standard deviation error bars, three input sizes for each. For each application and input size, the data shown include execution time at the phone alone, that of CloneCloud with WiFi (CC-WiFi), and that of CloneCloud with 3G (CC-3G). The partition choice is annotated with M for "monolithic" and O for "off-loaded," also indicating the relative improvement from the phone-alone execution.

Application	Input Size	Phone Exec. (sec) Mean (std)	Clone Exec. (sec) Mean (std)
VS	100KB	6.1 (0.32)	0.2 (0.01)
	1MB	59.3 (1.49)	2.2 (0.01)
	10MB	579.5 (20.76)	22.5 (0.08)
IS	1 img	22.1 (0.26)	0.9 (0.07)
	10 img	212.8 (0.44)	8.0 (0.03)
	100 img	2122.1 (1.27)	79.2 (0.44)
BP	depth 3	3.3 (0.10)	0.2 (0.01)
	depth 4	52.1 (1.45)	1.8 (0.07)
	depth 5	302.7 (3.76)	10.9 (0.19)

Table 1. Execution times of virus scanning (VS), image search (IS), and behavior profiling (BP) applications, three input sizes for each. For each application and input size, the data shown include execution time at the phone alone and execution time at the clone alone.

each next. The virus scanner scans the contents of the phone file system against a library of 1000 virus signatures, one file at a time. We vary the size of the file system between 100KB and 10 MB. The image search application finds all faces in images stored on the phone, using a face-detection library that returns the mid-point between the eyes, the distance in between, and the pose of detected faces. We only use images smaller than 100KB, due to memory limitations

of the Android face-detection library. We vary the number of images from 1 to 100. The privacy-preserving targeted-advertising application uses behavioral tracking across websites to infer the user's preferences, and selects ads according to a resulting model; by doing this tracking at the user's device, privacy can be protected (see Adnostic [Toubiana 2010]). We implement Adnostic's web page categorization, which maps a user's keywords to one of the hierarchical interest categories—down to nesting levels 3-5—from the DMOZ open directory [dmo]. The application computes the cosine similarity between user interest keywords and predefined category keywords.

For these applications, the static analysis with JChord took 23 seconds on average with sun jdk1.6.0_22 on a desktop machine; recall that this analysis need only be run once per application. Generating an optimizer (ILP) script from the profile trees, constraints and execution conditions, and solving the optimization problem, takes less than one second.

Table 1 shows the execution times of the applications at the phone alone or at the clone alone. The clone execution time is a lower-bound on execution time since, in practice, at least some part of the application must run on the phone. The difference between columns 3 and 4 captures the speedup opportunity due to the disparity between phone and cloud computation resources.

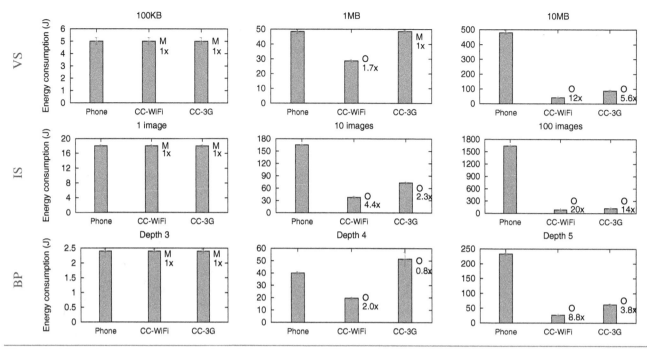

Figure 9. Mean phone energy consumption of virus scanning (VS), image search (IS), and behavior profiling (BP) applications with standard deviation error bars, three input sizes for each. For each application and input size, the data shown include execution time at the phone alone, that of CloneCloud with WiFi (CC-WiFi), and that of CloneCloud with 3G (CC-3G). The partition choice is annotated with M for "monolithic" and O for "off-loaded," also indicating relative improvement over phone-only execution.

Figures 8 and 9 shows execution times and phone energy consumption for the three applications, respectively. All measurements are the average of five runs. Each graph shows Phone, CloneCloud with WiFi (CC-WiFi), and CloneCloud with 3G (CC-3G). CC-Wifi and CC-3G results are annotated with the relative improvement and the partitioning choice, whether the optimal partition was to run monolithically on the phone (M) or to off-load to the cloud (O). In the experiments, WiFi had latency of 69ms and bandwidth of 6.6Mbps, and 3G had latency of 680ms, and bandwidth of 0.4Mbps.

CloneCloud chooses to keep local the smallest workloads from each application, deciding to off-load 6 out of 9 experiments with WiFi. With 3G, out of all 9 experiments, CloneCloud chose to off-load 5 experiments. For off-loaded cases, each application chooses to offload the function that performs core computation from its worker thread: scanning files for virus signature matching for VS, performing image processing for IS, and computing similarities for BP. CC-WiFi exhibits significant speed-ups and energy savings: 12x, 20x, and 10x speed-up, and 12x, 20x, and 9x less energy for the largest workload of each of the three applications, with a completely automatic modification of the application binary without programmer input. A clear trend is that larger workloads benefit from off-loading more: this is due to amortization of the migration cost over a larger computation at the clone that receives a significant speedup.

A secondary trend is that energy consumption mostly follows execution time: unless the phone switches to a deep sleep state while the application is off-loaded at the clone, its energy expenditure is proportional to how long it is waiting for a response. When the user runs a single application at a time, deeper sleep of the phone may further increase observed energy savings. We note that one exception is CC-3G, where although execution time decreases, energy consumption increases slightly for behavior profiling with depth 4. We believe this is due to our coarse energy cost model, and only occurs for close decisions. We hope to explore a finer-grained energy cost model that better captures energy consumption as future work.

CC-3G also exhibits 7x, 16x, and 5x speed-up, and 6x, 14x, and 4x less energy for the largest workload of each of the three applications. Lower gains can be explained given the overhead differences between WiFi and 3G networks. As a result, whereas migration costs about 15-25 seconds with WiFi, it shoots up to 40-50 seconds with 3G, due to the greater latency and lower bandwidth. In both cases, migration costs include a network-unspecific thread-merge cost— patching up references in the running address space from the migrated thread—and the network-specific transmission of the thread state. The former dominates the latter for WiFi, but is dominated by the latter for 3G. Our current implementation uses the DEFLATE compression algorithm to reduce the amount of data to send; we expect off-loading benefits to improve with other optimizations targeting the network overheads (in particular, 3G network overheads) such as redundant transmission elimination.

7. Related Work

CloneCloud is built upon previous research done in automatic partitioning, migration, and remote execution. Most related is the recent work on MAUI [Cuervo 2010]. Similar to MAUI [Cuervo 2010], CloneCloud partitions applications using a framework that combines static program analysis with dynamic program profiling and optimizes execution time or energy consumption using an optimization solver. For offloaded execution, MAUI performs method shipping with relevant heap objects, but CloneCloud migrates specific threads with relevant execution state on demand and can merge migrated state back to the original process.

Although MAUI and CloneCloud have similar workflows, a number of differences distinguish them. First, supporting native method calls was an important design choice we made, which increases its applicability. MAUI does not support remotely executing virtualized methods calling native functions (e.g., two methods that share native state). Second, CloneCloud requires little programmer help, whereas MAUI requires programmers to annotate methods as REMOTABLE. Third, MAUI does not focus on the details of method shipping, whereas we present a detailed design for state migration and merging, which is a major source of design challenges. Lastly, CloneCloud solves the optimization problem asynchronously, whereas MAUI requires a solver to be running at the server at runtime.

Further out, there is a much prior work on partitioning, migration, and remote execution, which we summarize.

Partitioning We first summarize work on partitioning of distributed systems. Coign [Hunt 1999] automatically partitions a distributed application composed of Microsoft COM components to reduce communication cost of partitioned components. The application must be structured to use COM components and partitioning points are coarse-grained COM boundaries. Coign focuses on static partitioning, which works better in a stable environment. Giurgiu [2009] takes a similar approach for applications designed on top of distribution middleware such as OSGi. Wishbone [Newton 2009] and Pleiades [Kothari 2007] compile a central program into multiple code pieces with stubs for communication. They are primarily intended for sensor networks, and they require programs to be written in special languages (a stream-processing language, and an extended version of C, respectively). J-Orchestra [Tilevich 2002] creates partitioned applications automatically by a compiler that classifies anchored unmodifiable, anchored modifiable, or mobile classes. After the analysis, it rewrites all references into indirect references (i.e., references to proxy objects) for a cluster of machines, and places classes with location constraints (e.g., ones with native state constraints) to proper locations. Fi-

nally, for distributed execution of partitioned applications, it relies on the RMI middleware. Chroma [Balan 2003] uses *tactics*, application-specific knowledge for remote execution in a high-level declarative form, for run-time partitioning.

There are also Java program partitioning systems for mobile devices, whose limitation is that only Java classes without native state can be placed remotely [Gu 2003, Messer 2002, Ou 2007]. The general approach is to partition Java *classes* into groups using adapted MINCUT heuristic algorithms to minimize the component interactions between partitions. Also, different proposals consider different additional objectives such as memory, CPU, or bandwidth. Besides disallowing native execution offloading, this previous work does not consider partitioning constraints like our work does, the granularity of partitioning is coarse since it is at class level, and it focuses on static partitioning.

On a related front, Links [Cooper 2006], Hops [Serrano 2006], and UML-based Hilda [Yang 2006] aim to statically partition a client-server program written in a high-level functional language or a high-level declarative language into two or three tiers. Yang [2007] examines partitioning of programs written in Hilda based on cost functions for optimizing user response time. Swift [Chong 2007] statically partitions a program written in the Jif programming language into client-side and server-side computation. Its focus is to achieve confidentiality and integrity of the partitioned program with the help of security labels in the program annotated by programmers.

Migration There has been previous work on supporting migration in Java. MERPATI [Suezawa 2000] provides JVM migration checkpointing the entire heap and all the thread execution environments (call stack, local variables, operand stacks) and resuming from a checkpoint. In addition, there have been different approaches on distributed Java virtual machines (DJVMs). They assume a cluster environment where homogeneous machines are connected via fast interconnect, and try to provide a single system image to users. One approach is to build a DJVM upon a cluster below the JVM. Jessica [Ma 1999] and Java/DSM [Yu 1997] rely on page-based distributed shared memory (DSM) systems to solve distributed memory consistency problems. To address the overhead of false sharing in page-based DSM, Jessica2 [Zhu 2002] is an object-based solution. cJVM [Aridor 1999] modifies the JVM to support method shipping to remote objects with proxy objects, creating threads remotely, and supporting distributed stacks. Object migration systems such as Emerald [Jul 1988] move objects to the sites running threads requesting to access the objects. In contrast, CloneCloud migration chooses partial threads to offload, moves only their relevant execution state (thread stack and reachable heap objects), and supports merging between existing state and migrated execution state.

Remote execution Remote execution of resource-intensive applications for resource-poor hardware is a well-known approach in mobile/pervasive computing. Most remote execution work carefully designs and pre-partitions applications between local and remote execution. Typical remote execution systems run a simple visual, audio output routine at the mobile device and computation-intensive jobs at a remote server [Balan 2002, Flinn 2001; 1999, Fox 1996, Rudenko 1998, Young 2001]. Rudenko [1998] and Flinn [1999] explore saving power via remote execution. Cyber foraging [Balan 2002; 2007] uses surrogates (untrusted and unmanaged public machines) opportunistically to improve the performance of mobile devices, similarly to data staging [Flinn 2003] and Slingshot [Su 2005]. In particular, Slingshot creates a secondary replica of a home server at nearby surrogates. ISR [Satyanarayanan 2005] provides the ability to suspend on one machine and resume on another machine by storing virtual machine (e.g., Xen) images in a distributed storage system.

8. Discussion and Future Work

CloneCloud is limited in some respects by its inability to migrate native state and to export unique native resources remotely. Conceptually, if one were to migrate at a point in the execution in which a thread is executing native code, or has native heap state, the migrator would have to collect such native context for transfer as well. However, the complexity of capturing such information in a portable fashion (and the complexity of integrating such captures after migration) is significantly higher, given processor architecture differences, differences in file descriptors, etc. As a result, CloneCloud focuses on migrating at execution points where no native state (in the stack or the heap) need be collected and migrated.

A related limitation is that CloneCloud does not virtualize access to native resources that are not virtualized already and are not available on the clone. For example, if a method accesses a camera/GPS on the mobile device, CloneCloud requires that method to remain pinned on the mobile device. In contrast, networking hardware or an unvirtualized OS facility (e.g., Android's image processing API) are available on both the mobile device and the clone, so a method that needs to access them need not be pinned. An alternative design would have been to permit migration of such methods, but enable access to the unique native resource via some RPC-like mechanism. We consider this alternative a complementary point in the design space, and plan to pursue it in conjunction with thread-granularity migration in the future.

The system presented in this paper allows only perfunctory concurrency between the unmigrated threads and the migrated thread; pre-existing state on the mobile device remains unmodifiable until the migrant thread returns. As long as local threads only read existing objects and modify only newly created objects, they can operate in tandem with the clone. Otherwise, they have to block. A promising direction, whose benefits may or may not be borne out by the associated complexity, lies in extending this architecture to support full concurrency between the mobile device and clones. To achieve this, we need to add thread synchronization, heap object synchronization, on-demand object paging to access remote objects, etc.

While in this paper we assume that the environment in which we run clone VMs is trusted, the future of roaming devices that use clouds where they find them demands a more careful approach. For instance, many have envisioned a future in which public infrastructure machines such as public kiosks [Garriss 2008] and digital signs are widely available for running opportunistically off-loaded computations. We plan to extend our basic system to check that the execution done in the remote machine is trusted. Automatically refactoring computation around trusted features on the clone is an interesting research question.

9. Conclusion

This paper takes a step towards seamlessly interfacing between the mobile and the cloud. Our system overcomes design and implementation challenges to achieve basic augmented execution of mobile applications on the cloud, representing the whole-sale transfer of control from the device to the clone and back. We combine partitioning, migration with merging, and on-demand instantiation of partitioning to address these challenges. Our prototype delivers up to 20x speedup and 20x energy reduction for the simple applications we tested, without programmer involvement, demonstrating feasibility for the approach, and opening up a path for a rich research agenda in hybrid mobile-cloud systems.

Acknowledgments

We would like to thank our shepherd Pradeep Padala and the anonymous reviewers for their insightful feedback.

References

[ADP] Android dev phone 1. `developer.android.com/guide/developing/device.html`.

[AOS] Android open source project. `source.android.com/`.

[Dal] Dalvik VM. `developer.android.com/guide/basics/what-is-android.html`.

[HPR] Hprof: A heap/cpu profiling tool in J2SE 5.0. `java.sun.com/developer/technicalArticles/Programming/HPROF.html`.

[JCh] JChord. `code.google.com/p/jchord`.

[lps] lp_solve. `lpsolve.sourceforge.net`.

[Mon] Monsoon power monitor. `www.msoon.com`.

[dmo] Open directory project. `www.dmoz.org`.

[Aridor 1999] Y. Aridor, M. Factor, and A. Teperman. cJVM: a single system image of a JVM on a cluster. In *ICPP*, 1999.

[Balan 2002] R. Balan, J. Flinn, M. Satyanarayanan, S. Sinnamohideen, and H.-I. Yang. The case for cyber foraging. In *ACM SIGOPS European Workshop*, 2002.

[Balan 2007] R. K. Balan, D. Gergle, M. Satyanarayanan, and J. Herbsleb. Simplifying cyber foraging for mobile devices. In *MobiSys*, 2007.

[Balan 2003] R. K. Balan, M. Satyanarayanan, S. Park, and T. Okoshi. Tactics-based remote execution for mobile computing. In *MobiSys*, 2003.

[Beccue 2009] Mark Beccue and Dan Shey. Mobile Cloud Computing. Research report, ABI Research, 2009.

[Cadar 2008] C. Cadar, D. Dunbar, and D. R. Engler. Klee: Unassisted and automatic generation of high-coverage tests for complex systems programs. In *OSDI*, 2008.

[Chong 2007] S. Chong, J. Liu, A. C. Myers, and X. Qi. Secure web applications via automatic partitioning. In *SOSP*, 2007.

[Chun 2009] B.-G. Chun and P. Maniatis. Augmented smartphone applications through clone cloud execution. In *HotOS*, 2009.

[Cooper 2006] E. Cooper, S. Lindley, P. Wadler, and J. Yallop. Links: Web programming without tiers. In *Proc. 5th International Symposium on Formal Methods for Components and Objects*, 2006.

[Cuervo 2010] E. Cuervo, A. Balasubramanian, D. Cho, A. Wolman, S. Saroiu, R. Chandra, and P. Bahl. MAUI: Making smartphones last longer with code offload. In *MobiSys*, 2010.

[Flinn 2001] J. Flinn, D. Narayanan, and M. Satyanarayanan. Self-tuned remote execution for pervasive computing. In *HotOS*, 2001.

[Flinn 1999] J. Flinn and M. Satyanarayanan. Energy-aware adaptation for mobile applications. In *SOSP*, 1999.

[Flinn 2003] J. Flinn, S. Sinnamohideen, N. Tolia, and M. Satyanarayanan. Data staging for untrusted surrogates. In *USENIX FAST*, 2003.

[Fox 1996] A. Fox, S. D. Gribble, E. A. Brewer, and E. Amir. Adapting to network and client variability via on-demand dynamic distillation. In *ASPLOS*, 1996.

[Garriss 2008] S. Garriss, R. Càceres, S. Berger, R. Sailer, L. van Doorn, and X. Zhang. Trustworthy and personalized computing on public kiosks. In *MobiSys*, 2008.

[Giurgiu 2009] I. Giurgiu, O. Riva, D. Juric, I. Krivulev, and G. Alonso. Calling the cloud: Enabling mobile phones as interfaces to cloud applications. In *Middleware*, 2009.

[Gu 2003] X. Gu, K. Nahrstedt, A. Messer, I. Greenberg, and D. Milojicic. Adaptive offloading inference for delivering applications in pervasive computing environments. In *PerCom*, 2003.

[Hunt 1999] G. C. Hunt and M. L. Scott. The Coign automatic distributed partitioning system. In *OSDI*, 1999.

[Jul 1988] E. Jul, H. Levy, N. Hutchinson, and A. Black. Fine-grained mobility in the emerald system. *ACM Trans. on Computer Systems*, 1988.

[Kothari 2007] N. Kothari, R. Gummadi, T. Millstein, and R. Govindan. Reliable and efficient programming abstractions for wireless sensor networks. In *PLDI*, 2007.

[Ma 1999] M. J. M. Ma, C.-L. Wang, and F. C. M. Lau. JESSICA: Java-enabled single-system-image computing architecture. *Journal of Parallel and Distributed Computing*, 1999.

[Messer 2002] A. Messer, I. Greenberg, P. Bernadat, D. Milojicic, D. Chen, T.J. Giuli, and X. Gu. Towards a distributed platform for resource-constrained devices. In *ICDCS*, 2002.

[Muthitacharoen 2001] A. Muthitacharoen, B. Chen, and D. Mazières. A low-bandwidth network file system. In *SOSP*, 2001.

[Newton 2009] R. Newton, S. Toledo, L. Girod, H. Balakrishnan, and S. Madden. Wishbone: Profile-based partitioning for sensornet applications. In *NSDI*, 2009.

[Ou 2007] S. Ou, K. Yang, and J. Zhang. An effective offloading middleware for pervasive services on mobile devices. *Journal of Pervasive and Mobile Computing*, 2007.

[Rudenko 1998] A. Rudenko, P. Reiher, G. J. Popek, and G. H. Kuenning. Saving portable computer battery power through remote process execution. *MCCR*, 1998.

[Satyanarayanan 2005] M. Satyanarayanan, Michael A. Kozuch, Casey J. Helfrich, and David R. O'Hallaron. Towards seamless mobility on pervasive hardware. *Journal of Pervasive and Mobile Computing*, 2005.

[Satyanarayanan 2009] M. Satyanarayanan, Bahl P., Caceres R., and Davies N. The case for vm-based cloudlets in mobile computing. *Pervasive Computing*, 8(4), 2009.

[Serrano 2006] M. Serrano, E. Gallesio, and F. Loitsch. HOP: a language for programming the web 2.0. In *Proc. 1st Dynamic Languages Symposium*, 2006.

[Su 2005] Y.-Y. Su and J. Flinn. Slingshot: Deploying stateful services in wireless hotspots. In *MobiSys*, 2005.

[Suezawa 2000] Takashi Suezawa. Persistent execution state of a java virtual machine. In *JAVA '00: Proceedings of the ACM 2000 conference on Java Grande*, 2000.

[Tilevich 2002] E. Tilevich and Y. Smaragdakis. J-orchestra: Automatic java application partitioning. In *ECOOP*, 2002.

[Tolia 2006] N. Tolia, M. Kaminsky, D. G. Andersen, and S. Patil. An architecture for Internet data transfer. In *NSDI*, 2006.

[Toubiana 2010] V. Toubiana, A. Narayanan, D. Boneh, H. Nissenbaum, and S. Barocas. Adnostic: Privacy preserving targeted advertising. In *NDSS*, 2010.

[Yang 2007] F. Yang, N. Gupta, N. Gerner, X. Qi, A. Demers, J. Gehrke, and J. Shanmugasundaram. A unified platform for data driven web applications with automatic client-server partitioning. In *WWW*, 2007.

[Yang 2006] F. Yang, J. Shanmugasundaram, M. Riedewald, and J. Gehrke. Hilda: A high-level language for data-driven web application. In *ICDE*, 2006.

[Young 2001] C. Young, Y. N. Lakshman, T. Szymanski, J. Reppy, and D. Presotto. Protium, an infrastructure for partitioned applications. In *HotOS*, 2001.

[Yu 1997] W. Yu and A. L. Cox. Java/DSM: A platform for heterogeneous computing. *Concurrency - Practice and Experience*, 1997.

[Zhu 2002] W. Zhu, C.-L. Wang, and F.C.M.Lau. JESSICA2: A distributed java virtual machine with transparent thread migration support. In *CLUSTER*, 2002.

Symbolic Crosschecking of Floating-Point and SIMD Code

Peter Collingbourne   Cristian Cadar   Paul H. J. Kelly

Department of Computing
Imperial College London
{peter.collingbourne03, c.cadar, p.kelly}@imperial.ac.uk

Abstract

We present an effective technique for crosschecking an IEEE 754 floating-point program and its SIMD-vectorized version, implemented in KLEE-FP, an extension to the KLEE symbolic execution tool that supports symbolic reasoning on the equivalence between floating-point values.

The key insight behind our approach is that floating-point values are only reliably equal if they are essentially built by the same operations. As a result, our technique works by lowering the Intel Streaming SIMD Extension (SSE) instruction set to primitive integer and floating-point operations, and then using an algorithm based on symbolic expression matching augmented with canonicalization rules.

Under symbolic execution, we have to verify equivalence along every feasible control-flow path. We reduce the branching factor of this process by aggressively merging conditionals, if-converting branches into `select` operations via an aggressive phi-node folding transformation.

We applied KLEE-FP to OpenCV, a popular open source computer vision library. KLEE-FP was able to successfully crosscheck 51 SIMD/SSE implementations against their corresponding scalar versions, proving the *bounded* equivalence of 41 of them (i.e., on images up to a certain size), and finding inconsistencies in the other 10.

Categories and Subject Descriptors   C.1.2 [*Multiple Data Stream Architectures (Multiprocessors)*]: Single-instruction-stream, multiple-data-stream processors (SIMD); D.2.4 [*Software/Program Verification*]: Reliability; D.2.5 [*Testing and Debugging*]: Symbolic execution

General Terms   Reliability, Verification

EuroSys'11, April 10–13, 2011, Salzburg, Austria.

1. Introduction

Single Instruction Multiple Data (SIMD) computing is an increasingly popular means of improving the performance of programs by exploiting their data level parallelism. A number of traditionally scalar architectures have been extended with SIMD support, such as the Streaming SIMD Extensions (SSE), 3DNow! and Advanced Vector Extensions (AVX) for x86, NEON for ARM, or AltiVec for PowerPC. Furthermore, GPU programming languages, such as OpenCL and CUDA, are based on the SIMD execution model.

SIMD processors exploit data level parallelism by providing instruction sets that operate on one-dimensional arrays of data called vectors. While automatic vectorization is an active area of research [Eichenberger 2004, Larsen 2000, Naishlos 2003], the difficulty of reasoning about data dependencies and arithmetic precision means that optimizing scalar code to use SIMD instructions is still a mostly manual process. Unfortunately, manually translating scalar code into an equivalent SIMD version is a difficult task, because any programming error may cause the hand-optimized SIMD code to act differently from the purportedly equivalent scalar version. In this paper, we propose a novel automatic technique for verifying that the SIMD version of a piece of code is equivalent to its (original) scalar version.

Our technique is based on symbolic execution [King 1975], which provides a systematic way for exploring all feasible paths in a program for inputs up to a certain size. On each explored path, our technique works by building the symbolic expressions associated with the scalar and respectively the SIMD version of the code, and proving their equivalence.

While symbolic crosschecking has been successfully employed in the past (e.g., in the context of block cipher implementations [Smith 2008]), we need to address a series of challenges to apply it to the verification of SIMD vectorizations. First, we need to model the semantics of a real SIMD instruction set, which the current generation of symbolic execution tools do not handle. Second, and more importantly, SIMD code makes intensive use of floating point operations. Due to the complexity of floating point semantics [IEEE Task P754 2008], it is extremely difficult — if not infeasible — to build a constraint solver for floating point, and as a result there are currently no such constraint solvers available. Thus, in this paper we take a different approach, in which we prove the equivalence of two symbolic floating point expressions by first applying a series of expression canonicalization rules, and then syntactically matching the two expressions. The key insight into why our approach works is that constructing two equivalent values from the same inputs in floating point can usually only be done reliably by performing the same operations.

This paper makes the following contributions:

1. We present a symbolic execution (SE) based technique for crosschecking SIMD vectorizations against their scalar implementations.

2. We implement our technique in a tool called KLEE-FP, an extension to the open source symbolic execution tool KLEE [klee.llvm.org].

3. We reason about floating-point values (which KLEE's constraint solver cannot handle), using expression matching augmented with canonicalization rules that express strict equivalences in floating-point and mixed FP-integer expressions. As far as we know, this is the first practical SE-based technique that can precisely handle IEEE 754 floating point arithmetic.

4. We evaluate our technique by applying KLEE-FP to OpenCV [Intel], a popular open source computer vision library. KLEE-FP was able to crosscheck a total of 51 SIMD/SSE implementations against their corresponding scalar versions, proving the bounded equivalence of 41 of them on images up to a certain size, and finding inconsistencies in the other 10 pairs.

To achieve this, the semantics for a substantial portion of the Intel SSE instruction set are implemented via translation to an intermediate representation. We improve the tractability of our technique by implementing an aggressive variant of if-conversion using phi-node folding [Chuang 2003, Lattner 2004], to replace control-flow forking with predicated `select` instructions, in order to reduce the number of paths explored by symbolic execution.

2. Overview

This section illustrates the main features of our technique by showing how it can be used to verify the equivalence between a scalar and an SIMD implementation of a simple routine. Our code example, shown in Figure 1, is based on one of the OpenCV benchmarks we evaluated (specifically `thresh(BINARY_INV, f32)`; see §5). The code defines a routine called `zlimit`, which takes as input a floating point array `src` of size `size`, and returns as output the array `dst` of the same size. Each element of `dst` is the greater of the corresponding elements of `src` and 0. The routine consists of both a scalar and an SIMD implementation; users choose

```
1   void zlimit (int simd, float *src, float *dst,
2               size_t  size ) {
3       if (simd) {
4           __m128 zero4 = _mm_set1_ps(0.f);
5           while ( size >= 4) {
6               __m128 srcv = _mm_loadu_ps(src);
7               __m128 cmpv = _mm_cmpgt_ps(srcv, zero4);
8               __m128 dstv = _mm_and_ps(cmpv, srcv);
9               _mm_storeu_ps(dst, dstv );
10              src += 4; dst += 4; size -= 4;
11          }
12      }
13      while ( size ) {
14          *dst = *src > 0.f ? *src : 0.f;
15          src++; dst++; size--;
16      }
17  }
18
19  int main(void) {
20      float  src [64], dstv [64], dsts [64];
21      uint32_t *dstvi = ( uint32_t *)dstv;
22      uint32_t *dstsi = ( uint32_t *)dsts;
23      unsigned i;
24      klee_make_symbolic(src, sizeof (src), "src");
25      zlimit (0, src, dsts, 64);
26      zlimit (1, src, dstv, 64);
27      for (i = 0; i < 64; ++i)
28          assert ( dstvi [i] == dstsi [i]);
29  }
```

Figure 1. Simple test benchmark.

between the two versions via the `simd` argument. The SIMD implementation makes use of Intel's SSE instruction set.

The first loop of the routine, at lines 5–11, contains the core of the SIMD implementation, and is a good illustration of how SIMD code is structured. Each iteration of the loop processes four elements of array `src` at a time. The variables `srcv`, `cmpv` and `dstv` are of type `__m128`, i.e., 128-bit vectors consisting of four `float`s each. The code first loads four values from `src` into `srcv` by using the SIMD instruction `_mm_loadu_ps()` (line 6). It then compares each element of `srcv` to the corresponding element of `zero4`, which was initialized on line 4 to a vector of four 0 values (line 7). The output vector `cmpv` contains the result of each comparison as a vector of four 32-bit bitmasks each consisting of all-ones (if the `srcv` element was > 0) or all-zeros (otherwise). Next it applies the `cmpv` bitmask to `srcv` by performing a bitwise AND of `cmpv` and `srcv` to produce `dstv`, a copy of `srcv` with values ≤ 0 replaced by 0 (line 8). Finally, it stores `dstv` into `dst` (line 9).

The second loop of the `zlimit` routine, at lines 13–16, is the scalar implementation, which is also used by the SIMD version to process the last few elements of `src` when the size is not an exact multiple of 4.

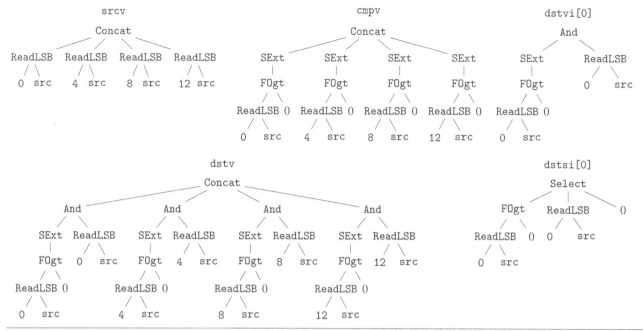

Figure 2. Symbolic expressions assigned to variables `srcv`, `cmpv`, `dstv` and to the array elements `dstvi[0]` and `dstsi[0]` of Figure 1. `src` represents the symbolic array `src`. The ReadLSB (Read Least Significant Byte first) node represents a 4-byte little-endian array read, FOgt floating point greater-than comparison, SExt sign extension, Select the equivalent of the C ternary operator and Concat bitwise concatenation.

The `main` function constitutes the test harness. In order to use KLEE-FP, developers have to identify the scalar and the SIMD versions of the code being checked, and the inputs and outputs to these routines. In our example, we have one input, namely the array `src`. Thus, the first step is to mark this array as *symbolic*, meaning that its elements could initially have any value (see §4.1 for more details). This is accomplished on line 24 by calling the function `klee_make_symbolic()` provided by KLEE, which takes three arguments: the address of the memory region to be made symbolic, its size in bytes, and a name used for debugging purposes only. Then, on line 25 we call the scalar version of the code and store the result in `dsts`, and on line 26 we call the SIMD version and store the result in `dstv`. Finally, on lines 27–28 each element of `dstv` is compared against the corresponding element of `dsts`. Note that we use bitcasting to integers via the pointers `dstvi` and `dstsi` for a bitwise comparison. As we will further discuss in Section 3, this is necessary because in the presence of NaN (*Not a Number*) values, the C floating point comparison operator `==` does not always return `true` if its floating-point operands are the same, as distinguished from a bitwise comparison.

To use KLEE-FP to run this benchmark, the user first compiles the code to LLVM bitcode [Lattner 2004], the low level representation on which KLEE and our extension KLEE-FP operate. The bitcode file can then be run directly by KLEE-FP.

Before KLEE-FP begins executing the input bitcode file, it first carries out a number of transformations. One of these is a lowering pass that replaces instruction-set specific SIMD operations with standard, instruction-set neutral instructions. Section 3.3 discusses this pass in more detail.

KLEE-FP interprets a program by evaluating the transformed bitcode instructions sequentially. During symbolic execution, values representing variables and intermediate expressions are manipulated. Both vector and scalar values are represented as bitvectors: concrete values by bitvector constants and symbolic ones by bitvector expressions. Vectors have bitwidth $s \times n$, where s is the bitwidth of the underlying scalar and n is the number of elements in the vector. Section 3 gives more details on our modeling approach.

For example, during the first iteration of the `zlimit` SIMD loop, the variables `srcv`, `cmpv` and `dstv` defined at lines 6–8 in Figure 1 are represented by the three expressions shown on the left hand side of Figure 2. Similarly, the results `dstvi[0]` and `dstsi[0]` are represented by the two expressions shown on the right side of Figure 2.

When KLEE-FP reaches an `assert` statement, it tries to prove that the associated expression is always `true`. For example, during the first iteration of the loop at lines 27–28 the expressions `dstvi[0]` and `dstsi[0]` are compared. To this end, KLEE-FP applies a series of expression rewrite rules, whose goal is to bring the expressions to a canonical normal form. As discussed in Section 4.4, one of our canonicalization rules transforms an expression tree of the form $\text{And}(\text{SExt}(P), X)$ into $\text{Select}(P, X, 0)$, where P is an arbitrary boolean predicate and X an arbitrary expression. For our example, this rule transforms the expression corresponding to `dstvi[0]` shown in Figure 2 to be identical to expression `dstsi[0]`, shown in the same figure. Once both expressions are canonicalized, we attempt to prove their equivalence by (1) using a simple syntactical matching for the floating-point subtrees, and (2) using a constraint solver for the integer subtrees. As highlighted in the introduction, the reason we are able to prove the equivalence of floating-point expressions by bringing them to canonical form and then syntactically matching them is that constructing two equivalent values from the same inputs in floating point can usually only be done reliably in a limited number of ways. As a consequence, we found that in practice we only need a relatively small number of expression canonicalization rules in order to apply our technique to real code (see §4.4).

One concern not covered by this simple example, which has a single execution path, is the number of proofs that are needed: under symbolic execution, every feasible program path is explored, and we have to conduct the proof on every path. Thus, an important optimization is to reduce the number of paths explored by merging multiple ones together. This optimization is discussed in detail in Section 4.2.

3. Modeling Floating Point and SSE Operations

This section discusses our approach for modeling floating point and SSE operations in KLEE-FP. In Section 3.1 we start by presenting our floating point extension to KLEE. Then, in Section 3.2 we describe our modeling of SSE vector operations, and in Section 3.3 we present our lowering pass that translates SSE intrinsics into standard LLVM operations. Finally, in Section 3.4 we discuss the way we handle LLVM atomic intrinsics.

3.1 Floating Point Operations

In order to add support for floating point, we extended KLEE's constraint language to include floating point types and operations. Floating point operation semantics are derived from those presented by LLVM, whose floating point instructions include $+$, $-$, $\times$, $\div$, remainder, conversion to and from signed or unsigned integer values (`FPToSI`, `FPToUI`, `UIToFP`, `SIToFP`), conversion between floating point precisions (`FPExt`, `FPTrunc`) and the relational operators $<$, $=$, $>$, $\leq$, $\geq$ and $\neq$. Of particular importance for our crosschecking algorithm (§ 4.3) is the fact that relational operators can occur in both *ordered* and *unordered* form. Ordered and unordered operators differ in the way they treat `NaN` values: if any operand is a `NaN`, ordered comparisons always evaluate to `false` while unordered ones to `true`.

A comparison of two floating point values x and y must have one of four mutually exclusive outcomes: $x < y$, $x = y$, $x > y$ or x `UNO` y (either or both of x and y

Shorthand	FCmp operation	Meaning
$\text{FOeq}(X, Y)$	$\text{FCmp}(X, Y, \{=\})$	Ordered $=$
$\text{FOlt}(X, Y)$	$\text{FCmp}(X, Y, \{<\})$	Ordered $<$
$\text{FOle}(X, Y)$	$\text{FCmp}(X, Y, \{<, =\})$	Ordered $\leq$
$\text{FUno}(X, Y)$	$\text{FCmp}(X, Y, \{\text{UNO}\})$	Unordered test

Table 1. Floating point predicate shorthand semantics.

are `NaN`). We establish a set $\mathbf{O} = \{<, =, >, \text{UNO}\}$ of these outcomes. Then, any floating point relational operator may be represented by a subset of $\mathbf{O}$: for example, ordered $\leq$ (`FOle`) is represented by $\{<, =\}$.

In KLEE-FP, all floating point relational operators are represented using a generic `FCmp` expression. The first two operands to `FCmp` are the comparison operands, whereas the third operand is a subset of $\mathbf{O}$, known as the *outcome set* (represented internally using a vector of four bits, based on the floating point predicate representation used by LLVM [Lattner 2004]). In this paper we normally refer to predicate operations using shorthand names rather than using `FCmp`. Table 1 gives a few examples of mappings between shorthand names and associated `FCmp` operations. In Section 4.4 we show how outcome sets can be used to simplify expressions involving floating-point comparisons.

In future work, we may also wish to store the rounding mode of each non-relational operation. However, we have not yet found this necessary, because none of the code we have worked with changes the rounding mode.

3.2 SSE Vector Operations

Intel's Streaming SIMD Extension operates on a set of eight 128-bit vector registers, called *XMM* registers. Each of these registers can be used to pack together either four 32-bit single-precision floats, two 64-bit double-precision floats, or various combinations of integer values (e.g., four 32-bit ints, or eight 16-bit shorts).

Since the same register set is used to operate on different data types, it is possible to perform an operation of a certain type on the result of an operation of a different type: e.g., one could perform a single-precision computation on the result of a double-precision, or even integer, computation. As a consequence, in order to capture the precise semantics of SSE vector operations, it is important to model SSE registers at the bit-level. Fortunately, KLEE already models its constraints with bit-level accuracy [Cadar 2008] by using the *bitvector* data type provided by its underlying constraint solver, STP [Ganesh 2007]. Thus, we model each XMM register as a 128-bit STP bitvector that can be treated as storing different data types, depending on the instruction that uses the register.

At the LLVM intermediate language level, SSE vectors are represented by 128-bit typed arrays. There are only three generic operations that operate on these arrays: `insertelement`, `extractelement` and `shufflevector`.

#	LLVM intrinsic (llvm.x86.)	# Occurrences in OpenCV	Instruction	Function
1	sse.cmp.ps	19	CMPPS	Compare Packed Single-Precision Floating-Point Values
2	sse.max.ps	4	MAXPS	Return Maximum Packed Single-Precision Floating-Point Values
3	sse.min.ps	6	MINPS	Return Minimum Packed Single-Precision Floating-Point Values
4	sse2.pslli.w	5	PSLLW	Shift Packed Data Left Logical
5	sse2.psubus.b	17	PSUBUSB	Subtract Packed Unsigned Integers with Unsigned Saturation
6	sse2.psubus.w	11	PSUBUSW	

Table 2. Examples of SSE intrinsics supported by KLEE-FP. The entire list consists of 37 intrinsics.

All other SSE instructions are implemented as LLVM intrinsics, as discussed in the next section.

The `extractelement` operation takes as arguments a 128-bit wide array (e.g., an eight element array of 16-bit integers) and an offset into this array, and returns the element at that offset. For example,

%res = **extractelement** <8 x **i16**> %a, **i32** 3

extracts the fourth element of the array a (which contains eight 16-bit shorts) and stores it in %res. Similarly,

%res = **insertelement** <8 x **i16**> %a, **i16** 10, **i32** 2

returns in %res an array with all values equal to those of the array %a except for the third element which receives the value 10.

The `shufflevector` instruction takes two vectors of the same type and returns a permutation of elements from those two vectors. The permutation is specified using an immediate vector argument whose elements represent offsets into the vectors. For example,

%res = **shufflevector** <4 x **float**> %a, <4 x **float**> %b, <4 x **i32**> <**i32** 0, **i32** 1, **i32** 4, **i32** 5>

returns in %res a vector with its 2 lower order elements taken from the 2 lower order elements of %a and its 2 higher order elements from the 2 lower order elements of %b.

In our implementation, we model these three operations using the bitvector extraction and concatenation primitives provided by STP. The modeling is straightforward. For example, if A is the 128-bit bitvector representing the array a, $\text{Extract}^{16}(A, 48)$ is the bitvector expression encoding the `extractelement` operation above, where $\text{Extract}^W(BV, k)$ extracts a bitvector of size W starting at offset k of bitvector BV.

3.3 SSE Intrinsic Lowering

Not all SSE instructions are implemented in terms of vector operations; most of them are represented using LLVM intrinsics. To enable comparison with scalar code, we implemented a pass that translates them into standard LLVM instructions by making use of the `extractelement` and `insertelement` operations presented in Section 3.2.

We added support for 37 SSE intrinsics; Table 2 shows a few examples. These 37 intrinsics were sufficient to handle the OpenCV benchmarks on which we evaluated our technique (§ 5). An example of a call to an SSE-specific intrinsic is shown below:

%res = **call** <8 x **i16**> @llvm.x86.sse2.pslli.w (<8 x **i16**> %arg, **i32** 1)

This instruction shifts every element of %arg left by 1 yielding %res. The lowering pass transforms this call into the following sequence of instructions:

%1 = **extractelement** <8 x **i16**> %arg, **i32** 0
%2 = **shl i16** %1, 1
%3 = **insertelement** <8 x **i16**> **undef**, **i16** %2, **i32** 0
%4 = **extractelement** <8 x **i16**> %arg, **i32** 1
%5 = **shl i16** %4, 1
%6 = **insertelement** <8 x **i16**> %3, **i16** %5, **i32** 1
...
%22 = **extractelement** <8 x **i16**> %arg, **i32** 7
%23 = **shl i16** %22, 1
%res = **insertelement** <8 x **i16**> %21, **i16** %23, **i32** 7

These instructions carry out the same task as the intrinsic but are expressed in terms of the standard LLVM instructions `insertelement`, `extractelement` and `shl`.

3.4 Atomic Intrinsics

LLVM provides a number of intrinsics which are used to represent atomic operations. Since our OpenCV benchmarks use atomic operations, we needed to add support for them to KLEE-FP.

An example of such an LLVM atomic intrinsic is the following:

%res = **call i32** @llvm.atomic.load.add.i32.p0i32 (**i32**∗ %ptr, **i32** 1)

This operation atomically loads a 32-bit integer from the given memory pointer %ptr, increments it, stores the result to %ptr and returns the value originally loaded from %ptr in %res.

Since KLEE does not support threading or signals, KLEE-FP uses a very simple work-around for atomic operations: it simply lowers them to equivalent sequences of non-atomic instructions. For example, the atomic operation shown above is translated to:

%res = **load i32**∗ %ptr
%1 = **add i32** %res, 1
store i32 %1, **i32**∗ %ptr

Our atomic lowering pass handles all 13 atomic intrinsics supported by LLVM 2.7, and was subsequently contributed to the main LLVM branch to be used by similar tools.

4. Crosschecking Algorithm

Crosschecking an SIMD routine against its scalar equivalent using our technique involves four main stages. First, we write a test harness that invokes the scalar and SIMD versions of the code on the same symbolic input, and asserts that their results are equal. For example, the `main()` function in Figure 1 represents the test harness for our simple `zlimit` benchmark. Second, we use *symbolic execution* to explore all the feasible paths in the code under test (§4.1). To increase the applicability of symbolic execution, we apply an aggressive version of *phi node folding* to statically merge paths (§4.2), which reduces the number of paths we have to track by an exponential factor on some benchmarks. Then, on each explored path, we try to prove that the symbolic expressions corresponding to the scalar and SIMD variants are equivalent. To do so, we first canonicalize the expressions through a series of expression rewrite rules and analyses (§4.4), and then use expression matching and constraint solving to prove that the resulting expressions are equivalent (§4.3).

4.1 Symbolic Execution

KLEE-FP uses symbolic execution [King 1975] to explore all the feasible paths in a program up to a certain input size. Symbolic execution runs the program on a *symbolic* input, whose value is initially unconstrained. As the program runs, it tracks the constraints on each symbolic memory location. If code uses a symbolic expression in a conditional, it follows both outcomes of the branch (if both are possible), constraining the conditional expression to be `true` on the true path and `false` on the other. Each of the two paths is explored in the same way, forking execution whenever both sides of a conditional expression are possible.

There are two fundamental limitations of symbolic execution which are relevant to this work:

1. It does not handle symbolically-sized objects. Thus, for code that uses arbitrarily-sized data structures, we can only verify the *bounded* equivalence of SIMD and scalar versions, i.e. we can verify they are equivalent up to a certain input size.

2. The number of paths in a program is in general exponential in the number of branches encountered during execution, thus for some programs, symbolic execution may fail to explore all feasible paths in a practical amount of time even for small input sizes. To reduce the number of explored paths, we discuss in Section 4.2 an approach for statically merging paths via phi-node folding.

In our work, we use the symbolic execution tool KLEE, which is built on top of the LLVM compiler infrastructure.

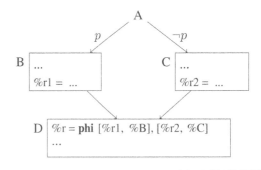

Figure 3. Diamond control flow pattern.

We found KLEE to be a good match for implementing our technique because it handles C/C++ code, tracks constraints with bit-level accuracy, and provides an easily extensible expression language [Cadar 2008].

4.2 Phi Node Folding

To reduce the number of explored paths, we apply a more aggressive variant of *phi-node folding* (also known as *if-conversion*) [Chuang 2003, Lattner 2004], which attempts to statically merge program paths.

Phi-node folding usually operates on the static single-assignment (SSA) form of a program [Alpern 1988] and targets branches with a control flow structure matching the diamond pattern shown in Figure 3, commonly associated with `if` statements and the C ternary operator. The beginning of block D contains one or more *phi* nodes, which select the correct register values (in our example, that of `%r`) depending on what block was previously executed.

We can reduce the amount of branching in a program by merging all four basic blocks in a diamond pattern into a single block. This is accomplished by unconditionally executing blocks B and C and using the branch predicate p to select the result via `select` instructions.

The traditional application of phi-node folding in compilers has both *safety* and *performance* restrictions. Because blocks B and C are executed unconditionally, it is only safe to perform the transformation if neither block contains an instruction that may throw an exception or cause any other side effects. Most arithmetic instructions satisfy these constraints. However, floating point instructions do not, because they may throw an exception if either operand is a `NaN`. Furthermore, the transformation is only performed when folding is cheap enough, in order to minimize the amount of unnecessary work done by the CPU.

Due to forking, the cost of not applying the optimization in a symbolic execution context is usually greater than that of applying it. Furthermore, since KLEE-FP's crosschecking algorithm (§4.3) and expression canonicalization rules (§4.4) do not interfere with the side effects associated with floating point expressions, is it is always safe to fold floating point instructions in KLEE-FP.